Environmental Isotopes in Biodegradation and Bioremediation

Edited by

C. Marjorie Aelion
University of Massachusetts, Amherst, USA
and University of South Carolina, Columbia, USA

Patrick Höhener
Université de Provence, Marseille, France

Daniel Hunkeler
Centre for Hydrogeology, University of Neuchâtel, Switzerland

Ramon Aravena
University of Waterloo, Ontario, Canada

CRC Press
Taylor & Francis Group
Boca Raton   London   New York

CRC Press is an imprint of the
Taylor & Francis Group, an **informa** business

CRC Press
Taylor & Francis Group
6000 Broken Sound Parkway NW, Suite 300
Boca Raton, FL 33487-2742

Library of Congress Cataloging-in-Publication Data

Environmental isotopes in biodegradation and bioremediation editors / C. Marjorie
    Aelion ... [et al.].
        p. cm.
    "A CRC title."
    Includes bibliographical references and index.
    ISBN 978-1-56670-661-2 (hardcover : alk. paper)
    1. Bioremediation--Technological innovations. 2.= Radioisotopes--Industrial
applications. 3. Radioisotopes--Scientific applications. 4. Radioisotopes in hydrology.
5. Radioisotopes in microbiology. 6. Biodegradation. I. Aelion, C. Marjorie (Claire
Marjorie) II. Title.

TD192.5.E955 2010
628.5--dc22   2009032735

Visit the Taylor & Francis Web site at
http://www.taylorandfrancis.com

and the CRC Press Web site at
http://www.crcpress.com

Contents

SECTION I   Isotope Fundamentals

SECTION II   Isotopes and Microbial Processes

SECTION III   Isotopes in Field Applications

SECTION IV   Isotope Emerging Areas

Preface

The application of isotopes has expanded from the field of geology to biogeochemistry, and on to many areas of environmental sciences. The use of isotopes in natural systems has been described in many excellent books covering marine systems, terrestrial biotic systems, geologic earth science systems, and the atmosphere. Similarly, the role of microorganisms in natural and contaminated environments has been discussed in several excellent books that describe microbial physiology and biochemistry in detail. This is the first book, to our knowledge, that brings the world of isotopes to the world of hydrology and microbial ecology in both natural and contaminated environments. Isotopes are a powerful tool to investigate this world and provide information on the interaction of microorganisms with natural compounds and pollutants in the area of biodegradation and its application in bioremediation and chemical cycling. The development of enhanced analytical capabilities and separation techniques, improvements in detection limits, and accessibility of instrumentation have opened up this world by allowing compound-specific isotope analysis.

This book began at the urging of Brian Kenet, who attended a session at an international meeting chaired by C. Marjorie Aelion in 1999 in which she presented a paper entitled "Soil Gas and Isotopic Assessment of Anaerobic Petroleum Biodegradation." Kenet urged again in 2001 after attending a similar session chaired by Aelion entitled, "Isotopic Approaches in Bioremediation." Kenet was convinced that there was significant and growing interest on the use of isotopes in biodegradation and environmental contaminant studies. Aelion agreed and the book took form; coauthors were garnered, the outline of the book was generated, and the book was contracted in 2002 with the expanded view of natural and contaminated environments. Since that time and during the writing of the book, the field has exploded. Initially in the domain of a relatively small group of specialized academic researchers, the field of isotopic applications to contaminant hydrology and microbial ecology has expanded to great numbers of researchers and to the commercial sector. This was due in part to increased access and availability to commercial analytical facilities and analytical developments. Continued development of the technique will expand the field even further to areas as yet untapped.

This book is intended to provide sufficient technical detail to be used as a textbook for advanced undergraduate students and graduate students in the fields of geology, environmental science, and microbial ecology who want to understand the role of isotopes in microbial ecology with more specific detail on biodegradation and bioremediation. The questions of what amount of a compound comes from anthropogenic release, whether these chemicals undergo degradation in the environment, and whether they persist and accumulate are of great concern and isotopes may help to answer these questions. Advanced students in the field need both theoretical and practical knowledge to thoroughly understand the complexities of the use of isotopes in these fields and an understanding of the potential for future expansion. Additionally the book will be useful to environmental practitioners in the field of bioremediation and scientists in environmental regulatory agencies who must understand, interpret, and evaluate the data presented to them by the environmental practitioners in the day-to-day realities of site remediation.

To this end in this first edition we have divided the book into four broad areas and the following four sections: "Isotope Fundamentals," "Isotopes and Microbial Processes," "Isotopes in Field Applications," and "Isotope Emerging Areas." The first section covers important background and theoretical information that is necessary for the understanding of later chapters. Chapter 1 defines isotopes, how isotopes were discovered, and how isotope data are reported with respect to international standards. Chapter 2 details how isotopes are analyzed and the importance of understanding the levels of sensitivity and accuracy that can be achieved for the various isotopes. Chapter 3 provides the theoretical basis for understanding how isotopes undergo fractionation in the environment, the extent of fractionation in different fractionation processes, and the fundamental governing equations for these transformations. Chapter 4 highlights the fundamental principles of isotope fractionation during the transformation of organic contaminants and compiles large tables with fractionation factors for stable isotopes in contaminants undergoing various transformation processes.

The second section covers different environmental redox conditions that dictate the predominant microbial processes that will occur. Chapter 5 focuses on aerobic processes in which molecular oxygen (O_2) is the electron acceptor and reactions that allow the highest energy yield for the degrading organisms. Aerobic degradation leads to carbon dioxide as an end product and is important to carbon recycling and the enhanced bioremediation of contaminants. Chapter 6 focuses on methanogenesis that recycles carbon in anaerobic environments such as landfills, lake sediments, and marshes producing methane as an end product. Methanogenesis leads to the largest of all carbon isotope shifts in the processed carbon and in the produced methane, so isotope methods may be particularly useful if sufficient methane is generated. Chapter 7 describes isotope studies focusing on the two stable isotopes ^{15}N and ^{34}S, often combined with ^{18}O in oxygen-bearing compounds. Both nitrogen and sulfur can serve as electron acceptors by nitrate- and sulfate-reducing microorganisms, respectively, to oxidize carbon and can be used to track more reduced forms such as ammonia, elemental sulfur, and sulfides.

The third section describes, more specifically, the transformation of anthropogenic pollutants and the application to field sites. Chapter 8 explains how transformation processes can be assessed using stable isotopes at field sites in completely water-saturated porous media and presents quantitative concepts to estimate the extent of degradation with isotope data. Chapter 9 summarizes those processes that fractionate isotopes in gases and volatile pollutants in the unsaturated zone, with a special focus on gas-phase diffusion that may confound the interpretation of the fractionation caused by biodegradation and may necessitate modeling of both fractionation by gas-phase diffusion and by biodegradation.

The fourth section covers expanding fields that may warrant books in their own right. Chapter 10 summarizes the use of compounds labeled with stable isotopes for degradation studies in natural and contaminated environments including the emerging area of stable isotope probing that combines stable isotope labeling with the extraction and analysis of molecular biological markers of organisms responsible for degradation. Chapter 11 highlights the use of radiocarbon at natural abundance as an extremely sensitive tool used in combination with stable carbon isotopes to examine carbon in natural and contaminated environments. Finally, Chapter 12 breaks away from the classical focus of many environmental scientists, often limited to the elements of carbon, nitrogen, oxygen, and sulfur, to examine the world of multiple elements. This chapter gives an overview of the use of novel stable isotopes for the elements Li, Cl, Ca, Se, Cr, and Fe, which have potential applications in a wide range of environmental studies, a few of which are highlighted.

Several individuals gave freely of their time to improve the content of this book. Chapters were reviewed by experts in the field and we thank each of the following reviewers: Johannes Barth, University of Erlangen; Stefano Bernasconi, ETH Zurich; Daniel Bouchard, SNC Lavalin; Thomas Boyd, Naval Research Laboratory; Jeffrey Chanton, Florida State University; Keith Hackley, Illinois State Geological Survey; Lawrence Miller, U.S. Geological Survey; Barbara Morasch, EPFL, Lausanne; Elizabeth Guthrie Nichols, North Carolina State University; Egbert Schwartz, Northern Arizona State University; Greg Slater, Woods Hole Oceanographic Institute; Leonard Wassenaar, NWRI, Environment Canada; Tim Buscheck, Chevron; Gwenaël Imfeld, Laboratoire d'Hydrologie et de Geochimie de Strasbourg; and Florian Breider, University of Neuchâtel. Melissa Engle provided extensive support for the illustrations in several chapters. We appreciate and acknowledge the patience and important contributions of Jennifer Ahringer, Kari Budyk, Matt Lamoreaux, and Rachael Panthier of Taylor & Francis, in the production of this book.

To complement this book, we have established a supporting website at www.unine.ch/chyn/isotopes where additional material is posted. We hope that this book and website are well used by a variety of people in a variety of disciplines, and will garner interest around the globe as is evidenced by the geographic distribution of the books' collaborators.

We look forward to receiving comments on new areas for isotopic applications as they develop in the future. Finally, we request that any omissions or errors be excused as unintentional on the part of the authors and we hope these will be brought to our attention.

C. Marjorie Aelion
Patrick Höhener
Daniel Hunkeler
Ramon Aravena

Acknowledgments

Many people contributed to this effort and the names of many of them are listed in the preface. In addition to these individuals we would like to thank Mr. Peter Stone of the S.C. Department of Health and Environmental Control who, as a consummate geochemist and inquisitive scientist, initiated a great deal of thought and research into the world of stable isotopes. Dr. Brian Kirtland was a contributor to this world, and Dr. Thomas Leatherman remains a long-term contributor to multiple efforts.

Editors

C. Marjorie Aelion is dean of the School of Public Health and Health Sciences at the University of Massachusetts, Amherst. She worked for the U.S. Geological Survey, Water Resources Division as a hydrologist for three years before beginning her academic career at the University of South Carolina in Columbia from 1991 to 2008. She was an assistant professor, an associate professor, and a professor in the Department of Environmental Health Sciences, and the associate dean for research for the Arnold School of Public Health. She obtained her SMCE from the Massachusetts Institute of Technology in Cambridge, Massachusetts in civil engineering, and her PhD in environmental sciences and engineering, environmental chemistry and biology from the School of Public Health at the University of North Carolina, Chapel Hill. She has been a Fulbright awardee to the Université de Bretagne Occidentale in France, and the University of Wageningen in the Netherlands. Her research is in the area of biodegradation of organic contaminants, tools for assessing remedial technologies, including stable isotopes and naturally occurring radiocarbon, and the application and development of enhanced remediation systems for contaminated groundwater. She has additional interests in the associations of metals in residential soils with negative health outcomes in children. C. Marjorie Aelion is on the editorial board of *Bioremediation Journal* and *Oceans and Oceanography*, and a managing editor for *Biodegradation*. She is the author of more than 70 peer-reviewed scientific articles and one edited book.

Patrick Höhener is a professor in hydrogeochemistry at the University of Provence, Marseille. He obtained a PhD in environmental sciences in 1990 from the Swiss Federal Institute of Technology, where he later had a position as lecturer at Zurich and Lausanne. His research interests lie in the management and remediation of soils and aquifers contaminated with organic chemicals. This includes the development of site investigation methods as well as the development of low-cost remediation techniques such as in situ bioremediation. He is also interested in the biogeochemical processes that control elemental cycling at contaminated sites and the use of partitioning tracers and environmental isotopes as tools for process identification. Patrick Höhener is editor of the *Journal of Contaminant Hydrology* and the journal *Grundwasser* and the author of more than 50 peer-reviewed articles.

Daniel Hunkeler is a professor for groundwater quality at the Centre for Hydrogeology at the University of Neuchâtel, Switzerland, and an adjunct professor at the University of Waterloo, Canada. He obtained a PhD from the Swiss Federal Institute for Technology, Zürich. His research focuses on the development of stable isotope methods and their application to gain insight into the contaminant behavior at the field scale. He also works on the development of methods for the characterization and remediation of contaminated sites and the use of geochemical and isotope tools to characterize the dynamics of groundwater flow systems. Daniel Hunkeler directs the Centre for Hydrogeology Stable Isotope and Hydrochemistry Laboratory and teaches graduate courses on groundwater chemistry, contaminant hydrogeology, stable isotope methods, and water resources management at the University of Neuchâtel as well as international short courses on isotope methods. He is an associate editor of *Organic Geochemistry* and the editor of *Grundwasser*.

Ramon Aravena is a research professor in the Department of Earth Sciences and Environmental Sciences at the University of Waterloo. Dr. Aravena's research has focused on the application of environmental isotopes (^{18}O, ^{2}H, ^{13}C, ^{14}C, ^{3}H, ^{34}S, and ^{15}N) in hydrology, geochemistry, and quaternary geology. He has been involved in numerous groundwater studies in Latin America, Canada, the United States, and Europe related to the evaluation of groundwater resources and groundwater contamination. His current research focuses on groundwater contamination caused by agricultural, urban, and industrial activities. Dr. Aravena consults as part of the expert pool of the International Atomic Energy Agency, Vienna, Austria, for its projects worldwide. He has also developed a strong research collaboration network with European and Latin American universities. His teaching involves isotope hydrology and geochemistry courses at the University of Waterloo, Canada, and courses on isotope hydrology in Latin America organized by the International Atomic Energy Agency, and at the National Groundwater Association professional course program. He has also been involved in teaching short courses on the application of environmental isotopes in contaminant hydrogeology in different European universities. He is the author and coauthor of more than 100 peer-reviewed articles.

Contributors

C. Marjorie Aelion
Department of Public Health
University of Massachusetts
Amherst, Massachusetts

and

Department of Environmental
  Health Sciences
University of South Carolina
Columbia, South Carolina

Ariel D. Anbar
School of Earth & Space Exploration
  and the Department of Chemistry &
  Biochemistry
Arizona State University
Tempe, Arizona

C. Fred T. Andrus
Department of Geological Sciences
University of Alabama
Tuscaloosa, Alabama

Ramon Aravena
Department of Earth and
  Environmental Sciences
University of Waterloo
Waterloo, Ontario, Canada

Brian Beard
Department of Geology & Geophysics
University of Wisconsin-Madison
Madison, Wisconsin

Stefano Bernasconi
Geological Institute
ETH Zürich
Zürich, Switzerland

Daniel Bouchard
SNC-Lavalin Environnement Inc.
Montréal, Quebec, Canada

Martin Elsner
Helmholtz Zentrum
  München
Institute of Groundwater
  Ecology
Neuherberg, Germany

Patrick Höhener
Université de Provence
Laboratoire Chimie Provence
Marseille, France

Edward R. C. Hornibrook
Department of Earth Sciences
Bristol Biogeochemistry Research
  Centre
University of Bristol
Bristol, United Kingdom

Daniel Hunkeler
Centre for Hydrogeology
University of Neuchâtel
Neuchâtel, Switzerland

Silvia A. Mancini
Golder Associates
Mississauga, Ontario, Canada

Bernhard Mayer
Department of Geoscience
University of Calgary
Calgary, Alberta, Canada

Barbara Morasch
Environmental Chemistry
  Laboratory
Swiss Federal Insitute of
  Technology Lausanne (EPFL)
Lausanne, Switzerland

R. Sean Norman
Department of Environmental
  Health Sciences
University of South Carolina
Columbia, South Carolina

Christopher S. Romanek
Department of Earth and
  Environmental Sciences
University of Kentucky
Lexington, Kentucky

Section I

Isotope Fundamentals

1 Fundamentals of Environmental Isotopes and Their Use in Biodegradation

Patrick Höhener and C. Marjorie Aelion

CONTENTS

1.1   WHAT ARE ENVIRONMENTAL ISOTOPES?

This chapter explains what isotopes are, what their average occurrence in nature is, how isotope measurements are reported with respect to international standards, and how isotope data can be used in biodegradation. The focus is on environmental isotopes, a loosely defined class of primary isotopes in biological and hydrological systems. Because hydrogen, carbon, oxygen, nitrogen, and sulfur are the major elements that make up organic matter, biologists are most interested in the isotopes of these elements. All of these elements are relatively light and, therefore, their isotopes have relatively large mass differences and are subject to fractionation by biochemical or physical processes—which makes them interesting tracers. However, the isotopes of chlorine, calcium, iron, and others are relevant to microbial processes and are candidates to figure among the environmental isotopes. The notion of environmental isotopes has been used in other textbooks (Fritz and Fontes 1980; Clark and Fritz

1997; Mook and De Vries 2000) that focus on tracing the hydrological cycle and its associated biogeochemical processes. It should be noted, however, that environmental isotopes does not signify that these isotopes were all produced by natural reactions (see ^{14}C as an example of an isotope that is of natural and anthropogenic origin). Also, some of the heavier isotopes that are used in geology for dating of rocks or tracing rock-forming processes may be called environmental isotopes. Therefore, there is no exact definition of what environmental isotopes are and this book does not make any attempt to put forward such a definition. We will focus on stable and radioactive isotopes of the elements: hydrogen, carbon, nitrogen, oxygen, sulfur, silica, and chlorine. Chapter 12 will give supplementary information on lithium, chlorine, calcium, selenium, chromium, and iron.

1.2   DISCOVERY OF ISOTOPES

Until the late nineteenth century, chemists were convinced that each chemical element has a distinct and integer molar mass. The elements were ordered in the periodic table of elements according to their molar masses, and minor deviations from integer molar masses were thought to be caused by analytical errors. Only after the discovery of radioactivity by Henry Becquerel and Marie and Pierre Curie at the end of the nineteenth century (Nobel Prize in physics in 1903), the ideas about the nature of elements began to change. The discovery of radioactivity soon led to the description of the different modes of radioactive decay, and it became recognized that decay causes elements to produce other elements of lower molecular weights. This led to the description of the uranium and thorium decay series and to the recognition that elements like uranium and thorium are composed of molecules with different molecular weights (e.g., for uranium, the mass could be either 235 or 238 g/mol). However, this and the fractional molecular masses found for some other elements remained somewhat of an enigma until the 1930s when the neutron was discovered. With the understanding of the role of neutrons in atoms and the help of modern analytical techniques, it became clear that an element can be composed of molecules with variable numbers of neutrons and that only the number of protons is invariable for each element. Molecules of different masses of the same element were thereafter called isotopes, a term first used by Frederick Soddy in 1913. The name isotope comes from the Greek *isos* meaning equal and *topos* meaning place and reflects the fact that isotopes are at the same place on the periodic table. Today, more than 2200 isotopes of the 92 naturally occurring elements are known. Not long after the discovery of neutrons came the first precise measurements of isotope abundance ratios, which were achieved using a mass spectrometer by Alfred Nier in 1936. This marked the start of a new research era in geosciences. As the mass spectrometric techniques became ready for routine analyses, scientists began to explore the natural variations of isotopes in air, water, soils, rocks, organic matter, and various other materials, and also began to investigate the fractionation processes that lead to small differences in isotope abundance ratios. Since then, isotope techniques have become standard tools to evaluate problems in paleoclimatology, hydrology, hydrogeology, oceanography, biochemistry, organic geochemistry (Galimov 2006), and many other branches of science. A number of excellent textbooks on isotope techniques are available also (Fritz and Fontes 1980; Clark and Fritz 1997; Kendall and McDonnell 1998; Cook and Herczeg 1999; Mook and De Vries 2000; Hoefs 2004; Johnson 2004; Flanagan, Ehleringer, and Pataki 2005).

It is no surprise that isotope techniques are also becoming attractive tools to study problems in environmental microbiology such as the microbial breakdown of organic matter or environmental pollutants, the biogenic formation of greenhouse gases, the cycling of nitrogen, the mobilization of iron, or the bioremediation of soils and aquifers. However, the samples that need to be analyzed to study such problems are often small, heterogeneous, or of complex chemical composition and need an advanced processing before being transferred to a mass spectrometer. The techniques of sample introduction to mass spectrometers evolved only very little during the 50 years after the introduction of mass spectrometry (see Chapter 2). Although it was possible, from the 1970s on, to couple elemental analyzers to mass spectrometers and measure stable isotopes (e.g., carbon, nitrogen, sulfur, and hydrogen) in various samples after combustion of the bulk material, the isotopic composition of single molecule species in the bulk matrix remained unresolved. A major analytical breakthrough was achieved in 1978 by coupling a gas chromatograph to a mass spectrometer (Matthews and Hayes 1978) and later by the invention of a combustion interface (Merritt et al. 1995). This technique called gas chromatography-combustion-isotope ratio mass spectrometry (GC-C-IRMS), allows routine measurements of stable isotope abundances in separated organic molecules. The technique profits from the massive progress that has been made in gas chromatography, as this is the most widely used technique for separation and analysis of organic pollutants since its invention in the late 1940s. Since the late 1990s, GC-C-IRMS is commercially available and widely applied in many laboratories. As a consequence, many stable isotope studies on the fate of pollutants in environmental systems appeared in literature within the last 10 years.

Although stable isotope biogeochemistry based on mass spectrometry is a recent branch of science, it must be noted that physical chemists were able to study the characteristics of isotope-labeled compounds by more than 50 years ago. These chemists manufactured isotope-labeled compounds (e.g., fully deuterated hydrocarbons) using organic synthesis and measured their physicochemical properties such as vapor pressure, molar volume, or viscosity relative to unlabeled counterparts in order to understand molecular interactions in solvents. By 1953, it was recognized that ^{13}C-labeled solvents (benzene, chloroform, and carbon tetrachloride) have a higher vapor pressure than ^{12}C-labeled solvents (Baertschi, Kuhn, and Kuhn 1953), and isotope vapor pressure effects from over 600 studies were known and reviewed more than 30 years ago (Jancso and Van Hook 1974). The measurement techniques for vapor pressure were based on distillation apparatuses and high-precision mercury manometers (Davis and Schiessler 1953), and the results obtained, for example on the vapor–liquid equilibrium effects of solvents, compare well with present-day GC-C-IRMS techniques (Wang and Huang 2003). However, the old physical-chemical literature is often not well known to present day environmental scientists.

1.3 WHAT ARE ELEMENTS, NUCLIDES, ISOTOPES, ISOTOPOMERS, AND ISOTOPOLOGUES?

Elements are defined by the number of protons (Z) in their nucleus. The arrangement of elements according to their proton number and the number of electron shells around the nucleus is depicted in the periodic table of elements. The mass number or atomic weight (A) of an element is equal to the sum of both protons (Z) and neutrons (N) in the nucleus, or

$$A = N + Z. \tag{1.1}$$

A single element can have two or more different atomic weights (A) due to differences in the number of neutrons that can exist in the nucleus. The resulting different atoms are called nuclides. The periodic table of elements may give information on some different nuclides of selected elements. Nuclides of the same number of protons (Z) but different atomic weights (A) are called isotopes. While protons have a positive charge, neutrons have no charge, so the number of neutrons does not affect the charge of an atom. Conventional short notation for nuclides and isotopes use a superscript of the atomic weight preceding the elemental symbol (e.g., ^{12}C, ^{13}C). An exception to this general notation rule is made for the hydrogen isotopes: ^{2}H is called deuterium (and noted sometimes like an element as D), and ^{3}H is called tritium (sometimes abbreviated as T). When the names of isotopes have to be given in full text (e.g., in manuscript titles), the element's full name is written, followed by a hyphen and the A number (e.g., carbon-14, radon-222).

To date, about 270 stable nuclides and about 2,000 radionuclides have been identified (Clark and Fritz 1997). Radionuclides can decay by different modes, including the α-, β-, and γ-decays, which are the most important. Radionuclides that emit alpha particles (helium nuclei) or beta particles (electrons) decay into an isotope of another element, which may or may not be stable. For example, the decay series of uranium-238 proceeds via 14 different decay reactions involving 8 elements and coming to a halt only when the stable isotope lead-206 is formed.

Charts of nuclides are available which display isotopes in horizontal rows (Figure 1.1). The charts have the atomic number Z as y-axis and the number of neutrons N as x-axis. In Figure 1.1, only the nuclide chart for the elements with 1 to 8 protons is shown. For heavier elements, the International Atomic Energy Agency (IAEA) publishes a nuclide chart (http://www-nds.iaea.org/nudat2/). As can be seen in Figure 1.1, the number of stable isotopes per element within this section of the nuclide chart ranges between 1 and 3, whereas the number of radioactive isotopes can be as high as 15 for oxygen. However, many of these radioisotopes have half-lives that are too short to be of use in environmental studies. Look at the 15 isotopes of nitrogen as an example: ^{10}N to ^{12}N and ^{16}N to ^{24}N have half-lives shorter than 10 seconds. If we wished to use these isotopes as tracers, we would need to synthesize the isotopes on our study site, fuel them into a process that we want to study, and measure them immediately. It becomes easily understandable that only isotopes with longer half-lives are potentially useful, particularly outside laboratories. In the case of nitrogen, there is only one such radioactive isotope, ^{13}N with

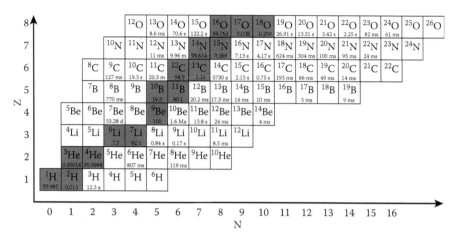

FIGURE 1.1 Nuclide chart of isotopes with proton numbers (Z) from 1 to 8 (*y*-axis) versus neutron numbers N (*x*-axis). Grey squares designate stable isotopes and the number below each isotope is its mean global natural abundance. White squares show radionuclides, with their half-life times. Half-life times are <1 ms when not given below isotope.

a half-life of about 10 minutes. Still, this half-life is at the limit of being a useful environmental tracer, although attempts have been made to study denitrification in soils (Tiedje et al. 1981), or sewage sludge (Kaspar, Tiedje, and Firestone 1981), or to study fertilizer uptake by plants in the field (Gerwing, Caldwell, and Goodroad 1979) by using ^{13}N. However, such ^{13}N-based studies remain unrepeated since they were too costly. Thus, for nitrogen, we are basically left with only the two stable isotopes ^{14}N and ^{15}N, which are very useful for environmental purposes. A similar problem is noted for oxygen: there are three stable isotopes (^{16}O to ^{18}O), but the 12 radioactive oxygen isotopes ^{12}O to ^{15}O and ^{19}O to ^{26}O are too short-lived to be of practical use for environmental studies (Figure 1.1). For hydrogen and carbon, however, we are fortunate, since both elements have two stable isotopes and radio-active isotopes (^{3}H, ^{14}C) with half-lives that are well suited for many purposes. Many inorganic or organic compounds labeled with ^{3}H or ^{14}C are commercially available and there is no need to use the short-lived ^{11}C or other radioactive carbon isotopes as tracers.

When looking at isotope distribution in chemical compounds, one must intro-duce the notions of isotopomers and isotopologues, which should not be con-fused. Isotopomers (isotopic isomers) are isomers having the same number of each isotopic atom but differ in their positions. For example, CH_3CHDCH_3 and $CH_3CH_2CH_2D$ form a pair of constitutional isotopomers of propane. Isotopomers undergo different processes in the environment, which could cause isotope frac-tionation. Isotopologues are chemical species that differ only in the isotopic composition of their molecules or ions. An example is water, where three of its hydrogen-related isotopologues are: HOH, HOD, and DOD. Simply, the isoto-pologue of a chemical species has at least one atom with a different number of neutrons.

1.4 NATURAL ABUNDANCE OF STABLE ISOTOPES

The original isotopic compositions of solar systems are a result of nuclear processes in stars and planets. Isotopic compositions in terrestrial environments change gradually by the processes of radioactive decay, cosmic ray interactions, mass-dependent fractionations that accompany inorganic and biological reactions, and anthropogenic activities such as the processing of nuclear fuels, reactor accidents, and the testing of nuclear weapons. Stable isotopes are nuclides that do not appear to decay to other isotopes on geologic time scales but may themselves be produced by the decay of radioisotopes. However, the rate of formation of the elements that are discussed in this book are far below any measurable rate, and the global mean abundance of stable isotopes of H, C, N, O, S, and Cl can be regarded as constant.

Elements usually have one common isotope and that is the form found in greatest abundance in nature. Table 1.1 summarizes the natural occurrence of the stable isotopes of interest in this book, giving values recommended by the International Union of Pure and Applied Chemistry (De Laeter et al. 2003). For instance, 98.93% of carbon in nature occurs as ^{12}C and only 1.07% of carbon occurs as ^{13}C. These numbers are depicted in the nuclide chart (Figure 1.1) in the squares for stable isotopes below the isotope name. Similarly, 99.9885% of hydrogen is found as ^{1}H, but a second stable isotope (^{2}H, or deuterium, D) is present in 0.0115% (Table 1.1). For sulfur, carbon, nitrogen, and oxygen the average terrestrial abundance ratio of the heavy to the most abundant light isotope ranges from 1:22 (sulfur) to 1:500 (oxygen), whereas the ratio $^{2}H/^{1}H$ is much lower at 1:6410. Note that for most elements discussed in this book, the lightest stable isotope is the most abundant. This is not the case, however, for some of the heavier elements like chromium or iron (Table 1.1).

Since the chemical behavior of an atom is largely determined by its electrons, isotopes exhibit nearly identical chemical behavior. The electronic structure of atoms

TABLE 1.1
Average Terrestrial Abundance of the Stable Isotopes of H, C, N, O, Si, S, Cl, Cr, and Fe

Element	Stable Isotopes and Natural Abundance (%)			
Hydrogen	^{1}H (99.9885)	^{2}H (0.0115)		
Carbon	^{12}C (98.93)	^{13}C (1.07)		
Nitrogen	^{14}N (99.636)	^{15}N (0.364)		
Oxygen	^{16}O (99.757)	^{17}O (0.038)	^{18}O (0.205)	
Silica	^{28}Si (92.223)	^{29}Si (4.685)	^{30}Si (3.092)	
Sulfur	^{32}S (94.99)	^{33}S (0.75)	^{34}S (4.25)	^{36}S (0.01)
Chlorine	^{35}Cl (75.76)	^{37}Cl (24.24)		
Chromium	^{50}Cr (4.345)	^{52}Cr (83.789)	^{53}Cr (9.501)	^{54}Cr (2.365)
Iron	^{54}Fe (5.845)	^{56}Fe (91.754)	^{57}Fe (2.119)	^{58}Fe (0.282)

Source: De Laeter, J. R., Bohlke, J. K., De Bievre, P., Hidaka, H., Peiser, H. S., Rosman, K. J. R., and Taylor, P. D. P., *Pure and Applied Chemistry,* 75:683–800, 2003.

and molecules is essentially independent of isotopic distribution of nuclear mass and within the limits established by the Born–Oppenheimer approximation (Jancso and Van Hook 1974). However, measurable and important small differences occur for a number of specific processes. Due to their larger masses, heavier isotopes tend to form shorter and more stable chemical bonds (Urey 1947; Bigeleisen and Wolfsberg 1958; Elsner et al. 2005), tend to occupy smaller molar volumes (Bartell and Roskos 1966), and tend to diffuse more slowly than lighter isotopes of the same element (Craig 1953) because molecular diffusion is inversely proportional to the square root of the molecular mass. This mass effect is most pronounced for ^1H and ^2H because ^2H has twice the mass of ^1H. For heavier elements, the relative mass difference between isotopes is much less but not always negligible. Environmental processes such as microbial transformation, photosynthesis, evaporation, and gas-phase diffusion have slightly different rates with respect to different isotopes. This leads to stable isotope fractionation—a phenomenon described in detail in Chapter 3. Fractionation processes lead to small but measurable differences in different pools of elements and compounds in nature. An example is given for the ranges of the isotope ratios in different pools of carbon (Figure 1.2). Isotope ratios are usually reported as ratios R, which is defined as the abundance of the heavy isotope relative to the abundance of the light isotope ($R = {}^{13}C/{}^{12}C$ for carbon). The mean global isotope ratio $^{13}C/^{12}C$ is $R = 0.011237$ (Table 1.1). According to Figure 1.2, deviations from this mean global ratio occur. Often, carbon is found to be somewhat enriched in ^{12}C ($R < 0.011237$), but marine bicarbonate and some carbonate rocks are also found to be enriched in ^{13}C

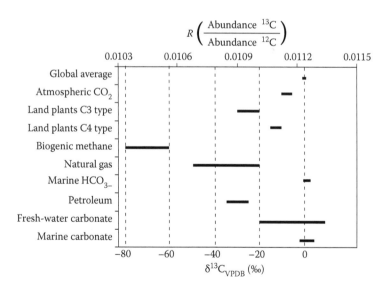

FIGURE 1.2 Ratios of isotope abundance $^{13}C/^{12}C$ in various carbonaceous materials (upper x-axis). Also shown are the values of the more convenient delta notation (lower x-axis). Data from (Mook, W. G., and J. J. De Vries, *Environmental isotopes in the hydrological cycle: Principles and applications. Vol. 1 Introduction—Theory, methods review, IAEA*, Isotope Hydrology Section, Vienna, 2000; Clark, I., and P. Fritz, *Environmental isotopes in hydrogeology*. CRC Press, Boca Raton, FL, 1997.)

(Figure 1.2). The variations in R are large in comparison to the analytical uncertainty of modern mass spectrometers and these variations can be traced reproducibly. As a consequence, it is important to note that the natural abundances of stable isotopes in Table 1.1 are not universal constants but global means. The analytical detection of slight changes of the natural abundance of stable isotopes and the comparison of local measurements with global means is thus a powerful tool to trace specific biologic or chemical physical processes in the environment.

Reporting isotope abundances as in Figure 1.2 as R values is unpractical since small and uneven numbers have to be compared to each other. A much more convenient way is to compare measured R values to an international standard and report only the deviation from this standard. To do so, standards—to which everybody relates the measurement—and an internationally accepted notation to report isotopic data of samples relative to these standards are needed. This will be introduced in the next section.

1.5   STABLE ISOTOPES: DELTA-NOTATION AND INTERNATIONAL STANDARDS

The measurement of absolute isotope ratios or abundances is not easily done and requires very sophisticated mass spectrometric equipment (see Chapter 2). However, for most studies, the monitoring of relative variations in stable isotope concentrations rather than absolute abundances provides the necessary information about the processes. By measuring a known reference material under the same conditions as our sample, we can then express isotopic concentrations as the difference between sample and standard.

Stable isotope variations at or near natural abundance levels are usually reported in the delta (δ) notation, a value given in parts per thousand (‰, per mil). Isotope ratios (R) are measured for samples and standard by mass spectrometry and the relative measure delta (δ) is calculated as written here for the ^{13}C and ^{12}C (Coplen 1996):

$$\delta^{13}C = \frac{R_{sample} - R_{standard}}{R_{standard}} \cdot 1000 \qquad (1.2)$$

with R_{sample} being the $^{13}C/^{12}C$-ratio determined by mass spectrometry and $R_{standard}$ is the abundance ratio of the international standard (Table 1.2). Delta values are not absolute isotope abundances but differences between sample and the international standard that, by definition, has a delta value of 0‰. For instance, if a biomass sample is found to have a $^{13}C/^{12}C$-ratio R_{sample} greater than the $R_{standard}$ by five parts per mil, this value is reported as $\delta^{13}C = +5$‰. A positive δ value means that the isotopic ratio of the sample is higher than that of the standard; a negative δ value means that the isotopic ratio of the sample is lower than that of the standard. Many isotope geochemists advocate always prefacing the δ value with a sign, even when the value is positive, to distinguish between a true positive δ value and a δ value that is merely missing its sign (a frequent occurrence with users unfamiliar with isotope terminology).

As a basis for global comparison of stable isotope data, international standards have been introduced. Two organizations are taking the lead in distributing these standards: the National Institute of Standards and Technology (NIST, www.nist.gov) in the

TABLE 1.2
The International References and their Abundance Ratios for Most of the Stable Isotopes Discussed in this Book

Element	Ratio of Isotopes	Standard(s)	Abbreviation	Abundance Ratio $R_{standard}$[a]
Hydrogen	$^2H/^1H$	Vienna Standard Mean Ocean Water	VSMOW	1.5575×10^{-4}
Helium	$^3He/^4He$	Atmospheric helium	AIR He	1.38×10^{-6}
Lithium	$^7Li/^6Li$[b]	Lithium carbonate from Spodumene deposits in North Carolina	L-SVEC	12.1735[b]
Boron	$^{11}B/^{10}B$	Boric acid	NBS 951	4.04362
Carbon	$^{13}C/^{12}C$	Carbonate from Vienna Pee Dee Belemnite	VPDB	0.011237
Nitrogen	$^{15}N/^{14}N$	Atmospheric nitrogen	AIR N_2	3.677×10^{-3}
Oxygen	$^{18}O/^{16}O$	Vienna Standard Mean Ocean Water	VSMOW	2.0052×10^{-3}
		Carbonate from Vienna Pee Dee Belemnite	VPDB	2.0672×10^{-3}
Silicon	$^{30}Si/^{28}Si$	Silica sand NBS-18 (= RM8546)	NBS-18	0.03365761[c]
		Silicon crystals	IRMM-018a	0.03365602[c]
Sulfur	$^{34}S/^{32}S$	H_2S in Vienna Ca–on Diablo Troilite	VCDT	4.5005×10^{-2}
Chlorine	$^{37}Cl/^{35}Cl$	Chloride ion in ocean water Standard Mean Ocean Chlorine	SMOC	0.324

[a] From Clark, I., and P. Fritz, *Environmental isotopes in hydrogeology*. CRC Press, Boca Raton, FL, 1997; see Chapter 12 for information on calcium, selenium, chromium, strontium, iron, and molybdenum isotopes and http://www.ciaaw.org/ for reference materials.

[b] Reporting of $^6Li/^7Li$ should be discontinued (Coplen, T. B., J. K. Bohlke, P. De Bievre, T. Ding, N. E. Holden, J. A. Hopple, H. R. Krouse, et al., *Pure and Applied Chemistry*, 74, 1987–2017, 2002).

[c] Valkiers, S., K. Ruße, P. Taylor, T. Ding, and M. Inkret, *International Journal of Mass Spectrometry*, 242, 321–23, 2005.

United States and the International Atomic Energy Agency (IAEA, www.iaea.org). Reference material is available through either of these sources. A complete list of reference materials and the recommendations for the reporting of isotopic compositions of the various elements is also given by the Commission of the International Union of Pure and Applied Chemistry (IUPAC, http://www.ciaaw.org/). Different protocols for reporting isotope data created some confusion before 1996. Since that time, the isotope community has adopted the IUPAC recommendations for reporting isotope data (Coplen 1996). The international standards for C, H, O, N, S, Cl, and Si are briefly introduced in the following and standards of other elements supplementary information is given in Chapter 12.

$^{13}C/^{12}C$: The international standard for $^{13}C/^{12}C$ introduced in 1957 (Craig 1957) was the internal calcite structure (rostrum) of the fossil *Belemnitella americana* from the Cretaceous Pee Dee Formation in South Carolina, in many publications referred to as the Pee Dee Belemnite (or PDB). The short notation $\delta^{13}C_{PDB}$ refers to this standard. It had an abundance ratio $R_{standard}$ $^{13}C/^{12}C$ of 0.011237 (Table 1.2). The IAEA in Vienna subsequently defined the hypothetical VPDB scale (Vienna PDB, considered as identical to the PDB) as reference for stable carbon analysis. Before the limited PDB supply was exhausted, a crushed white marble designated as NBS-19 was calibrated and forms the basis for the definition of VPDB. The material has a slightly positive $\delta^{13}C$:

$$\delta^{13}C_{NBS-19} = +1.95\text{‰ } \delta^{13}C_{VPDB} = +1.95\text{‰ } \delta^{13}C_{PDB}. \qquad (1.3)$$

The $\delta^{13}C$ values of various carbonaceous reference materials are reported in the new guidelines for $\delta^{13}C$ measurements (Coplen et al. 2006b). Furthermore, a normalization procedure for reporting $\delta^{13}C$ data is described (Coplen et al. 2006a).

$^2H/^1H$ and $^{18}O/^{16}O$: The international standard for $^2H/^1H$ and $^{18}O/^{16}O$ used to be the Standard Mean Ocean Water (SMOW), as defined by Craig (1961). In fact, Craig's SMOW never existed. It was a hypothetical water calibrated to the isotopic content of NBS-1, a water sample from the Potomac River (Clark and Fritz 1997). The IAEA prepared standard water from distilled seawater that has an isotopic composition close to SMOW. The reference is identified as Vienna Standard Mean Ocean Water (VSMOW) and has been used for about four decades as the accepted reference. The abundance ratios are given in Table 1.2. Although there is some debate whether VSMOW = SMOW (Clark and Fritz 1997), δ^2H_{VSMOW} and $\delta^{18}O_{VSMOW}$ are defined as 0. In order to supply another water that is highly depleted compared to ocean water, NIST distributes rainwater (Standard Light Antarctic Precipitation, SLAP) with a δ^2H_{VSMOW} of -428.0‰ and a $\delta^{18}O_{VSMOW}$ of -55.50‰ (Coplen 1994).

For $\delta^{18}O$ in carbonates, the isotope community refers to the $\delta^{18}O$ that had been measured in the Pee Dee Belemnite, noted $\delta^{18}O_{VPDB}$. The reference material is a white marble NBS-19, with $\delta^{18}O_{NBS-19} = -2.20$‰ $\delta^{18}O_{VPDB}$.

$^{15}N/^{14}N$: The international standard for $^{15}N/^{14}N$ is atmospheric nitrogen from air (AIR), with $\delta^{15}N_{AIR} = 0$ and an abundance ratio as given in Table 1.2. Reference materials are available as gas (N_2) or as ammonium or nitrate salts.

$^{34}S/^{32}S$: For several decades, troilite (FeS), from the Cañon Diablo meteorite, (CDT) has been employed as the international standard for $^{34}S/^{32}S$. This material had a $^{34}S/^{32}S$ abundance ratio of 0.0450, but is not available anymore. Reference materials are silver sulfides IAEA-S-1 ($\delta^{34}S_{VCDT} = -0.3$‰), IAEA-S-2 ($\delta^{34}S_{VCDT} = +22.67$‰) and barium sulfate NBS-127 ($\delta^{34}S_{VCDT} = +21.1$‰). Since 1995, the IUPAC (Krouse and Coplen 1997) recommends that abundances relative to $^{34}S/^{32}S$ ratios of all sulfur bearing substances be expressed on the VCDT scale, defined by assigning an exact $\delta^{34}S$ value of -0.3‰ (relative to VCDT) to the silver sulfide reference material IAEA-S-1. Confusions concerning the absolute values of some sulfur reference materials have been clarified by measurements (Qi and Coplen 2003) in comparison to elemental sulfur (IAEA-S-4-Soufre de Lacq, $\delta^{34}S_{VCDT} = +16.90$‰). Reporting of $\delta^{34}S$ measurements relative to CDT should be discontinued.

$^{37}Cl/^{35}Cl$: The reference for chlorine isotopes is Standard Mean Ocean Chloride (SMOC), as the isotopic composition in seawater chloride is very constant. The $^{37}Cl/^{35}Cl$ ratio of this standard is 0.324. Two reference materials are available, IRMM-641 and IRMM-642, both made from NaCl in pure water at different concentrations (Ostermann et al. 2001). It should be noted that Cl$^-$ has to be transformed to gaseous chloroform (or to cesium chloride) to be measured using a mass spectrometer. This is done by precipitation from the water sample by adding $AgNO_3$. The resulting AgCl is purified and reacted with CH_3I to methylchloride, CH_3Cl. This gas is suitable for measuring the $^{37}Cl/^{35}Cl$ ratio by normal IRMS.

$^{29}Si/^{28}Si$ and $^{30}Si/^{28}Si$: The international standard for stable silica isotopes is NBS28 silica sand (also named NIST RM 8546) with $\delta^{29}Si = 0‰$, $\delta^{30}Si = 0‰$ and $\delta^{18}O = +9.6‰$ (Valkiers et al. 2005). A new material (SiO_2 crystals, IRMM-018a) with an isotopic composition that is identical to NBS28 is distributed now (Valkiers et al. 2005).

1.6   NOTATIONS USED IN TRACER STUDIES

For environmental studies using tracers enriched in isotopes, labeled compounds are available with the heavy isotope making up to 99% of the tracer element (e.g., 99 atom % ^{13}C-acetate, 99 atom % ^{15}N-ammonium sulfate). To biologists the principal advantage of stable isotopes is that they are not radioactive and are therefore unrestricted in use and do not generate radioactive waste. A restriction occurs only for laboratories that routinely analyze the natural abundance of rare isotopes (e.g., ^{2}H, ^{13}C). For these laboratories, a certain danger of contamination of the measurement system or of sampling equipment exists when samples with very high abundance ratios are measured. Results from tracer studies are usually reported in units of atom percent (atom %). This value gives the absolute number of atoms of one isotope in 100 atoms of total element, as in the example for ^{13}C:

$$\text{atom } \% = \frac{^{13}C}{^{13}C + {^{12}C}} \times 100 \tag{1.4}$$

for the ^{13}C atom % calculation, the trace amounts of naturally present ^{14}C are usually negligible and the sum of ^{12}C and ^{13}C is taken to be total C.

Medical tracer studies of human physiology are most often reported in units of atom percent excess (APE), which is the enrichment level of a sample above a given background reading considered as zero. The background reading in atom % is subtracted from the experimental value to give APE:

$$\text{atom } \% \text{ excess (APE)} = (\text{atom } \% \text{ sample} - \text{atom } \% \text{ baseline}) \times 100. \tag{1.5}$$

1.7   RADIOISOTOPES

Radioisotopes are unstable nuclides that decay with specific half-life times to other nuclides under the emission of radiation. The most important decay modes are α-decay (emission of a helium atom), β-decay (emission of an electron β^- or a positron β^+), and γ-decay (emission of electromagnetic energy). Before exploring the tracer potential of radionuclides, let us first review the rules governing radioactive decay. The disappearance of radioactive atoms is a statistical logarithmic process in which a subset of atoms undergoes decay in a given time period. The activity, A, decreases according to the law of radioactive decay (Equation 1.6)

$$A_{(t)} = A_{(0)} \exp^{(-\lambda t)}. \tag{1.6}$$

For carbon-14, 1 atom out of 8,266 atoms decay each year, and the decay constant λ thus, is 1.2097×10^{-4} yr^{-1}. However, most books and tables on radioisotopes do not give decay constants λ, but rather half-life times $T_{1/2}$, which is related to λ via Equation 1.7:

$$T_{1/2} = \ln 2 / \lambda. \tag{1.7}$$

In our example, the half-life time for ^{14}C is 5730 years. The most commonly employed radioisotopes, their half-life times, and decay modes are listed in Table 1.3. Environmental concentrations of radionuclides are determined as radioactive activities and are given in Becquerel (Bq) per volume or mass. One Becquerel equals one decay per second. The unit used before the Bq was Curie, with 1 Ci = 3.7×10^{10} Bq.

Carbon-14 takes a special place in the naturally occurring radioisotopes because of its long half-life time. Willard Frank Libby described the use of this nuclide as a tool for archeological dating in his Nobel Prize–winning work (1952). However, ^{14}C is equally important as a tool to study the mixing of oceans or to discriminate old and recent carbon sources in microbial studies. A typical application is to distinguish the origin of carbon in an ecosystem from either fossil sources (petroleum or natural gas) or recent sources (biogas). Natural ^{14}C is produced in the upper atmosphere when cosmic neutrons hit nitrogen atoms and knock out one proton (the nitrogen isotope ^{14}N with 7 protons +7 neutrons is thus transformed into the carbon isotope ^{14}C with 6 protons and 8 neutrons, Figure 1.3). The production of ^{14}C in the atmosphere is balanced by its decay. However, explosion of nuclear weapons from 1950 to 1963 significantly disturbed that balance and in 1964 the ^{14}C atmospheric activities had almost doubled.

The ^{14}C activities are referenced to an international standard known as modern carbon (mC). Its activity is defined as 95% of the ^{14}C activity, in 1950, of the NBS oxalic acid standard. This is close to the activity of wood grown in 1890 in a fossil CO_2-free environment and equals 226 Bq kg^{-1} carbon (Clark and Fritz 1997). Measured ^{14}C activities are expressed in percent modern carbon (pMC). Like stable isotopes, ^{14}C undergoes fractionation and to maintain universality for dating purposes, pMC must be normalized to a common δ^{14}C value of $-25\%o$ (Clark and Fritz 1997).

TABLE 1.3

Radioisotopes of Some Elements of Interest in Biodegradation and Bioremediation

Isotope	Half-Life	Decay Mode	Sources
^{3}H (Tritium)	12.43 years	β^-	Cosmogenic, weapons testing
^{13}N	9.965 minutes	β^+	Synthesized tracer
^{11}C	20.3 minutes	β^+	Synthesized tracer
^{14}C	5,730 years	β^-	Cosmogenic, weapons testing
^{35}S	0.239 years	β^-	Industrial use
^{36}Cl	301,000 years	β^-	Cl$^-$

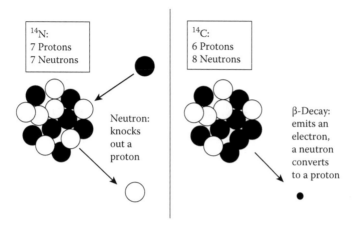

FIGURE 1.3 Atmospheric production of ^{14}C from ^{14}N by cosmic rays (neutron impact, causing a loss of a proton, left-hand side), and radioactive decay of ^{14}C to ^{14}N by β-decay (emission of an electron, generating a proton from a neutron, right-hand side).

After carbon, tritium is the next radioisotope of environmental relevance in the context of this book. Tritium activities are usually expressed as absolute concentrations, using tritium units (TU) as reporting units. So no reference standard is needed. One TU corresponds to one ^3H atom per 10^{18} atoms of hydrogen, which is 0.118 Bq/L of water. Due to thermonuclear bomb testing, tritium concentrations in the atmosphere and in rainwater peaked in 1963 worldwide, and tritium concentrations have been used to check the age of groundwater in aquifers or pore water in sediment cores to understand the local water cycle.

1.8 WHY SHOULD ONE USE ISOTOPES IN BIODEGRADATION AND BIOREMEDIATION?

Biodegradation is perhaps the most important process on earth. Without biodegradation, the organic matter synthesized by plants and other lithotrophic organisms would accumulate and the carbon dioxide in the atmosphere would not be regenerated. Biodegradation and biotransformation include the cycling of important elements, including carbon and nitrogen that are both essential molecular building blocks and components of contaminants. Without biodegradation, nutrients fixed in organic matter would not be recycled and chemicals produced by humans and released carelessly into the environment would not be transformed, mineralized, and detoxified. Bioremediation is applied biodegradation, put into action by scientists and engineers, most often with the aim to decontaminate soil or groundwater polluted by anthropogenic chemicals and has become a business of considerable economic importance. Thus the study of biodegradation is of great economic importance and for the sustainable development of people on earth.

As more tools are developed to study biodegradation, more scientific questions can be answered. Is soil organic matter respired to CO_2? Will chlorinated solvents be degraded in groundwater before migrating to a well for drinking water? Will sulfate-reducing microorganisms oxidize methane that is seeping through sediments at the seafloor and prevent volatilization of this greenhouse gas into the atmosphere? Will nitrate that runs off of agricultural lands into groundwater be reduced to the harmless nitrogen gas by denitrifying microorganisms? There is often a need to assign responsibility for contamination, for example whether a pollutant found downgradient of an industrial site in groundwater is actually originating from this site or from another source? Additional areas of interest are whether nitrate contamination originates from the use of chemical fertilizer or application of manure, and whether it can be guaranteed that 95% of the mass of a spilled chemical is removed by a certain remediation technique. Fundamental questions include: Is a microbial degradation reaction initiated by a certain biochemical reaction? Do the carbon atoms from a certain pollutant end up in microbial biomass, proteins, lipids, or even in the DNA of microbes?

To study these types of questions, scientists historically relied on measurements of concentrations of the compounds of interests and their reaction partners in the specific environment. This was not always successful and often led to uncertainty.

Stable or radioactive isotopes offer a great added value in studies on biodegradation and bioremediation for many reasons. One is that concentration measurements may ambiguously explain certain processes especially at field sites. For example, dilution of dissolved compounds decreases concentrations that are often misinterpreted as being caused by degradation. Furthermore, improper sampling techniques may cause loss of the target compounds and result in misinterpretation, which is of particular concern for gases and volatile pollutants and that can be lost during the sampling or during the transport and storage of samples before analysis. Less volatile compounds also can be lost when adsorbed onto materials used for sampling and storage, like tubing, pumps, vials, and stoppers.

The measurement of isotopes is less affected by the above mentioned problems. Stable isotope data are reported as isotope ratios and are ratios of concentrations of two compounds, one of which carries a rare stable isotope. Isotope ratios in gaseous or aqueous samples are therefore not affected by dilution or by sampling artifacts such as volatilization or sorption since those processes affect both compounds almost equally and thus the ratio remains stable. Ratios of stable isotopes are often an excellent tool to check the consistency of interpretations that are based on concentration data. In addition, radioisotopes contribute the factor of time as additional information to prove or falsify hypotheses. For example, carbon-14 allows one to distinguish between fossil and modern carbon. However, only carbon and hydrogen have natural radioactive isotopes with half-life times well suited for practical studies; unfortunately oxygen, nitrogen, and sulfur either do not have such natural radioactive isotopes or the radioactive isotopes have half-lives that are too short for practical purposes.

Historically, the limitation of using stable isotopes as a tool to study biodegradation was due to difficulties in introducing sample materials into an isotope mass spectrometer. The key development made in the last 10–20 years was the direct coupling of isotope mass spectrometers to classical analytical systems used in laboratories such as gas chromatographs and elemental analyzers that made isotope analyses faster and cheaper. Also, the detection limits for the analysis of isotope ratios in compounds or pollutants in air or water samples have been drastically lowered by developments of new injection techniques like solid-phase microextraction (SPME) or the Purge & Trap technique.

1.9   TYPE OF INFORMATION OBTAINED BY ISOTOPES

The variations in the abundance of stable isotopes in natural compounds or anthropogenic pollutants can be used in different ways for assessing different types of information. These types can be grouped into at least seven classes. The first possibility is to identify sources of compounds by direct analysis of the compound in a given environment and compare the isotopic abundance to a number of potential sources that have distinct but constant isotope abundances. A prerequisite for this approach is that the concentration of a compound in the environment only changes owing to dilution or phase transfer processes. Hence, its isotopic composition will remain largely constant and thus can be used to elucidate the source of the compound. Preferably, two or three different isotopes are measured, for example ^{15}N and ^{18}O in nitrate, or ^{2}H, ^{13}C, and ^{37}Cl in a chlorinated solvent. For example, one might compare the multi-isotope pattern of hydrogen, carbon, and chlorine in dissolved chloroethene in groundwater to those of the solvent used by a client suspected to be the polluter (Poulson and Drever 1999). Additionally, one might trace the origin of nitrate pollution in groundwater by comparing its ^{15}N and ^{18}O analysis with those of frequently applied fertilizers and with atmospheric nitrogen.

A second type of information that can be gained by stable isotope analysis is the origin of a compound of interest that was formed in a given specific environment and by comparing its isotope signal to the ones of potential precursors. A typical application is determining whether CO_2 found either in soil air or dissolved in groundwater originates from pollutant degradation from a natural process. Radiocarbon can help here most often, since pollutants are often derived from petroleum, which is ancient and therefore free of ^{14}C, whereas natural CO_2 contains modern ^{14}C signals (see Chapter 11). However, stable carbon isotopes also can help to track the origin of dissolved CO_2 in ground water. Stable carbon isotope ratios have been used to determine the percentages of total dissolved CO_2 in groundwater derived from degradation of heating oil versus dissolution of carbonates (Bolliger et al. 1999), or from sugar refinery wastewater versus CO_2 from carbonates (Wersin, Abrecht, and Höhener 2001). Stable isotopes also are used to identify the substrate in methanogenesis because the two major substrates, acetate and CO_2, react with hydrogen gas and are easily distinguished based on stable carbon and hydrogen analysis of methane (see Chapter 5).

A third type of information that is obtained with stable isotope data concerns the assessment of transformation pathways. This is now often applied to resolve the question of whether a contaminant at a given site undergoes aerobic or anaerobic transformation. The initial biochemical reaction in aerobic or anaerobic contaminant transformation is different, and this leads to different kinetic isotope effects that are observable in the remaining unreacted contaminant, provided that the transformation progressed to a certain extent. Initial reactions under aerobic conditions often involve a mono- or dioxygenase and leads to the opening of carbon-carbon or carbon-hydrogen bonds as is discussed in Chapters 4 and 7. In contrast, under anaerobic conditions, the initial reactions are different and may include a fumarate addition, a reaction that is associated with quite a large carbon isotope effect. Under field conditions, often mixed aerobic–anaerobic conditions prevail, for example in oxic soil with soil aggregates that are anoxic microniches. Several studies have successfully shown that it is possible to assign the prevailing degradation pathways for environmental pollutants in mixed aerobic–anaerobic environments using a dual isotope approach, for example with toluene (Vogt et al. 2008), with methyl tertiary-butyl ether (MTBE) in groundwater (Zwank et al. 2005), with the explosive hexahydro-1,3,5-trinitro-1,3,5-triazine (RDX) in soil (Bernstein et al. 2008), and with nitrobenzene (Hofstetter et al. 2008). Based solely on carbon isotope analysis, it was possible to distinguish between two aerobic degradation processes for 1,2-dichloroethane in the laboratory, since the oxidation creates small fractionation, whereas hydrolytic dehalogenation creates large isotope fractionation (Hirschorn et al. 2004).

The fourth type of information that can be obtained by isotope data is used to distinguish between abiotic and biotic processes that have acted on a contaminant along its migration from a source to a receptor or during a specific treatment, especially to distinguish between biochemical reactions and abiotic transformation or phase transfer processes. This is useful for verifying mineralization of contaminants at sites undergoing natural attenuation or biostimulation. Isotope data often can show that the pollutants were not simply transferred to another phase, since phase transfer processes discriminate rather poorly between isotopes compared to biodegradation. For example, sorption of dissolved contaminants to geosorbents causes fractionation of stable hydrogen or carbon isotopes that is either below detection limits (Slater et al. 2000; Schuth et al. 2003) or that is at the edge of detection limits (Kopinke et al. 2005). Nevertheless, there are cases where phase-transfer processes fractionate isotopes significantly. Recent work summarized in Chapter 9 presented evidence that gas-phase diffusion and volatilization through soil discriminates, quite importantly, for light isotopes in volatile pollutants of low molecular weight, (e.g., in gasoline hydrocarbons) and that the remaining fraction of compounds in liquid gasoline gets progressively heavier during volatilization.

The fifth class of application of stable isotope techniques is the quantification of the extent of transformation. The classical Rayleigh equation (Chapter 3) has always been in the center of such quantitative assessments, classically to describe variations of isotopes during distillation. Progress has been made to explore the use of the Rayleigh equation in groundwater and its limitations when applied to complex groundwater systems, for example, where dilution and biodegradation are physically separated (Abe and Hunkeler 2006; Van Breukelen 2007; Van Breukelen and Prommer 2008). Chapter 8 of this book is devoted to the Rayleigh equation and its limitations with respect to applications in groundwater. It gives the reader a tool for a quantitative assessment of the percentages of biodegradation versus dilution in ideal aquifers using stable isotope measurements in dissolved contaminants along plumes. Chapter 9 addresses how volatilization of pollutants can be quantitatively assessed with stable isotope measurements in the liquid pollution source that has not yet volatilized.

The sixth type of information that can be obtained by exploiting stable isotope data is the verification of reactive transport models. These models are used frequently for the prediction of the length, duration, and environmental impact of contaminant plumes in groundwater or vapor plumes in soil air. Such models can be quite complex, may contain many parameters, and need to be validated, which is usually done using concentration data. Isotope data can provide a further data set for model validation, with the advantage that isotope data may be less biased by sampling errors as discussed previously. A recent comparison of two different numerical models predicting the extension of a plume of chloroethenes at Dover Air Force Base in Delaware suggested that both models were able to correctly predict the concentration contours of chloroethenes, but only one model closely predicted stable carbon isotope ratios in the ethenes (Atteia, Franceschi, and Dupuy 2008). The other model gave completely different shapes of isotope contours. Isotope data identified that one model apparently predicted the concentration data correctly but the incorporated transport processes incorrectly.

The final type of information derived from the use of stable isotopes is associated with their use as both a chemical and a molecular tracer in radioisotope and stable isotope labeling studies that are useful for examining both natural and contaminated environments. Stable isotope labeling of nitrogen has enhanced our ability to quantify nitrogen cycling on time scales of minutes, and to identify the incorporation of carbon into macromolecular organic matter identified in different sediment organic fractions. Finally, as with the advancement of analytical instrumentation for the analysis of isotope ratios, significant advancements have been made in the extraction, purification, and analysis of isotopically labeled molecular biomarkers such as DNA and RNA. The combination of isotopic and bimolecular analyses has opened up future areas of research not only in biodegradation studies but also in the broader field of microbial ecology.

ACKNOWLEDGMENT

Funding was provided kindly by Jean-Dominique Meunier from CEREGE (Aix-en-Provence) through a grant from French CNRS (National Center for Scientific Research). This manuscript benefited from reviews by Stefano Bernasconi (ETH Zurich) and Barbara Morasch (University of Neuchâtel).

REFERENCES

Abe, Y., and D. Hunkeler. 2006. Does the Rayleigh equation apply to evaluate field isotope data in contaminant hydrogeology? *Environmental Science & Technology* 40:1588–96.

Atteia, O., M. Franceschi, and A. Dupuy. 2008. Validation of reactive model assumptions with isotope data: Application to the Dover case. *Environmental Science & Technology* 42:3289–95.

Baertschi, P., W. Kuhn, and H. Kuhn. 1953. Fractionation of isotopes by distillation of some organic substances. *Nature* 171:1018–20.

Bartell, L. S., and R. R. Roskos. 1966. Isotope effects on molar volume and surface tension—simple theoretical model and experimental data for hydrocarbons. *Journal of Chemical Physics* 44:457.

Bernstein, A., Z. Ronen, E. Adar, R. Nativ, H. Lowag, W. Stichler, and R. U. Meckenstock. 2008. Compound-specific isotope analysis of RDX and stable isotope fractionation during aerobic and anaerobic biodegradation. *Environmental Science Technology* 42:7772–77.

Bigeleisen, J., and M. Wolfsberg. 1958. Theoretical and experimental aspects of isotope effects in chemical kinetics. *Advances in Chemical Physics* 1:15–76.

Bolliger, C., P. Höhener, D. Hunkeler, K. Häberli, and J. Zeyer. 1999. Intrinsic bioremediation of a petroleum hydrocarbon-contaminated aquifer and assessment of mineralization based on stable carbon isotopes. *Biodegradation* 10:201–17.

Clark, I., and P. Fritz. 1997. *Environmental isotopes in hydrogeology.* Boca Raton, FL: CRC Press.

Cook, P. G., and A. L. Herczeg. 1999. *Environmental tracers in subsurface hydrology.* Boston, MA: Kluwer.

Coplen, T. B. 1994. Reporting of stable hydrogen, carbon, and oxygen isotopic abundances (Technical Report). *Pure and Applied Chemistry* 66:273–76.

Coplen, T. B. 1996. New guideline for reporting stable hydrogen, carbon, and oxygen isotope-ratio data. *Geochimica Cosmochimica Acta* 60:3359–60.

Coplen, T. B., J. K. Bohlke, P. De Bievre, T. Ding, N. E. Holden, J. A. Hopple, H. R. Krouse, et al. 2002. Isotope-abundance variations of selected elements (IUPAC Technical Report). *Pure and Applied Chemistry.* 74:1987–2017.

Coplen, T. B., W. A. Brand, M. Gehre, M. Groning, H. A. J. Meijer, B. Toman, and R. M. Verkouteren. 2006a. After two decades a second anchor for the VPDB delta C-13 scale. *Rapid Communications in Mass Spectrometry* 20:3165–66.

Coplen, T. B., W. A. Brand, M. Gehre, M. Groning, H. A. J. Meijer, B. Toman, and R. M. Verkouteren. 2006b. New guidelines for delta C-13 measurements. *Analytical Chemistry* 78:2439–41.

Craig, H. 1953. The geochemistry of stable carbon isotopes. *Geochimica Cosmochimica Acta* 3:53–92.

Craig, H. 1957. Isotopic standards for carbon and oxygen and correction factors for mass-spectrometric analysis. *Geochimica Cosmochimica Acta* 12:133–49.

Craig, H. 1961. Standards for reporting concentrations of deuterium and oxygen-18 in natural water. *Science* 133:1833–34.

Davis, R. T., and R. W. Schiessler. 1953. Vapor pressures of perdeuterobenzene and of perdeuterocyclohexane. *Journal of Physical Chemistry* 57:966.

De Laeter, J. R., Bohlke, J. K., De Bievre, P., Hidaka, H., Peiser, H. S., Rosman, K. J. R., and Taylor, P. D. P., Atomic weights of the elements: Review 2000—(IUPAC technical report). *Pure and Applied Chemisry,* 75:683-800, 2003.

Elsner, M., L. Zwank, D. Hunkeler, and R. Schwarzenbach. 2005. A new concept linking observable stable isotope fractionation to transformation pathways of organic pollutants. *Environmental Science & Technology* 39:6896–916.

Flanagan, L. B., J. R. Ehleringer, and D. E. Pataki, eds. 2005. *Stable isotopes and biosphere—Atmosphere interactions: Processes and biological controls.* Amsterdam: Elsevier.

Fritz, P., and J. C. Fontes. 1980. *Handbook of environmental isotope geochemistry.* Amsterdam: Elsevier.

Galimov, E. M. 2006. Isotope organic geochemistry. *Organic Geochemistry* 37:1200–62.

Gerwing, J. R., A. C. Caldwell, and L. L. Goodroad. 1979. Fertilizer nitrogen distribution under irrigation between soil, plant, and aquifer. *Journal Environmental Quality* 8:281–84.

Hirschorn, S. K., M. J. Dinglasan, M. Elsner, S. A. Mancini, G. Lacrampe-Couloume, E. A. Edwards, and B. S. Lollar. 2004. Pathway dependent isotopic fractionation during aerobic biodegradation of 1,2-dichloroethane. *Environmental Science & Technology* 38:4775–81.

Hoefs, J. 2004. *Stable isotope geochemistry.* 5th ed. Berlin: Springer.

Hofstetter, T. B., J. C. Spain, S. F. Nishino, J. Bolotin, and R. P. Schwarzenbach. 2008. Identifying competing aerobic nitrobenzene biodegradation pathways by compound-specific isotope analysis. *Environmental Science & Technology* 42:4764–70.

IUPAC. 1998. Isotopic compositions of the elements, 1997. *Pure and Applied Chemistry* 70:217–35.

Jancso, G., and W. A. Van Hook. 1974. Condensed phase isotope effects (especially vapor pressure isotope effects). *Chemical Reviews* 74:689–750.

Johnson, C. M. 2004. *Geochemistry of non-traditional stable isotopes.* Washington, DC: Mineralogical Society of America.

Kaspar, H. F., J. M. Tiedje, and R. B. Firestone. 1981. Denitrification and dissimilatory nitrate reduction to ammonium in digested sludge. *Canadian Journal of Microbiology* 27:878–85.

Kendall, C., and J. J. McDonnell. 1998. *Isotope tracers in catchment hydrology.* Amsterdam: Elsevier.

Kopinke, F. D., A. Georgi, M. Voskamp, and H. H. Richnow. 2005. Carbon isotope fractionation of organic contaminants due to retardation on humic substances: Implications for natural attenuation studies in aquifers. *Environmental Science & Technology* 39:6052–62.

Krouse, H. R., and T. B. Coplen. 1997. Reporting of relative sulfur isotope-ratio data (Technical Report). *Pure and Applied Chemistry* 69:293–95.

Libby, W. F. 1952. *Radiocarbon dating.* Chicago, IL: University of Chicago Press.

Matthews, D. E., and J. M. Hayes. 1978. Isotope-ratio-monitoring gas chromatography-mass spectrometry. *Analytical Chemistry* 50:1465–73.

Merritt, D. A., K. H. Freeman, M. P. Ricci, S. A. Studley, and J. M. Hayes. 1995. Performance and optimization of a combustion interface for isotope ratio monitoring gas-chromatography mass spectrometry. *Analytical Chemistry* 67:2461–73.

Mook, W. G., and J. J. De Vries. 2000. *Environmental isotopes in the hydrological cycle: Principles and applications.* Vol. 1 *Introduction—Theory, methods, review.* Vienna: IAEA, Isotope Hydrology Section.

Ostermann, M., M. Berglund, P. D. Taylor, and M. Mariassy. 2001. Certification of the chlorine content of the isotopic reference materials IRMM-641 and IRMM-642. *Fresenius' Journal of Analytical Chemistry* 371:721–25.

Poulson, S. R., and J. I. Drever. 1999. Stable isotope (C, Cl, and H) fractionation during vaporization of trichloroethylene. *Environmental Science & Technology* 33:3689–94.

Qi, H. P., and T. B. Coplen. 2003. Evaluation of the S-34/S-32 ratio of Soufre de Lacq elemental sulfur isotopic reference material by continuous flow isotope-ratio mass spectrometry. *Chemical Geology* 199:183–87.

Schuth, C., H. Taubald, N. Bolano, and K. Maciejczyk. 2003. Carbon and hydrogen isotope effects during sorption of organic contaminants on carbonaceous materials. *Journal of Contaminant Hydrology* 64:269–81.

Slater, G. F., J. M. E. Ahad, B. S. Lollar, R. Allen-King, and B. Sleep. 2000. Carbon isotope effects resulting from equilibrium sorption of dissolved VOCs. *Analytical Chemistry* 72:5669–72.

Tiedje, J. M., R. B. Firestone, M. K. Firestone, M. R. Betlach, H. F. Kaspar, and J. Sorensen. 1981. Use of N-13 in studies of denitrification. *Advances in Chemistry Series* 295–315.

Urey, H. C. 1947. The thermodynamic properties of isotopic substances. *Journal of Chemical Society* 1:562–81.

Valkiers, S., K. Ruße, P. Taylor, T. Ding, and M. Inkret. 2005. Silicon isotope amount ratios and molar masses for two silicon isotope reference materials: IRMM-018a and NBS28. *International Journal of Mass Spectrometry* 242:321–23.

Van Breukelen, B. M. 2007. Extending the Rayleigh equation to allow competing isotope fractionating pathways to improve quantification of biodegradation. *Environmental Science. &. Technology* 41:4004–10.

Van Breukelen, B. M., and H. Prommer. 2008. Beyond the Rayleigh equation: Reactive transport modeling of isotope fractionation effects to improve quantification of biodegradation. *Environmental Science & Technology* 42:2457–63.

Vogt, C., E. Cyrus, I. Herklotz, D. Schlosser, A. Bahr, S. Herrmann, H.-H. Richnow, and A. Fischer. 2008. Evaluation of toluene degradation pathways by two-dimensional stable isotope fractionation. *Environmental Science & Technology* 42:7793–800.

Wang, Y., and Y. Huang. 2003. Hydrogen isotopic fractionation of petroleum hydrocarbons during vaporization: Implication for assessing artificial and natural remediation of petroleum contamination. *Applied.. Geochemistry* 18:1641–51.

Wersin, P., J. Abrecht, and P. Höhener. 2001. Large scale redox plume in glaciofluvial deposits due to sugar-factory wastes and waste water at Aarberg, Switzerland. *Hydrogeology Journal* 9:282–96.

Zwank, L., M. Berg, M. Elsner, T. C. Schmidt, R. P. Schwarzenbach, and S. B. Haderlein. 2005. New evaluation scheme for two-dimensional isotope analysis to decipher biodegradation processes: Application to groundwater contamination by MTBE. *Environmental Science & Technology* 39:1018–29.

2 Analysis of Stable Isotopes

Daniel Hunkeler and Stefano Bernasconi

CONTENTS

2.1 INTRODUCTION

This chapter discusses the most important methods to measure light stable isotopes in compounds that are of interest when investigating biogeochemical cycles and contaminated sites. A particular emphasis is given to methods for isotope analysis of specific compounds such as organic contaminants that rely on separation techniques such as gas chromatography. However, methods for analysis of bulk samples where the entire sample is analyzed without prior separation are covered as well. The general principles of the methods are outlined. For detailed procedures, the reader is referred to the specific publications and protocols developed by different laboratories. The procedures for calculation of isotope ratios from raw data of the instruments are also covered here although this is automatically performed by most software packages.

2.2   PRINCIPLES OF ISOTOPE ANALYSIS

The isotope ratios of light elements are most commonly measured with gas-source magnetic sector mass spectrometers often denoted as isotope ratio mass-spectrometers (IRMS). Their measurement principle is different from that of mass-spectrometers commonly used for identification of organic compounds. The first type of these instruments was developed in 1947 by Alfred Nier. With these instruments, the molecules of interest are introduced in gaseous form into the ion source, are ionized, and travel through a magnetic field (Figure 2.1). The ions are deflected with a different radius depending on their mass/charge ratio making it possible to detect them in separate collectors, usually Faraday cups. Thus, what is measured is not directly the abundance of isotopes of an element but the abundance of small molecules with different mass (denoted as isotopologues) containing the element of interest. To be able to calculate the isotope ratio from the ion ratios, the measurement molecules should contain as few atoms as possible and at the same time it should be as easy to transform the molecules of interest into the measurement gas.

The types of molecules that are more commonly used for analysis of different isotopes are listed in Table 2.1. For carbon and oxygen, molecules with three different masses are generally measured using a universal triple collector setup to be able to calculate isotope ratios despite of the existence of multiple combinations of isotopes giving the same mass (see 2.9 for details). For hydrogen, nitrogen, and sulfur, two different masses are sufficient because the abundance of other isotopologues is very low. To be able to measure the isotopic composition of an element (e.g., carbon) in a compound (e.g., benzene), the compound has to be first transformed to the corresponding molecule of measurement (e.g., CO_2). With this approach, the mass spectrometer can be optimized for a few molecules allowing for a high precision of measurement. The measurement of isotope ratios always takes place relative to the isotope ratio of a reference gas (same type of molecules with a known isotopic composition) and thus, potential mass discrimination during transfer, ionization, or counting of the molecules cancels out. Although isotope ratios are usually expressed relative to an international standard using the delta notation (see Chapter 1), the international standard is not usually used as reference gas because it is not available in sufficient quantity. Rather, a laboratory working standard (e.g., a tank of CO_2 gas) is calibrated against the international standard. Thus, the measured isotope ratios relative to the laboratory standard have to be transformed to isotope ratios relative to the international standard.

More recently, new method for isotope analysis based on laser spectroscopy rather than mass spectrometry have been developed (e.g., Lis, Wassenaar, and Hendry 2008). The method relies on differences in vibrational frequencies of molecules depending on the isotope that is present in the chemical bond. The higher the mass of the isotope, the lower the vibrational frequency. By measuring the absorption of electromagnetic radiation at different wavelengths, the abundance of different isotopes can be quantified. The method is available to measure isotope ratios in H_2O, CO_2, and CH_4 and generally less expensive equipment is required than for mass spectrometry. While first instruments coupled to

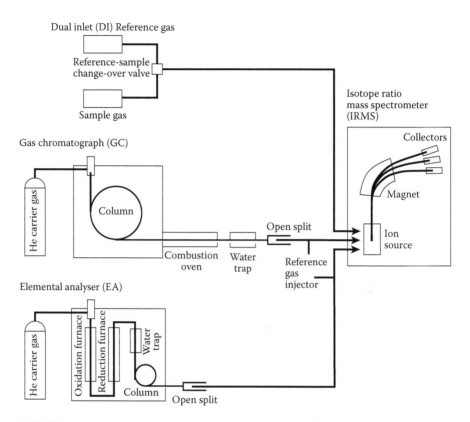

FIGURE 2.1 Schematic setup of dual inlet isotope ratio mass spectrometer (DI-IRMS) and continuous flow systems consisting of a gas chromatograph (GC) connected via a combustion interface to an IRMS or an elemental analyzer (EA) connected to an IRMS.

elemental analysers (EA) are emerging, these instruments cannot (yet) be linked to gas chromatography (GC) systems.

2.3 DUAL INLET AND CONTINUOUS-FLOW SYSTEMS

There are two different approaches of introducing the gaseous molecules into the ion source of an IRMS. The classic dual inlet approach dual inlet-isotope ratio mass spectrometer (DI-IRMS) uses two reservoirs containing pure reference and sample gas, respectively (Figure 2.1) at the same pressure to avoid pressure-dependant mass discrimination effects. Several pulses of sample and reference gas are injected alternately and the isotope ratio of the sample is calculated relative to the reference gas. While the dual inlet method allows for very precise measurements, it requires a time-consuming transformation of the compounds of interest to the measurement gas using preparation lines. Beginning in the 1970s, a new way of introducing sample and reference gas has been developed whereby the measurement molecules are carried to the ion source as pulses of measurement gas in

TABLE 2.1

Measurement Gas and Isotopologues Evaluated When Analysing Isotope Ratio of Different Elements. Minimal Amount of Compound Required for GC-IRMS Applications

Element	Isotope Ratio	Measurement Gas	Minimal Amount nmol of Element for GC-IRMS Analysis and Typical Uncertainty	Isotopologues and Thier Mass	
H	$^2H/^1H$	H_2	30 (5‰)	2	$^1H^1H$
				3	$^1H^2H$
C	$^{13}C/^{12}C$	CO_2	1 (0.5‰)	44	$^{12}C^{16}O^{16}O$
				45	$^{13}C^{16}O^{16}O$, $^{12}C^{16}O^{17}O$
				46	$^{12}C^{16}O^{18}O$, $^{13}C^{16}O^{17}O$, $^{12}C^{17}O^{17}O$
N	$^{15}N/^{14}N$	N_2	3 (0.5‰)	28	$^{14}N^{14}N$
				29	$^{14}N^{15}N$
				30	$^{15}N^{15}N$
O	$^{18}O/^{16}O$	O_2	5 (0.8‰)	32	$^{16}O^{16}O$
				33	$^{16}O^{17}O$
				34	$^{16}O^{18}O$, $^{17}O^{17}O$
		CO_2		See above	
		CO		28	$^{12}C^{16}O$
				29	$^{13}C^{16}O$, $^{12}C^{17}O$
				30	$^{12}C^{18}O$, $^{13}C^{17}O$
S	$^{34}S/^{32}S$	SO_2		64	$^{32}S^{16}O^{16}O$
				66	$^{32}S^{18}O^{16}O$
		SF_6		146	$^{32}S\ ^{19}F_6$
				148	$^{34}S\ ^{19}F_6$

a continuous stream of helium (Figure 2.1). This method is denoted as continuous-flow isotope ratio mass spectrometry (CF-IRMS). These instruments can be directly coupled with other instruments that convert the substance of interest into the measurement molecules (e.g., CO_2). Thus, the preparation of the measurement gas and analysis of its isotopic composition can be performed in one single process, which greatly reduces the time necessary for sample analysis. Furthermore, the required sample amount is much smaller compared to DI-IRMS making it possible to analyze isotope ratios of environmental compounds that can be present at relatively low concentrations.

The development of CF-IRMS for an increasing number of elements and coupled to an increasing number of instruments has led to a much wider application of isotope methods in many disciplines. The most common combinations are: The coupling of a GC via a conversion interface to the IRMS to measure isotope ratios of individual compounds in mixtures (e.g. benzene in gasoline), the coupling of an EA to an IRMS to quantify the average isotope ratio of solid or liquid samples (bulk samples), the coupling of an IRMS with a preparation system that makes it possible to generate the measurement gas with simple chemical reactions (e.g., liberation of CO_2 from carbonates by addition of phosphoric acid), and, more recently, the coupling of a high performance liquid chromatography (HPLC) system with an IRMS. These combinations are commercially available from several manufacturers. Other combinations have been developed as well by research laboratories such as the combination of a dissolved organic carbon/total organic carbon (DOC/TOC) analyzer with an IRMS (St-Jean 2003).

In CF-IRMS systems, the sample gas travels a relatively long distance being subject to separation and transformation processes, while the reference gas is usually injected immediately before the helium stream enters the IRMS via a reference gas injection box (Figure 2.1). As a result of the different pathways, the reference and sample gas peaks have a different shape and often a different height, in contrast to dual inlet systems, where both the reference and sample gas produce a signal of similar intensity and similar rectangular peak shape (Figure 2.2). Sample peak shapes of CF-IRMS systems are typically Gaussian instead of square shaped due to dispersion effects and the intensity of the sample peak is generally variable depending of the amount of sample introduced for example into the GC or EA (Figure 2.2). However, newly developed mass spectrometers provide accurate isotope ratio values despite these differences in peak shape and intensity between sample and reference gas.

A particular challenge for CF-IRMS systems is the measurement of H_2 and HD with a mass of 2 and 3 in a stream of helium with a mass of 4. Due to the collision of He ions, some of them will deviate from their trajectory and fall into the mass 3 detectors complicating accurate measurements of HD to H_2 ratios. Electrostatic filters added in front of the mass 3 collector were developed to eliminate the contribution of helium to mass 3. A further difficulty of H_2 isotope analysis, common to dual inlet and continuous-flow systems, is that in the ion source H_2 molecules react with H_2^+ to produce H_3^+ ions, which have the same mass as HD (Sessions, Burgoyne, and Hayes 2001). The amount of produced H_3^+ is proportional to the square of the pressure of H_2 in the source. Fortunately, the H_3^+ production is sufficiently consistent and stable so that it can be corrected for. The correction factor (usually denoted as H_3^+-factor) is usually determined by injecting reference gas H_2 at a different pressure and carrying out a linear regression between apparent mass 3 and signal intensity.

2.4 COMPOUND-SPECIFIC ISOTOPE ANALYSIS (CSIA)

2.4.1 Coupling with Gas Chromatography (GC)

The coupling of gas chromatographs with an IRMS makes it possible to analyze isotope ratios of individual compounds in complex mixtures (Matthews and Hayes 1978; Meier-Augenstein 1999). Such methods are usually referred to as compound-specific isotope analysis (CSIA). The instrument is frequently denoted as isotope-ratio-monitoring gas chromatography-mass spectrometry (IRM-GCMS) or more commonly gas chromatography isotope-ratio mass spectrometry (GC-IRMS). The latter abbreviation may be modified depending on the type of conversion interface for example into GC-C-IRMS if a combustion interface is used. The basic components of

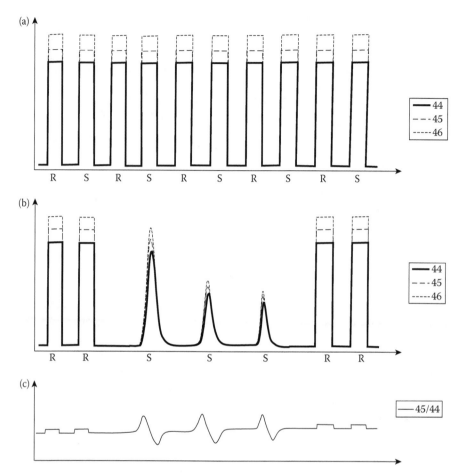

FIGURE 2.2 Signals for different CO_2 masses (see Table 2.1) recorded (a) by dual inlet IRMS, (b) by continuous flow IRMS, and (c) ratio between mass 45 and 44 for continuous flow IRMS. Signals for different masses are scaled to yield peaks of comparable height. R = Reference gas. S = Sample gas.

these analytical systems are illustrated in Figure 2.1. A key component of the coupled system is the interface that converts the compounds eluting from the GC to the measurement gas. The conversion interface is connected to IRMS via an open split. The open split consists of a capillary that is connected to the vacuum of the IRMS and draws in a part of the sample gas stream (typically 0.2–0.5 ml/min) while the remainder is vented. For carbon, the conversion consists of combustion of the compounds to CO_2 using a narrow bore reactor tube (0.5–1 mm bore) that is either packed with Cu-oxide pellets (850°C) or a combination of copper/nickel/platinum wires (940°C). The wire system has a smaller oxidant capacity and therefore needs to be reoxidized regularly with O_2, which can be done automatically. The water from combustion has to be removed because it would lead to protonation of CO_2 to produce HCO_2^+ in the ion source preventing accurate measurements. Furthermore, condensation of water in the mass spectrometer can shorten electronic contacts. Water removal takes place either by a cryogenic trap with liquid nitrogen or a Nafion membrane tube in a counterstream of helium.

For nitrogen isotopes, usually the same type of combustion reactor is used as for carbon. The combustion leads to the formation of nitrous oxides that are reduced to N_2 in a separate downstream reactor packed with copper and held at 650°C, followed by removal of water and CO_2. During nitrogen isotope analysis there is a higher contamination risk during sample processing and analysis than for CO_2 due to the high concentration of N_2 in the atmosphere. For analysis of hydrogen isotopes, the organic compounds are converted to H_2 either in an empty ceramic tube at a high temperature (typically 1400°C) or at a lower temperature in a tube filled with chromium as a catalyst. The first procedure has been shown to lead to complete conversion of organic compounds to H_2 (Burgoyne and Hayes 1998). For analysis of oxygen isotopes, the organic compounds are transformed to CO in an empty tube with a platinum inner surface at 1280°C. For chlorine isotopes, the common measurement gas is methyl chloride. While continuous-flow methods involving a GC (Shouakar-Stash, Drimmie, and Frape 2005) or EA (Wassenaar and Koehler 2004) were developed, no interface is available yet for on-line conversion of chlorinated hydrocarbons to methyl chloride. However, recently, it was demonstrated that the chlorine isotope ratio of chlorinated ethenes can be measured by directly introducing them into the ion source of an IRMS with a special collector setup after GC separation (Shouakar-Stash et al. 2006). For tetrachloroethene (PCE) the fragment ions 94/96 are measured, for trichloroethene (TCE) the fragment ions 95/97, and for dichloroethene (DCE) the molecular ions 96/98. The method requires only around 4 nmol Cl on the GC column. An alternative method consists of the coupling of a GC with a multicollector-inductively coupled plasma mass-spectrometer (ICP-MS) (Van Acker et al. 2006). While any organic compound that passes through a GC and contains Cl can be analyzed with this approach, the required mass is much higher and the instrumentation more expensive. Furthermore, it was recently shown that Cl isotope ratio can also be derived from ion patterns measured by quadrupole mass spectrometers (Sakaguchi-Soder et al. 2007). However, the precision of the measurement is lower than for IRMS systems.

Some of the main differences of GC-IRMS compared to other GC methods (e.g., GC-MS) are that a larger amount of compound is required and that the compounds should be baseline separated in the GC. The main reason for the second condition is that during separation of the compounds in the GC column, the molecules with heavy isotopes travel slightly faster than molecules with light isotopes due to a small isotope effect associated with partitioning between mobile and stationary phase. As a result, the initial part of the peak is slightly enriched in the heavy isotope (e.g., ^{13}C) while the peak end is depleted in the heavy isotopes. This trend is most apparent when the ratios of the ion traces (e.g., 45/44) are examined (Figure 2.2). The ratio trace first increases as slightly more CO_2 with ^{13}C arrives and then decreases. Precise measurements are only obtained if the entire peak is integrated. Baseline separation can be more challenging for GC-IRMS than for other GC systems because some additional dispersion occurs within the combustion interface and during transfer to the IRMS, widening the peaks.

Approaches to optimizing the GC resolution are not discussed in detail here since they are analogous to those used for any other GC methods and the reader is referred to the extensive literature on the subject. Important considerations are to inject the compounds in a short pulse (= narrow band), to use narrower and longer columns, or the selective purification of the samples before injection. Narrow bands can be reached by using retention gaps and low initial GC temperature, by condensing the sample at the column head by cooling it down (cryogenic refocusing) or by split injection, which ever leads to a loss of mass. When using narrow diameter columns, care has to be taken not to overload the column given that GC-IRMS requires a larger amount of compound. Furthermore, when considering longer columns it has to be kept in mind that the resolution only increases with the square root of column length (e.g., by 1.4 when the column length is doubled from 30 to 60 m).

With GC-IRMS systems, isotope ratios in almost any organic compound that can be introduced on a GC column can be analyzed as long as the requirements regarding amount and resolution described above are fulfilled and the accuracy and precision of the measurement is carefully evaluated. In some cases, the presence of other elements can interfere with the conversion in the interface. For example, the presence of Cl in organic molecules complicates the analysis of H isotopes due to the formation of HCl in the high temperature interface or Cr-chloride in the catalytic interface. However, for compounds containing only C, H, O, and N such interferences are not expected.

2.4.2 Coupling with Liquid Chromatography

The first coupling of liquid chromatography with isotope ratio mass-spectrometers (IRMS) was described by (Mohammadzadeh et al. 2005) and brought on the market in 2005, much later than GC-IRMS. The method opens a wide range of new applications of IRMS because many types of molecules (e.g., pharmaceutical) or many biomolecules (e.g., amino acids, carbohydrates, RNA, DNA) are too polar or thermo-labile to easily pass through a GC. So far only systems for carbon are available. With these instruments, the organic molecules eluting from the HPLC column are transformed to CO_2 by chemical oxidation in the liquid phase. The produced CO_2 is transferred to the gas phase, dried, and introduced into the IRMS. Because wet oxidation is used, only water can be used as mobile phase during HPLC separation restricting the possibilities for optimizing chromatographic resolution. The method requires slightly more mass of compound (around 4 nmol C) than GC-IRMS because the peaks are broader.

2.4.3 Sample Preparation

For environmental samples, a major challenge is often to introduce a sufficient mass of compound for analysis since compound concentrations in the environment can be low. In the following, several approaches for aqueous sample extraction and injection for GC-IRMS analysis are discussed with a special focus on compounds that are of particular interest in environmental studies including CO_2, methane, monoaromatic hydrocarbons (benzene, toluene, ethylbenzene, xylene = BTEX), polyaromatic hydrocarbons (PAH), and chlorinated solvents. For environmental applications, it is often desirable that the analytical method is sufficiently sensitive to analyze concentrations at least as low as regulatory limits or even lower, for example, to evaluate if low concentrations are a result of dilution or degradation. The most common methods for extracting aqueous samples are compared in Table 2.2. If the necessary equipment is available, the Purge & Trap technique is probably the method of choice for volatile compounds, although the system is susceptible to carry over effects if samples with highly variable concentrations are measured. When installing a purge and trap system (Zwank et al. 2003), it has to be ensured that the entire amount of trapped compounds are transferred to the GC column

TABLE 2.2
Methods for Sample Extraction in GC-IRMS Applications

Method References	Advantage	Disadvantage
Purge and trap	• Large fraction of compound in water can be extracted yielding low detection limit	• Required expensive equipment • Risk of carry over
Solid phase microextraction (SPME)	• Simple, fast, standard injector can be used • Low detection limit for compounds with elevated K_{ow}	• Some sorption competition occur when most sensitive carboxen-PDMS fibers are used for complex mixtures • High detection limit for compounds with low K_{ow} (e.g., vinyl chloride) • Fiber where sorption takes place is fragile
Headspace	• Simple, standard injector can be used • Suited for very volatile compounds such as methane, ethane, ethene	• Only small portion of mass is transferred, leading to high detection limit unless compounds are very volatile
In-needle extraction methods (e.g., ITEX, SPDE)	• Higher sorption capacity than SPME • Simple and robust set-up • Lower detection limits than SPME and headspace	• Less sensitive than purge and trap
Liquid–liquid extraction	• Low detection limit if sufficient volume is extracted • Standard injector can be used	• Time consuming • Separation of very volatile compounds from solvent can be difficult • Only small fraction of what has been extracted can be injected unless large volume injector used • Large volume injector suited for high molecular weight compounds but lead to loss of more volatile compounds possibly associated with fractionation
Twister	• Simple and fast • Higher sorption capacity than SPME	• Requires special desorption unit • Little choice of sorbent material

without losses within a split injector. For very volatile compounds such as vinyl chloride, this may require cryofocusing after desorption. For compounds with an elevated octanol-water partition coefficient, solid phase micro extraction (SPME) is a good alternative (Dias and Freeman 1997; Hunkeler and Aravena 2000; Zwank et al. 2003). With this method, a small volume of sorbent attached to a fiber is exposed to the gas phase or aqueous phase of the sample, followed by desorption directly in the GC injector. The method has the advantage that it can be used with standard injectors and different types of sorbents are available. However, the volume of sorbent quite small and the fiber quite fragile. Several recently developed systems have tried to overcome these limitations, for example, by coating a magnetic stirrer with a larger mass of sorbent (Gerstel-Twister, Mülheim an der Ruhr, Germany) for extraction of dissolved compounds or by packing a syringe needle with sorbent for extracting compounds from the headspace of samples (e.g., In tube extraction, ITEX; CTC Analytics, Switzerland or Solid Phase Dynamic Extraction, SPDE; Chromtech, Germany). For dissolved gases such as methane, ethane, and ethene, headspace extraction is the best suited method. To obtain narrow bands, split injection has to be used, which leads to some loss in sensitivity. Lower detection limit can be reached using a sample loop or by trapping a larger volume of headspace gas using a cryogenic system. Using standard purge and trap systems with 25 ml sample volume, detection limits for carbon isotopes in the range of a few $\mu g/L$ can be reached for chlorinated hydrocarbons and around one $\mu g/L$ or lower for BTEX compounds (Jochmann et al. 2006; Zwank et al. 2003). For hydrogen isotopes, the detection limits for petroleum hydrocarbons are about a factor 30 higher, while chlorinated hydrocarbons are difficult to analyze due to the liberation of Cl during pyrolysis.

Sample extraction generally involves one or several phase transfer processes (e.g., from the aqueous phase to the gas phase or from the aqueous phase to the SPME fiber), which can be associated with small isotope fractionation (Hunkeler and Aravena 2000). As long as extraction is performed under identical conditions for all samples, isotope fractionation is usually reproducible and shows little dependence on the amount of compound present in the sample. Further isotope fractionation can occur especially during injection. For example, Schmitt, Glaser, and Zech (2003) observed isotope fractionation during split injection of CO_2 and gaseous BTEX compounds, whereby the amount of fractionation depended not only on the split ratio but also on the amount of compound injected, which complicates corrections. A particularly large isotope fractionation was observed for CO_2 injection at a low split ratio. However, in other studies (Miyajima et al. 1995; Salata, Roelke, and Cifuentes 2000) using a different brand of GC, such fractionation for CO_2 was not observed suggesting that it may depend on the injector design and indicating that results obtained with one analytical system can not necessarily be extrapolated to others. Fractionation during injection can be minimized by using splitless injection, which is not always possible for example when analyzing volatile compounds.

2.4.4 SAMPLING AND SAMPLE PRESERVATION

During sampling, the same precautions have to be taken as with sampling for concentration analysis. Loss of compound due to sorption and degassing has to be avoided even though the effect of these losses may be less critical for isotope then for concentration analysis since they are generally only associated with very small isotope fractionation. Larger shifts can potentially occur due to biodegradation and should be avoided by adding preservatives directly in the field. The most common approach is to decrease the pH below 2 by adding hydrochloric acid or to increase it above 10 by adding tribasic sodium phosphate (Kovacs and Kampbell 1999) or sodium hydroxide. Both approaches are adapted for a wide range of volatile organic compounds with a few exceptions: methyl tertiary-butyl ether (MTBE) can hydrolyze under acidic conditions and chlorinated ethanes under basic conditions. When adding base, carbonates precipitate, which usually will not interfere with sample extraction. When adding acid, some small gas bubbles may form. Similar to concentration analysis carryover from sampling points with high concentration to those with low concentration should be avoided.

2.5   ISOTOPE ANALYSIS OF BULK SAMPLES

There are two main types of analytical devices used for bulk analyses. Bulk analysis refers to the analyses of the average isotope ratio of the complete sample without prior separation into different fractions. The first device to be developed was the coupling of an elemental analyzer to an IRMS (EA-IRMS) to measure the carbon, nitrogen, and sulfur isotopic composition of bulk samples in liquid or solid form (Giesemann et al. 1994; Matthews and Hayes 1978). A more recent development is the coupling of pyrolysis devices (Gehre and Strauch 2003) to analyze oxygen and hydrogen isotopes by high temperature (generally around 1450°C) thermal decomposition of the substances of interest in an inert helium atmosphere (sometimes called pyrolysis IRMS (PY-IRMS) or thermal conversion IRMS (TC-IRMS)).

2.5.1   ELEMENTAL ANALYZER-ISOTOPE RATIO MASS SPECTROMETER (EA-IRMS) FOR C, N, AND S ISOTOPE ANALYSIS

The EA-IRMS systems allow C, N, and S isotopes to be measured in all substances that can be combusted quantitatively to CO_2, N_2, and SO_2 in an oxidizing atmosphere. This is accomplished by interfacing a standard elemental analyzer to an IRMS via an open split. The open split is necessary, as the high flow of helium of 80 to 100 ml/min from the elemental analyzer cannot be accommodated by the mass spectrometer pumping system. The basic schematic of such a system is presented in Figure 2.1. The sample of interest is filled in tin or silver capsules, with or without the addition of other substances to improve the combustion, depending on the sample type. The samples are dropped via an autosampler in a furnace kept at 1030–1060°C. The drop of the sample in the furnace is timed with an oxygen pulse that reaches the combustion chamber immediately after the sample to ensure rapid and complete combustion. The combustion furnace is filled with chromium oxide and silvered cobaltous oxide to remove halogens. The gas is then passed over copper at 650°C in a second furnace to quantitatively reduce nitrogen oxides to N_2 and remove excess oxygen. For sulfur isotope analysis only one furnace is used. It is filled with WO_3 in the hottest zone and elemental copper to convert any SO_3 to SO_2 toward the bottom of the oven. The gases are subsequently passed across a water trap, separated on a packed GC column and transferred into the IRMS via an open split. The reference gas, to which the sample is compared when calculating the isotope ratios, is injected either via a reference gas injection box or alternatively via the second inlet of a DI-IRMS. Standardization and calibration of the system is accomplished by the inclusion of substances of known isotope composition that are treated exactly as the samples in each measurement batch. This allows correcting for possible isotope fractionations induced by the preparations system. When measuring small samples it is also necessary to consider the carbon blank deriving from the tin and silver capsule. Depending on the producer, it generally amounts to a few micrograms, which can significantly influence the results. For the measurement of sulfur isotopes in sulfates and sulfides, it is necessary to add vanadium pentoxide to the sample, generally in a 1:1 ratio as a catalyst to improve combustion.

2.5.2   PY-IRMS for O and H Isotope Analysis

Pyrolysis units for stable isotope analyses have rapidly evolved in the last few years. The principle is similar to that of an elemental analyzer, with the exception that the samples are heated at temperatures of up to 1450°C in absence of oxygen. At a high temperature, the materials of interest decompose to hydrogen gas and by adding a source of carbon to carbon monoxide. These gases are subsequently separated by GC in a packed GC column and transferred to the IRMS for measurement. The initial developments of PY-IRMS for the measurement of oxygen isotopes in cellulose were carried out with standard EA where the temperature of the furnace is limited to 1100°C. The furnace was packed with either glassy carbon or nickelized carbon to convert the oxygen liberated from the cellulose to carbon monoxide. Although these methods produced reliable results for cellulose $\delta^{18}O$, it has proven difficult to obtain good results with other substances. A decisive improvement came with the introduction of high temperature furnaces that allowed a 100% conversion of the pyrolyzed substances to H_2 and CO without the production of CH_4 and CO_2, which is observed at lower temperatures (Burgoyne and Hayes 1998). In addition, improvements in the furnace design have reduced the memory effects that were often observed in the first systems. With the high temperature conversion, oxygen isotopes in solids such as nitrates, sulfates, cellulose, or other organic substances as well as in water and other oxygen containing liquids can be obtained. Pyrolysis units can also be used for hydrogen isotope determination of waters, organic compounds, and hydrous minerals. Similar to elemental analyzer-isotope ratio mass spectrometer (EA-IRMS), calibration and standardization is accomplished by measurement of standards of known composition. Cross-calibrated standards for oxygen isotopes are available for several substances of interest, such as nitrates and sulfates (Brand et al. 2009).

2.5.3   Sample Preparation

For carbon and nitrogen isotopes of solid organic matter, the samples are best freeze-dried or dried in an oven at 60°C. The samples are then finely powdered, homogenized with a mortar and pestle, and filled into the tin capsules for measurement. No additional preparation is needed unless the sample contains carbonate minerals.

In this case carbonates have to be removed either with diluted acid or by fumigation over concentrated hydrochloric acid. For oxygen and hydrogen isotopes, organic matter has to be cleaned of any other substance containing either hydrogen or oxygen (e.g., water, carbonates, and clay minerals), as those would contaminate the measurement by decomposing as well. Particular care has to be taken with many organic substances that are hygroscopic. Cellulose, for instance, readily adsorbs moisture from the laboratory air and thus, exposure to air has to be minimized. For hydrogen isotopes in organic substances, it also has to be considered that oxygen-bound hydrogens such as those in hydroxyl groups readily exchange with ambient water in contrast to carbon-bound hydrogen. Thus, oxygen-bound hydrogens have to be eliminated by substitution with another functional group such as is done by nitration of cellulose or by equilibration with water vapor of known isotopic composition (Schimmelmann, Lewan, and Wintsch 1999). Organic liquids, if pure, can also be measured without previous separation during preparation. Precautions have to be taken with volatile organic compounds as they can evaporate. Special care has to be taken when they are introduced in the chamber of the autosampler and are waiting to be dropped because the chamber is relatively hot and is flushed with helium. This can lead to partial or complete loss of compound with a low boiling point and will change their isotopic composition.

Both nitrate and ammonia can be separated from diluted freshwaters by ion exchange columns. Nitrate is eluted from the ion exchange column and subsequently precipitated as silver nitrate (Silva et al. 2000). Silver nitrate can be measured for N isotopes by EA-IRMS and oxygen isotopes by PY-IRMS. Ammonium can also be isolated with ion exchange resins from diluted waters (Lehmann, Bernasconi, and McKenzie 2001). By using a nitrogen–free resin, ammonium can be measured by direct combustion of the resin without additional preparation steps. Ion exchange resins cannot be used in saline waters and soil extracts as the high ion concentrations in these solutions saturate the resins rendering them useless. For saline waters, ammonia can be collected as ammonium sulfate by diffusion into an acidified glass fiber filter enclosed in polytetrafluoroethylene (PTFE) tape (Sigman et al. 1997). Nitrate can be reduced to ammonium with Devarda's alloy (an alloy of aluminium, copper, and zinc) and subsequently be collected as ammonium sulfate. This method, however, does not allow the determination of oxygen isotopes in nitrate. Recently, a powerful method using denitrifying bacteria, which lack the enzyme to completely convert nitrate to N_2, has been developed. With these bacterial cultures, both the oxygen and nitrogen isotope composition of nitrate can be determined on the same sample—by analyzing the N_2O produced by the bacteria (Casciotti et al. 2002; Coplen, Böhlke, and Casciotti 2004). Sample size requirements vary with nitrate and ammonium concentrations. The denitrifier method has been tested for concentrations as low as 1 μM NO_3, the method by Lehmann, Bernasconi, and McKenzie 2001 to concentrations of 40 μM NH_4.

Sulfate and sulfide can be precipitated from the solution to determine their isotope composition. When both species are present, sulfide can be precipitated first by adding silver nitrate, cadmium acetate, or zinc acetate directly to the sample. The only precaution is that oxidation of the sulfide is avoided. The precipitated sulfide can be separated by filtration or centrifugation and sulfate can be subsequently precipitated by the addition of barium chloride. To avoid co-precipitation of barium carbonate, the sample should be acidified to a pH of approximately 5. The precipitated pure barium sulfate can be analyzed for sulfur and oxygen isotopes. Care should be taken not to reduce the pH too low or oxygen isotope exchange with the water could take place.

2.6   OTHER METHODS

The methods described above involve the transformation of the compounds of interest to the measurement gas at high temperature. A number of compounds can be analyzed using simpler methods, especially isotope ratios of dissolved inorganic carbon (DIC), carbonates, and dissolved oxygen. For analysis of DIC, the water samples are dispensed into a closed vessel, a headspace is introduced while avoiding contamination with atmospheric CO_2, followed by acid—normally phosphoric acid. The acid will transform all DIC to carbonic acid and dissolved CO_2. After a period of equilibration, the CO_2 in the headspace is measured either using a GC-IRMS system (Miyajima et al. 1995; Salata, Roelke, and Cifuentes 2000) or injected to an IRMS via an automated preparation system. Carbonates can be analyzed similarly. The carbonate is dispensed into a vial, the gas phase is replaced by helium, and phosphoric acid is added to evolve CO_2. Usually the carbonate measurement is carried out with an automated preparation system connected to an IRMS. CO_2 from the soil and unsaturated zone can be injected directly into a GC-IRMS system.

For O_2 analysis, no reaction at all is necessary since the isotope ratio can directly be measured in the form of O_2. Usually, the O_2 is equilibrated between the aqueous phase and an initially O_2 free headspace and then the O_2 injected into a modified EA (Wassenaar and Koehler 1999) or a GC-IRMS system (Roberts, Russ, and Ostrom 2000). In both cases, H_2O and CO_2 are removed by traps and the O_2 is separated on a column from other gases before entering the IRMS.

2.7   METHOD DEVELOPMENT AND QUALITY CONTROL

The development and testing of a new method for CF-IRMS involves several elements:

1. Verification of the linearity of the IRMS.

   It has to be verified regularly that the measured isotope ratio is independent of the amount of reference gas that is injected and hence independent of the signal intensity. Otherwise isotope ratios of samples of a different concentration or size cannot be compared. For this purpose, reference gas peaks with different intensity covering the whole measurement range are injected. This procedure is usually denoted as a linearity test although the term may be somewhat confusing. In concentration analysis, linearity usually refers to a linear increase of a signal with concentration. In contrast, in the IRMS world, a high linearity indicates that the obtained isotope ratio is independent of the amount of reference gas or sample injected. In manufacturer's specifications, the linearity is usually reported as the maximal variation per nA or V of signal.

2. Evaluation of isotope fractionation and nonlinearity effects associated with sample processing.

A good linearity for reference gas does not imply a good linearity for the sample because the sample undergoes a number of transformation and separation steps and because the peak shapes of reference and sample gas pulses are different. Furthermore, it has to be verified that the measured and "true" isotope ratio agree (accuracy) and the variability during repeat measurements has to be quantified (precision), which is usually different when measuring specific compounds than when injecting reference gas. Therefore, the whole analytical chain has to be tested with compounds of known isotopic composition. For this purpose, certified reference material or, if not available, in-house reference material is analyzed in different quantities analogous to samples covering the whole measurement range. This test has to be carried out exhaustively during method development and to a lesser degree each time the method is used. Depending on the outcome, different types of correction procedures are applied assuming a satisfactory precision:

- High linearity and accuracy. This is the desired outcome and no corrections are needed.
- High linearity but occurrence of systematic offset between measured value and actual value of reference material. Such an offset can be due to reproducible isotope fractionation during one of the preparation steps (e.g., SPME extraction). If the offset is constant and statistically significant, a simple correction can be done.
- Limited linearity. Especially for EA/PY-IRMS and for hydrogen isotopes on some GC-IRMS systems, a significant dependence of the isotope ratio on the sample size is observed. If the relationship is stable during the measurement period, a correction can be carried out. The correction factor is commonly calculated from a linear regression of size versus isotope composition for a number of standards of variable size, which are interspersed with the samples in a measurement sequence.

3. Evaluation of detection limits.

Especially for contaminant studies where concentrations in some zones may be low, a good knowledge of the detection limit is necessary. The evaluation of the detection limit is usually carried out together with the second step by progressively diminishing the amount of substance analyzed. Below a certain signal intensity (e.g., 0.5 V), the signal may become strongly nonlinear. Jochmann and colleagues (2006) proposed a procedure for determining the detection limit. According to their method, the detection limit is reached when the mean value of a standard ($n = 3$) deviates by more than a predefined interval (e.g., 0.5‰ for carbon) from the mean value of a series of standards at higher concentration levels.

During routine analysis a number of quality control measures should be taken including

- Daily verification of the precision and linearity of the system.
- Daily analysis of reference material at different amounts covering the measurement range to verify the linearity and accuracy for samples.
- Maintaining a control chart with results of the analysis of reference material to rapidly identify deviations from normal operation conditions. Control charts are a standard tool in analytical chemistry consisting of a chart in which the results of analysis of reference compounds are plotted as a function of time.

2.8   REPORTING UNCERTAINTY

As with other analytical methods, isotope data should be reported with the corresponding measurement uncertainty. When analyzing a sample or reference compound several times under exactly the same conditions, usually a high precision can be obtained, which varies depending on the element. However, it does not reflect the overall uncertainty of the method. If the same sample were analyzed again at different concentrations, the obtained values may be again precise but slightly offset from the first value due to some nonlinearity. Since samples are usually measured at different signal intensities, the reported uncertainty should reflect the overall uncertainty over the whole measurement range including the uncertainty due to slight nonlinearity effects (i.e., a slight amount dependence of the isotope ratio). The overall uncertainty can be characterized by the variability observed during repeated injection of different amounts of reference material. Typical uncertainties for GC-IRMS are given in Table 2.1. The uncertainty is larger for hydrogen isotopes than for other isotopes due to the low abundance of ^2H and the low ion yield in the source of the IRMS. The uncertainties obtained by repeated sample injections are larger than those obtained by multiple reference injections directly into the IRMS because separation and transformation steps add additional errors. The IAEA and the U.S. EPA have planned a detailed guideline for CSIA of organic contaminants to be published. Additional details on quality control measures and the quantification of uncertainty can be found in an U.S. EPA guideline (U.S. EPA, 2009).

2.9   CALCULATION OF ISOTOPE RATIOS

The IRMS system provides information about ratios of isotopologues (identical molecules with a different isotopic composition) relative to the reference gas from which the isotope ratio of elements in the molecules are calculated. The calculation procedure is discussed in detail for CO_2, probably the most commonly measured molecule. The basic information obtained from the IRMS are electrical signals proportional to the amount of ions with specific masses (44, 45 and 46 in case of CO_2) reaching the detectors (ion currents). In a first step, the true isotopologue ratio has to be calculated from the measured ion-current ratios based on the known true isotopologue ratio of the reference gas:

$$^{45}R_{sample} = \frac{^{45}I_{sample}}{^{45}I_{reference}} \cdot {}^{44}R_{reference}. \tag{2.1}$$

Where

$^{45}R_{sample}$ is the ratio between isotopologues with mass 45 and 44, respectively in the sample gas.

$^{45}R_{reference}$ is the ratio between isotopologues with mass 45 and 44, respectively in the reference gas that can be calculated from the known isotope ratio of the reference gas.

$^{45}I_{sample}$ is the measured ratio between the ion current for mass 45 and 44, respectively for the sample gas.

$^{45}I_{reference}$ is the measured ratio between the ion currents for mass 45 and 44, respectively for the reference gas.

A direct calculation of the $^{13}C/^{12}C$ isotope ratio from isotopologue ratio of CO_2 with mass 44 and 45 is not possible. While the mass 44 only contains CO_2 molecules with $^{12}C(^{12}C^{16}O_2)$, mass 45 contains CO_2 with $^{13}C(^{13}C^{16}O_2)$ as well as a smaller amount of CO_2 with $^{12}C(^{12}C^{16}O^{17}O)$. Thus the mass 45 has to be corrected for the contribution of $^{12}C^{16}O^{17}O$ for accurate $^{13}C/^{12}C$ calculations. Similarly when calculating $^{18}O/^{16}O$ it has to be taken into account that mass 46 also contains $^{12}C^{17}O_2$ and $^{13}C^{17}O^{16}O$ apart from $^{12}C^{18}O^{16}O$. When performing the correction, one is confronted with the problem that there are two measured parameters, the 45/44 and 46/44 ratio and three unknowns, the $^{13}C/^{12}C$, $^{18}O/^{16}O$, and $^{17}O/^{16}O$ ratio that leads to a no unique solution. A further equation is required. For that purpose usually a relationship between ^{17}O and ^{18}O content is used, which relies on the assumption that in all processes these two isotopes are fractionated proportionally whereby ^{17}O shows less fractionation than ^{18}O. The classical Craig correction (1957) uses the following relationship between ^{17}O and ^{18}O: $(^{17}R/^{17}R_{st}) = (^{18}R/^{18}R_{st})^a$ with $a = 0.5$ where ^{17}R and $^{17}R_{st}$ are the $^{17}O/^{16}O$ of the sample and standard, respectively and ^{17}R and $^{17}R_{st}$ the $^{18}O/^{16}O$ ratios of the sample and standard, respectively. However, it was shown later that a is not constant but varies between 0.5 and 0.53. Based on measured data for ^{18}O and ^{17}O, Santrock, Studley, and Hayes (1985) proposed the relationship $^{17}R = k \cdot ^{18}R^a$ where k is a constant characteristic of the relationship between ^{17}R and ^{18}R in the terrestrial oxygen pool ($K = 0.0099235$). For a, they selected an average value of 0.516.

To calculate isotope ratios, equations that relate the isotope ratios of elements to the isotopologue ratios are derived. The equations rely on the assumption that the probability of a certain isotope to be at a certain position in the molecule corresponds to its abundance and is independent of other isotopes present at other positions. Taking ^{13}C in CO_2 as an example, the probability that a ^{13}C occurs in the molecule is independent of which oxygen isotopes are present in the molecule and corresponds to the abundance of ^{13}C in the total CO_2 pool. The fractional abundance of an isotope corresponds to the amount of the isotope divided by the total amount of the corresponding element:

$$^{n}F_{i} = \frac{\text{amount of isotope with mass } n \text{ of element } i}{\text{amount of element } i}. \qquad (2.2)$$

Similarly the fractional abundance of an isotopologue with a certain mass is given by the amount of the isotopologue (e.g., CO_2 with mass 45) divided by the total amount of all isotopologues (e.g., CO_2).

$$^nM = \frac{\text{amount of isotopologue with mass } n}{\text{total amount of all isotopologues}}. \tag{2.3}$$

Using a simple probability function, the fractional abundance of the various isotopologues is then related to the fractional abundance of the isotope by:

$$^{44}M = {}^{12}F_C \cdot {}^{16}F_O \cdot {}^{16}F_O. \tag{2.4}$$

$$^{45}M = {}^{13}F_C \cdot {}^{16}F_O \cdot {}^{16}F_O + 2 \cdot {}^{12}F_C \cdot {}^{16}F_O \cdot {}^{17}F_O. \tag{2.5}$$

$$^{46}M = 2 \cdot {}^{12}F_C \cdot {}^{16}F_O \cdot {}^{18}F_O + 2 \cdot {}^{13}F_C \cdot {}^{16}F_O \cdot {}^{17}F_O + {}^{12}F_C \cdot {}^{17}F_O \cdot {}^{17}F_O. \tag{2.6}$$

By dividing Equation 2.5 and 2.6 by Equation 2.4, the equations can be expressed in terms of molecular and isotope ratios:

$$^{45}R = {}^{13}R + 2 \cdot {}^{17}R. \tag{2.7}$$

$$^{46}R = 2 \cdot {}^{18}R + 2 \cdot {}^{13}R \cdot {}^{17}R + {}^{17}R^2. \tag{2.8}$$

Using these two equations and the third equation relating ^{17}O and ^{18}O by Craig (1957) or Santrock, Studley, and Hayes (1985), the $^{13}C/^{12}C$ and $^{18}O/^{17}O$ ratios can be calculated. The approach of Craig involves some simplifications that lead to the following two equations where all molecular and isotope ratios are expressed in the delta notation:

$$\delta^{13}C = 1.0676 \cdot \delta^{45}R - 0.0338 \cdot \delta^{46}R. \tag{2.9}$$

$$\delta^{18}O = 1.0010 \cdot \delta^{46}R - 0.0021 \cdot \delta^{45}R. \tag{2.10}$$

The approach of Santrock, Studley, and Hayes (1985) avoids simplifications but requires a numerical procedure to calculate isotope ratios. The Craig (1957) correction is more commonly used because of its long tradition and simplicity. However, in some of the newer software packages the more accurate Santrock correction can be selected as well. All modern instruments calculate the isotope ratios directly. For isotopes other than carbon and oxygen in CO_2, analogous calculation procedures were developed. For molecules containing only one element (N_2 or O_2) the calculation is straightforward while for molecules composed of two or three elements it is more complicated. For SO_2, the correction of the contribution of $^{32}S^{16}O^{18}O$ in addition to $^{34}S^{16}O^{16}O$ to mass 66 required particular attention. A special calculation procedure is also required for evaluation of data for chlorinated compounds directly injected into an IRMS without prior conversion because they usually contain several Cl atoms and because the heavy isotope occurs at a high abundance (Elsner and Hunkeler 2008).

REFERENCES

Brand, W. A., T. B. Coplen, A. T. Aerts-Bijma, J. K. Bohlke, M. Gehre, H. Geilmann, M. Groning, H. G. Jansen, H. A. J. Meijer, S. J. Mroczkowski, H. P. Qi, K. Soergel, H. Stuart-Williams, S. M. Weise, and R. A. Werner. 2009. Comprehensive inter-laboratory calibration of reference materials for delta O-18 versus VSMOW using various on-line high-temperature conversion techniques. *Rapid Communications in Mass Spectrometry* 23 (7):999-1019.

Burgoyne, T. W., and J. M. Hayes. 1998. Quantitative production of H_2 by pyrolysis of gas chromatographic effluents. *Analytical Chemistry* 70:5136–41.

Casciotti, K. L., D. M. Sigman, M. G. Hastings, J. K. Böhlke, and A. Hilkert. 2002. Measurement of the oxygen isotopic composition of nitrate in seawater and freshwater using the denitrifier method. *Analytical Chemistry* 74 (19):4905–12.

Coplen, T. B., J. K. Böhlke, and K. L. Casciotti. 2004. Using dual-bacterial denitrification to improve delta N-15 determinations of nitrates containing mass-independent [17]O. *Rapid Communications in Mass Spectrometry* 18 (3):245–50.

Craig, H. 1957. Isotopic standards for carbon and oxygen and correction factors for mass-spectrometric analysis of carbon dioxide. *Geochimica Cosmochimica Acta* 12:133–49.

Dias, R. F., and K. H. Freeman. 1997. Carbon isotope analyses of semivolatile organic compounds in aqueous media using solid-phase microextraction and isotope ratio monitoring GC/MS. *Analytical Chemistry* 69:944–50.

Elsner, M., and D. Hunkeler. 2008. Evaluating chlorine isotope effects from isotope ratios and mass spectra of polychlorinated molecules. *Analytical Chemistry* 80:4731–40.

Gehre, M., and G. Strauch. 2003. High-temperature elemental analysis and pyrolysis techniques for stable isotope analysis. *Rapid Communications in Mass Spectrometry* 17 (13):1497–1503.

Giesemann, A., H. J. Jager, A. L. Norman, H. P. Krouse, and W. A. Brand. 1994. Online sulfur-isotope determination using an elemental analyzer coupled to a mass-spectrometer. *Analytical Chemistry* 66 (18):2816–19.

Hunkeler, D., and R. Aravena. 2000. Determination of stable carbon isotope ratios of chlorinated methanes, ethanes and ethenes in aqueous samples. *Environmental Science & Technology* 34:2839–44.

Jochmann, M. A., M. Blessing, S. B. Haderlein, and T. C. Schmidt. 2006. A new approach to determine method detection limits for compound-specific isotope analysis of volatile organic compounds. *Rapid Communications in Mass Spectrometry* 20 (24):3639–48.

Kovacs, D. A., and D. H. Kampbell. 1999. Improved method for the storage of groundwater samples containing volatile organic analytes. *Archives of Environmental Contamination Toxicology* 36:242–47.

Lehmann, M. F., S. M. Bernasconi, and J. A. McKenzie. 2001. A method for the extraction of ammonium from freshwaters for nitrogen isotope analysis. *Analytical Chemistry* 73 (19):4717–21.

Lis, G., L. I. Wassenaar, and M. J. Hendry. 2008. High-precision laser spectroscopy D/H and $^{18}O/^{16}O$ measurements of microliter natural water samples. *Analytical Chemistry* 80 (1):287–93.

Matthews, D. E., and J. M. Hayes. 1978. Isotope-ratio-monitoring gas chromatography-mass spectrometry. *Analytical Chemistry* 50 (11):1465–73.

Meier-Augenstein, W. 1999. Applied gas chromatography coupled to isotope ratio mass spectrometry. *Journal of Chromatography A* 842:351–71.

Miyajima, T., Y. Yamada, Y. T. Hanba, K. Yoshii, T. Koitabashi, and E. Wada. 1995. Determining the stable isotope ratio of total dissolved inorganic carbon in lake water by GC/C/IRMS. *Limnology and Oceanography* 40 (5):994–1000.

Mohammadzadeh, H., I. Clark, M. Marschner, and G. St-Jean. 2005. Compound specific isotope analysis (CSIA) of landfill leachate DOC components. *Chemical Geology* 218:3–13.

Roberts, B. J., M. E. Russ, and N. E. Ostrom. 2000. Rapid and precise determination of the $\delta^{18}O$ of dissolved and gaseous dioxygen via gas chromatography-isotope ratio mass spectrometry. *Environmental Science & Technology* 34:2337–41.

Sakaguchi-Soder, K., J. Jager, H. Grund, F. Matthaus, and C. Schuth, 2007. Monitoring and evaluation of dechlorination processes using compound-specific chlorine isotope analysis. *Rapid Communications in Mass Spectrometry* 21 (18):3077–84.

Salata, G. G., L. A. Roelke, and L. A. Cifuentes. 2000. A rapid and precise method for measuring stable carbon isotope ratios of dissolved inorganic carbon. *Marine Chemistry* 69:153–61.

Santrock, J., S. A. Studley, and J. M. Hayes. 1985. Isotopic analyses based on the mass spectrum of carbon dioxide. *Analytical Chemistry* 57:1444–48.

Schimmelmann, A., M. D. Lewan, and R. P. Wintsch. 1999. D/H isotope ratios of kerogen, bitumen, oil, and water in hydrous pyrolysis of source rocks containing kerogen types I, II, IIS, and III. *Geochimica et Cosmochimica Acta* 63 (22):3751–66.

Schmitt, J., B. Glaser, and W. Zech. 2003. Amount-dependent isotopic fractionation during compound-specific isotope analysis. *Rapid Communications in Mass Spectrometry* 17 (9):970–77.

Sessions, A. L., T. W. Burgoyne, and J. M. Hayes. 2001. Determination of the H_3 factor in hydrogen isotope ratio monitoring mass spectrometry. *Analytical Chemistry* 73:200–7.

Shouakar-Stash, O., R. J. Drimmie, and S. K. Frape. 2005. Determination of inorganic chlorine stable isotopes by continuous flow isotope ratio mass spectrometry. *Rapid Communications in Mass Spectrometry* 19:121–27.

Shouakar-Stash, O., R. J. Drimmie, M. Zhang, and S. K. Frape. 2006. Compound-specific chlorine isotope ratios of TCE, PCE and DCE isomers by direct injection using CF-IRMS. *Applied Geochemistry* 21:766–81.

Sigman, D. M., M. A. Altabet, R. Michener, D. C. McCorkle, B. Fry, and R. M. Holmes. 1997. Natural abundance-level measurement of the nitrogen isotopic composition of oceanic nitrate: An adaptation of the ammonia diffusion method. *Marine Chemistry* 57 (3–4):227–42.

Silva, S. R., C. Kendall, D. H. Wilkison, A. C. Ziegler, C. C. Y. Chang, and R. J. Avanzino. 2000. A new method for collection of nitrate from fresh water and the analysis of nitrogen and oxygen isotope ratios. *Journal of Hydrology* 228 (1–2):22–36.

St-Jean, G., 2003. Automated quantitative and isotopic (^{13}C) analysis of dissolved inorganic carbon and dissolved organic carbon in continuous-flow using a total organic carbon analyser. *Rapid Communications in Mass Spectrometry* 17:419–28.

US EPA. 2009, Hunkeler, D., R. U. Meckenstock, B. Lollar, T. C. Schmidt, and J. T. Wilson. A guide for assessing biodegradation and source identification of organic groundwater contaminants using compound specific isotope analysis (CSIA). EPA/600/R-08/148

Van Acker, M., A. Shahar, E. D. Young, and M. L. Coleman. 2006. GC/Multiple collector-ICPMS method for chlorine stable isotope analysis of chlorinated aliphatic hydrocarbons. *Analytical Chemistry* 78:4663–67.

Wassenaar, L. I., and G. Koehler. 1999. An on-line technique for the determination of the $\delta^{18}O$ and $\delta^{17}O$ of gaseous and dissolved oxygen. *Analytical Chemistry* 71:4965–68.

Wassenaar, L. I., and G. Koehler. 2004. On-line technique for the determination of the $\delta^{37}Cl$ of inorganic and total organic Cl in environmental samples. *Analytical Chemistry* 76:6384–88.

Zwank, L., M. Berg, T. C. Schmidt, and S. B. Haderlein. 2003. Compound-specific carbon isotope analysis of volatile organic compounds in the low-microgram per liter range. *Analytical Chemistry* 20:5575–83.

3 Principles and Mechanisms of Isotope Fractionation

Daniel Hunkeler and Martin Elsner

CONTENTS

3.1 PRINCIPLES AND MECHANISMS OF ISOTOPE FRACTIONATION

The fate of organic compounds in the environment is controlled by a number of processes. These include transport processes in the aqueous and gas phase, phase transfer processes, and transformation processes. During the first two types of processes, the molecules remain intact. The third process—chemical or biological transformation—involves breaking of bonds. The presence of a heavy isotope in molecules can have a particularly large effect on such transformation processes. Therefore, the origin and magnitude of isotope fractionation during transformation processes will be discussed in detail below. However, transport processes in particular diffusion in the gas phase can also lead to isotope fractionation. Before the effect of different types of processes on isotope ratios is discussed, some general definitions and concepts related to isotope fractionation are introduced.

3.2   EXPRESSING AND QUANTIFYING ISOTOPE FRACTIONATION

The abundance of different isotopes in a system are usually reported as isotope ratio R

$$R_A = \frac{^H A}{^L A} \tag{3.1}$$

or more commonly in the delta notation relative to the isotope ratio of reference material (see Chapter 2)

$$\delta_A = \left(\frac{R_A}{R_{std}} - 1 \right) \tag{3.2}$$

where $^H A$ is the amount of heavy isotope in the system, $^L A$ the amount of the light isotopes, and R_{std} the isotope ratio of the reference material. Since variations of isotope ratios are often small, the δ-values are usually expressed on a per mill scale by multiplication with 1000. For example a value of −0.025 calculated according to Equation 3.2 corresponds to −25‰. Although according to rules of metrology this conversion must not be incorporated into the equation, the delta notation is frequently defined in the literature as

$$\delta_A(‰) = \left(\frac{R_A}{R_{std}} - 1 \right) \cdot 1000 \tag{3.3}$$

For nonreactive processes, isotope fractionation corresponds to an uneven distribution of isotopes among molecules of a compound present in different phases (aqueous, gaseous, and solid) or at different locations in a system. In other words, the isotope ratio R of molecules of a compound in different phases or locations varies. Similarly, for reactive processes, isotope fractionation refers to a difference in R between reactants and products. Isotope fractionation can occur under equilibrium conditions (equilibrium isotope effect [EIE]) or as a result of a kinetic process that proceeds at different rates depending on the isotope that is present (kinetic isotope effect [KIE]). Strictly speaking, fractionation does not take place between different isotopes, but between molecules having a different isotopic composition denoted as isotopologues. We could, therefore, also use a concept of isotopologue fractionation. However, since we usually measure the average isotopic composition of molecules and not the abundance of different isotopologues, we usually discuss fractionation in terms of isotopes. The magnitude of isotope fractionation can be characterized by the isotope fractionation factor α. For equilibrium conditions, α corresponds to the ratio between R of the compound in the two phases or, in case of reactive processes, to the ratio between R of reactant and product. For an equilibrium process

$$A \leftrightarrow B$$

the isotope fractionation factor is given by

$$\alpha_{BA}^{eq} = \frac{R_B}{R_A} \qquad (3.4)$$

with

$$R_A = \frac{^H A}{^L A} \quad \text{or} \quad R_B = \frac{^H B}{^L B}$$

where
R_A, R_B is the isotope ratio of compound in phase A or B, respectively or isotope ratio of reactant and product, respectively.
$^H A$, $^H B$ is the total amount of the heavy isotope in compound A or B.
$^L A$, $^L B$ is the total amount of the light isotope in compound A or B.
The fractionation factor can also be expressed in terms of isotope ratios in the delta notation using $R_A = (\delta_A + 1) \times R_{std}$ (Equation 3.2)

$$\alpha_{BA}^{eq} = \frac{\delta_B + 1}{\delta_A + 1}$$

Or when using the δ-values converted into per mill

$$\alpha_{BA}^{eq} = \frac{\delta_B(\permil) + 1000}{\delta_A(\permil) + 1000} \qquad (3.5)$$

The isotope fractionation factor characterizes the overall preference of the process for the light isotope and is sometimes also denoted as bulk isotope fractionation factor.
In case of a kinetic process

$$A \rightarrow B$$

the isotope fractionation factor is defined analogously to Equation 3.4. However, R_B now corresponds to the isotope ratio of product that is formed at a moment in time, usually denoted as instantaneous product

$$R_B = \frac{d^H B}{d^L B} \tag{3.6}$$

The isotope ratio of the instantaneous product is different from the average isotope ratio of accumulated product because the latter consists of a mixture of product that has been produced throughout the process. Often isotope fractionation is expressed using the isotope enrichment factor that is defined as

$$\varepsilon_{BA} = \alpha_{BA} - 1 \tag{3.7}$$

Similarly as for the δ-notation, the obtained value is frequently transformed in per mill (‰) by multiplying with a 1000 and again the equation that includes the transformation factor can frequently be found in the literature

$$\varepsilon_{BA}(‰) = (\alpha_{BA} - 1) \times 1000 \tag{3.8}$$

For equilibrium isotope fractionation, quantification of isotope fractionation is straightforward. The isotope fractionation factor corresponds to the ratio of R in different phases, or between reactant and product of an equilibrium reaction as defined by Equation 3.4. For kinetic isotope fractionation, quantification of α is more demanding because the isotope ratio of an instantaneously formed product usually cannot be determined directly except at the beginning of the reaction. The isotope fractionation factor is usually quantified based on the change of the isotope ratio of the reactant in time. Therefore, an equation is required that relates α to the evolution of the reactant isotope ratio during the progress of reaction easily be measured. To develop such an equation we start from the definition of α for kinetic processes

$$\alpha_{BA}^{kin} = \frac{d^H B / d^L B}{{}^H A / {}^L A} \tag{3.9}$$

For mass balance reasons

$$d^H B = -d^H A$$
$$d^L B = -d^L A \tag{3.10}$$

Combining Equations 3.9 and 3.10 with rearrangement leads to

$$\frac{d^H A}{{}^H A} = \alpha_{BA}^{kin} \times \frac{d^L A}{{}^L A} \tag{3.11}$$

Equation 3.4 above can also be derived based on two first-order rate expressions for light and heavy isotopes, respectively

$$d^H A = {}^H k_{\text{bulk}} \times {}^H A$$
$$d^L A = {}^L k_{\text{bulk}} \times {}^L A \tag{3.12}$$

whereby

$$\alpha_{\text{bulk}}^{\text{kin}} = \frac{{}^H k_{\text{bulk}}}{{}^L k_{\text{bulk}}} \tag{3.13}$$

where ${}^H k_{\text{bulk}}$ and ${}^L k_{\text{bulk}}$ are the average rate constants at which heavy and light isotopes, respectively, are transferred from the reactant to the product pool. Hence, the fractionation factor can be considered to represent the ratio between first order reaction rate constants for disappearance of heavy and light isotopes from the substrate pool. We will later see that $\alpha_{\text{bulk}}^{\text{kin}}$ is also expected to be constant for reactions that are nonfirst order. For simplicity, in the following the kinetic isotope fractionation factor will be expressed as α.

Integration of Equation 3.11 from ${}^L A_0$ to ${}^L A$ and ${}^H A_0$ to ${}^H A$, respectively gives

$$\ln \frac{{}^H A}{{}^H A_0} = \alpha \cdot \ln \frac{{}^L A}{{}^L A_0} \qquad \text{or}$$

$$\frac{{}^H A}{{}^H A_0} = \left(\frac{{}^L A}{{}^L A_0} \right)^{\alpha} \tag{3.14}$$

dividing both sides by ${}^L A / {}^L A_0$ yields

$$\frac{R}{R_0} = \left(\frac{{}^L A}{{}^L A_0} \right)^{(\alpha - 1)} \tag{3.15}$$

where R and R_0 is the isotope ratio of reactant at a given time t and at time zero, respectively. The fraction of substrate that has not reacted yet, f, at time t is given by

$$f = \frac{C}{C_0} = \frac{{}^L A + {}^H A}{{}^L A_0 + {}^H A_0} = \frac{{}^L A \times (1 + R)}{{}^L A_0 \times (1 + R_0)} \tag{3.16}$$

If either the concentration of the heavy isotope ${}^H A$ and ${}^H A_0$ is small, as for most elements (H,C,N, and O) at natural abundance, or if the change in isotope ratio is small and hence $1 + R \approx 1 + R_0$, ${}^L A / {}^L A_0$ can be approximated by f and Equation 3.15 simplifies to

$$\frac{R}{R_0} = f^{(\alpha - 1)} \tag{3.17}$$

Equation 3.17 is the most fundamental equation to describe the isotope behavior usually denoted as a Rayleigh equation. It is usually considered to describe the isotope evolution of the reactant for a well-mixed, open system. The term open system simply indicates that the formed product does not interact any more with the reactant as is typically the case in irreversible transformation processes, even if they take place in a closed container. However, the equation also applies to phase transfer processes as long as the compound in the second phase is constantly removed. A classical example is the condensation of water in a cloud where the formed rain drops are immediately removed from the system. An equation for the accumulated product can be derived from Equation 3.17 using an isotope mass balance equation that links isotope ratio of reactant and accumulated product and assuming that all atoms of an element are transferred to the product

$$R_0 = R \times f + (1-f) \times \bar{R}_B \qquad (3.18)$$

where

$\bar{R}_B$ is the Isotope ratio of the accumulated product.

R_0 is the Initial isotope ratio of the reactant.

Inserting Equation 3.17 into Equation 3.18 leads to

$$\bar{R}_B = R_0 \times \frac{1 - f \times \alpha}{1 - f} \qquad (3.19)$$

Equation 3.17 is often expressed using the delta notation for isotope ratios and the isotope enrichment factor instead of the isotope fractionation factor. Given that $R_A = (\delta_A + 1) \times R_{std}$ and $\alpha = \varepsilon + 1$, on a logarithmic scale, Equation 3.17 for carbon isotopes transforms to

$$\ln \frac{\delta^{13}C + 1}{\delta^{13}C_0 + 1} = \varepsilon \times \ln f$$

or when inserting values on the per mill scale

$$\ln \frac{\delta^{13}C(\text{‰}) + 1000}{\delta^{13}C_0(\text{‰}) + 1000} = \varepsilon \times \ln f \qquad (3.20)$$

Since δ values are generally much smaller than 1 (e.g., 50‰ = 0.05), and since for such cases $\ln(1 + \delta)$ is approximately equal to δ as can be shown using a Taylor series expansion, Equation 3.20 can be further simplified to

$$\delta^{13}C = \delta^{13}C_0 + \varepsilon \times \ln f \qquad (3.21)$$

By inserting this equation into the following isotope mass balance equation and rearrangement

$$\delta^{13}C_0 = f \times \delta^{13}C + (1-f) \times \delta^{13}C_B \qquad (3.22)$$

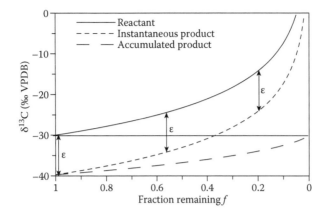

FIGURE 3.1 Evolution of isotope ratio of reactant, instantaneous product, and accumulated product as a function of the progress of reaction for a compound with an initial isotope ratio $\delta^{13}C = -30‰$ undergoing a reaction with an isotope enrichment factor of $\varepsilon = -10‰$.

again an equation for the accumulating product $\delta^{13}C_B$ is obtained

$$\delta^{13}C_B = \delta^{13}C_0 - \varepsilon \times \frac{f \times \ln f}{1-f} \tag{3.23}$$

Figure 3.1 illustrates the isotope evolution of reactant, instantaneous product and accumulated product as a function of the progress of reaction expressed by f. The graph illustrates some fundamental features of the behavior of isotopes during an irreversible process: The rate of change of the isotope ratio of the reactant increases as the reaction progresses and hence isotope fractionation becomes more apparent at a high degree of transformation. The instantaneous product is always offset by approximately ε compared to the $\delta^{13}C$ of the reactant, while the accumulated product approaches the isotope ratio of the reactant as the reaction reaches completion. Hence at the beginning of the transformation, the difference between product and reactant provides an indication of the magnitude of isotope fractionation. In some reactions, an element can partition between different products such as chlorine during reductive dechlorination of PCE to TCE and Cl⁻. In this case, it is the weighted average isotope ratio of all products that follow the product lines shown in Figure 3.1, while the individual products may show a constant offset relative to each other depending on the isotope effects associated with the respective product formation. (Hunkeler, van Breukelen and Elsner, 2009)

To determine fractionation factors from isotope data, occasionally the approximate Equation 3.21 is used. More accurate results are obtained, however, if Equations 3.15 or 3.17 are linearized. In particular, this is necessary when isotope fractionation leads to large shifts in isotope ratio as frequently observed for H isotopes. Noting that $L_A / L_{A_0} = f / [1 + R / 1 + R_0]$ according to Equation 3.16, the linearized form of Equation 3.15 corresponds to

$$\ln \frac{R}{R_0} = (\alpha - 1) \times \ln \frac{f}{(1+R)/(1+R_0)} \tag{3.24}$$

which must also be used with isotopically labeled compounds (Hunkeler 2002). The linearized form of Equation 3.17 is

$$\ln \frac{R}{R_0} = (\alpha - 1) \times \ln f \qquad (3.25)$$

An alternative linearized form of Equation 3.17 is given by Scott and colleagues (2004)

$$\ln R = (\alpha - 1) \times \ln C + \ln \left[\frac{R_0}{(C_0)^{\alpha-1}} \right] \qquad (3.26)$$

As long as the precision of the isotope analysis corresponds to the range of uncertainty of the standard gas that is used, it is recommended to quantify isotope fraction using Equation 3.25 or the corresponding equation in the delta notation (Equation 3.20) without forcing the regression through zero. The corresponding plots of $\ln(R/R_0)$ versus $\ln f$ are often referred to as Rayleigh plots.

In laboratory studies to determine isotope fractionation factors, often replicate experiments are carried out raising the question of how the data should be combined to obtain a representative fractionation factor. The unweighted average fractionation factor should only be used if the replicates consist of a similar number of observations spread over a similar range in f. Otherwise, Scott and colleagues (2004) propose a method based on linear regression with a dummy variable and a method based on a Pitman estimator.

3.3 ISOTOPE FRACTIONATION DURING TRANSPORT PROCESSES

Under saturated conditions, transport of dissolved compounds occurs by advection and dispersion (e.g., Freeze and Cherry 1979). Advection corresponds to the displacement of molecules with the movement of the water. In the classical advection-dispersion theory, it is usually assumed that the advective displacement of molecules takes place with the average groundwater flow velocity within some representative aquifer volume. However, not all molecules cover the same distance in a given amount of time for two reasons. The groundwater flow velocity varies around the average value at different scales, from the pore scale to the formation scale. As a result, some molecules are ahead of the average migration distance while others trail behind. This phenomenon is usually referred to as mechanical dispersion. Secondly, in the direction of strong concentration gradients diffusion will occur. Both effects lead to a smoothing of concentration gradients in both the flow direction and lateral to the flow direction. The effect of the two processes is usually lumped into a dispersion coefficient according to the following equation

$$D = \alpha_{Disp} \times v + D_{diff} \qquad (3.27)$$

where

D is the Dispersion coefficient

D_{Diff} is the Effective diffusion coefficient

α_{Disp} is the Dispersivity

v is the Flow velocity.

The first term in the dispersion coefficient corresponds to the variability of advective transport that is not included in the average velocity, while the second term corresponds to the contribution of diffusion. The dispersivity is usually much larger in the longitudinal direction and reflects the heterogeneity of the porous medium. Unless the groundwater flow velocity is very slow, the first term is usually much larger than the second term and hence transport is usually dominated by advective processes. However, exceptionally (e.g., in case of relatively homogeneous sand), the contribution of diffusion to transversal dispersion can be in the same order of magnitude as the contribution of mechanical dispersion and hence diffusion cannot always be completely neglected when considering transport in the aqueous phase. Furthermore, in heterogeneous formation, molecules may diffuse into stagnant zones at different scales.

Under unsaturated conditions, transport in the gas phase becomes relevant for volatile organic compounds, which are an important contaminant class (e.g., Conant, Gillham, and Mendoza 1996). This is due to the fact that velocities of diffusion are much larger in the gas phase than in the aqueous phase, because collision between contaminant molecules and the molecules of the given phase are much less frequent. Also, for gaseous compounds, advective transport can be important. It can be induced by density differences between vapor containing contaminants and the surrounding gas, barometric pressure variations, or changes in the volume of gases due to consumption or production of gaseous compounds (e.g., Falta et al. 1989).

ISOTOPE FRACTIONATION DURING ADVECTION AND DIFFUSION

Advection is not expected to lead to isotope fractionation since molecules are simply carried along with the moving aqueous phase independent of their mass. Diffusion occurs due to the random motion of molecules (Brownian motion) leading to a net transport of molecules along concentration gradients. The average velocity of a molecule can be calculated based on the kinetic theory of gases. According to this theory, the average energy of a molecule is only a function of the temperature and independent of its mass and corresponds to $3/2 \times k \times T$. This energy corresponds to the kinetic energy of the molecule $1/2\ mv^2$. Hence we obtain

$$\frac{1}{2} m \times v^2 = \frac{3}{2} \times k \times T$$

or

$$v = \sqrt{\frac{3 \times k \times T}{m}} \qquad (3.28)$$

where k is the Boltzmann constant, T the absolute temperature and m the mass of the molecule.

This equation illustrates that the velocity of a molecule in the gas phase is the smaller the higher its mass is. In addition to the mass, also the molecular volume influences the diffusion coefficient, because larger molecules are more likely to collide and hence are slowed down.

If a molecule diffuses through another gas, the diffusion coefficient depends also on the mass of this gas. Using the Enskog kinetic theory (e.g., Wittko and Kohler 2006), the diffusion coefficient of molecule I in a mixture of molecules I and j can be expressed as

$$D_i = \frac{3}{8 \times n \times \sigma_{ij}^2} \times \left[\frac{k \times T \times (m_i + m_j)}{2 \times \pi \times m_1 \times m_2} \right]^{1/2} \propto \left[\frac{m_i + m_j}{m_i \times m_j} \right]^{1/2} = \mu^{-1/2} \qquad (3.29)$$

where n is the particle number density, m_i the mass of compound i, m_j the mass of compound j, σ_{ij} the average diameter of the two compounds and μ the reduced mass. Hence the diffusion coefficient is proportional to the inverse of the square root of the reduced mass. In order to evaluate the potential for isotope fractionation by diffusion, we are interested in the ratio of diffusion coefficients for different isotopologues, which can be considered as a fractionation factor for diffusion. Given that the molecule diameter varies little with isotope substitution, the ratio between the diffusion coefficients of two isotopologues for diffusion through air can be expressed as

$$\frac{{}^H D_g}{{}^L D_g} = \sqrt{\frac{\mu_L}{\mu_H}} = \sqrt{\frac{m_L \times (m_H + m_{\text{air}})}{m_H \times (m_L + m_{\text{air}})}} \qquad (3.30)$$

where ${}^H D_g$ is the diffusion coefficient of the heavy isotopologue with mass m_H, ${}^L D_g$ the diffusion coefficient of the light isotopologue with mass m_L, and m_{air} is the average mass of air. This relationship is sometimes referred to as the "inverse square root of mass law." Table 3.1 illustrates ratios of diffusion coefficients for common organic contaminants, CO_2 and CH_4 diffusing through air.

As indicated by the obtained values in Table 3.1, in systems where diffusion is an important transport process, isotope fractionation in the low per mille range may occur, especially for small molecules where the effect of isotopic substitution on molecular weight is greatest. Isotope fractionation due to gas-phase diffusion was indeed observed for compounds such as hydrocarbons (Bouchard, Höhener, and Hunkeler 2008; Bouchard et al. 2008) and CO_2 (Cerling et al. 1991), and the observed isotope patterns were successfully reproduced by mathematical modeling based on Equation 3.30.

Equation 3.30 has also been applied to simulate isotope fractionation in the liquid phase. However, theoretical considerations and experimental data indicate that it overestimates isotope fractionation when applied to liquids (e.g., Bourg 2008). In liquids, the motion of molecules is much more restricted and hence cannot be appropriately described by the kinetic theory of gases. Calculations and experiments on diffusion of ions and noble gases suggested that the mass dependence of the diffusion coefficient in liquids can be described by a power law

$$D \propto m^{-\beta} \qquad (3.31)$$

TABLE 3.1

Effect of Presence of Heavy Isotope on Diffusion Coefficient of Organic Compounds and CO_2 in Gas Phase Calculated According to Equation 3.30. The Average Mass of Air is 28.8 g/mol

	m_L	m_H	Heavy Isotope	$^H D_g / ^L D_g$	$(^H D_g / ^L D_g - 1) \times 1000$
					Gas
CO_2	44	45	One ^{13}C	0.99559	−4.4
CH_4	18	19	One ^{13}C	0.98367	−16.3
Benzene	78	79	One ^{13}C or ^2H	0.99829	−1.71
Toluene	92	93	One ^{13}C or ^2H	0.99872	−1.28
MTBE	88	89	One ^{13}C	0.99858	−1.42
Trichloroethene	130	131	One ^{13}C or ^2H	0.99931	−0.69
Trichloroethene	130	132	One ^{37}Cl	0.99863	−1.37
Tetrachloroethene	164	165	One ^{13}C or ^2H	0.99955	−0.45
Tetrachloroethene	164	166	One ^{37}Cl	0.99907	−0.93

with an exponent β that is significantly lower (< 0.2) than for gas phase diffusion (Bourg and Sposito 2008).

Nevertheless, mathematical simulations of the behavior of *tert*-butyl-alcohol (TBA) and methyl *tert*-butyl ether (MTBE) suggested that diffusive isotope fractionation may occur for compounds migrating through higher permeability zones bordered by lower permeability zones (LaBolle et al. 2008). Diffusion into low permeability zones is expected to be more rapid for light molecules and hence heavy isotopes may become enriched in the plume migrating through the higher permeability zone. However, the relevance of this process has not yet been demonstrated by experimental studies.

3.4 ISOTOPE FRACTIONATION DURING PHASE TRANSFER PROCESSES

During phase transfer processes, isotope fractionation may occur under equilibrium conditions or due to the transfer of different isotopologues from one phase to another at a different rate. A classical example is the evaporation of water that leads to a depletion of heavy isotopes in the vapor phase under equilibrium and kinetic conditions (Clark and Fritz 1997). Relevant phase transfer processes for organic contaminants are dissolution and vaporization of organic liquids, sorption to solids, and air–water partitioning. The isotope fractionation during phase transfer processes is generally small. Therefore, experiments that measure isotope ratios in one of the two phases after one-step equilibration often fail to detect any isotope fractionation. A more sensitive approach to detect isotope fractionation is to replace one of the phases with material devoid of the molecules of interest and repeat the equilibration several times, which leads to a magnification of isotope changes. In case of volatilization, it is possible to continuously renew the gas phase whereby fractionation may occur under equilibrium or kinetic conditions depending on how rapidly the gas phase is removed.

For vaporization of organic compounds, it is frequently observed that molecules with heavy isotopes are slightly more volatile than molecules with light isotopes (Table 3.2a), which is usually denoted as an inverse isotope effect. This counterintuitive phenomenon can be explained by Van-der-Waals interactions of molecules in the condensed phase, which slow down molecular vibrations (i.e., the periodical movement of atoms inside the molecules around their common center of mass) shifting their energy to lower frequencies compared to the gas phase. Concomitantly, also the energetic difference between molecules with light and heavy isotopes respectively is decreased in the condensed phase (see the next section on transformation processes) compared to energetic differences in the gas phase. As a result, it takes slightly less energy for a molecule with a heavy isotope to go to the gas phase so that molecules with heavy isotopes become enriched in the gas phase. The opposite trend is observed in the presence of hydrogen bonds. Here, the loss of intermolecular hydrogen bonds in the gas phase leads to a normal isotope effect (i.e., light isotopes evaporate preferentially) in a similar way as during bond breakage in chemical reactions. A controlled field experiment under unsaturated conditions and column studies suggest that the carbon isotope fractionation during vaporization is negligible compared to isotope fractionation during diffusion and biodegradation (see Chapter 9). However, a notable hydrogen isotope effect was observed during vaporization of MTBE in a column experiment (Kuder et al. 2009). For dissolution from the organic phase, the little data that are available suggests that isotope fractionation is negligible. For sorption, one-step equilibration experiments failed to detect any isotope effect. However, multistep and column experiments demonstrated a small carbon isotope fractionation, whereby the heavy isotopes are slightly depleted in the organic phase (see Table 3.2b). This observation may be attributable to yet another phenomenon, namely that bonds to heavy isotopes are slightly shorter than to light isotopes. Consequently, heavy isotopologues have a smaller molecular volume so that less energy is required for cavity formation in the water and their dissolution is slightly preferred. Regarding sorption, theoretical considerations suggest that the process might influence isotope ratios at the front of migrating plumes. However, the relevance of the process at the field scale has not been demonstrated yet.

3.5 ISOTOPE FRACTIONATION DURING TRANSFORMATION PROCESSES

For reactive processes, the rate constant or equilibrium constant frequently depends on which isotope is present at the reactive position, which is denoted as an isotope effect. As a result of this isotope effect, isotopes become unevenly distributed among reactant and products. In other words, isotope fractionation occurs. This section discusses the origin of isotope effects during transformation processes and illustrates in semiquantitative terms the factors that control the magnitude of isotope fractionation. A more detailed treatment of the theory of isotope fractionation can be found in specific textbooks on isotope effects (e.g., Melander and Saunders 1980; Cook 1991; Kohen and Limbach 2006). At first, equilibrium isotope effects (EIEs) are treated, then kinetic isotope effects (KIEs). It is always assumed that the heavy isotope is located at the

TABLE 3.2
Isotope Fractionation During Phase Transfer Processes

Compound	Conditions	ε_C	ε_H	ε_{Cl}	References
(a) Vaporization					
TCE	Equilibrium, 1 step, 5 to 15°C	+0.7			(Huang et al. 1999)
	Equilibrium, 1 step, 35°C	+0.1			(Poulson and Drever 1999)
	Continuous removal of vapor, 22 ± 2°C	+0.09	+8.9 ± 5.9	−0.52	
		+0.35 ± 0.02		−1.64 ± 0.13	
	Continuous removal of vapor, 24 ± 1°C	+0.24 ± 0.06		−1.82 ± 0.22	(Huang et al. 1999)
Dichloromethane	Equilibrium, 1 step, 25 ± 0.1°C	+0.31 ± 0.04		−0.13	(Huang et al. 1999)
		+1.33			(Huang et al. 1999)
	Continuous removal of vapor, 24 ± 1°C	+0.65 ± 0.02		−1.48 ± 0.06	(Huang et al. 1999)
(b) Sorption					
Benzene, Toluene	Graphite, Activated carbon	<±0.5			(Slater et al. 2000)
	Equilibrium, 1 step, ?? °C				
Benzene, Toluene, p-Xylene	Lignite, Lignite coke, Activated carbon	<±0.5	<±8		(Schüth et al. 2003)
	Equilibrium, 1 step, 20°C				
Benzene	Humic acid	+0.44 ± 0.10			(Kopinke et al. 2005)
	Equilibrium, multistep, ?? °C				
Benzene	Humic acid coated column	+0.17			(Kopinke et al. 2005)
Toluene	Humic acid	+0.60 ± 0.10			(Kopinke et al. 2005)
	Equilibrium, multistep, ?? °C				

(Continued)

TABLE 3.2 (Continued)

Compound	Conditions	ε_C	ε_H	ε_{Cl}	References
o-Xylene	Humic acid coated column	+ 0.92			(Kopinke et al. 2005)
2,4-dimethylphenol	Humic acid coated column	+ 0.35			(Kopinke et al. 2005)
TCE, PCE	Graphite, Activated carbon Equilibrium, 1 step, ?? °C	< ±0.5[a]			(Slater et al. 2000)
TCE, c-DCE, VC	Lignite, Lignite coke, Activated carbon Equilibrium, 1 step, 20°C	< ±0.5			(Schüth et al. 2003)
MTBE	Humic acid Equilibrium, multistep, ?? °C	Not detected			(Kopinke et al. 2005)

[a] In some experiments (PCE-activated carbon, TCE-graphite) a small enrichment of ^{13}C in the dissolved phase was observed.

reacting position. In larger molecules, this is not always the case. The heavy isotope can be located at a position that is not attacked and hence has no influence on the rate constant leading to a dilution of the isotope effect. How this dilution effect can be taken into account will be discussed further below.

3.5.1 ENERGIES INHERENT IN MOLECULES

To understand the origin of isotope effects associated with transformation processes, the energies inherent in molecules have to be considered. Figure 3.2 illustrates the simplest case, the potential energy well of a diatomic molecule A-B as a function of the distance of the atoms to each other. Electrons act like the glue that keeps the positively charged atomic nuclei together in their molecular structure. Figure 3.2 illustrates that if the atoms are too close together, there is electrostatic repulsion by the positively charged nuclei, whereas if they are too far apart, the attraction by the electrons becomes too weak and the chemical bond is broken. To displace the atoms from their optimum bond distance, energy must be invested in either direction resulting in the surface of electron energy shown in Figure 3.2. Electron energy is therefore the most important molecular energy form for chemical bonding. Of all molecular energies, however, it is the one that is not sensitive to isotopic substitution. The reason is the Born–Oppenheimer approximation: Electron movement within binding wave functions is so fast that the electrons sense isotopic nuclei as resting bodies and are, therefore, not influenced if these nuclei have different isotopic masses. This explains why there is generally no relationship between isotope effects and reaction rates or chemical equilibria.

Molecules can take up energy in other, more subtle ways besides excitation of their electrons, however. When receiving small amounts of energy, molecules may start to move faster through space (translational energy). Adding larger amounts of energy can make them rotate more quickly around their molecular axes (rotational energy). At about room temperature, groups of atoms within the molecule may start to vibrate more strongly with respect to their common center of mass (vibrational energy). All of these energies are mass sensitive and, therefore, contribute to isotope effects. According to quantum mechanics, the uptake of energy in molecules does not occur continuously, but is quantized meaning that discrete quantum energy levels of the molecule are occupied if energy is supplied. This is sketched for vibrational energies of the diatomic molecule A-B in Figure 3.2a. Figure 3.2b shows moreover that the lowest vibrational energy level does not coincide with the bottom of the electronic energy well meaning that even at 0 K the molecular vibrational energy is not zero. This phenomenon is called zero point energy (ZPE) and may be explained by Heisenberg's uncertainty principle. Since the precise location and momentum of a particle cannot be determined at the same time, and since at 0 K the location of the atomic nuclei is constrained to the lowest vibrational energy state, it follows that their momentum cannot be zero and that they must possess a finite vibrational energy. This ZPE that is quantified relative to the lowest point of the potential energy well (Figure 3.2b) is given by

$$E_{ZPE} = \frac{1}{2} \times h \times v \qquad (3.32)$$

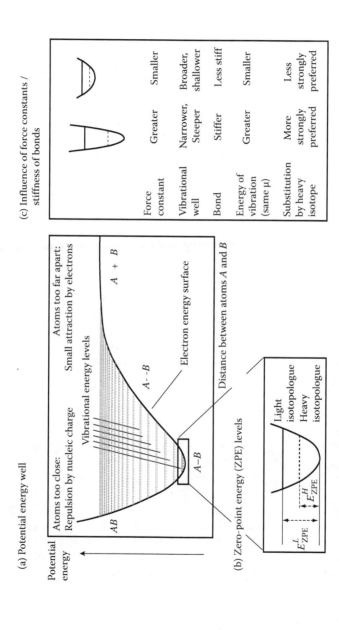

FIGURE 3.2 (a) Potential energy of a diatomic molecule A–B, (b) location of the lowest energy levels as a function of isotope that is present in the bond, and (c) effect of force constant on properties of the molecule.

where

 h is the Planck constant ($6.6 \cdot 10^{34}$ Js)

 v is the Frequency of vibration (s^{-1}).

The vibrational frequency corresponds to

$$v = \frac{1}{2} \cdot \pi \cdot \sqrt{f/\mu} \qquad\qquad (3.33)$$

with

$$\mu = (m_A \times m_B)/ (m_A + m_B) \qquad\qquad (3.34)$$

where

 f is the force constant

 μ is the reduced mass.

 m_A, m_B are the masses of atoms A and B, respectively

In contrast to the force constant, which does not depend on the isotopes present in the bond, the reduced mass is sensitive to the mass of the respective atoms. This also means that the vibrational frequency is influenced by isotopic substitution (Equation 3.33). As a result, the ZPE varies depending on the presence of different isotopes (Figure 3.2a). Since electron energies are not mass-dependent (see above), the ZPE is left as an important, if not often, dominant contributor to isotope effects (Huskey 1991). Zero point energies provide us with a key to understanding and predicting qualitative trends in isotope effects. Instead of considering Gibbs reaction enthalpies for prediction isotope effects, it is instructive to compare vibrations of reactants and products. Similarly, instead of looking at the strength (i.e., the total energy) of chemical bonds, it is expedient to consider their stiffness, that is, the force constant of the associated molecular vibration (see Figure 3.2c).

While the force constants are independent of the isotope present in a bond, they vary between different bonds corresponding to changes in the characteristics of the potential energy wells (Figure 3.3c). A bond is said to be stiffer if the vibrational force constant is greater, meaning that the well itself is narrower, the associated vibration has a greater frequency, and ZPE for different isotopes are less closely spaced (Figure 3.2c). In the same way a bond is said to be less stiff, if the vibration has a lower force constant, meaning that the well is wider, the frequency of the vibration is smaller and the ZPE are more closely spaced. As we will see below, the changes of the stiffness of bonds and consequently the spacing of ZPE levels during reactions is a key factor contributing to isotope effects. In order to understand how these changes lead to isotope effects we next consider a simple reaction.

3.5.2 Changes in Molecular Energy During Reaction

The energy inherent in molecules undergoing reaction can be illustrated using a potential energy surface as shown in Figure 3.3 for a simple reaction

$$AB + C \rightarrow A + BC$$

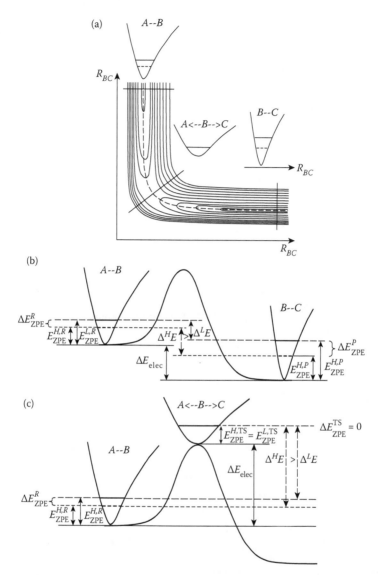

FIGURE 3.3   (a) Potential energy surface corresponding to a reaction of $AB + C \rightarrow A + BC$ with cross sections and illustration of zero point energies (ZPEs). (b) Potential energy profile with cross sections and ZPEs for reactant and product. (c) Potential energy profile with cross sections and ZPEs for reactant and transition state. The cross section for the transition state corresponds to the symmetrical vibration mode. In the symmetrical vibration mode the isotopic atom B is not in motion and hence the ZPE of the transition state does not depend on the isotopic substitution. The full lines correspond to the ZPE of molecules with a light isotope at the reactive site, the dashed line to molecules with a heavy isotope. For clarity, the ZPE is magnified compared to the change in electronic energy. Subscripts "R" indicated reactants, "P" products, "TS" transition state, "L" light isotopes and "H" heavy isotopes. ΔE_{ZPE} is the difference in ZPE between molecules with light and heavy isotopes for reactant (superscript R) and product (superscript P). ΔE_{elec} is the difference in electronic energy.

The cross sections A--B and B--C correspond to the potential energy wells known from Figure 3.2, where only the ZPE levels for isotopic substitution at the atom B are sketched. Figure 3.3 illustrates that the path of minimum activation energy first brings all atoms in close proximity before B is transferred between A and C. The saddle point A--B--C is the transition state of the reaction. Such cases of simultaneous bond cleavage and formation are called group transfer reactions and are encountered with hydrogen (i.e., $B = H$), as well as in second-order nucleophilic substitution (S_N2) reactions.

3.5.3 EQUILIBRIUM ISOTOPE EFFECT (EIE)

The example of Figure 3.3 may be taken to illustrate the types of isotope effects on isotopes of B: (i) the EIE that can be determined if the compounds A--B and B--C are in equilibrium; (ii) the KIE that is observable in A--B, if the compound reacts to B--C in an irreversible fashion. For EIEs the reaction path of Figure 3.3 is not important so that it is sufficient to consider just reactants and products. Since both are stable and their structures are well defined, they may be taken to consider the contributions of molecular energies to isotope effects in an exemplary way.

As discussed above, when considering the baseline energy that is present inside molecules even at 0 K, contributions to isotope effects stem only from vibrational zero point energies, often indicated by ZPE. At room temperature it must also be considered that molecules take up thermal energy so that besides their lowest energy level, other levels are occupied as well. Such energy levels are sketched for vibrations in Figure 3.2a and similar energy ladders may be drawn for rotations and translations. The adequate averaging and accounting over all possible states of a system is done in terms of partition functions. The canonical partition function Q weighs the number of accessible states in a chemical sample (sharply increasing with temperature) with the probability of encountering this state (sharply decreasing with temperature) so that it arrives at a specific energy for a given temperature. These principles of statistical mechanics link microscopic to macroscopic thermodynamic properties. The equilibrium constant for molecules containing B as light isotope can therefore be expressed as follows

$$^LK = \frac{\prod Q^L_{\text{products}}}{\prod Q^L_{\text{reactants}}} \cdot \exp(-\Delta^LE/kT) \tag{3.35}$$

where

Q is the canonical partition function of the molecules participating in the reaction.

Δ^LE is the energy difference between product and reactant molecules at 0 K (see Figure 3.3b).

As illustrated in Figure 3.3b and (note that in the example Δ^LE and ΔE_{elec} are negative,) the energy difference between product and reactants Δ^LE is related to the ZPEs by

$$\Delta^L E = \Delta E_{\text{elec}} + E_{\text{ZPE}}^{L,P} - E_{\text{ZPE}}^{L,R} \tag{3.36}$$

where

$E_{\text{ZPE}}^{L,R}$ is the ZPE of the reactant AB

$E_{\text{ZPE}}^{L,P}$ is the ZPE of the product BC.

Inserting Equation 3.36 into Equation 3.35 leads to

$$^L K = \frac{\Pi Q_{\text{products}}^L}{\Pi Q_{\text{reactants}}^L} \cdot \exp\left[-\Delta E_{\text{elec}} / kT\right] \cdot \exp\left[\left(E_{\text{ZPE}}^{L,R} - E_{\text{ZPE}}^{L,P}\right) / kT\right] \tag{3.37}$$

In this equation, the ΔE_{elec} term corresponds to the energy difference between the hypothetical vibrationless states of products and reactants (i.e., the bottom levels of the electron energy surfaces), the ZPE term reflects the difference in vibrational ZPE between products and reactants as discussed above, and the partition functions Q account for excitation of translations, rotations, and vibrations at higher energies than the ZPE level.

To obtain an equation for the EIE, we can write two reaction equations depending on the isotope present in the reacting bond

$$A^L B + C \rightarrow A + {}^L BC$$

$$A^H B + C \rightarrow A + {}^H BC.$$

Taking into account that partition functions for all compounds without isotope substitution and the ΔE_{elec} term cancel, the ratio of equilibrium constant for light and heavy isotopes, respectively, is given by

$$\text{EIE} = \frac{^L K}{^H K} = \frac{Q_{^L BC}}{Q_{^H BC}} \cdot \frac{Q_{A^H B}}{Q_{A^L B}} \cdot \exp\left[\left(\left(E_{\text{ZPE}}^{L,R} - E_{\text{ZPE}}^{H,R}\right) - \left(E_{\text{ZPE}}^{L,P} - E_{\text{ZPE}}^{H,P}\right)\right) / kT\right] \tag{3.38}$$

The partition function can be subdivided into the translational, rotational, and vibrational contribution. The isotope effect is therefore often subdivided into a lumped contribution from translations and rotations called "mass and moments of inertia" (MMI), a second part that takes into account excitation of higher-energy vibrational levels (EXC) and a third part that describes the differences in ZPE.

$$\text{EIE} = \text{MMI} \times \text{EXC} \times \text{ZPE} \tag{3.39}$$

Table 3.3 gives an example of calculated EIEs illustrating the preference of isotopes for being in different chemical compounds. The table 3.3 illustrates that isotope effects are indeed in most, but not in all, cases dominated by the ZPE contribution. The ZPE approximation—that is, if the contributions of thermal excitation (MMI and EXC) are neglected and only the ZPE term is considered—provides therefore a

TABLE 3.3
Carbon Equilibrium Isotope Effects for the Reaction $CH_3X \rightarrow CH_3CH_3$,
Computed From Theoretical Force Constants

X	MMI	EXC	ZPE	Total
F	0.9985	0.9999	1.0099	1.0083
Cl	0.9961	1.0021	0.9904	0.9885
Br	0.9920	1.0042	0.9892	0.9855
I	0.9909	1.0062	0.9813	0.9784
CN	0.9939	1.0012	1.0041	0.9991

Source: Hartshorn, S. R., and V. J. Shiner, *Journal of the American Chemical Society,* 94 (26), 9002–12, 1972.

qualitative, if not often even semiquantitative, approach to interpreting observable isotope effects. For semiquantitative treatments Equation 3.39 may be simplified to

$$EIE = ZPE = \exp\left[\left(\left(E_{ZPE}^{L,R} - E_{ZPE}^{H,R}\right) - \left(E_{ZPE}^{L,P} - E_{ZPE}^{H,P}\right)\right)/kT\right]$$

$$= \exp\left[(\Delta E_{ZPE}^{R} - \Delta E_{ZPE}^{P})/kT\right]. \tag{3.40}$$

The equation illustrates that the EIE depends on differences in ZPE between reactants and products. The differences in ZPE in turn depend on stiffness of the bond as illustrated in Figure 3.2c. Hence the EIE reflects changes in the properties of the bonds of the molecules when passing from the reactant to the product state. In the simple example considered, there is only one vibrational mode for reactant and product corresponding to the stretching of the bond A--B and B--C, respectively. The ZPE difference (ΔE_{ZPE}) is smaller for the reactants than the products because the potential energy well is narrower for the product (Figure 3.3b). Hence, the term in the exponential function of Equation 3.40 is smaller than zero and accordingly the EIE smaller than one. In other words, the equilibrium constant is larger for the heavy isotope and thus the heavy isotope accumulates in the product characterized by the tighter potential energy well. As illustrated by this example, the heavy isotope is generally more strongly preferred in bonds characterized by a tighter potential energy. This aspect will come into play again when considering KIEs.

The stiffness of bonds (and hence the vibrational frequencies) do not only change during reactions but, to a much smaller extent, can also change during phase transfer processes as was touched upon in the section on vaporization. Remember that for vaporization of organic molecules, vibrational frequencies are often slightly lower in the condensed phase due to interactions between molecules than in the gas phase (Section 3.3). Lower vibrational frequencies correspond to a closer spacing between ZPE levels. Hence, if we consider the solvent as the reactant and the vapor as the product, the exponential term in Equation 3.40 is negative and the EIE becomes smaller than one (inverse isotope effect). In other words, the heavy isotopes become slightly enriched in the gas phase.

So far we considered the smallest possible molecule that has only one vibrational mode. For larger molecules we have to sum the ZPE changes for all bonds that change in stiffness and all vibrational modes

$$\text{EIE} = \exp\left[-\left(\sum_i \Delta E^R_{\text{ZPE},i} - \sum_i \Delta E^P_{\text{ZPE},i}\right)/kT\right] \quad (3.41)$$

where

$\Delta E^{R\,\text{or}\,P}_{\text{ZPE},i}$ is the difference in ZPE for a vibrational mode i.

Thus, in order to quantify EIE we need to know the ZPE differences for reactants and products that depend on the vibrational frequencies and hence the force constants of the bonds. Since the force constants and vibrational frequencies are accessible for stable molecules from infrared spectroscopy or computations, the determination of isotope effects is relatively straightforward for an equilibrium reaction.

3.5.4 KINETIC ISOTOPE EFFECT (KIE)

Chemical kinetics is frequently rationalized using the transition state theory (TST). According to the TST, the reaction rate depends on the abundance of molecules in the transition state and the tendency of molecules in the transition state to go forward. The former can be characterized by an equilibrium constant relating transition state to reactants while the latter depends on some universal constants. If both are put together, the reaction rate can be expressed as follows

$$k_r = \frac{k \times T}{h} \times K_{\text{TS}} \quad (3.42)$$

where

k_r is the reaction rate
k is the Boltzmann constant
h is the Planck constant
T is the absolute temperature
K_{TS} is the equilibrium constant for the formation of transition state from reactants.

Similar as for the EIE, the KIE can be expressed in terms of partition functions. Using the following two reactions for the two isotopes

$$A^L B + C \leftrightarrow {}^L\text{TS}$$

$$A^H B + C \leftrightarrow {}^H\text{TS}$$

we obtain

$$\text{KIE} = \frac{{}^L k_r}{{}^H k_r} = \frac{Q_{{}^L\text{TS}}}{Q_{{}^H\text{TS}}} \times \frac{Q_{A^H B}}{Q_{A^L B}} \quad (3.43)$$

   The KIE expression given above is also denoted as a semiclassical rate ratio because it does not take into account tunneling. Tunneling describes the phenomenon that, molecules can react even though they do not have sufficient energy to move over the potential energy barrier. This phenomenon is particularly strong if the energy barrier is very steep. If tunneling is neglected and again the ZPE approximation is used, the KIE can be expressed in terms of vibrational frequencies in analogy to Equation 3.41. There are two notable differences compared to the equilibrium treatment, however. First, it is not as straightforward to obtain vibrational frequencies of transition states. They cannot be measured experimentally since transition states are not stable and their computations is challenging. Secondly, the vibrational mode corresponding to the dissociation of the bond during reaction is not included in the summation of the transition state because the vibration is no longer a periodic movement that is treated quantum mechanically, but a nonperiodic decomposition motion whose classical description leads to the term $k \times T/h$ in Equation 3.42. While for reactants, the number of vibrational modes is 3N-6 for nonlinear molecules and 3N-5 for linear molecules (with N being the number of atoms in the molecule), for the transition state these numbers are therefore by 1 smaller.

$$\text{KIE} = \frac{^L k_r}{^H k_r} = \exp\left[\left(\sum_i^{3N-6} \Delta E_{ZPE,i}^R - \sum_i^{3N-7} \Delta E_{ZPE,i}^{TS}\right)/kT\right] \tag{3.44}$$

   During passage from reactant to transition state, the force constants and hence the vibrational frequencies change significantly. To what extent such changes lead to a KIE depends on where the isotopic atom is located. A significant so-called primary KIE occurs if the isotopic atom is located in a bond that is broken during the reaction. In general, bonds are looser or even broken in the transition state. This generally implies that the ZPE levels for molecules with different isotopes are therefore closer together (Figure 3.3c) and the heavy isotopes less strongly preferred in the transition state. Consequently, light isotopes react preferentially, a phenomenon commonly encountered in (bio)chemical reactions. Or, when expressed in mathematical terms, the ΔE_{ZPE} summation in Equation 3.44 for the transition state is smaller than the summation for the reactant state leading to a positive value for the exponential term and thus a KIE larger than one. From this it follows that the KIE becomes largest, if the term of the transition state becomes zero. This situation may occur under two circumstances: (i) The reacting bond is almost completely broken in the transition state so that the vibrational potential is very shallow and the spacing between the isotopic ZPE levels becomes negligible, or (ii) Alternatively, the vibration may become insensitive to isotopic substitution if the isotope does not experience any movement in the vibration. Specifically, we consider the transition state A--B--C of Figure 3.3c. If A-B-C were a stable molecule, two vibrational stretching modes would be expected. In the asymmetric stretch, a periodic movement first bring B and C together and move A away, then A and B would come together and C would move away, and so on.

Conversely, in the symmetric stretch *A* and *C* would swing periodically toward and away from *B*, while *B* itself would remain stationary (Figure 3.3a). Since the transition state *A--B--C* is not a stable molecule, one of the two stretches is not a periodic vibration, but corresponds to the nonperiodic decomposition motion discussed earlier. This can easily be identified as the asymmetric stretch, where *B* and *C* approach each other and *A* moves away corresponding to the movements of atoms in the reaction along the dashed line in Figure 3.3a. The periodic stretching mode, which corresponds to a movement perpendicular to the reaction path at the highest point of the energy curve (Figure 3.3a), remains as intact vibration. Since *B* does not move within the symmetric stretch, isotopic substitution in *B* does not have any consequences on the frequency of the vibration. Consequently, the vibrational ZPE levels of molecules with and without isotopic substitution are the same (Figure 3.3c). Consequently, very large contributions by the ZPE term can be expected in such group transfer reactions with symmetric transition state.

KIE do not only occur if the isotope of interest is substituted in the reacting bond but also when the substitution is in a position adjacent to the reaction center. The reason is that also the stiffness of adjacent bonds is slightly influenced by the (bio)chemical reaction, which gives rise to secondary isotope effects. Predicting qualitative trends of secondary isotope effects in α-position is facilitated by the following rule of thumb. If the number of substituents at a reacting center decreases (e.g., from sp^3 to sp^2), this corresponds to a less cramped coordination environment that allows more space for bending vibrations of the remaining bonds. Consequently, the frequency of these vibrations decreases, the bonds are less stiff in the context of Figure 3.2c, and the secondary isotope effects of the involved atoms are normal (i.e., greater than unity). Conversely, if the number of substituents increases (e.g., from sp^2 to sp^3), the frequency of the bending vibrations becomes greater, the corresponding bonds are stiffer, and secondary isotope effects tend to be inverse (i.e., smaller than unity). However, in general secondary isotope effects are at least one order of magnitude smaller than primary isotope effects.

The considerations above illustrate that the KIE mainly originates from a difference in ZPE, which is a phenomenon of quantum mechanics. The KIE does not depend on the exact shape of the potential energy profile but is determined by the geometry of the transition state. For this reason, KIE measurements have been used for decades in organic chemistry to gain insight into the structure of transition states. Since the KIE is independent of the exact shape of the electron energy profile, the magnitude of the isotope effect is not linked to the magnitude of the absolute reaction rate, but depends instead very strongly on changes in molecular vibrations.

3.5.4.1 Magnitude of the Kinetic Isotope Effect (KIE)

As shown above, KIEs tend to be greatest, if under certain conditions (e.g., symmetrical transition state) differences in $\Delta E_{\mathrm{ZPE}}^{\mathrm{TS}}$ of the transition state disappear completely so that only the contributions from the reactants remain in Equation 3.44. This makes it possible to calculate ballpark estimates for maximum KIEs. If we only consider the vibrational energy of the reacting bond in the reactant and use $E_{\mathrm{ZPE}} = 1/2 \times h \times v$ (Equation 3.32), the expression for the KIE simplifies to

$$\text{KIE} = \frac{{}^L k_r}{{}^H k_r} = \exp\left[\left(\Delta E_{ZPE}^R\right)/k \times T\right] = \exp\left[\left(E_{ZPE}^{L,R} - E_{ZPE}^{H,R}\right)\right]/k \times T$$

$$= \exp\left[\frac{h}{2 \times k \times T} \times \left(v_{ZPE}^{L,R} - v_{ZPE}^{H,R}\right)\right] \tag{3.45}$$

Using Equation 3.33, the ratio between the vibrational frequency of light and heavy isotopes, respectively, can be expressed as

$$\frac{v_{ZPE}^{L,R}}{v_{ZPE}^{H,R}} = \sqrt{\frac{\mu_H}{\mu_L}} \tag{3.46}$$

Substitution of Equation 3.46 into Equation 3.45 leads to

$$\text{KIE} = \exp\left[\frac{h \times v_{ZPE}^{L,R}}{2 \times k \times T} \times \left(1 - \sqrt{\frac{\mu_L}{\mu_H}}\right)\right] \tag{3.47}$$

or expressed in terms of wave numbers

$$\text{KIE} = \exp\left[\frac{h \times c \times \tilde{v}_{ZPE}^{L,R}}{2 \times k \times T} \times \left(1 - \sqrt{\frac{\mu_L}{\mu_H}}\right)\right] \tag{3.48}$$

where

$\tilde{v}_{ZPE}^{L,R}$ is the wave number (cm^{-1})

c is the speed of light ($3 \times 10^8 \times m \times s^{-1}$).

Using Equation 3.48, the KIE can be calculated when the mass of the atoms in the bond and the vibrational frequency (i.e., its wave number) are given (Table 3.4). The obtained values are known as Streitwieser semiclassical limits (Huskey 1991).

Based on the values given in Table 3.4 some general statements about KIEs can be made. (1) A larger KIE can be expected, the larger the relative mass difference

TABLE 3.4

Streitwieser Semiclassical Limits for KIE Due to Breakage of Different Bonds During Reactions at 25°C

Bond	Frequency (cm^{-1})	Isotope	KIE
C—H	2900	$^{12}C/^{13}C$	1.021
		$^{1}H/^{2}H$	6.44
C—C	1000	$^{12}C/^{13}C$	1.049
C—Cl	750	$^{12}C/^{13}C$	1.057
C—Cl	750	$^{35}Cl/^{37}Cl$	1.013
C—N	1150	$^{12}C/^{13}C$	1.060
C—N	1150	$^{14}N/^{15}N$	1.044
C—O	1100	$^{12}C/^{13}C$	1.061
C—O	1100	$^{16}O/^{18}O$	1.067

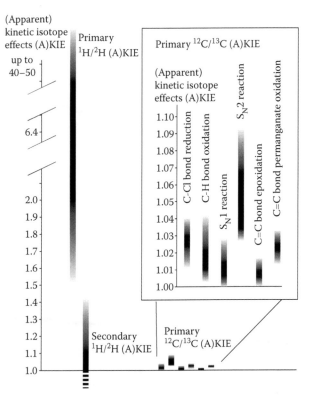

FIGURE 3.4 Typical ranges for kinetic isotope effects (KIEs) for common reactions (Based on data from Elsner, M., L. Zwank, D. Hunkeler, and R. P. Schwarzenbach, *Environmental Science & Technology,* 39, 6896–916, 2005).

between the two isotopes is. Hence, a particularly large KIE is expected for hydrogen, while it is smaller for carbon and even smaller for examples for chlorine (see Figure 3.4). (2) For a given element (e.g., C), the magnitude of the KIE depends on the bonding partner, with larger KIEs for heavier bonding partners. However, the values in Table 3.4 have to be considered as approximate ballpark values. The actual KIE can be different from these values for the following reasons:

- In the transition state, the bond may not be completely broken or the transition state not symmetrical. Thus some of the ZPE difference of the reactant state cancels out and the KIE is smaller than the given values.
- The summation of vibrational energies and frequencies has to take into account all vibrational modes. During passage from reactant state to transition state the strength of adjacent bonds may vary, too, and hence the KIE may be smaller or larger compared to the Streitwieser limits.
- For hydrogen isotopes, tunneling may occur leading to a larger KIE than given above since the probability of tunneling is higher for the lighter isotope.

For a more accurate calculation of KIEs, the structure of the transition state has to be simulated using methods of computational chemistry. Based on the obtained force constants and other factors contributing to the isotope effect (e.g., tunneling), the KIE can then be calculated.

Figure 3.4 summarizes experimental KIE reported for important reaction mechanisms (Elsner et al. 2005). In general, the values are in the same order of magnitude as predicated by Streitwieser Limits (Table 3.4) and some of the expected trends can be observed. A very large KIE is observed, for example, for the light element hydrogen, whereas KIE of carbon are much smaller. Also, within the range of carbon KIE relatively large values are observed if C is bound to heavy bonding partners such as in Cl bond reduction and S_N2 reactions. It is interesting to note that the KIE for S_N1 is smaller than for S_N2, which is due to the fact that the loss of bonding is partly compensated by an increased strength of adjacent bonds in the case of S_N1 reactions. Similarly, also the KIE for oxidation of a double bond is relatively small because the loss in bond strength is partly compensated by the formation of new bonds.

3.5.4.2 Effect of Different Rate Laws

According to the derivation of Equation 3.43, the KIE relates first-order rate constants. In the following it is demonstrated that the KIE expression is also valid for higher-order rate constants. The probability that two molecules with isotope substitution react with each other is very low and therefore, the ratio of reaction rates for a higher-order reaction is the same as for a first-order reaction. This becomes apparent by considering the simple reaction $A + B \rightarrow M$ as an example and assuming that the reaction is of n-th order with respect to B. The ratio between the reaction rates for light and heavy isotope is given by

$$\frac{r_L}{r_H} = \frac{{}^L k \times [B_L]^n \times [A]}{{}^H k \times [B_L]^n \times [A]} = \frac{{}^L k \times [B_L] \times [B_L]^{n-1} \times [A]}{{}^H k \times [B_H] \times [B_L]^{n-1} \times [A]} = \frac{{}^L k \times [B_L]}{{}^H k \times [B_H]} \qquad (3.49)$$

A more complicated situation arises if the reaction is catalyzed by an enzyme. In this case, the reaction is preceded by a binding to the enzyme and may involve several activation steps prior to the irreversible step. In the simplest case, the reaction mechanism can be described as follows

$$A + E \underset{k_{-1}}{\overset{k_1}{\rightleftharpoons}} EA \xrightarrow{k_2} E + P \qquad (3.50)$$

Once a steady state is reached, meaning that the concentration of the intermediate substrate-enzyme complex is constant, the kinetics of the reaction can be characterized by two entities, the limiting overall conversion rate V and the Michaelis constant K_m. The two parameters are related to the rate constants k_1, k_{-1} and k_2 as follows

$$V = k_2 \times e_0$$

$$K_m = \frac{k_{-1} + k_2}{k_1} \qquad (3.51)$$

where

e_0 is the total enzyme concentration.

It is generally assumed that the binding step is not sensitive to isotope substitution while isotope fractionation frequently occurs during the chemical reaction step. The isotope effect associated with the reaction step is denoted as *intrinsic isotope effect*. The question is now how the intrinsic isotope effect is related to the KIE that we observe when carrying out isotope studies with enzymes, here denoted as apparent kinetic isotope effect (AKIE).

To evaluate the relationship between AKIE and intrinsic isotope effect, we have to evaluate how the enzymatic transformation rate varies as a result of isotope substitution. We can conceptualize the simultaneous transformation of molecules with and without isotope substitution as enzyme inhibition by a competing substrate. The ratio of reaction rate in case of competitive inhibition is given by Cornish-Bowden (1995)

$$\frac{r_L}{r_H} = \frac{(V^L / K_m^L) \times [A_L]}{(V^H / K_m^H) \times [A_H]} \qquad (3.52)$$

Hence, the observed AKIE reflects how the ratio V/K_m varies as a result of isotope substitution

$$\text{AKIE} = \frac{V^L / K_m^L}{V^H / K_m^H} \qquad (3.53)$$

The ratio V/K_m is given by

$$\frac{V}{K_m} = \frac{k_2 \cdot k_1}{k_{-1} + k_2} \qquad (3.54)$$

Insertion of Equation 3.54 for light and heavy isotope into Equation 3.53 and assuming that the rate of the binding step is not affected by isotope substitution one obtains

$$\text{AKIE} = \frac{k_2^L}{k_2^H} \times \frac{k_{-1} + k_2^H}{k_{-1} + k_2^L} = \frac{\dfrac{k_2^L}{k_2^H} + \dfrac{k_2^L}{k_{-1}}}{1 + \dfrac{k_2^L}{k_{-1}}} \qquad (3.55)$$

The ratio k_2^L / k_{-1} characterizes the tendency of the substrate-enzyme complex EA to react on to form P, versus its tendency to dissociate again from the enzyme. This ratio is usually denoted as commitment to catalysis C. The ratio k_2^L / k_2^H corresponds to the intrinsic KIE associated with the actual reaction step. Equation 3.55 can be rewritten as

$$\text{AKIE} = \frac{\dfrac{k_2^L}{k_2^H} + C}{1 + C} \qquad (3.56)$$

Equation 3.56 illustrates how the intrinsic KIE is related to the observed AKIE. If the commitment to catalysis is large (i.e., if every molecule that binds to the enzyme immediately reacts independently of its isotopic substitution), AKIE approaches 1. In other words, the enzyme does not discriminate between molecules with different isotopes. In the other extreme, if the commitment to catalyses is very low, the intrinsic KIE is fully expressed. Hence, depending on the relative ratio of the different rate constants of the enzyme, the AKIE varies between no KIE and the intrinsic KIE that would be expected for an analogous chemical reaction. In the derivation above only the most simple enzyme mechanism was considered. It can be shown that in case of even more reaction steps, the AKIE reflects the KIEs of all steps leading up to and including the first irreversible one. The equation given above strictly only applies once a steady state is reached (Paneth 1985). Especially for large commitment to catalysis, it may take some time until a steady state is reached. Hence, initially during an enzymatic process, the AKIE or the fractionation factor may not be constant. Furthermore, during a transformation process, the substrate concentration steadily decreases and hence the substrate-enzyme complex concentration needs to be continuously readjusted to a new steady state level. However, for most reactions this readjustment is rapid, and if transformation is followed over several orders of magnitude of concentration decrease usually a constant isotope fractionation is observed.

3.5.5 CORRECTING FOR HEAVY ISOTOPES AT POSITIONS THAT DO NOT PARTICIPATE IN REACTION

When using the common Rayleigh equation to evaluate isotope data, the fractionation factor that we obtain represents the average tendency of heavy isotopes to becoming enriched in the reactant, independent of their location in the molecule. Such bulk isotopic fractionation factors are very useful for interpretation of field data. However, if we want to gain insight into the factors that control the magnitude of isotope fractionation or unravel reaction mechanisms from isotope data, we need information about the KIE values specific to the reactive position. This section summarizes a method to derive AKIE values from measured isotope data that was developed in earlier work (Elsner et al. 2005).

The observed isotope fractionation based on changes in the average isotope ratio of a whole molecule is smaller than the actual KIE because the heavy isotope may be located at sites other than where the reaction takes place. This is illustrated in Figure 3.5 for hydrolytic dehalogenation of 1,3-dichloropropane. We can distinguish three different isotopologues: those that do not contain a ^{13}C inside their structure (case 1),

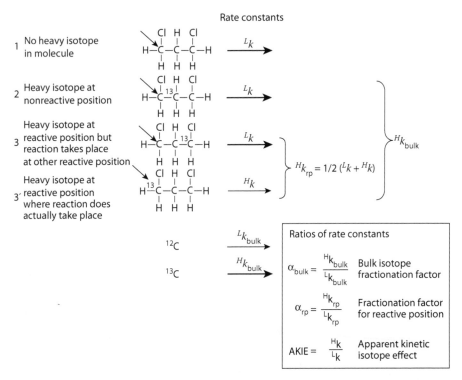

FIGURE 3.5 Isotopologues of 1,3-Dichloropropane with different isotopes at reactive and nonreactive positions assuming transformation by hydrolytic dehalogenation. Molecules with more than one heavy isotope were neglected due to their low abundance. ^{L}k and ^{H}k: Rate constants for molecules with light and heavy isotope, respectively at reacting position, $^{L}k_{Bulk}$ and $^{H}k_{Bulk}$: Average rate constants at which heavy and rate isotopes, respectively, are transformed independent of their location in the molecule.

those that contain ^{13}C in the C-2 position, which is nonreactive (case 2), and those that contain ^{13}C in the reactive C-1 position (case 3 and 3′; note that both describe two possible reactions of the same isotopomer). Dilution of the isotope effect may occur due to two different phenomena: (i) the heavy isotope is located at a nonreactive position and hence does not have an effect on the reaction rate (case 2, Figure 3.5), or (ii) the molecule contains several identical reactive positions. Although in the second case, the heavy isotope is present at a reactive position, it is in intramolecular competition with the other potentially reactive position so that it does not always get the change to react. As will be explained in the following, a correction for these two factors can be carried out if the reaction mechanism is known.

Step 1: Correction for nonreactive positions

Let us assume that a molecule has n atoms of an element of which x are at reactive positions. (In the case of 1,3-dichloropropane, $n = 3$ and $x = 2$.) In a first step, we need to subtract the dilution of the isotope effect by nonreactive positions (case 2, Figure 3.5) in order to evaluate how quickly the heavy isotopes located at any of the reacting sites (case 3 and 3', Figure 3.5) become enriched as the reaction proceeds. If we only consider the reactive positions in the molecules, we expect that heavy isotopes become enriched following a Rayleigh trend analogous to Equation 3.17

$$\frac{R_{rp}}{R_{0,rp}} = f^{(\alpha_{rp}-1)} \tag{3.57}$$

where

R_{rp} is the isotope ratio of atoms at reactive positions only

α_{rp} is the average fractionation factor for all x reactive positions (Figure 3.5).

In contrast, the isotope ratio at the nonreactive site will not change because the reaction rate is not sensitive to what isotope is present at this location (that is, if secondary isotope effects are absent). This position is simply carried along during the reaction and conserves its isotope ratio. Hence

$$\frac{R_{nrp}}{R_{0,nrp}} = 1 \tag{3.58}$$

where

R_{nrp} is the isotope ratio of atoms at nonreactive position.

What we measure is the average isotope ratio R of all positions in the molecule. We can calculate the isotope ratio of the reactive positions from R if we know the initial isotope ratio of the nonreactive positions $R_{0,nrp}$. In absence of other information, it is reasonable to assume that their initial isotope ratio is equivalent to the initial isotope ratio of the whole molecule, that is $R_{0,rp} = R_{0,nrp} = R_0$. Otherwise, an accurate calculation can be carried out for any other initial intramolecular isotope distribution if known. Using an isotope mass balance equation we can write an equation that links the isotope ratio of reactive and nonreactive positions to the measured isotope ratio

$$R \times n = R_{rp} \times x + R_{nrp} \times (n-x) = R_{rp} \times x + R_0 \times (n-x). \tag{3.59}$$

Rearrangement of Equation 3.59 and insertion into Equation 3.57 leads to

$$\frac{R_{rp}}{R_0} = \frac{\frac{n}{x} \times R - \frac{n-x}{x} \times R_0}{R_0} = \frac{R_0 + \frac{n}{x} \times (R - R_0)}{R_0} = f^{(\alpha_{rp}-1)} \tag{3.60}$$

In other words, the shift of the measured isotope ratio from its original ratio has to be scaled by a factor n/x before the Rayleigh equation is applied. In the delta notation using carbon as an example, Equation 3.60 corresponds to

$$\ln\frac{1 + \delta^{13}C_0 + \frac{n}{x} \times (\delta^{13}C - \delta^{13}C_0)}{1 + \delta^{13}C_0} = \varepsilon_{rp} \times \ln f. \tag{3.61}$$

A linearized form of Equations 3.60 or 3.61 can be used to quantify α_{rp} or ε_{rp}, respectively, based on measured isotope data if the number of reactive sites x is known.

Step 2: Correcting for multiple reactive sites

In a second step, the α_{rp} is transformed to the AKIE. Let us assume that the molecule has z indistinguishable reactive positions that compete with each other to be the reacting site (intramolecular competition). In the case of 1,3-dichloropropane, $z = 2$. In case of primary isotope effects and assuming that the reaction involves one of the reactive position, all of them are in competition for reaction and hence $x = z$. In contrast, if the reaction is concerted meaning that several atoms of the element of interest react simultaneously, or in case of secondary isotope effects, there is no intramolecular competition and $z = 1$. These situations will be illustrated in Chapter 4 with some examples.

α_{rp} corresponds to the ratio between the reaction rate if a heavy isotope is present at any of the reactive positions, $^H k_{rp}$, to the reaction rate if all reactive positions are occupied by light isotopes, $L_{k_{rp}}$ (see Figure 3.5). Assuming that $^L k$ is the reaction rate for a light isotope at the attacked site and $^H k$ if a heavy isotope is present, α_{rp} can be expressed by

$$\alpha_{rp} = \frac{^H k_{rp}}{L_{k_{rp}}} = \frac{^H k + (z-1) \times L_k}{z \times L_k} \tag{3.62}$$

Equation 3.62 can be rearranged to

$$\text{AKIE} = \frac{^L k}{^H k} = \frac{1}{z \times (a_{rp} - 1) + 1} \tag{3.63}$$

The calculation yields AKIE values that may be smaller than the intrinsic KIE if nonfractionating rate-limiting steps are present, as discussed for enzymatic reactions. This difference has to be taken into account when calculated AKIE values are compared to expected KIE values. The calculation of AKIE values is useful to gain insight into factors that control the magnitude of observed isotope fractionation. The obtained AKIE values also make it possible to compare isotope fractionation for different molecules transformed by the same reaction mechanism.

3.5.6 CASE STUDY

The calculation procedure is illustrated for transformation of different chlorinated alkanes by *Xanthobacter autotrophicus* (Abe, Zopfi, and Hunkeler 2009). The initial transformation step of the haloalkane dehalogenase enzyme consists of a nucleophilic substitution of Cl by an S_N2 mechanism. The obtained bulk isotope fractionation factor decreases with increasing chain length from $1/\alpha = 1.0301$ to 1.0135 (Table 3.5). In a first step, the isotope data were corrected for nonreactive positions by carrying out a linear regression with the transformed data according to Equation 3.61. Then the

TABLE 3.5

Bulk Fractionation Factor, Reactive-Position Specific Fractionation Factor and Apparent Kinetic Isotope Effect (AKIE) for Transformation of Chlorinated Alkanes by Hydrolytic Dehalogenation (S_N2 Nucleophilic Substitution). To Simplify the Comparison of Fractionation Factor with AKIE, the Former is Reported as an Inverse Ratio

Compound	n	x		$1/\alpha$	$1/\alpha_{rp}$	AKIE
1,2-Dichloroethane 1,2-DCA	2	2	Cl Cl | | H — C — C — H | | H H	1.0301		1.0567
1,3-Dichloropropane 1,3-DCP	3	2	Cl H Cl | | | H — C — C — C — H | | | H H H	1.0204		1.0567
Chloropropane CP	3	1	Cl H H | | | H — C — C — C — H | | | H H H	1.0174	1.0484	1.0484
Chlorobutane CB	4	1	Cl H H H | | | | H — C — C — C — C — H | | | | H H H H	1.0135	1.0487	1.0487

Source: Data from Abe, Y., J. Zopfi., and D. Hunkeler, *Isotopes in Environmental and Health Studies,* 45 (1), 18–26, 2009.

position-specific fractionation factor was transformed to the AKIE using Equation 3.63. The AKIE values are in a similar range. The observed value is close to the Streitwieser semiclassical limit (1.057, Table 3.4), which can be expected because S_N2 reactions have a symmetrical transition state. A slight difference in AKIE can be observed between substrates with one and two reactive sites, respectively. These differences may be caused by differences in the structure of the transition state or differences in the commitment to catalysis. The example illustrates that once isotope data are corrected for dilution by nonreactive and multiple reactive sites, the obtained AKIE can be close to the expected KIE.

3.5.7 ESTIMATION OF ISOTOPE FRACTIONATION FACTORS FOR REACTIONS WITH KNOWN MECHANISM

Above it was demonstrated how AKIE values can be derived from measured isotope data as average of the whole molecule. The reverse procedure, approximate estimation of the expected observable isotope fractionation for a transformation according to a given mechanism, is also of interest if new compounds are explored. Equation 3.62 can be used to transform known KIE into position-specific isotope fractionation factors α_{rp}. Then α_{rp} can be transformed further using the following approximate equation (Elsner et al, 2005)

$$(\alpha_{rp} - 1) \approx \frac{n}{x} \times (\alpha - 1) \tag{3.64}$$

For many reactions $z = x$ and Equations 3.62 and 3.64 can be combined to the following approximate equation that provides a direct link between the observed fractionation factor and the AKIE

$$\text{AKIE} = \frac{1}{1 + n \times (\alpha - 1)} \tag{3.65}$$

The obtained α values have to be considered as maximal values that can be expected for a reaction. The actual values may be smaller, especially for enzyme catalyzed reactions, depending on the structure of the transition state and the commitment to catalysis.

REFERENCES

Abe, Y., J. Zopfi, and D. Hunkeler. 2009. Effect of molecule size on carbon isotope fractionation during biodegradation of chlorinated alkanes by *Xanthobacter autotrophicus* GJ10. *Isotopes in Environmental and Health Studies* 45 (1):18–26.

Bouchard, D., P. Höhener, and D. Hunkeler. 2008. Carbon isotope fractionation during volatilization of a petroleum hydrocarbon source and diffusion across a porous medium: A column experiment. *Environmental Science & Technology* 42 (21):7801–6.

Bouchard, D., D. Hunkeler, P. Gaganis, R. Aravena, P. Höhener, and P. Kjeldsen. 2008. Carbon isotope fractionation during migration of petroleum hydrocarbon vapors in the unsaturated zone: Field experiment at Værløse Airbase, Denmark, and modeling. *Environmental Science & Technology* 42:596–601.

Bourg, I. C. 2008. Comment on "Modeling sulfur isotope fractionation and differential diffusion during sulfate reduction in sediments of the Cariaco Basin" by M. A. Donahue, J. P. Werne, C. Meile, and T. W. Lyons. *Geochimica Cosmochimica Acta* 72 (23):5852–54.

Bourg, I. C., and G. Sposito. 2008. Isotopic fractionation of noble gases by diffusion in liquid water: Molecular dynamics simulations and hydrologic applications. *Geochimica Cosmochimica Acta* 72 (9):2237–47.

Cerling, T. E., D. K. Solomon, J. Quade, and J. R. Bowman. 1991. On the isotopic composition of carbon in soil carbon dioxide. *Geochimica Cosmochimica.. Acta* 55:3403–5.

Clark, I. D., and P. Fritz. 1997. *Environmental isotopes in hydrogeology*. Boca Raton, FL: Lewis Publishers.

Conant, B. H., R. W. Gillham, and C. A. Mendoza. 1996. Vapor transport of trichloroethylene in the unsaturated zone: Field and numerical modeling investigations. *Water Resources Research* 32 (1):9–22.

Cook, P. F., ed. 1991. *Enzyme mechanism from isotope effects*. Boca Raton, FL: CRC Press.

Cornish-Bowden, A. 1995. *Fundamentals of enzyme kinetics.* London: Portland Press.

Elsner, M., L. Zwank, D. Hunkeler, and R. P. Schwarzenbach. 2005. A new concept linking observable stable isotope fractionation to transformation pathways of organic groundwater contaminants. *Environmental Science & Technology* 39:6896–916.

Falta, R. W., I. Javandel, K. Pruess, and P. A. Witherspoon. 1989. Density-driven flow of gas in the unsaturated zone due to the evaporation of volatile organic compounds. *Water Resources Research* 25 (10):2159–69.

Freeze, R. A., and J. A. Cherry. 1979. *Groundwater.* Upper Saddle River, NJ: Prentice Hall.

Hartshorn, S. R., and V. J. Shiner. 1972. Calculation of H/D, C-12/C-13, and C-12/C-14 fractionation factors from valence force fields derived for a series of simple organic-molecules. *Journal of the American Chemical Society* 94 (26):9002–12.

Huang, L., N. C. Sturchio, T. Abrajano, L. J. Heraty, and B. D. Holt. 1999. Carbon and chlorine isotope fractionation of chlorinated aliphatic hydrocarbons by evaporation. *Organic Geochemistry* 30:777–85.

Hunkeler, D. 2002. Quantification of isotope fractionation in experiments with deuterium-labeled substances. *Applied and Environmental Microbiology* 68 (10):5205–7.

Hunkeler, D., B. M. Van Breukelen, and M. Elsner. 2009. Modeling chlorine isotope trends during sequential transformation of chlorinated ethenes, *Environmental Science and Technology*, in press.

Huskey, W. P. 1991. Origin and interpretation of heavy-isotope effects. In *Enzyme mechanism from isotope effects,* ed. P. F. Cook, pp. 37–72, Boca Raton: FL: CRC Press.

Kohen, A., and H. H. Limbach, eds. 2006. *Isotope effects in chemistry and biology.* Boca Raton: FL: CRC Press.

Kopinke, F. D., A. Georgi, M. Voskamp, and H. H. Richnow. 2005. Carbon isotope fractionation of organic contaminants due to retardation on humic substances: Implications for natural attenuation studies in aquifers. *Environmental Science & Technology* 39 (16):6052–62.

Kuder, T., P. Philp, and J. Allen. 2009. Effects of volatilization on carbon and hydrogen isotope ratios of MTBE. *Environmental Science and Technology* 43 (6): 1763–1768.

LaBolle, E. M., G. E. Fogg, J. B. Eweis, J. Gravner, and D. G. Leaist. 2008. Isotopic fractionation by diffusion in groundwater. *Water Resources Research* 44, W07405.

Melander, L., and W. H. Saunders. 1980. *Reaction rates of isotopic molecules.* New York: John Wiley.

Paneth, P. 1985. On the application of the steady state to kinetic isotope effects. *Journal of the American Chemical Society* 107:7070–71.

Poulson, S. R., and J. I. Drever. 1999. Stable isotope (C, Cl, and H) fractionation during vaporization of trichloroethylene. *Environmental Science & Technology* 33 (20):3689–94.

Schüth, C., H. Taubald, N. Bolano, and K. Maciejczyk. 2003. Carbon and hydrogen isotope effects during sorption of organic contaminants on carbonaceous materials. *Journal Contaminant Hydrology* 64:269–81.

Scott, K. M., X. Lu, C. M. Cavanaugh, and J. S. Liu. 2004. Optimal methods for estimating kinetic isotope effects from different forms of the Rayleigh distillation equation. *Geochimica Cosmochimica Acta* 68:433–42.

Slater, G. F., J. M. E. Ahad, B. Sherwood Lollar, R. Allen-King, and B. Sleep. 2000. Carbon isotope effects resulting from equilibrium sorption of dissolved VOCs. *Analytical Chemistry* 72:5669–72.

Wittko, G., and W. Kohler. 2006. Influence of isotopic substitution on the diffusion and thermal diffusion coefficient of binary liquids. *European Physical Journal E* 21 (4):283–91.

4 Isotope Fractionation during Transformation Processes

Daniel Hunkeler and Barbara Morasch

CONTENTS

4.1 INTRODUCTION

This chapter discusses the occurrence and magnitude of stable isotope fractionation during biotic and abiotic transformation of common organic contaminants. It provides an overview of isotope enrichment factors (ε), which are necessary for the quantification of contaminant degradation based on stable isotope data from field sites and can help to characterize degradation mechanisms. This chapter also discusses why the magnitude of isotope fractionation varies between different substances. Because the observed isotope fractionation is caused by isotope effects during the initial irreversible transformation step (see Chapter 3), the reaction mechanism of this step is discussed for common organic contaminants. We will illustrate how the isotope effect can be derived from observed isotope data and compared to theoretically expected values for a given reaction. While these calculations are useful to gain insight into factors that control the magnitude of isotope fractionation and sometimes to identify reaction mechanisms, for practical application, isotope fractionation factors can also directly be used for the evaluation of field isotope data (as discussed in Chapter 8) provided that the reaction mechanism is known. If not, the measurement of isotope ratios of several elements may provide insight into the reaction mechanism. Different mechanisms frequently involve bonds between different elements as illustrated below for methyl tertiary-butyl ether (MTBE) and lead to a characteristic isotope fractionation pattern among different elements. Hence the measurement of isotope ratios of multiple elements can help to reveal the dominant reaction mechanisms in complex environments or even to identify the contribution of different mechanisms to the biodegradation of environmental contaminants.

The chapter covers the most common organic contaminants starting with chlorinated hydrocarbons, followed by (poly-) aromatic and aliphatic hydrocarbons, and finally gasoline additives. The reported isotope fractionation factors originate from numerous laboratory studies and were determined as illustrated in Chapter 3.

4.2   KEY FACTORS THAT INFLUENCE THE MAGNITUDE OF ISOTOPE FRACTIONATION

As discussed in detail in Chapter 3, biotic and abiotic processes often differ in their rate of reaction depending on whether a heavy or light isotope of an element is present in the bond that is broken during the first irreversible transformation step. This phenomenon is denoted as a kinetic isotope effect (KIE). The magnitude of a KIE can be quantified by comparing the rate constants for molecules with and without a heavy isotope at the reactive site (e.g., a ^{13}C-H bond vs a ^{12}C-H bond that is broken)

$$\text{KIE} = \frac{k_{^{12}\text{C-H}}}{k_{^{13}\text{C-H}}} = \frac{^{L}k}{^{H}k} \qquad (4.1)$$

where ^{L}k and ^{H}k are the rate constant for molecules with a light and heavy isotope, respectively at the reactive position. The KIE is the isotope effect that would be detected if the fractionating step was investigated in isolation and only the isotope ratio at the reactive position was considered. However, in experiments with compounds labeled at natural abundance and when using isotope ratio mass spectrometers (IRMS) for the analysis, one cannot determine the isotope fractionation at the reactive site in isolation because the compounds are transformed into a measurement gas (e.g., CO_2 for carbon isotope analysis) prior to analysis (see Chapter 2). Due to this transformation, the measured isotope ratio can no longer be attributed to a specific position in the molecule but corresponds to the average of all positions. There are other methods to obtain isotope ratios for each position in the molecule such as site-specific natural isotope fractionation—nuclear magnetic resonance (SNIF-NMR). However, they require a large amount of molecules and are not yet suited for typical environmental applications. Based on the average isotope ratios obtained by IRMS, one can quantify the average isotope fractionation irrespective of the position of the heavy isotope in the molecule (see Chapter 3). The obtained property is usually denoted as the bulk isotope fractionation factor (Elsner et al. 2005) and for first-order kinetics is given by

$$\alpha_{bulk} = \frac{^{13}k_{bulk}}{^{12}k_{bulk}}, \qquad (4.2)$$

where $^{13}k_{bulk}$ and $^{12}k_{bulk}$ are the average rate constants at which ^{13}C and ^{12}C, respectively, react independent of their position in the molecule, referred to as bulk isotope rate constants. Since bulk isotope fractionation factors are frequently close to one, it is more convenient to express them as bulk isotope enrichment factors on a per mill scale according to

$$\varepsilon_{bulk}(\%o) = (\alpha_{bulk} - 1) \times 1000. \qquad (4.3)$$

Note that KIEs are generally larger than one because they correspond to the rate constant of the light isotope that reacts faster over that of the heavy isotope, while α_{bulk} values correspond to the inverse ratio and hence are smaller than one. The bulk isotope fractionation factor (or more precisely the inverse of it) is usually smaller than the KIE for several reasons; two important factors, "dilution" and "masking," will be addressed in the following section and their effect illustrated with examples. A more fundamental discussion of the factors that control the magnitude of isotope fractionation can be found in Chapter 3.

4.2.1 DILUTION OF ISOTOPE FRACTIONATION

Most molecules contain several atoms of the element of interest that is investigated by stable isotope analysis. When a heavy isotope is present in the molecule but not located at the right (= reactive) position it will have no effect on the reaction rate. This means, the molecule reacts at the fast rate (^{L}k) although a heavy isotope is present. The higher the number of atoms of the element of interest, the larger the probability that a heavy isotope is located at a wrong (= nonreactive) position. Hence, the isotope effect experienced by molecules with a heavy isotope at the reactive positions becomes diluted by molecules with a heavy isotope present at other positions. As a result, the observed isotope effect is usually smaller than the theoretical KIE (Equation 4.1) and decreases with increasing molecule size. With some simple calculation (see Chapter 3), the dilution effect by nonreactive positions can be eliminated, which makes it possible to compare isotope fractionation data from molecules of different sizes. Let us consider the aerobic oxidation of hexane. The initial enzymatic transformation by alkane hydroxylases generally takes place at either of the two terminal positions (C_α) while the four intermediate carbons are not involved. Hence, two of the six carbons are (potentially) reactive positions (Figure 4.1). The first correction of the isotope fractionation factor concerns the presence of nonreactive positions

$$\left(\alpha_{\text{reactive position}} - 1\right) \approx (\alpha_{\text{bulk}} - 1) \times \frac{n}{x}. \tag{4.4}$$

Or in terms of isotope enrichment factors

$$\varepsilon_{\text{reactive position}}(\text{‰}) \approx \varepsilon_{\text{bulk}}(\text{‰}) \times \frac{n}{x}, \tag{4.5}$$

where α_{bulk} and $\varepsilon_{\text{bulk}}$ are bulk isotope fractionation/enrichment factors, $\alpha_{\text{reactive position}}$ and $\varepsilon_{\text{reactive position}}$ the isotope fractionation/enrichment factors if only atoms located at the potentially reactive positions of the molecule are taken into account, n the

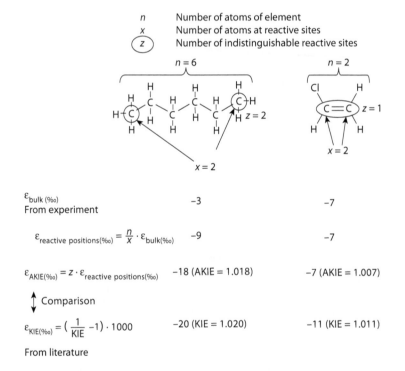

FIGURE 4.1   Calculation of the apparent kinetic isotope effect (AKIE) from measured isotope fractionation data ($\varepsilon_{\text{bulk}}$) and comparison with literature KIE values for aerobic oxidation of hexane (Bouchard, D., D. Hunkeler, and P. Höhener, *Organic Geochemistry*, 39, 23–33, 2008.) and vinyl chloride (Chu, K.-H., S. Maendra, D. L. Song, M. E. Conrad, and L. Alvarez-Cohen, *Environmental Science & Technology*, 38, 3126–30, 2004; Chartrand, M. M. G., A. Waller, T. E. Mattes, M. Elsner, G. Lacrampe-Couloume, J. M. Gossett, E. A. Edwards, and B. Sherwood Lollar, *Environmental Science & Technology*, 39, 1064–70, 2005.)

number of atoms of an element ($n = 6$, in the example of hexane) and x number of atoms at reactive positions ($x = 2$, in the example of hexane).

The isotope enrichment factor for the reactive position can also be obtained by transforming the isotope data followed by linear regression according to equation given in Chapter 3. The calculated $\alpha_{\text{reactive position}}$ does not fully reflect the KIE (Equation 4.1) because for our example hexane, just one of the two indistinguishable reactive positions (C_α) will be transformed by the enzyme in the initial step. Only when present at this position will the heavy isotope affect the reactive rate. Thus, a second correction is made for the presence of multiple indistinguishable reactive positions (for hexane: $z = 2$) to obtain the apparent KIE, AKIE (= KIE, which is potentially affected by masking as discussed below)

$$\text{AKIE} = \frac{1}{1 + z \times \left(\alpha_{\text{reactive position}} - 1 \right)} = \frac{1}{1 + z \times \varepsilon_{\text{reactive position}} (\text{‰})/1'000} \quad (4.6)$$

where z is the number of indistinguishable reactive positions.

In analogy to the bulk isotope enrichment factors we can define the AKIE as an enrichment factor on a per mill scale

$$\varepsilon_{\text{AKIE}} (\text{‰}) = \left(\frac{1}{\text{AKIE}} - 1 \right) \times 1'000 = z \times \varepsilon_{\text{reactive position}} (\text{‰}) \quad (4.7)$$

These values can be compared to the theoretical KIE expressed on a per mill scale

$$\varepsilon_{\text{KIE}} (\text{‰}) = \left(\frac{1}{\text{KIE}} - 1 \right) \times 1000 \quad (4.8)$$

Note that x is not always equivalent to z as in the case of hexane. The example of the oxidation of vinyl chloride (VC) illustrates the difference between x and z (Figure 4.1). During the initial transformation of VC to an epoxide, both carbons are involved and together represent the reactive position. However, given that both carbons concurrently participate in the initial transformation, they are not in intramolecular competition. There is only one indistinguishable reactive position ($z = 1$) containing two atoms ($x = 2$). In other words, the difference between our two examples is that the initial reaction in hexane degradation only involves one out of two carbon atoms at the reactive positions ($x = z = 2$), while in the case of VC, two carbons together make up one reactive position ($x = 2$; $z = 1$). For reactions with $x = z$ Equation 4.6 transforms to

$$\text{AKIE} = \frac{1}{1 + n \times \varepsilon_{\text{bulk}} (\text{‰})/1,000}. \quad (4.9)$$

Or on the per mill scale

$$\varepsilon_{AKIE}(\text{‰}) = n \times \varepsilon_{bulk}(\text{‰}) \qquad (4.10)$$

4.2.2 MASKING OF ISOTOPE FRACTIONATION

In many cases, an AKIE for biological processes or for reactions involving the solid phase is smaller than the expected KIE for the corresponding reaction in a homogeneous liquid solution. This difference comes from the fact that the KIE corresponds to an—often theoretical—scenario where the reaction step is studied in isolation, whereas in reality it is preceded by binding steps (e.g., interaction with a reactive surface or an enzyme). Although binding steps are usually not associated with significant isotope fractionation, they can partly mask the actual KIE. How much smaller the AKIE is compared to the KIE, depends on the rate of the binding step relative to the rate of the initial reaction step (see Equation 3.56). If each molecule that reaches the reactive site after a binding step reacts immediately, no isotope fractionation will be detectable given that the binding step is not selective to the isotope at the reactive site. In contrast, if the reaction step is slow compared to binding and dissociation steps, isotope fractionation will become detectable in the substrate pool. The rates of binding and dissociation steps are usually not known as they are difficult to determine. The comparison of AKIE values with KIE values reveals to what extent binding and dissociation are rate limiting during a sequence of steps (e.g., Melander and Saunders 1980).

To illustrate the influence of dilution and masking on the observed isotope fractionation, let us consider the aerobic biodegradation of C_3–C_{10} alkanes. They are all degraded by a similar reaction mechanism (hydroxylation by an alkane monooxygenase, see Section 4.4.3) and hence degradation should be associated with a similar KIE (Table 4.1). However, the bulk isotope enrichment factors (ε_{bulk}) decrease over nearly two orders of magnitude with increasing number of carbon atoms. At a first glance, this trend follows the dilution effect discussed before. However, after correction for the number of carbon atoms and reactive positions per molecule using Equations 4.5 and 4.7, $\varepsilon_{reactive\ positions}$ and ε_{AKIE} still decrease over one order of magnitude from C_3 to C_{10}. Thus, the decrease cannot be explained by dilution alone. In contrast, for a chemical transformation of alkanes by OH radicals in the gas phase, the obtained values of ε_{AKIE} and $AKIE_{chemical}$ are similar for all compounds. This suggests that the observed decrease of the KIE for biodegradation is related to transport and binding steps, such as substrate uptake into the cells or formation of the substrate–enzyme complex, which play a role in biodegradation but not during a chemical reaction in the gas phase. Intuitively, the larger the alkane, the longer it may take to transport it into the cell. As the isotope effect is caused inside the cells, isotope fractionation of larger molecules is less apparent outside of the cell compared to smaller substrate molecules.

TABLE 4.1

Carbon Isotope Enrichments during Aerobic Biodegradation and Chemical Transformation of *n*-alkanes

Compound	Number of C Atoms	Biodegradation				Chemical Transformation	
		ε_{bulk}(‰)	$\varepsilon_{reactive\ positions}$(‰)	ε_{AKIE}(‰)	$AKIE_{biodeg}$	ε_{AKIE}(‰)	$AKIE_{chemical}$
Propane	3	−11.0	−16.5	−33.0	1.033	−13.8	1.014
n-Butane	4	−5.6	−11.2	−22.4	1.022	−12.8	1.013
n-Pentane	5	−3.8	−9.5	−19.0	1.019	−13.8	1.014
n-Hexane	6	−2.3	−6.9	−13.8	1.013	−20.6	1.021
n-Heptane	7	−1.4	−4.9	−9.8	1.010	−16.7	1.017
n-Octane	8	−0.9	−3.6	−7.2	1.007	−16.7	1.017
n-Decane	10	−0.2	−1.0	−2.0	1.002	n.d.	n.d.

Sources: Values adapted from Anderson, R. S., L. Huang, R. Iannone, A. E. Thompson, and J. Rudolph, *Journal of Physical Chemistry*, 108, 11537–44, 2004; Bouchard, D., D. Hunkeler, and P. Höhener, *Organic Geochemistry*, 39, 23–33, 2008.

4.2.3 DIFFERENT REACTION MECHANISMS

A number of compounds can be degraded by different reaction mechanisms and isotope fractionation will often vary depending on the mechanism. A good example is MTBE (Kuder et al. 2005; Zwank, Berg et al. 2005; Elsner, McKelvie et al. 2007). Under oxic conditions, MTBE is biodegraded via cleavage of a C–H bond at the methoxy group by a monooxygenase (Figure 4.2a), while under anoxic conditions a C-O bond is likely cleaved by an S_N2 mechanism (Figure 4.2b). Furthermore, under acidic conditions, MTBE can become hydrolyzed chemically likely via an S_N1 mechanism (Figure 4.2c). For aerobic biodegradation, generally a small carbon and large hydrogen bulk isotope enrichment factor is observed, while for anaerobic biodegradation carbon isotope, fractionation is larger and hydrogen isotope fractionation smaller than for aerobic biodegradation (Table 4.2). These differences are due to the type of bond that is broken. Cleavage of a bond to a heavier element tends to lead to a larger isotope effect than when a lighter bonding partner is present (see Chapter 3). Hence carbon isotope fractionation is larger for splitting a C–O bond than for splitting a C–H bond. The difference in hydrogen isotope fractionation can be explained by the presence of H in the reactive bond for aerobic oxidation, while no bond to H is directly involved in the case of an S_N2 reaction. For acid hydrolysis, the carbon isotope enrichment factor is intermediate while the hydrogen isotope enrichment factor is in the same range as for aerobic biodegradation. The quite large hydrogen isotope fractionation may be surprising given that the S_N1 reaction does not directly involve changes to bonds with hydrogen. However, due to the high relative mass difference between the two isotopes, ^1H and ^2H,

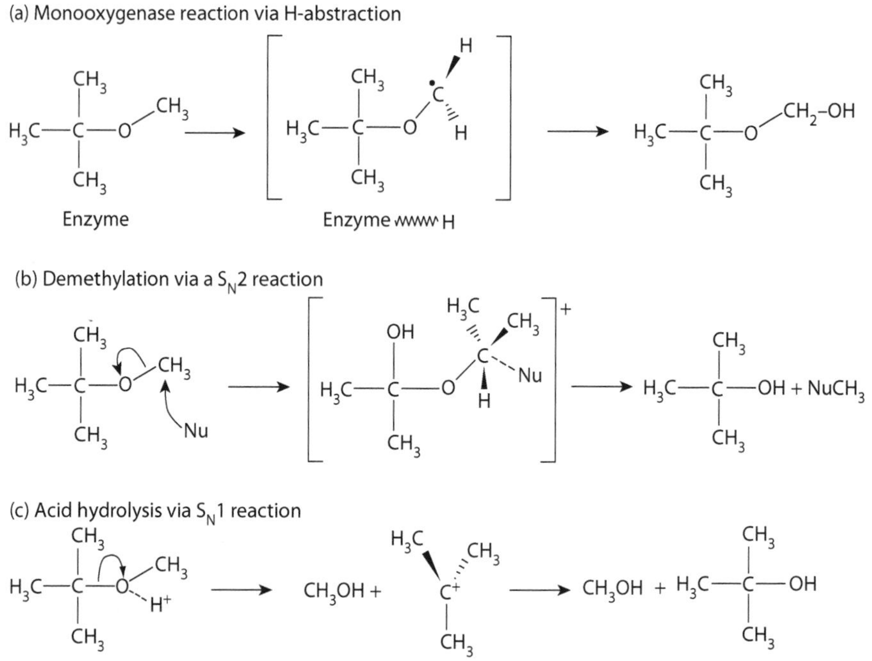

(a) Monooxygenase reaction via H-abstraction

(b) Demethylation via a S_N2 reaction

(c) Acid hydrolysis via S_N1 reaction

FIGURE 4.2   Initial biotic and abiotic transformation mechanisms of MTBE.

hydrogen can also influence the reaction rate when present in a position adjacent to the reactive center, a phenomena denoted as secondary isotope effect.

In order to evaluate in more detail whether the observed isotope enrichment factors are consistent with the expected reaction mechanisms, we can translate them again into AKIE values and compare them to KIE values (Table 4.2). Depending on the reaction scenario, different values for n, x, and z have to be used. In other words, in order to be able to calculate AKIE we need to know—or to postulate—the reaction mechanism. In general, the calculated AKIE correspond well to the expected ranges, although for aerobic degradation they are in the lower range, possibly due to the effect of masking.

In summary, the calculation of AKIE values provides insight into factors controlling the magnitude of isotope fractionation. By calculating AKIE values for known reactions, isotope fractionation data from molecules having different sizes can be compared to each other and with theoretical KIE data. AKIE calculations can also be used to identify initial reactions by comparing the resulting AKIE with theoretical KIE values for different reaction mechanisms.

While these calculations provide insight into factors controlling isotope fractionation, AKIE values are not needed for a quantitative evaluation of isotope data at field sites. Here, laboratory-derived isotope enrichment factors (ε_{bulk}) can directly be used to quantify contaminant degradation based on isotope measurements. Care has to be taken to select the appropriate isotope enrichment factor for the reactions taking place. If the reaction mechanism is not known, it can be characterized by comparing the shifts in isotope ratios for different elements as illustrated above for

TABLE 4.2

Carbon and Hydrogen Isotope Fractionation During Transformation of MTBE via Different Pathways. The Bulk Isotope Enrichment Factors are Transformed into Apparent Kinetic Isotope Effects Taking into Account the Number of Each Element in the Molecule (n), the Number of Atoms at Reactive Positions (x), and the Number of Indistinguishable Reactive Positions (z), which Vary Depending on the Element and Reaction Mechanism

Pathway		n	x	z	ε_{bulk}(‰)	ε_{AKIE} (‰)	AKIE	KIE
Aerobic	C	5	1	1	–2.0	–9.9	1.01	1.01–1.03
biodegradation	H	12	3	3	–41.4	–497	1.99	2–50
Anaerobic	C	5	1	1	–13.4	–67.0	1.07	1.03–1.09
biodegradation	H	12	9	1	–16.0	–21.3	1.02	0.95–1.05
(S_N2)								
Acid hydrolysis	C	5	1	1	–4.9	–24.5	1.03	1.0–1.03
S_N1	H	12	3	1	–55.0	–220.0	1.28	1.0–1.15

Sources: The Bulk Isotope Enrichment Factors Correspond to Average Values from Laboratory Studies by Hunkeler, D., B. J. Butler, R. Aravena, and J. F. Barker, *Environmental Science & Technology*, 35, 676–81, 2001; Gray, J. R., G. Lacrampe-Couloume, D. Gandhi, K. M. Scow, R. D. Wilson, D. M. Mackay, and B. Sherwood Lollar *Environmental Science & Technology*, 36, 1931–38, 2002; Kuder, T., J. T. Wilson, P. Kaiser, R. V. Kolhatkar, P. Philp, and J. Allen, *Environmental Science & Technology*, 39, 213–20, 2005; Somsamak, P., H. H. Richnow, and M. M. Häggblom, *Environmental Science & Technology*, 39, 103–9, 2005; Somsamak, P., H. H. Richnow, and M. M. Häggblom, *Applied Environmental Microbiology*, 72, 1157–63, 2006; Rosell, M., D. Barcelo, T. Rohwerder, U. Breuer, M. Gehre, and H. H. Richnow, *Environmental Science & Technology*, 41, 2036–43, 2007.

the example of MTBE. Once the reaction mechanism identified, the correct isotope enrichment factor can then be selected for quantifying the reaction. Strategies for quantifying reactions at the field scale will be discussed in Chapter 8.

In the following paragraphs, the occurrence and magnitude of stable isotope effects are discussed, structured by types of contaminants starting with chlorinated hydrocarbons, followed by aromatic and aliphatic hydrocarbons, and finally by gasoline additives. Isotope fractionation is generally reported as bulk isotope enrichment factor (ε_{bulk}). Where helpful for linking the observed isotope fractionation to reaction mechanisms, AKIE values, usually expressed as ε_{AKIE}, are calculated and compared to expected KIE values.

4.3 ISOTOPE FRACTIONATION DURING TRANSFORMATION OF CHLORINATED HYDROCARBONS

Chlorinated hydrocarbons, particularly tetrachloroethene (PCE), trichloroethene (TCE), and their degradation products, belong to the most abundant organic pollutants in industrialized countries (Squillace et al. 1999; Squillace, Moran, and

Price 2004). They derive from numerous sources such as industrial sites, landfills, or dry-cleaning facilities and are of great concern because of their toxicity and because they frequently cause extended groundwater contaminations at spill sites (Guilbeault, Parker, and Cherry 2005). The range of isotope enrichment factors for different reactions is summarized in Table 4.3. Values from individual studies are given in Table 4.6 at the end of the chapter.

4.3.1 CHLORINATED ETHENES

Isotope fractionation for this group of compounds has been studied extensively due to their importance as groundwater contaminants. The investigated processes correspond to the main reactions by which these compounds are transformed under natural conditions and during engineered remediation (i.e., anaerobic and aerobic biodegradation, abiotic reductive transformation by Fe(0), and chemical oxidation).

Under oxic conditions, only the less chlorinated compounds such as VC or dichloroethene (DCE) can serve as a carbon and energy source. In the presence of another carbon source (e.g., methane, methanol, or aromatic hydrocarbons), some bacteria can also transform the more chlorinated compounds (TCE or PCE) cometabolically. In either case, transformation generally occurs via epoxidation of the carbon double bond (Figure 4.3c, Habets-Crützen et al. 1984). Since only two carbon atoms are present in the molecule and both are involved in the initial transformation step, ε_{bulk} corresponds directly to $\varepsilon_{AKIE,C}$ (see Figure 4.1). The $\varepsilon_{AKIE,C}$ values (–3.2 to –9.8‰) correspond quite well to the $\varepsilon_{KIE,C}$ from a reference experiment (–11‰, Singleton et al. 1997) except for TCE for which a larger value was observed. The relatively small carbon isotope effect can be explained by the absence of bond breaking during the initial transformation step (Figure 4.3c). Furthermore, only a very small chlorine isotope fractionation occurs (–0.3‰; Abe et al. 2009) since the carbon-chlorine bond is not involved in the initial transformation step.

Under anoxic conditions, transformation primarily occurs via reductive dechlorination, also denoted as hydrogenolysis, whereby C–Cl bonds are sequentially cleaved leading to formation of lower chlorinated ethenes (Scholz-Muramatsu et al. 1995; Maymo-Gatell et al. 1997). Transformation can occur fortuitously or the chlorinated ethenes serve as terminal electron acceptors, a process known as dehalorespiration. Numerous investigations revealed that the cis-1,2-dichloroethene (cDCE) isomer is the predominant product of the reduction of TCE. Trans-1,2-dichloroethene (tDCE) as well as 1,1-dichloroethene (1,1-DCE) are produced in minor quantities (Bradley 2003; Smidt and De Vos 2004). Typically, dehalogenation rates decrease with a decreasing number of chlorine atoms in the molecule. The observed carbon isotope fractionation is large for VC, cDCE, and tDCE while it is somewhat smaller for TCE, 1,1-DCE, and especially for PCE (Table 4.3). These differences may be due to differences in the respective reaction mechanism. The exact mechanism for reductive dechlorination of chlorinated ethenes is not known yet. Possible mechanisms have been postulated mainly based on experiments with model systems with reductants such as cobalamines (Glod, Angst et al. 1997; Glod, Brodmann et al. 1997; Pratt and van der Donk 2006; Bühl, Vinković Vrček, and Kabrede 2007). The initial step may involve a dissociative one-electron transfer (Figure 4.3a), a mechanism, in particular, proposed for TCE and PCE (Glod, Angst et al. 1997). Such a

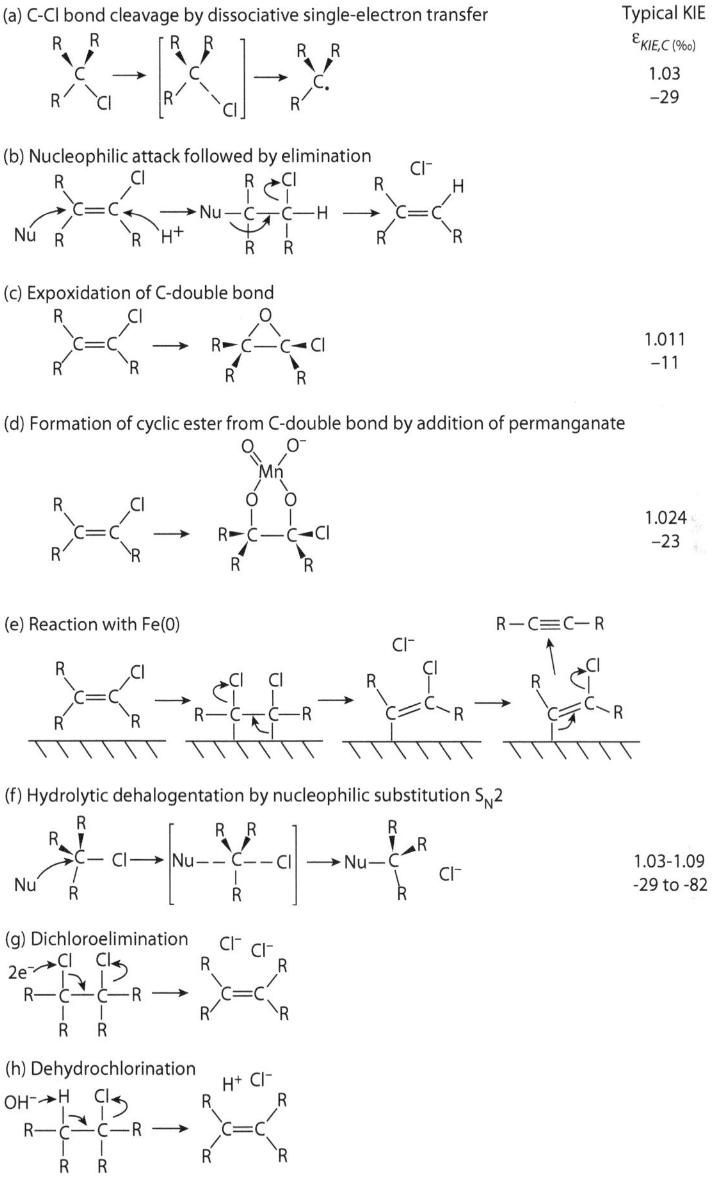

(a) C-Cl bond cleavage by dissociative single-electron transfer

Typical KIE

$\varepsilon_{KIE,C}$ (‰)

1.03
−29

(b) Nucleophilic attack followed by elimination

(c) Expoxidation of C-double bond

1.011
−11

(d) Formation of cyclic ester from C-double bond by addition of permanganate

1.024
−23

(e) Reaction with Fe(0)

(f) Hydrolytic dehalogentation by nucleophilic substitution S_N2

1.03-1.09
-29 to -82

(g) Dichloroelimination

(h) Dehydrochlorination

FIGURE 4.3 Initial transformation steps of chlorinated hydrocarbons by different mechanisms, typical carbon kinetic isotope effect (KIE), and corresponding $\varepsilon_{KIE,C}$ if no dilution by nonreactive positions occurred. Where known, the transition state is shown in brackets. Whenever known, typical carbon and hydrogen KIE from the literature are given (Shiner, V. J., and F. P. Wilgis, *Isotopes in organic chemistry.* Vol. 8 *Heavy isotope effects,* Elsevier, New York, 1992; Singleton, D. A., S. R. Merrigan, J. Liu, and K. N. Houk, *Journal of the American Chemical Society,* 119 (14), 3385–86, 1997; Houk, K. N., and T. Strassner, *Journal Organic Chemistry,* 64 (3), 800–2, 1999; Elsner, M., L. Zwank, D. Hunkeler, and R. P. Schwarzenbach, *Environmental Science & Technology,* 39, 6896–916, 2005.)

TABLE 4.3

Range of Carbon Isotope Enrichment Factors (‰) for the Biotic and Abiotic Transformation of Chlorinated Hydrocarbons. Values of the Individual Studies are given in Table 4.5

Compound	Biotic Aerobic		Biotic Anaerobic Reductive	Reductive dechlorination by Fe(0)	Abiotic
	Oxidation	Cometabolic oxidation	Reductive dechlorination		Oxidation by permanganate
(a) Chlorinated ethenes					
PCE			−0.4 to −16.4[a]	−5.7 to −25.3	−17.0
TCE	−18	−1.1	−2.5 to −18.5	−7.5 to −13.5	−21.4 to −26.8
cDCE	−7.1 to −9.8	n.d.	−14.1 to −29.7	−6.9 to −16.0	−21.1
tDCE		−3.5 to −6.7	−21.4 to −30.3		
1,1-DCE			−5.8 to −23.9		
VC	−5.7 to −8.2	−3.2 to −4.8	−21.0 to −31.1	−6.9 to −19.3	
(b) Chlorinated benzenes	Oxidation		Reductive dechlorination		
1,2,3-TCB			−2.4 to −3.5		
1,2,4-TCB	n.d.		−3.2		

	Oxidative cleavage of C-H	Hydrolytic dehalogenation S_N2	Dichloroelimination	Reductive dechlorination by Fe(0) or Cr(III)	Dichloroelimination by Fe(0) or Cr(III)	Dehydrochlorination
(c) Chlorinated ethanes						
1,2-DCA	-3.0 to -5.3	-21.5 to -33.0	-32.1			
1,1,1-TCA				-13.6 to -15.8		
1,1,2-TCA			-2.0			
1,1,2,2-TeCA					-12.7 to -19.3	-25.6 to -27.1
PCA					-14.7	-27.9 to -30.5
HCA					-10.4	
(d) Chlorinated methanes		Nucleophilic substitution S_N2	Nucleophilic substitution S_N2	Reduction on Fe(II) or iron oxides		
DCM		-42.4 to -66.3	-45.8 to -61.0			
CT				-25.8 to -32.0 -15.9[b]		

n.d. no significant isotope fractionation.

[a] *Desulfitobacterium* sp. Viet1 (Cichocka, D., G. Imfeld, H. H. Richnow, and I. Nijenhuis, *Chemosphere*, 71, 639–48, 2008).

[b] Reduction by mackinawite (FeS).

mechanism would likely lead to primary carbon and chlorine isotope effects due to the cleavage of a C–Cl bond in the initial step diluted by the carbon and chlorines present at the other position(s). Alternatively (Figure 4.3b), the cobalamin may act as a nucleo-phile forming an ethylcobalamin with simultaneous protonation of the second carbon (Glod, Brodmann et al. 1997; Pratt and van der Donk 2006; Bühl, Vinković Vrček, and Kabrede 2007). The ethylcobalamin then reductively dissociates into a dechlorinated compound and free chlorine. For this mechanism, the observed carbon isotope fraction-ation would likely be larger than in the first case since both carbon atoms are involved and hence no dilution occurs, provided that the first step is rate limiting.

The considerable variability of carbon isotope fractionation effects for different chlorinated ethenes suggests that different compounds react by a different mecha-nism. Given the large $\varepsilon_{bulk,C}$ for VC and cDCE (–14.1 to –31.1‰), these compounds may be transformed by the second mechanism (Figure 4.3b), which does not involve a dilution of the isotope effect while for other substrates the first mechanism may occur (Figure 4.3a). Alternatively, rate limiting transport steps preceding the actual reactive step might mask the isotope effect to different degrees. Indeed, in experi-ments on reductive dechlorination of PCE, the observed carbon isotope fractionation was larger in experiments with ruptured cells than in those with intact cells suggest-ing that uptake is partly rate limiting (Nijenhuis et al. 2005). However, such an effect was not observed for TCE (Cichocka et al. 2007).

Reductive transformation of chlorinated ethenes by Fe(0) was postulated to occur via a di-sigma-bonded surface-bound intermediate (Figure 4.3e), which is both car-bon positions of the molecule are transiently bound to the zero-valent iron surface (Arnold and Roberts 2000). Afterward, one or two C–Cl bonds are reductively cleaved leading to an ethene or acethylene product. A carbon isotope effect can occur during formation of the bound intermediate, which involves the conversion of a π-bond to a σ-bond or during the subsequent bond cleavage. Significant carbon iso-tope fractionation was indeed observed for transformation of all chlorinated ethenes (Table 4.3). The observed carbon isotope enrichment factors vary over a large range (e.g., from $\varepsilon_{bulk,C} = -5.7$ to $\varepsilon_{bulk,C} = -25.3‰$ for PCE) suggesting that the rate limiting step may vary depending on the state of the surface.

The initial rate limiting step during oxidation of chlorinated solvents by perman-ganate consists of the generation of a cyclic ester, which involves transformation of a π-bond to a σ-bond and formation of two carbon-oxygen bonds (Figure 4.3d, Nelson and Henley 1995). It involves both carbons of the chlorinated ethene and hence no dilution of the isotope effect occurs. Substantial carbon isotope fractionation is observed for all chlorinated ethenes ranging from $\varepsilon_{bulk,C} = -17.0‰$ for PCE to $\varepsilon_{bulk,C} = -26.8‰$ for TCE (Table 4.3). Since both carbons are involved in the initial step, these values correspond to $\varepsilon_{AKIE,C}$ and are close to the expected KIE of $\varepsilon_{KIE,C} = -23‰$ (Figure 4.3d, Houk and Strassner 1999). It is worth mentioning that the reaction with permanganate leads to a larger KIE than epoxidation ($\varepsilon_{KIE,C} = -11‰$, Singleton et al. 1997) although in both cases the double bond is oxidized (Figure 4.3c and d). However, one particular difference is that in the five-ring configuration of a cyclic ester, bonds are less constrained than in the three-ring configuration of the epoxide. In the transitions state of the cyclic ester, the frequency of vibrations is likely lower than in the epoxide and hence the isotope effect is larger (see Chapter 3).

4.3.2 Chlorinated Ethanes

Under oxic conditions, 1,2-dichloroethane (1,2-DCA) can be biodegraded via two different pathways. One pathway involves an oxidation likely catalyzed by a monooxygenase leading to the cleavage of a C-H bond and subsequent formation of a hydroxyl group. The second pathways consists of hydrolytic dehalogenation via a S_N2 nucleophilic substitution (Figure 4.3f, Hage and Hartmans 1999). The formed intermediates, 1,2-dichloroethanol and 2-chloroethanol, respectively, are further transformed to chloroacetaldehyde that is degraded via chloroacetate to glycolate. The magnitude of isotope fractionation strongly varies depending on the pathway. Oxidation by a monooxygenase leads to a small ^{13}C enrichment compared to a nucleophilic substitution (Table 4.3). This pattern can be explained by the influence of the atomic mass of the bonding partner on the carbon isotope effect (Huskey 1991). The cleavage of a C-H bond (molecular weight of hydrogen = 1) leads to a smaller carbon isotope effect compared to the cleavage of a C–Cl bond (molecular weight of chlorine = 35; see Chapter 3). If the observed isotope fractionation is corrected for intramolecular competition between the two identical reactive positions, an $\varepsilon_{AKIE,C}$ of −6 to −9.6‰ for oxidation (C-H bond cleavage) and −46 to −64.8‰ for hydrolytic dehalogenation (C–Cl bond cleavage) is obtained, which is within the expected range of $\varepsilon_{AKIE,C}$.

Under reductive conditions, 1,2-DCA can be biodegraded by dichloroelimination (Figure 4.3g) to ethene (Belay and Daniels 1987; Egli et al. 1987), which is associated with a significant carbon isotope fractionation. The relatively large $\varepsilon_{bulk,C}$ of −32.1‰ (Table 4.3) suggests again that the reaction mechanism involves both carbons and hence no dilution of the isotope effect occurs. Similar to 1,2-DCA, the initial reaction in anaerobic 1,1,2-trichloroethane (TCA) degradation is dichloroelimination. Surprisingly, the dechlorination of TCA was found to be associated with significantly smaller isotope enrichments of $\varepsilon_{bulk,C} = -2‰$. The reason for this variation is not yet known.

Especially highly chlorinated ethanes can also be transformed by abiotic reactions. During dehydrohalogenation (Figure 4.3h) of 1,1,2,2-tetrachloroethane (1,1,2,2-TeCA) and pentachloroethane (PCA) a large $\varepsilon_{bulk,C}$ of −25.6‰ to −30.5‰ was observed again indicating that both positions are involved in the rate limiting step without a dilution effect (Elsner, Cwiertny et al. 2007). In contrast, reductive dechlorination of 1,1,1-trichloroethane by Cr(II), Fe(0), or Cu-plated iron gave smaller $\varepsilon_{bulk,C}$ values (−13.6 to −15.6‰) due to dilution by the second carbon atom that does not participate in the reaction (Elsner, Cwiertny et al. 2007; Hofstetter et al. 2007). Similarly, the relatively small carbon isotope enrichment factors for dichloroelimination of 1,1,2,2-TeCA, PCA, and hexachloroethane (HCA) indicate that only one position is involved in the initial transformation step (Hofstetter et al. 2007). However, for 1,1,2,2-TeCA, another study reported larger carbon isotope enrichment factors (Elsner, Cwiertny et al. 2007).

4.3.3 Chlorinated Methanes

Chlorinated methanes are important environmental pollutants that can be degraded by bacteria cometabolically or serve as a carbon and energy source. Various initial reaction mechanisms have been described, for example, reduction, nucleophilic substitution, or oxygenation (Wackett et al. 1992). Under oxic conditions, dichloromethane (DCM) can be transformed to formaldehyde by methylotrophic bacteria via a biologically mediated nucleophilic substitution (Wackett et al. 1992). Since only one carbon atom is present, the isotope effect is undiluted by nonreactive positions. The observed carbon isotope effects are very pronounced (Table 4.3) and correspond to $\varepsilon_{AKIE,C}$ of -42.2 to $-66.3‰$, which are typical for S_N2 reactions (Figure 4.3f). For abiotic reductive transformation of carbon tetrachloride (CT) by various iron (hydr)oxides and iron sulfides, the observed magnitude of carbon isotope fractionation was reported to be between $\varepsilon_{bulk,C} = -21.6‰$ and $-32‰$ (Table 4.3) except for reduction by mackinawite (Zwank, Elsner et al. 2005). These values correspond again to $\varepsilon_{AKIE,C}$ and are in the typical range for the reductive cleavage of a C–Cl bond (Figure 4.3a).

4.3.4 CHLORINATED BENZENES

Chlorinated benzenes are widely used as solvents, degreasers, pesticides, and for chemical syntheses. In the presence of oxygen, chlorinated benzenes are generally attacked by a ring dioxygenase that introduces two hydroxyl groups in neighboring positions in the aromatic ring (van der Meer et al. 1991). The reaction mechanism corresponds to toluene dioxygenation (Figure 4.4a), which is discussed in more detail in the paragraph on monoaromatic hydrocarbon (Section 4.4.1). No significant carbon isotope fractionation was observed during aerobic biodegradation of 1,2,4-trichlorobenzene (TCB) by *Pseudomonas* sp. P51 (Table 4.3) similarly as for toluene dioxygenation. Reductive dechlorination during degradation of 1,2,3-TCB and 1,2,4-TCB by *Dehalococcoides* sp. CBDB1 causes significant but relatively small carbon isotope fractionation (Griebler, Adrian et al. 2004; Table 4.3). During the reaction, a chlorine substituent is replaced by hydrogen in an analogy to reductive dechlorination of PCE or TCE. The small carbon isotope fractionation compared to chlorinated ethenes can be explained by the higher number of carbon atoms of the benzene skeleton and thus stronger dilution of the isotope effect.

4.4 ISOTOPE FRACTIONATION DURING TRANSFORMATION OF PETROLEUM HYDROCARBONS

Similar to chlorinated solvents, petroleum hydrocarbons are of environmental concern. Petroleum hydrocarbons may enter the environment via accidental fuel spills or at industrial sites. Among them, the monoaromatic hydrocarbons benzene, toluene, ethylbenzene, and xylene isomers (BTEX) belong to the list of priority pollutants of the U.S. Environmental Protection Agency (US-EPA 1999). These putatively mutagenic and carcinogenic substances make up the bulk (>50 wt %) of the water-soluble fraction of gasoline (Coleman et al. 1984). In the following sections, transformation mechanisms and isotope fractionation will be summarized for BTEX, low-molecular

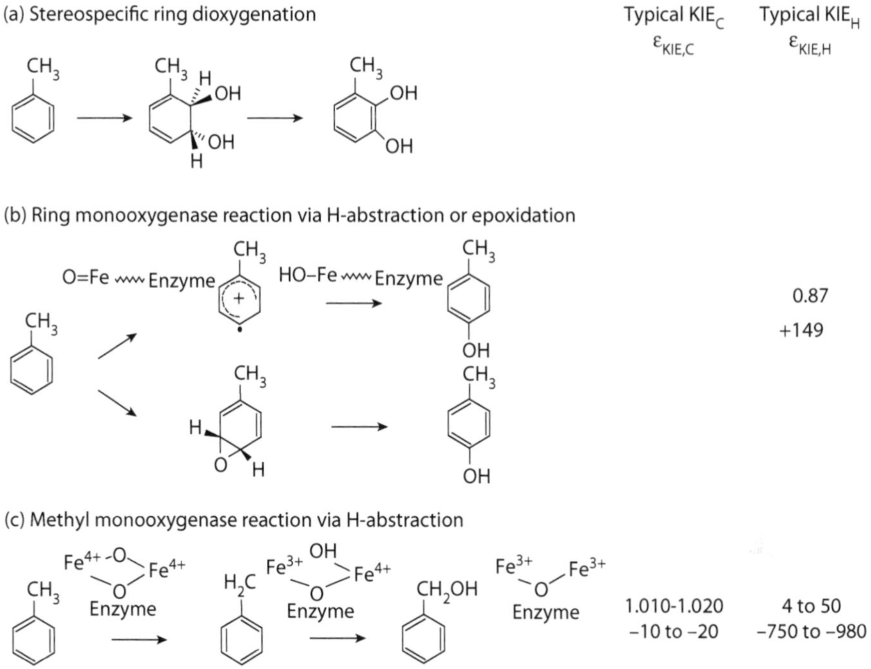

(a) Stereospecific ring dioxygenation

Typical KIE$_C$ Typical KIE$_H$
$\varepsilon_{KIE,C}$ $\varepsilon_{KIE,H}$

(b) Ring monooxygenase reaction via H-abstraction or epoxidation

0.87

+149

(c) Methyl monooxygenase reaction via H-abstraction

1.010-1.020   4 to 50
−10 to −20   −750 to −980

FIGURE 4.4 (Continued) Initial transformation steps of mono- and polyaromatic hydrocarbons by different mechanisms. Unstable intermediates are shown in brackets. Reaction mechanisms were adapted from (a) Gibson, D. T., G. E. Cardini, F. C. Maseles, and R. E. Kallio, *Biochemistry,* 9, 1631–35, 1970; Jeffrey, A. M., H. J. C. Yeh, D. M. Jerina, T. R. Patel, J. F. Davey, and D. T. Gibson, *Biochemistry,* 14, 575–84, 1975; Karlsson, A., J. V. Parales, R. E. Parales, D. T. Gibson, H. Eklund, and S. Ramaswamy, *Science,* 299, 1039–42, 2003, (b) Whited, G. M., and D. T. Gibson, *Journal of Bacteriology,* 173 (9), 3010–16, 1991; McClay, K., B. G. Fox, and R. J. Steffan, *Applied and Environmental Microbiology,* 66, 1877–82, 2000; Bertrand, E., R. Sakai, E. Rozhkova-Novosad, L. Moe, B. G. Fox, J. T. Groves, and R. N. Austin, *Journal of Inorganic Biochemistry,* 99, 1998–2006, 2005, (c) Nesheim, J. C., and J. D. Lipscomb, *Biochemistry,* 35, 10240–47, 1996; Jin, Y., and J. D. Lipscomb, *Biochemistry,* 3, 6178–86, 1999, (d) Frey, P. A, *Annual Review of Biochemistry,* 70, 121–48, 2001; Selmer, T., A. J. Pierik, and J. Heider, *Biological Chemistry,* 386, 981–88, 2005, (e) Ulrich, A. C., H. R. Beller, and E. A. Edwards, *Environmental Science & Technology,* 39, 6681–91, 2005; Safinofski, M., and R. U. Meckenstock, *Environmental Microbiology,* 8, 347–52, 2006. Wherever known, typical carbon and hydrogen KIE effects from the literature are given. (Nesheim, J. C., and J. D. Lipscomb, *Biochemistry,* 35, 10240–47, 1996; Elsner, M., L. Zwank, D. Hunkeler, and R. P. Schwarzenbach, *Environmental Science & Technology,* 39, 6896–916, 2005; Brecker, L., M. F. Kögl, C. E. Tyl, R. Kratzer, and B. Nidetzky, *Tetrahedron Letters,* 47, 4045–49, 2006; de Visser, S. P., K. Oh, A.-R. Han, and W. Nam, *Inorganic Chemistry,* 46, 4632–41, 2007.)

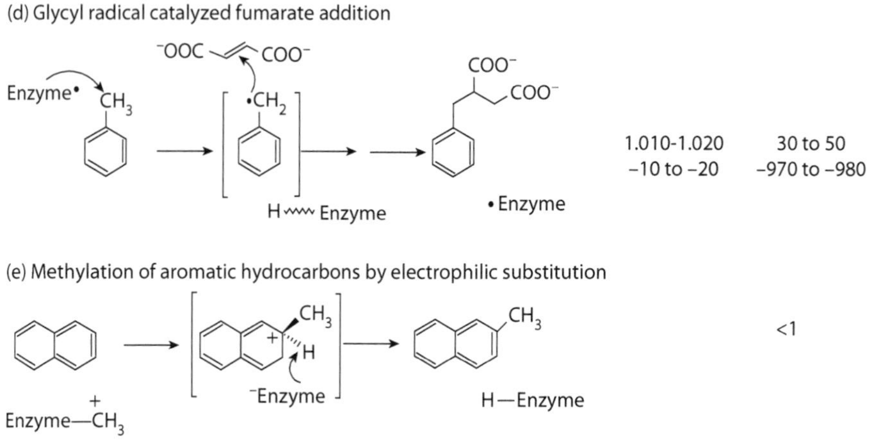

(d) Glycyl radical catalyzed fumarate addition

1.010-1.020   30 to 50
−10 to −20   −970 to −980

(e) Methylation of aromatic hydrocarbons by electrophilic substitution

<1

FIGURE 4.4   (Continued)

weight Polycyclic aromatic hydrocarbons (PAHs), and alkanes. Ranges of isotope enrichment factors for these compounds can be found in Table 4.4, while a comprehensive list of values from individual studies can be found in Table 4.7.

4.4.1   MONOAROMATIC HYDROCARBONS

Aerobic and anaerobic degradation of monoaromatic hydrocarbons has been thoroughly studied. For most of the known biochemical reactions isotope fractionation factors have been determined. As monoaromatic hydrocarbons are stabilized by the aromatic ring system and high activation energies have to be overcome, chemical reactions are not relevant for their turnover under environmental conditions.

Under oxic conditions, O_2 is the cosubstrate for biodegradation of aromatic hydrocarbons by either mono- or dioxygenase enzymes as will be discussed using toluene as an example. As other dioxygenases, the toluene dioxygenase reaction (TDO) uses one complete O_2 molecule and catalyzes the formation of two OH-groups at the positions 2 and 3 of the ring skeleton (Yeh, Gibson, and Liu 1977). This reaction leads—via a short-time intermediate with broken aromatic character—to the formation of a catechol (Figure 4.4a). Only small carbon isotope fractionation was observed for this reaction with $\varepsilon_{bulk,C} = -0.4$‰, corresponding to $\varepsilon_{AKIE,C} = -1.4$‰. The small AKIE reflects the absence of bond breaking during the reaction.

The toluene monooxygenases introduce only one oxygen atom of the cosubstrate O_2 into the aromatic substrate while the second is released as H_2O (Nozaki and Hayaishi 1984). An initial monooxygenation step can target either carbons of the aromatic ring at any position (Figure 4.4b) or the methyl carbon (Figure 4.4c). For the ring-monooxygenation of toluene, two initial reaction mechanisms have been postulated (Figure 4.4b). Either enzyme-bound oxygen forms a radical by subtraction of a hydrogen atom followed by attachment of a OH group to yield a cresol (Harayama, Kok, and Neidle 1992; Lange and Que 1998), or an unstable epoxide is formed, a reaction which includes loosening of a C=C bond and the formation of C–O–C (Figure 4.4b, Whited and Gibson 1991). Carbon isotope fractionation upon monooxygenation is somewhat larger ($\varepsilon_{bulk,C} = -1.1‰$) than for the dioxygenase reaction.

Finally, the reaction at the methyl group proceeds via H-abstraction and formation of a radical (Figure 4.4c). This reaction is associated with the largest carbon isotope fractionation ($\varepsilon_{bulk,C} = -3.3‰$) in aerobic toluene degradation. The calculated $\varepsilon_{AKIE,C} = -23.1‰$ agrees well with typical values expected for cleavage of a C–H bond ($\varepsilon_{KIE,C} = -10$ to $-20‰$). Carbon isotope fractionation caused by the same methyl monooxygenation was also investigated for degradation of m- and p-xylene and resulted in slightly smaller isotope enrichment factors of $\varepsilon_{bulk,C} = -1.7$ and $-2.3‰$, respectively (Table 4.7).

Isolation of anaerobic toluene degrading microorganisms in the late 1980s and 1990s opened the way for the systematic study of their degradation pathways (Grbic-Galic and Vogel 1987). As an initial reaction, a fumarate molecule is added to the methyl group of toluene to form a succinyl adduct, called benzyl succinate (Biegert, Fuchs, and Heider 1996). Ethylbenzene, xylenes, alkyl-substituted polyaromatic compounds, and alkanes are degraded by the same mechanism (Spormann and Widdel 2000). During this initial reaction, a radical-bearing enzyme cleaves the C-H bond of the toluene methyl group and the formed methyl radical reacts with the cosubstrate fumarate (Figure 3.3d; Selmer, Pierik, and Heider 2005). Fractionation of stable carbon isotopes during anaerobic toluene degradation under various redox conditions resulted in enrichment factors ranging from $\varepsilon_{bulk,C} = -0.5$ to $-6.2‰$ (Table 4.4). These values correspond to $\varepsilon_{AKIE,C}$ values of -3.5 to $-31.0‰$ (Equation 4.10). Especially the higher values that are in the expected range for cleavage of a C-H bond ($\varepsilon_{KIE,C} = -10$ to $-20‰$). Experiments with a mixed culture that degraded toluene under methanogenic conditions gave hydrogen isotope enrichment factors of $\varepsilon_{bulk,H} = -12$ to $-69‰$ (Ward et al. 2000) corresponding to $\varepsilon_{AKIE,H} = -96$ to $-522‰$ or $AKIE_H$ values of 1.106 and 2.08. The reported isotope effects are in the lower range of values expected for cleavage of a C–H bond.

4.4.2 POLYAROMATIC HYDROCARBONS

Under oxic conditions, carbon isotope fractionation during biodegradation of polyaromatic hydrocarbons such as naphthalene, 2-methylphenanthrene, and fluoranthene was studied (Table 4.4). For these substrates, the initial reaction generally consists of a dioxygenation and the mechanism is very similar to TDO discussed above. For naphthalene dioxygenation, a very small carbon isotope enrichment factor of $\varepsilon_{bulk,C} = -0.1‰$ was obtained while no significant ^{13}C enrichment occurred for larger PAHs. Compared to toluene with seven carbon atoms ($\varepsilon = -0.4‰$ during TDO reaction), naphthalene has 10 carbon atoms and the dilution therefore is higher. However, dilution alone does not explain the difference in $\varepsilon_{bulk,C}$. Corresponding AKIEs expressed as $\varepsilon_{AKIE,C}$ are smaller for naphthalene ($-0.5‰$) than for toluene ($-1.4‰$) possibly due to the increasing importance of rate limiting transport steps for larger molecules. A similar trend to smaller AKIE for larger molecules has been observed for alkanes (see section 4.2.2). Using the AKIE for naphthalene, maximal values of $\varepsilon_{bulk,C}$ for larger PAH can be predicted. These calculations demonstrate that no significant carbon isotope fractionation can be expected for PAHs consisting of 11 or more carbons.

Data on stable isotope fractionation caused by anaerobic biodegradation of PAHs is very limited. This is a direct consequence of the small number of bacterial cultures available that can degrade PAHs without O_2. In analogy for anaerobic degradation, the isotope enrichment is expected to become smaller with increasing numbers of carbon atoms. Indeed, carbon isotope enrichment factors for the degradation of naphthalene ($\varepsilon_{bulk,C} = -1.1‰$) and methylnaphthalene ($\varepsilon_{bulk,C} = -0.9‰$) under sulfate-reducing conditions are smaller than most enrichment factors observed for anaerobic biodegradation of monoaromatic compounds (Table 4.7). The initial step in anaerobic naphthalene degradation has not yet been proven unambiguously but is postulated to be a direct methylation of the aromatic ring. The enzymatic methyl group transfer and its binding to the aromatic ring lead to an intermediary positive charge in the molecule. The aromatic ring is reestablished after the enzyme has abstracted an H-radical in the following step (Figure 4.4e). The product of this reaction is 2-methylnaphthalene. The 2-methylnaphthalene itself is degraded via fumarate addition and the formation of a methyl-naphtyl succinate, analogous to anaerobic toluene degradation (Figure 4.4d).

4.4.3 N-ALKANES AND CYCLIC ALKANES

n-Alkanes are saturated, unbranched chains of hydrocarbons that originate from petroleum and other thermally mature deposits of organic carbon. It is now recognized that enzymes capable of alkane oxygenation are widespread in prokaryotic and in eukaryotic cells (for a review see Widdel and Rabus 2001; van Beilen and Funhoff 2005). A large number of aerobic bacteria have the potential to degrade n-alkanes and their principal degradation pathways have been studied in detail. Using n-octane as an example, the initial step is a hydroxylation by an alkane monooxygenase. This enzyme catalyzes a monooxygenation, which introduces one oxygen as OH-group into the substrate, typically at the terminal or subterminal positions C_α or C_β of the hydrocarbon chain. The first intermediate is n-octanol or 2-octanol depending on the exact type of enzyme employed. It is not yet known whether carbon isotope fractionation varies depending on the position at which the reaction takes place. For the oxidation of alkanes by OH-radicals (Anderson et al. 2004), a smaller carbon isotope effect was observed for secondary carbons (carbons connected to two other carbons) compared to primary and tertiary carbons (carbons connected to one and three carbons, respectively). In the microcosm study presented in 4.2.2, the AKIE decreased with increasing chain length from $\varepsilon_{AKIE,C} = -33$ (C3) to $-2‰$ (C10, Table 4.2) similarly as for PAH likely due to the increasing importance of rate limiting transport steps for larger molecules.

In absence of oxygen, alkanes are biodegradable via fumarate addition, which follow the same reaction mechanism as discussed before for mono- and polyaromatic hydrocarbons with alkyl residues (Figure 4.4d). Degradation of n-hexane by a denitrifying pure culture resulted in a ^{13}C enrichment of $\varepsilon_{bulk,C} = -2.3‰$ (Vieth and Wilkes 2006). The calculated $\varepsilon_{AKIE,C}$ for this reaction is $-14‰$ and corresponds well to the expected $\varepsilon_{KIE,C} = -10$ to $-20‰$ (Figure 4.4d) for radical catalyzed fumarate addition reactions.

TABLE 4.4

Ranges of Carbon and Hydrogen Isotope Enrichment Factors (‰) **for the Transformation of Mono- and Polyaromatic Hydrocarbons and Alkanes. Values of the Individual Studies are Given in Table 4.7**

Compound	Biotic Aerobic Degradation (Oxygenases)		Biotic Anaerobic Degradation	
	Carbon	Hydrogen	Carbon	Hydrogen
(a) Aromatic hydrocarbons				
Benzene	−0.5 to −4.3	−11.0 to −17.0	−0.8 to −3.6	−26 to −79
m-,p-Cresol			−1.6 to −3.9	
Ethylbenzene			−2.1	
o-Ethyltoluene			−3.7	
Toluene	0 to −3.3	−2 to −159	−0.5 to −6.2	−12 to −98.4
Styrene	−1.2			
o-,m-,p-Xylene	−0.6 to −2.3		−1.1 to −3.2	−29.6
(b) Polyaromatic hydrocarbons				
Naphthalene	−0.1		−1.1	
2-Methylnaphthalene			−0.9	
2-Methylphenanthrene	n.d.			
Fluoranthene	n.d.			
(c) Aliphatic hydrocarbons				
Methane	−26.5			
Ethane	−8.0			
n-Alkanes C_3–C_6	−2.2 to −10.8		−1.6 to −5.7	
n-Alkanes C_7–C_{10}	−0.2 to −1.4		−1.9 to 2.1	
n-Alkanes C_{12}–C_{30}	n.d.			

n.d. No Significant Isotope Fractionation

4.5 ISOTOPE FRACTIONATION DURING TRANSFORMATION OF GASOLINE ADDITIVES

After the ban of tetra-ethyl lead, methyl *tert*-butyl ether (MTBE) has become the most common antiknocking agent in gasoline. Due to ubiquitous spills of gasoline, MTBE is among the most frequently detected groundwater contaminants in the United States and has also been detected in groundwater in many European countries. Further compounds used as fuel oxygenates are *tert*-amyl methyl ether (TAME) and ethyl *tert*-butyl ether (ETBE). The carbon and hydrogen isotope enrichment factors for biodegradation of MTBE and ETBE are summarized in Table 4.5 together with values for abiotic transformation of MTBE by acid hydrolysis and oxidation with permanganate. The origin of the variability for carbon isotope fractionation is linked to differences in the reaction mechanisms as discussed in detail in Section 4.2.3.

TABLE 4.5

Ranges of Carbon and Hydrogen Isotope Enrichment Factors (‰) Published for the Biotic and Abiotic Transformation of Gasoline Additives. Results of the Individual Studies Are Given in Table 4.8

Compound		Biotic		Abiotic		
		Aerobic	Anaerobic	Acid Hydrolysis	Oxidation by Permanganate	Sorption/ Partitioning
MTBE	ε_C	−0.3 to −2.4	−7.4 to −15.6	−4.9		n.d. to −0.7
	ε_H	−29 to −66	−11.5 to −16.0	−55	−109	
ETBE	ε_C	−0.7 to −0.8				
	ε_H	−11 to −14				
TAME	ε_C		−11.2 to −13.7			
	ε_H					
TBA	ε_C	−4.2	n.d.			n.d. to −1.5
	ε_H					

n.d. No Significant Isotope Fractionation

4.6 SUMMARY AND CONCLUSIONS

In this chapter, we illustrated that many biotic and abiotic transformation reactions of common environmental contaminants are accompanied by a specific enrichment of the heavy isotopes in the residual, nondegraded compound. This enrichment can be especially pronounced for small molecules or if the element of interest is only present at a small number in the molecule. For molecules containing larger numbers of the element of interest in a molecule, the isotope effect becomes increasingly diluted because the probability that the heavy isotope is located at the reactive site (and is influencing the transformation rate) is decreasing. Furthermore, as illustrated by the example of alkanes, binding and dissociation steps preceding the actual transformation may become more rate limiting, too, for larger molecules and increasingly mask the isotope effect. For compound, the isotope fractionation pattern between different elements can vary depending on specific chemical reaction mechanisms. These specific isotope enrichment patterns of multiple elements allow characterizing reaction mechanisms in laboratory studies as well as in complex field settings. Once the reaction mechanism is identified, an appropriate isotope enrichment factor can be selected for a quantitative evaluation of isotope data. However, even for the same mechanism and the same element some variation in the isotope enrichment factor can exist as illustrated in the tables above. The calculated degree of transformation will be associated with a range of uncertainty.

In addition to the substance groups discussed above, isotope fractionation data become increasingly available for pesticides, nitro-aromatic compounds, and others (Ewald et al. 2007; Bernstein et al. 2008; Hofstetter et al. 2008; Penning, Cramer, and Elsner 2008). In some studies, additional elements such as N and O are included.

TABLE 4.6

Comprehensive Overview on Carbon Isotope Enrichment Factors (‰) for Biotic and Abiotic Transformations of Chlorinated Hydrocarbons

Compound	Transformation Type	Pathway	ε_C	Comment	Reference
PCE	Anaerobic	Biotic reductive dechlorination	−5.5	Mixed culture KB-1	(Slater et al. 2001)
			−5.5		
			−2.6		
PCE	Anaerobic	Biotic reductive dechlorination	−5.4	Enrichment culture	(Slater et al. 2001)
			−5.1		
PCE	Anaerobic	Biotic reductive dechlorination	−0.42	*Sulfurospirillum multivorans*	(Nijenhuis et al. 2005)
PCE	Anaerobic	Biotic reductive dechlorination	−5.2	*Desulfitobacterium* sp. PCE-S	(Nijenhuis et al. 2005)
PCE	Anaerobic	Biotic reductive dechlorination	−0.5	*Sulfurospirillum halorespirans*	(Cichocka et al. 2007)
PCE	Anaerobic	Biotic reductive dechlorination	−16.4	*Desulfitobacterium* sp. Viet1	(Cichocka et al. 2008)
PCE	Anaerobic	Biotic reductive dechlorination	−6.0	*Dehalococcoides* ethenogenes 195	(Cichocka et al. 2008)
PCE	Abiotic	Chemical oxidation by permanganate	−17		(Poulson and Naraoka 2002)
PCE	Abiotic	Dechlorination by Fe(0)	−25.3		(Dayan et al. 1999)
PCE	Abiotic	Dechlorination by Fe(0)	−5.7	Peerless iron	(VanStone et al. 2004)
			−9.2		
PCE	Abiotic	Dechlorination by Fe(0)	−9.7	Connelly iron	(VanStone et al. 2004)
			−15.5		
PCE	Abiotic	Reduction by cyanocobalamin	−13.0		(Nijenhuis et al. 2005)
TCE	Anaerobic	Reductive dechlorination	−7.1	Mixed culture (Pinellas)	(Sherwood Lollar et al. 1999)
TCE	Anaerobic	Biotic reductive dechlorination	−2.5	Mixed culture KB-1	(Bloom et al. 2000)
			−6.6		

(Continued)

TABLE 4.6 (Continued)

Compound	Transformation Type	Pathway	ε_c	Comment	Reference
TCE	Anaerobic	Biotic reductive dechlorination	-14.3 -13.4 -13.9 -15.2	Mixed culture KB-1	(Slater et al. 2001)
TCE	Anaerobic	Biotic reductive dechlorination	-9.6	Dehalococcoides ethenogenes 195	(Lee, Conrad, and Alvarez-Cohen 2007)
TCE	Anaerobic	Biotic reductive dechlorination	-16.4	Sulfurospirillum multivorans	(Lee, Conrad, and Alvarez-Cohen 2007)
TCE	Anaerobic	Biotic reductive dechlorination	-3.3	Dehalobacter restrictus PER-K23	(Lee, Conrad, and Alvarez-Cohen 2007)
TCE	Anaerobic	Biotic reductive dechlorination	-16.0	Dehalococcoides-containing enrichment culture ANAS	(Lee, Conrad, and Alvarez-Cohen 2007)
TCE	Anaerobic	Biotic reductive dechlorination	-18.5	Sulfurospirillum halorespirans	(Cichocka et al. 2007)
TCE	Anaerobic	Biotic reductive dechlorination	-18.4	Sulfurospirillum multivorans	(Cichocka et al. 2007)
TCE	Anaerobic	Biotic reductive dechlorination	-12.1	Desulfitobacterium sp. PCE-S	(Cichocka et al. 2007)
TCE	Anaerobic	Biotic reductive dechlorination	-3.5	Desulfuromonas michiganensis	(Cichocka et al. 2008)
TCE	Anaerobic	Biotic reductive dechlorination	-8.4	Geobacter lovleyi SZ	(Cichocka et al. 2008)
TCE	Anaerobic	Biotic reductive dechlorination	-3.3	Dehalobacter restrictus PER-K23	(Cichocka et al. 2008)
TCE	Anaerobic	Biotic reductive dechlorination	-13.5	Dehalococcoides ethenogenes 195	(Cichocka et al. 2008)
TCE	Aerobic	Oxidation	-18.2	Burkholderia cepacia G4	(Barth et al. 2002)
TCE	Aerobic	Cometabolic oxidation	-1.1	Methylosinus trichosporium OB3b	(Chu et al. 2004)
TCE	Abiotic	Reduction by cyanocobalamin	-15.2		(Nijenhuis et al. 2005)
TCE	Abiotic	Oxidation by permanganate	-25.1 -26.8		(Hunkeler et al. 2003)

Compound	Condition	Process	System/Culture	δ value	Reference
TCE	Abiotic	Oxidation by permanganate		−21.4	(Poulson and Naraoka 2002)
TCE	Abiotic	Dechlorination by Fe(0)		−8.6	(Dayan et al. 1999)
TCE	Abiotic	Dechlorination by Fe(0)		−10.1	(Schuth et al. 2003)
TCE	Abiotic	Dechlorination by Fe(0)	Peerless iron	−13.9, −13.0	(VanStone et al. 2004)
TCE	Abiotic	Dechlorination by Fe(0)	Connelly iron	−10.5, −7.5	(VanStone et al. 2004)
cDCE	Anaerobic	Reductive dechlorination	Mixed culture KB-1	−14.1, −16.1	(Bloom et al. 2000)
cDCE	Anaerobic	Reductive dechlorination	Mixed culture KB-1	−21.9, −25.5, −18.8, −18.9	(Slater et al. 2001)
cDCE	Anaerobic	Reductive dechlorination	Lab Microcosm	−19.2	(Hunkeler, Aravena, and Cox 2002)
cDCE	Anaerobic	Biotic reductive dechlorination	*Dehalococcoides* ethenogenes 195	−21.1	(Lee, Conrad, and Alvarez-Cohen 2007)
cDCE	Anaerobic	Biotic reductive dechlorination	*Dehalococcoides* sp. BAV1	−16.9	(Lee, Conrad, and Alvarez-Cohen 2007)
cDCE	Anaerobic	Biotic reductive dechlorination	*Dehalococcoides*-containing enrichment culture ANAS	−29.7	(Lee, Conrad, and Alvarez-Cohen 2007)
cDCE	Aerobic	Cometabolic oxidation	*Methylosinus trichosporium* OB3b	n.d.	(Chu et al. 2004)
cDCE	Aerobic		Enrichment culture, 12–14°C	−9.8, −8.8	(Tiehm et al. 2008)
cDCE	Aerobic		Enrichment culture, 22–24°C	−7.1, −8.2	(Tiehm et al. 2008)

(Continued)

TABLE 4.6 (Continued)

Compound	Transformation Type	Pathway	ε_c	Comment	Reference
cDCE	Aerobic	Oxidation	-8.5	*Polaromonas* J666	(Abe et al. 2009)
cDCE	Abiotic	Oxidation by permanganate	-21.1		(Poulson and Naraoka 2002)
cDCE	Abiotic	Dechlorination by Fe(0)	-14.4		(Dayan et al. 1999)
cDCE	Abiotic	Dechlorination by Fe(0)	-15.9	Peerless iron	(VanStone et al. 2004)
			-16.0		
cDCE	Abiotic	Dechlorination by Fe(0)	-11.9	Connelly iron	(VanStone et al. 2004)
			-6.9		
tDCE	Anaerobic	Biotic reductive dechlorination	-30.3	Lab Microcosm	(Hunkeler et al. 2002)
tDCE	Anaerobic	Biotic reductive dechlorination	-21.4	*Dehalococcoides* sp. BAV1	(Lee et al. 2007)
tDCE	Anaerobic	Biotic reductive dechlorination	-28.3	*Dehalococcoides*- containing enrichment culture ANAS	(Lee et al. 2007)
tDCE	Aerobic	Cometabolic oxidation	-3.5	*Methylomonas methanica*	(Brungard et al. 2003)
tDCE	Aerobic	Cometabolic oxidation	-6.7	*Methylosinus trichosporium* OB3b	(Brungard et al. 2003)
11-DCE	Anaerobic	Reductive dechlorination	-7.3	Lab Microcosm	(Hunkeler, Aravena, and Cox 2002)
11-DCE	Anaerobic	Reductive dechlorination	-5.8	*Dehalococcoides* ethenogenes 195	(Lee, Conrad, and Alvarez-Cohen 2007)
11-DCE	Anaerobic	Reductive dechlorination	-8.4	*Dehalococcoides* sp. BAV1	(Lee, Conrad, and Alvarez-Cohen 2007)
11-DCE	Anaerobic	Reductive dechlorination	-23.9	*Dehalococcoides*-containing enrichment culture ANAS	(Lee, Conrad, and Alvarez-Cohen 2007)
VC	Anaerobic	Reductive dechlorination	-26.6	Mixed culture KB-1	(Bloom et al. 2000)
			-21.5		
VC	Anaerobic	Reductive dechlorination	-21.0	Mixed culture KB-1	(Slater et al. 2001)
			-26.0		
			-22.2		

VC	Anaerobic	Reductive dechlorination	-31.1	Lab Microcosm	(Hunkeler, Aravena, and Cox 2002)
VC	Anaerobic	Reductive dechlorination	-24.0	Dehalococcoides sp. BAV1	(Lee, Conrad, and Alvarez-Cohen 2007)
VC	Anaerobic	Reductive dechlorination	-22.7	Dehalococcoides-containing enrichment culture ANAS	(Lee, Conrad, and Alvarez-Cohen 2007)
VC	Aerobic	Oxidation	-8.2	Mycobacterium sp. JS60	(Chartrand et al. 2005)
VC	Aerobic	Oxidation	-7.1	Mycobacterium sp. JS61	(Chartrand et al. 2005)
VC	Aerobic	Oxidation	-7.0	Mycobacterium sp. JS617	(Chartrand et al. 2005)
			-7.1		
VC	Aerobic	Oxidation	-7.6	Nocardioides sp. JS614	(Chartrand et al. 2005)
VC	Aerobic	Oxidation	-5.7	Mycobacterium aurum L1	(Chu et al. 2004)
VC	Aerobic	Cometabolic oxidation	-3.2	Methylosinus trichosporium OB3b	(Chu et al. 2004)
VC	Aerobic	Oxidation	-7.2	Nocardioides sp. JS614	(Abe et al. 2009)
			-7.3		
VC	Aerobic	Cometabolic oxidation	-4.8	Mycobacterium vaccae JOB5	(Chu et al. 2004)
VC	Aerobic		-4.5	Enrichment culture (Alameda)	(Chu et al. 2004)
VC	Aerobic		-5.5	Enrichment culture (Travis)	(Chu et al. 2004)
VC	Aerobic		-6.3	Enrichment culture, 12–14°C	(Tiehm et al. 2008)
			-6.5		
VC	Aerobic		-5.5	Enrichment culture, 22–24°C	(Tiehm et al. 2008)
			-5.4		
			-5.4		
VC	Abiotic	Dechlorination by Fe(0)	-16.6	Peerless iron	(VanStone et al. 2004)
			-19.3		
VC	Abiotic	Dechlorination by Fe(0)	-10.4	Connelly iron	(VanStone et al. 2004)
			-6.9		

(Continued)

TABLE 4.6 (Continued)

Compound	Transformation Type	Pathway	ε_c	Comment	Reference
HCA	Abiotic	Dichloroelimination by Cr(II)	−10.4		(Hofstetter et al. 2007)
PCA	Abiotic	Dichloroelimination by Cr(II)	−14.7		(Hofstetter et al. 2007)
1,1,2,2-TeCA	Abiotic	Dehydrohalogenation to TCE	−25.6		(Elsner, Cwiertny et al. 2007)
1,1,2,2-TeCA	Abiotic	Dehydrohalogenation to TCE	−27.1		(Hofstetter et al. 2007)
1,1,2,2-TeCA	Abiotic	Dichloroelimination by Cr(II)[a], Fe(0)[b], Cu/Fe[c]	−18.0[a] −19.3[b] −17.0[c]		(Elsner, Cwiertny et al. 2007)
1,1,2,2-TeCA	Abiotic	Dichloroelimination n by Cr(II)	−12.7		(Hofstetter et al. 2007)
1,1,1-TCA	Abiotic	Reductive dehalogenation by Cr(II)[a], Fe(0)[b], Cu/Fe[c]	−15.8[a] −13.6[b] −13.7[c]		(Elsner, Cwiertny et al. 2007)
1,1,2-TCA	Anaerobic	Dichloroelimination	−2.0	Lab Microcosm	(Hunkeler, Aravena, and Cox 2002)
1,2-DCA	Aerobic	Hydrolytic dehalogenation	−32.0	Xanthobacter autotrophicus GJ10	(Hunkeler and Aravena 2000)
1,2-DCA	Aerobic	Hydrolytic dehalogenation	−31.7 −33.0	Xanthobacter autotrophicus GJ10	(Hirschorn et al. 2004)
1,2-DCA	Aerobic	Hydrolytic dehalogenation	−32.4 −31.9	Ancylobacter aquaticus AD20	(Hirschorn et al. 2004)
1,2-DCA	Aerobic	Hydrolytic dehalogenation	−23.5 −24.1 −21.5	Enrichment culture (T)	(Hirschorn et al. 2004)

Compound	Condition	Process	Value	Culture/Description	Reference
1,2-DCA	Aerobic	Hydrolytic dehalogenation	-31.6, -30.5, -30.4, -32.1, -32.6, -32.3	Microcosm (EL)	(Hirschorn et al. 2004)
1,2-DCA	Aerobic	Hydrolytic dehalogenation	-25.0, -26.2	Microcosm (WL)	(Hirschorn et al. 2004)
1,2-DCA	Aerobic	Oxidation	-3.0	Pseudomonas sp. DCA1	(Hirschorn et al. 2004)
1,2-DCA	Aerobic	Oxidation	-3.3	Enrichment culture (T)	(Hirschorn et al. 2004)
1,2-DCA	Aerobic	Oxidation	-3.5, -4.2, -5.3	Enrichment culture (EL)	(Hirschorn et al. 2004)
1,2-DCA	Anaerobic	Dichloroelimination	-32.1	Lab Microcosm	(Hunkeler, Aravena, and Cox 2002)
DCM	Aerobic	Hydrolytic dehalogenation	-42.4	Hyphomicrobium MC8b	(Heraty et al. 1999)
DCM	Aerobic	Nuclephilic substitution	-58.4	Methylobacterium dichloromethanicum DM4	(Nikolausz et al. 2006)
DCM	Aerobic	Nuclephilic substitution	-61.9	Methylopila helvetica DM1	(Nikolausz et al. 2006)
DCM	Aerobic	Nuclephilic substitution	-60.2	Methylophilus leisingeri DM11	(Nikolausz et al. 2006)
DCM	Aerobic	Nuclephilic substitution	-61.9	Hyphomicrobium denitrificans	(Nikolausz et al. 2006)
DCM	Aerobic	Nuclephilic substitution	-66.3	Hyphomicrobium sp. KDM2	(Nikolausz et al. 2006)
DCM	Aerobic	Nuclephilic substitution	-62.8	Hyphomicrobium sp. KDM4	(Nikolausz et al. 2006)
DCM	Aerobic	Nuclephilic substitution	-62.8	Hyphomicrobium sp. DM2	(Nikolausz et al. 2006)
DCM	Aerobic	Nuclephilic substitution	-46.7	Hyphomicrobium sp. GJ21	(Nikolausz et al. 2006)
DCM	Aerobic	Nuclephilic substitution	-44.0	Methylorhabdus multivorans DM13	(Nikolausz et al. 2006)
DCM	Denitrifying	Nuclephilic substitution	-61.0	Hyphomicrobium sp. KDM2	(Nikolausz et al. 2006)
DCM	Denitrifying	Nuclephilic substitution	-59.3	Hyphomicrobium sp. KDM4	(Nikolausz et al. 2006)

(Continued)

TABLE 4.6 (Continued)

Compound	Transformation Type	Pathway	ε_C	Comment	Reference
DCM	Denitrifying	Nuclephilic substitution	−58.4	*Hyphomicrobium* sp. DM2	(Nikolausz et al. 2006)
DCM	Denitrifying	Nuclephilic substitution	−45.8	*Hyphomicrobium* sp. GJ21	(Nikolausz et al. 2006)
CT	Abiotic	Reductive dehalogenation by Fe(II) and different iron minerals	−15.9		(Zwank, Elsner et al., 2005)
			−26.1		
			−25.8		
			−32.0		
			−29.5		
			−26.7		
1,2,4-TCB	Aerobic		n.d.	*Pseudomonas* sp. P51	(Griebler, Adrian et al. 2004)
1,2,4-TCB	Anaerobic	Reductive dechlorination	−3.2	*Dehalococcoides* sp. CBDB1	(Griebler, Adrian et al. 2004)
1,2,3-TCB	Anaerobic	Reductive dechlorination	−3.5	*Dehalococcoides* sp. CBDB1	(Griebler, Adrian et al. 2004)
			−3.4	*Dehalococcoides* sp. CBDB1	

n.d. No Significant Isotope Fractionation.

TABLE 4.7

Overview on Isotope Enrichment Factors for Biotic Transformation of Aromatic and Aliphatic Hydrocarbons including Values for Replicates

Compound	Transformation Type	Pathway	Isotope Enrichment ε_C	Isotope Enrichment ε_H	Comment	Author
Benzene	Aerobic	?	-0.5		Mixed culture, enrichment factor estimated from publication	(Stehmeier et al. 1999)
			-0.8			
			-1			
			-0.5			
Benzene	Aerobic	?	-1.47	-12.8	Acinetobacter sp.	(Hunkeler, Andersen et al. 2001)
			-1.44			
Benzene	Aerobic	?	-3.41	-11.2	Burkholderia sp.	(Hunkeler, Andersen et al. 2001)
			-3.65			
Benzene	Aerobic	?	-2.1		Unsaturated sand microcosms	(Bouchard et al. 2008)
			-1.4			
Benzene	Aerobic	Dihydroxylation	-1.3		Rhodococcus opacus B-4	(Fischer et al. 2008)
Benzene	Aerobic	Dihydroxylation	-0.7		Pseudomonas putida ML2	(Fischer et al. 2008)
Benzene	Aerobic	Monohydroxylation	-1.7	-11	Ralstonia pickettii PKO1	(Fischer et al. 2008)
Benzene	Aerobic	Monohydroxylation	-4.3	-17	Cupriavidus necator ATCC 17967	(Fischer et al. 2008)
Benzene	Aerobic	Unknown	-2.6	-16.0	Alicycliphilus denitrificans BC	(Fischer et al. 2008)
Benzene	Denitrifying	Unknown, Methylation?	-2.4	-31	Enrichment culture Swamp NO$_3$-2b	(Mancini et al. 2008)
			-2.4	-26		
			-2.3	-39		
			-2.0	-41		
			-2.4	-32		

(Continued)

TABLE 4.7 (Continued)

Compound	Transformation Type	Pathway	Isotope Enrichment		Comment	Author
Benzene	Denitrifying	Unknown, Methylation?	-2.6	-47	Enrichment culture Swamp NO$_3$-1b	(Mancini et al. 2008)
Benzene	Denitrifying	Unknown, Methylation?	-2.8	-47	Enrichment culture Cart NO$_3$- PW1	(Mancini et al. 2008)
Benzene	Denitrifying	Unknown, Methylation?	-1.9	-31	Enrichment culture Swamp NO$_3$-Cons	(Mancini et al. 2008)
Benzene	Sulfate-reducing	Unknown, Hydroxylation?	-3.6	-79	Enrichment culture Cart SO$_4$ 1a	(Mancini et al. 2008)
			-3.4	-78		
			-2.7	-79		
Benzene	Sulfate-reducing	Unknown	-1.9	-59	Enrichment culture	(Fischer et al. 2008)
Benzene	Methanogenic	Unknown, Hydroxylation?	-1.9	-58	Enrichment culture Cart CH$_4$-1	(Mancini et al. 2008)
			-1.9	-64		
			-1.8	-58		
			-2.0	-54		
			-2.3	-72		
			-2.1	-58		
Benzene	Methanogenic	Unknown, Hydroxylation?	-0.8	-34	Mixed culture or CH$_4$-1b	(Mancini et al. 2008)
Benzene	Methanogenic	Unknown, Hydroxylation?	-1.1	-38	Enrichment culture Cart CH$_4$-1	(Mancini et al. 2008)
Ethylbenzene	De-nitrifying	Hydroxylation of ethyl side chain	-2.1		Strain EBN1	(Meckenstock et al. 2004)
Ethylbenzene	Anaerobic		-2.1		Field experiment	(Richnow, Annweiler et al. 2003)
o-Ethyltoluene	Sulfate-reducing		-3.7		Enrichment culture	(Wilkes et al. 2000)
Toluene	Aerobic		n.d.		Lab microcosm	(Sherwood Lollar et al. 1999)
Toluene	Aerobic	Methyl monooxygenation	-2.6	-159	Pseudomonas putida mt-2	(Morasch et al. 2002)
				-115		
				—		

Compound	Condition	Reaction			Organism	Reference
Toluene	Aerobic	Methyl monooxygenation	-3.3		Pseudomonas putida mt-2	(Morasch et al. 2002)
Toluene	Aerobic	Methyl monooxygenation	-2.5	-159	Pseudomonas putida mt-2, low iron conditions	(Morasch et al. 2002)
Toluene	Aerobic	Methyl monooxygenation	-1.9	-115	Pseudomonas putida mt-2, high iron conditions	(Morasch et al. 2002)
			-1.8	-97		
Toluene	Aerobic	Methyl monooxygenation	-2.8	-140	Pseudomonas putida mt-2	(Morasch et al. 2002)
Toluene	Aerobic	Methyl monooxygenation	-0.4	-8.6	Cladosporium sphaerospermum	(Vogt et al. 2008)
Toluene	Aerobic	Ring monooxygenation	-1.1		Ralstonia pickettii PKO1	(Morasch et al. 2002)
Toluene	Aerobic	Ring dioxygenation	-0.4		Pseudomonas putida F1	(Morasch et al. 2002)
Toluene	Aerobic		-0.7		Unsaturated sand microcosms	(Bouchard, Hunkeler, and Höhener 2008)
			-0.8			
Toluene	Aerobic	Ring dioxygenation	-1.8	-2	Rhodococcus opacus B-4	(Vogt et al. 2008)
Toluene	De-nitrifying	Fumarate addition	-1.7		Thauera aromatica	(Meckenstock et al. 1999)
Toluene	De-nitrifying	Fumarate addition	-2.7	-35	Thauera aromatica	(Vogt et al. 2008)
Toluene	De-nitrifying	Fumarate addition	-5.6	-79	Azoarcus sp. T	(Vogt et al. 2008)
			-6.2	-79		
			-5.7	-78		
Toluene	De-nitrifying	Fumarate addition	-3.0	-45	Azoarcus sp. EbN1	(Vogt et al. 2008)
			-3.8	-58		
			-2.9	-50		
Toluene	Anoxygenic phototrophic	Fumarate addition	-4.0	-23	Blastochloris sulfoviridis	(Vogt et al. 2008)
Toluene	Fe(III)-reducing	Fumarate addition	-1.8		Geobacter metallireducens	(Meckenstock et al. 1999)
Toluene	Sulfate-reducing	Fumarate addition	-1.5		Column experiment	(Meckenstock et al. 1999)

(Continued)

TABLE 4.7 (Continued)

Compound	Transformation Type	Pathway	Isotope Enrichment		Comment	Author
Toluene	Fe(III)-reducing	Fumarate addition	-1.3 -1.3 -1.0 -1.0	-34.6 -37.2	*Geobacter metallireducens* with solid Fe(III)	(Tobler et al. 2008)
Toluene	Fe(III)-reducing	Fumarate addition	-1.0 -1.0 -1.0		*Geobacter metallireducens* with solid Fe(III)	(Tobler, Hofstetter, and Schwarzenbach 2008)
Toluene	Fe(III)-reducing	Fumarate addition	-2.9 -3.0 -3.1 -3.6	-98.4	*Geobacter metallireducens* with solid Fe(III)-citrate	(Tobler, Hofstetter, and Schwarzenbach 2008)
Toluene	Sulfate-reducing	Fumarate addition	-1.7		Strain TRM1	(Meckenstock et al. 1999)
Toluene	Sulfate-reducing	Fumarate addition	-2.0	-66	Strain TRM1	(Vogt et al. 2008)
Toluene	Sulfate-reducing	Fumarate addition	-2.2		*Desulfosarcina cetonica*	(Morasch et al. 2001)
Toluene	Sulfate-reducing	Fumarate addition	-2.6 -2.3 -2.4	-80 -68 -74	*Desulfosarcina cetonica*	(Vogt et al. 2008)
Toluene	Sulfate-reducing	Fumarate addition	-0.8		Enrichment culture	(Ahad et al. 2000)
Toluene	Sulfate-reducing	Fumarate addition	-2.7 -2.8 -2.8	-88 -81 -87	Enrichment culture Zz. 53-56	(Ahad et al. 2000)
Toluene	Methanogenic	Fumarate addition	-0.5		Enrichment culture	(Ahad et al. 2000)
Toluene	Methanogenic	Fumarate addition		-12 -65	Mixed consortium	(Ward et al. 2000)
Toluene	Abiotic	Fenton-like oxidation	n.d.			(Ahad and Slater 2008)

Compound	Redox condition	Process	Value		Experiment / Strain	Reference
m-Xylene	Aerobic	Methyl monooxygenation	-1.7		Pseudomonas putida mt-2	(Morasch et al. 2002)
m-Xylene	Aerobic		-0.8		Unsaturated sand microcosms	(Bouchard et al. 2008)
			-0.6			
m-Xylene	Sulfate-reducing	Fumarate addition	-1.8		Strain OX39	(Morasch et al. 2004)
m/p-Xylene	Anaerobic	Fumarate addition	-1.5		Field experiment	(Richnow, Annweiler et al. 2003)
p-Xylene	Aerobic	Methyl monooxygenation	-2.3		Pseudomonas putida mt-2	(Morasch et al. 2002)
o-Xylene	Sulfate-reducing	Fumarate addition	-3.2		Enrichment culture	(Wilkes et al. 2000)
o-Xylene	Sulfate-reducing	Fumarate addition	-1.1		Column experiment	(Richnow, Meckenstock et al. 2003)
o-Xylene	Sulfate-reducing	Fumarate addition	-1.5		Strain OX39	(Morasch et al. 2004)
o-Xylene			-2.6	-29.6	Field experiment	(Steinbach et al. 2004)
m-Cresol	Sulfate-reducing	Fumarate addition	-3.9		Desulfobacterium cetonicum	(Morasch et al. 2004)
p-Cresol	Sulfate-reducing	Fumarate addition	-1.6		Desulfobacterium cetonicum	(Morasch et al. 2004)
Styrene	Aerobic		-1.2		Mixed culture, enrichment factor estimated from publication	(Stehmeier et al. 1999)
Naphthalene	Aerobic	Ring dioxygenation	-0.1		Pseudomonas putida NCIMB9816	(Morasch et al. 2002)
Naphthalene	Sulfate-reducing	Ring methylation	-1.1		Enrichment culture	(Griebler, Safinofski et al. 2004)
2-Methyl-naphthalene	Sulfate-reducing	Fumarate addition	-0.9		Enrichment culture	(Griebler, Safinofski et al. 2004)
2-Methyl-phenanthrene	Aerobic		n.d.		Sphingomonas sp. 2MPII	(Mazeas and Budzinski 2002)
Fluoranthene	Aerobic		n.d.		Sphingomonas paucimobilis	(Trust Hammer et al. 1995)
Methane	Aerobic		-26.5		Marine sediment	(Kinnaman, Valentine, and Tyler 2007)
Ethane	Aerobic		-8.0		Marine sediment	(Kinnaman, Valentine, and Tyler 2007)

(Continued)

TABLE 4.7 (Continued)

Compound	Transformation Type	Pathway	Isotope Enrichment	Comment	Author
Propane	Aerobic		−4.8	Marine sediment	(Kinnaman, Valentine, and Tyler 2007)
Propane	Aerobic	H-abstraction?	−10.8	Unsaturated sand microcosms	(Bouchard et al. 2008)
Propane	Sulfate-reducing	Fumarate addition	−5.7	Enrichment culture	(Kniemeyer et al. 2007)
n-Butane	Aerobic		−2.9	Marine sediment	(Bouchard et al. 2008)
n-Butane	Aerobic	H-abstraction?	−5.6	Unsaturated sand microcosms	(Bouchard et al. 2008)
n-Butane	Sulfate-reducing	Fumarate addition	−1.6	Enrichment culture	(Bouchard et al. 2008)
n-Pentane	Aerobic	H-abstraction?	−3.8	Unsaturated sand microcosms	(Bouchard et al. 2008)
			−2.4		
n-Pentane	Anaerobic	Fumarate addition	−2.3	Oilfield reservoir	(Vieth and Wilkes 2006)
n-Pentane	Anaerobic	Fumarate addition	−2.7	Oilfield reservoir	(Wilkes et al. 2008)
iso-Pentane	Anaerobic	Fumarate addition	−1.6	Oilfield reservoir	(Vieth and Wilkes 2006)
iso-Pentane	Anaerobic	Fumarate addition	−1.6	Oilfield reservoir	(Wilkes et al. 2008)
n-Hexane	Aerobic	H-abstraction?	−2.2	Unsaturated sand microcosms	(Bouchard et al. 2008)
			−2.4		
			−2.3		

n-Hexane	Denitrifying	Fumarate addition	-2.3		Strain HxN1	(Vieth and Wilkes 2006)
n-Hexane	Anaerobic	Fumarate addition	-1.6		Oilfield reservoir	(Vieth and Wilkes 2006)
n-Heptane	Aerobic	H-abstraction?	-1.4		Unsaturated sand microcosms	(Bouchard et al. 2008)
n-Heptane	Anaerobic	Fumarate addition	-1.9		Oilfield reservoir	(Vieth and Wilkes 2006)
			-2.1			
n-Octane	Aerobic	H-abstraction?	-0.7		Unsaturated sand microcosms	(Bouchard et al. 2008)
			-1.1			
			-0.9			
n-Decane	Aerobic	H-abstraction?	-0.3		Unsaturated sand microcosms	(Bouchard et al. 2008)
			-0.2			
n-Alkanes (C$_{14}$-C$_{30}$)	Aerobic		n.d.		Enrichment culture	(Mazeas, Budzinski, and Raymond 2002)
Alkanes (C$_{12}$-C$_{24}$)	Aerobic		n.d.	n.d.	Strain GIM2.5 Chrysosporium 1767	(Peng et al. 2004)

n.d.) No significant isotope fracionation.

TABLE 4.8

Overview on Isotope Enrichment During Biotic and Abiotic Transformation of Gasoline Additives including Values for Replicates

Compound	Transformation Type	Pathway	Value C	Value H	Comment	Author
MTBE	Abiotic	Acid hydrolysis	-4.9	-55		(Elsner, McKelvie et al. 2007)
MTBE	Abiotic	Oxidation by permanganate	—	-109		(Elsner, McKelvie et al. 2007)
MTBE	Aerobic	Methyl monooxygenase	-2.0 -1.9 -1.6 -1.8		Enrichment culture	(Hunkeler, Butler et al. 2001)
MTBE	Aerobic	Methyl monooxygenase	-1.5		Enrichment culture, 3-methylpentane as cosubstrate	(Hunkeler, Butler et al. 2001)
MTBE	Aerobic	Methyl monooxygenase	-2.0 -2.2 -2.4	-36 -33 -33 -37	Methylibium petroleiphilum PM1	(Gray et al. 2002)
MTBE	Aerobic	Methyl monooxygenase	-1.5 -1.4 -1.8	-66 -29	Enrichment culture	(Gray et al. 2002)
MTBE	Aerobic	Methyl monooxygenase	-2.4	-42	Methylibium sp. R8	(Rosell et al. 2007)
MTBE	Aerobic	Hydrolysis?	-0.48	n.d.	β-Proteobacterium L108	(Rosell et al. 2007)
MTBE	Aerobic	Hydrolysis?	-0.28	n.d.	Rhodococcus ruber IFP2001	(Rosell et al. 2007)
MTBE	Anaerobic, methanogenic?		-8.1		Field	(Kolhatkar et al. 2002)

Compound	Process	Mechanism			Experiment	Reference
MTBE	Anaerobic?		-7.4 -10.2	-11.5	Field	(Kuder et al. 2002)
MTBE	Anaerobic		-8.9		Microcosm experiment	(Kuder et al. 2002)
MTBE	Anaerobic		-13.0	-16.0	Microcosm experiment	(Kuder et al. 2005)
MTBE	Anaerobic, methanogenic?		-15.6		Microcosm experiment	(Somsamak, Richnow, and Häggblom 2005)
MTBE	Anaerobic, inhibition of methanogenesis		-14.6		Microcosm experiment	(Somsamak, Richnow, and Häggblom 2005)
MTBE	Anaerobic			-15.6	Field, εH calculated based on MTBE concentration. based on carbon isotopes and assuming anaerobic degradation	(Zwank, Berg et al. 2005)
MTBE	Sulfate-reducing		-14.5 -13.9 -13.7 -14.4		Enrichment culture	(Somsamak, Richnow, and Häggblom 2006)
MTBE	Methanogenic		-14.0–-14.4		Enrichment culture	(Somsamak, Richnow, and Häggblom 2006)
TAME	Anaerobic, methanogenic?		-13.7		Microcosm experiment	(Somsamak, Richnow, and Häggblom 2005)
TAME	Anaerobic, inhibition of methanogenesis		-11.2		Microcosm experiment	(Somsamak, Richnow, and Häggblom 2005)
ETBE	Aerobic	Acidic hydrolysis?	-0.7	-14	β-Proteobacterium L108	(Rosell et al. 2007)
ETBE	Aerobic	Acidic hydrolysis?	-0.8	-11	Rhodococcus ruber IFP2001	(Rosell et al. 2007)
TBA	Aerobic	Methyl monooxygenase	-0.8 -4.2	-11	Enrichment culture, 3-methylpentane as cosubstrate	(Hunkeler, Butler et al. 2001)
TBA	Anoxic		n.d.		Field	(Kuder et al. 2005)

In contrast to transformation processes, physicochemical partitioning and transport processes are, in general, only associated with very small or no isotope fractionation. An exception is diffusion, where isotope effects play an important role for volatile compounds in the unsaturated zone (see Chapter 3 and 9). Since transformation of many common organic contaminants is associated with significant isotope fractionation, compound-specific isotope analysis is a valuable tool to address the fate of these contaminants during natural attenuation and engineered remediation. Strategies for evaluating and quantifying reactions at the field scale are discussed in Chapter 8.

ACKNOWLEDGMENT

We would like to thank Patrick Höhener for his contributions to this chapter and Gwenaël Imfeld for the helpful comments.

REFERENCES

Abe, Y., R. Aravena, J. Zopfi, O. Shouakar-Stash, E. Cox, J. D. Roberts, and D. Hunkeler. 2009. Carbon and chlorine isotope fractionation during aerobic oxidation and reductive dechlorination of vinyl chloride and cis-1,2-dichloroethene. *Environmental Science & Technology* 43:101–7.

Ahad, J. M. E., B. Sherwood Lollar, E. A. Edwards, G. F. Slater, and B. E. Sleep. 2000. Carbon isotope fractionation during anaerobic biodegradation of toluene: Implications for intrinsic bioremediation. *Environmental Science & Technology* 34:892–96.

Ahad, J. M. E., and G. F. Slater. 2008. Carbon isotope effects associated with Fenton-like degradation of toluene: Potential for differentiation of abiotic and biotic degradation. *Science of the Total Environment* 401:194–98.

Anderson, R. S., L. Huang, R. Iannone, A. E. Thompson, and J. Rudolph, 2004. Carbon kinetic isotope effects in the gas phase reactions of light alkanes and ethene with the OH radical at 296 +/- 4 K. *Journal of Physical Chemistry* 108:11537–44.

Arnold, W. A., and A. L. Roberts. 2000. Pathways and kinetics of chlorinated ethylene and chlorinated acetylene reaction with Fe(0) particles. *Environmental Science & Technology* 34:1794–805.

Barth, J. A. C., G. Slater, C. Schueth, M. Bill, A. Downey, M. Larkin, and R. M. Kalin. 2002. Carbon isotope fractionation during aerobic biodegradation of trichloroethene by *Burkholderia cepacia* G4: A tool to map degradation mechanisms. *Applied and Environmental Microbiology* 68:1728–34.

Belay, N., and L. Daniels. 1987. Production of ethane, ethylene, and acetylene from halogenated hydrocarbons by methanogenic bacteria. *Applied Microbiology and Biotechnology* 53:1604–10.

Bernstein, A., Z. Ronen, E. Adar, R. Nativ, H. Lowag, W. Stichler, and R. U. Meckenstock. 2008. Compound-specific isotope analysis of RDX and stable isotope fractionation during aerobic and anaerobic biodegradation. *Environmental Science & Technology* 42:7772–77.

Bertrand, E., R. Sakai, E. Rozhkova-Novosad, L. Moe, B. G. Fox, J. T. Groves, and R. N. Austin. 2005. Reaction mechanisms of non-heme diiron hydroxylases characterised in whole cells. *Journal of Inorganic Biochemistry* 99:1998–2006.

Biegert, T., G. Fuchs, and J. Heider. 1996. Evidence that anaerobic oxidation of toluene in the denitrifying bacterium *Thauera aromatica* is initiated by formation of benzylsuccinate from toluene and fumarate. *European Journal of Biochemistry* 238: 661–8.

Bloom, Y., R. Aravena, D. Hunkeler, E. A. Edwards, and S. K. Frape. 2000. Carbon isotope fractionation during microbial dechlorination of trichloroethene, cis-1,2-dichloroethene and vinyl chloride: Implication for assessment of natural attenuation. *Environmental Science & Technology* 34:2768–72.

Bouchard, D., D. Hunkeler, and P. Höhener. 2008. Carbon isotope fractionation during aerobic biodegradation of *n*-alkanes and aromatic compounds in unsaturated sand. *Organic Geochemistry* 39:23–33.

Bradley, P. M. 2003. History and ecology of chloroethene biodegradation: A Review. *Bioremediation Journal* 7:81–109.

Brecker, L., M. F. Kögl, C. E. Tyl, R. Kratzer, and B. Nidetzky. 2006. NMR study on ^{13}C-kinetic isotope effects at ^{13}C natural abundance to characterize oxidations and an enzyme-catalyzed reduction. *Tetrahedron Letters* 47:4045–49.

Brungard, K. L., Munakata-Marr, J., Johnson, C. A. and Mandernack, K. W., 2003. Stable carbon isotope fractionation of trans-1, 2-dichloroethylene during Co-metabolic degradation by methanotrophic bacteria. *Chemical Geology*, 195(1–4): 59–67.

Bühl, M., I. Vinković Vrček, and H. Kabrede. 2007. Dehalogenation of chloroalkenes at cobalt centers. A model density function study. *Organometallics* 26:1494–504.

Chartrand, M. M. G., A. Waller, T. E. Mattes, M. Elsner, G. Lacrampe-Couloume, J. M. Gossett, E. A. Edwards, and B. Sherwood Lollar. 2005. Carbon isotopic fractionation during aerobic vinyl chloride degradation. *Environmental Science & Technology* 39:1064–70.

Chu, K.-H., S. Maendra, D. L. Song, M. E. Conrad, and L. Alvarez-Cohen. 2004. Stable carbon isotope fractionation during aerobic biodegradation of chlorinated ethenes. *Environmental Science & Technology* 38:3126–30.

Cichocka, D., G. Imfeld, H. H. Richnow, and I. Nijenhuis. 2008. Variability in microbial carbon isotope fractionation of tetra- and trichloroethene upon reductive dechlorination. *Chemosphere* 71:639–48.

Cichocka, D., M. Siegert, G. Imfeld, J. Andert, K. Beck, G. Diekert, H. H. Richnow, and I. Nijenhuis. 2007. Factors controlling the carbon isotope fractionation of tetra- and trichloroethene during reductive dechlorination by *Sulfurospirillum* ssp and *Desulfitobacterium* sp strain PCE-S. *FEMS Microbiology Ecology* 62:98–107.

Coleman, W. E., J. W. Munch, R. P. Streicher, H. P. Ringhand, and F. C. Knopfler. 1984. The identification and measurement of components in gasoline, kerosene, and No. 2 fuel oil that partition into the aqueous phase after mixing. *Archives of Environmental Contamination and Toxicology* 13:171–78.

Dayan, H., T. Abrajano, N. C. Sturchio, and L. Winsor. 1999. Carbon isotopic fractionation during reductive dehalogenation of chlorinated ethenes by metallic iron. *Organic Geochemistry* 30:755–63.

de Visser, S. P., K. Oh, A.-R. Han, and W. Nam. 2007. Combined experimental and theoretical study on aromatic hydroxylation by mononuclear nonheme iron(IV)-oxo complexes. *Inorganic Chemistry* 46:4632–41.

Egli, C., R. Scholtz, A. M. Cook, and T. Leisinger. 1987. Anaerobic dechlorination of tetrachloromethane and 1,2-dichloroethane to degradable products by pure cultures of Desulfobacterium sp. and Methanobacterium sp. *FEMS Microbiology Letters* 43:257–61.

Elsner, M., D. M. Cwiertny, A. L. Roberts, and B. Sherwood Lollar. 2007.1,1,2,2-tetrachloroethane reactions with OH-, Cr(II), granular iron, and a copper-iron bimetal: Insights from product formation and associated carbon isotope fractionation. *Environmental Science & Technology* 41:4111–17.

Elsner, M., J. McKelvie, G. L. Couloume, and B. Sherwood Lollar. 2007. Insight into methyl tert-butyl ether (MTBE) stable isotope fractionation from abiotic reference experiments. *Environmental Science & Technology* 41:5693–700.

Elsner, M., L. Zwank, D. Hunkeler, and R. P. Schwarzenbach. 2005. A new concept to link observed stable isotope fractionation to degradation pathways of organic groundwater contaminants. *Environmental Science & Technology* 39:6896–916.

Ewald, E. M., A. Wagner, I. Nijenhuis, H. H. Richnow, and U. Lechner. 2007. Microbial dehalogenation of trichlorinated dibenzo-p-dioxins by a Dehalococcoides-containing mixed culture is coupled to carbon isotope fractionation. *Environmental Science & Technology* 41:7744–51.

Fischer, A., I. Herklotz, S. Herrmann, M. Thullner, S. A. B. Weelink, A. J. M. Stams, M. Schlomann, H. H. Richnow, and C. Vogt. 2008. Combined carbon and hydrogen isotope fractionation investigations for elucidating benzene biodegradation pathways. *Environmental Science & Technology* 42:4356–63.

Frey, P. A. 2001. Radical mechanisms of enzymatic catalysis. *Annual Review of Biochemistry* 70:121–48.

Gibson, D. T., G. E. Cardini, F. C. Maseles, and R. E. Kallio. 1970. Incorporation of oxygen-18 into benzene by *Pseudomonas putida*. *Biochemistry* 9:1631–35.

Glod, G., W. Angst, C. Holliger, and R. Schwarzenbach. 1997. Corrinoid-mediated reduction of tetrachloroethene, trichloroethene, and trichlorofluoroethene in homogeneous aqeous solution: Reaction kinetics and reaction mechanisms. *Environmental Science & Technology* 31:253–60.

Glod, G., U. Brodmann, C. Holliger, and R. Schwarzenbach. 1997. Cobalamin-mediated reduction of *cis-* and *trans-*dichloroethene, 1,1-dichloroethene, and vinyl chloride in homogeneous aqueous solution: Reaction kinetics and mechanistic considerations. *Environmental Science & Technology* 31:3154–60.

Gray, J. R., G. Lacrampe-Couloume, D. Gandhi, K. M. Scow, R. D. Wilson, D. M. Mackay, and B. Sherwood Lollar. 2002. Carbon and hydrogen isotopic fractionation during biodegradation of methyl *tert-*butyl ether. *Environmental Science & Technology* 36:1931–38.

Grbic-Galic, D., and T. M. Vogel. 1987. Transformation of toluene and benzene by mixed methanogenic cultures. *Applied and Environmental Microbiology* 53:254–60.

Griebler, C., L. Adrian, R. U. Meckenstock, and H. H. Richnow. 2004. Stable carbon isotope fractionation during aerobic and anaerobic transformation of trichlorobenzene. *FEMS Microbiology Ecology* 48:313–21.

Griebler, C., M. Safinofski, A. Vieth, H. H. Richnow, and R. U. Meckenstock. 2004. Combined application of stable carbon isotope analysis and specific metabolites determination for assessing in situ degradation of aromatic hydrocarbons in a tar oil-contaminated aquifer. *Environmental Science & Technology* 38:617–31.

Guilbeault, M. A., B. L. Parker, and J. A. Cherry. 2005. Mass and flux distributions from DNAPL zones in sandy aquifers. *Ground Water*, 43:70–86.

Habets-Crützen, A. Q. H., L. E. S. Brink, C. G. van Ginkel, J. A. M. de Bont, and J. Tramper. 1984. Production of epoxides from gaseous alkenes by resting-cell suspensions and immobilized cells of alkene-utilizing bacteria. *Applied Microbiology and Biotechnology* 20:245–50.

Hage, J. C., and S. Hartmans. 1999. Monooxygenase-mediated 1,2,-dichloroethane degradation by *Pseudomonas* sp. strain DCA1. *Applied and Environmental Microbiology* 65:2466–70.

Harayama, S., M. Kok, and E. L. Neidle. 1992. Functional and evolutionary relationships among diverse oxygenases. *Annual Review of Microbiology* 46:565–601.

Heraty, L. J., M. E. Fuller, L. Huang, T. Abrajano, and N. C. Sturchio. 1999. Isotope fractionation of carbon and chlorine by microbial degradation of dichloromethane. *Organic Geochemistry* 30:793–99.

Hirschorn, S. K., M. J. Dinglasan, M. Elsner, S. A. Mancini, G. Lacrampe-Couloume, E. A. Edwards, and B. Sherwood Lollar. 2004. Pathway dependent isotopic fractionation during aerobic biodegradation of 1,2-dichloroethane. *Environmental Science of Technology* 38:4775–81.

Hofstetter, T. B., A. Neumann, W. A. Arnold, A. E. Hartenbach, J. Bolotin, C. J. Cramer, and R. P. Schwarzenbach. 2008. Substituent effects on nitrogen isotope fractionation during abiotic reduction of nitroaromatic compounds. *Environmental Science & Technology* 42:1997–2003.

Hofstetter, T. B., C. M. Reddy, L. J. Heraty, M. Berg, and N. C. Sturchio. 2007. Carbon and chlorine isotope effects during abiotic reductive dechlorination of polychlorinated ethanes. *Environmental Science & Technology* 41:4662–68.

Houk, K. N., and T. Strassner. 1999. Establishing the (3+2) mechanism for the permanganate oxidation of alkenes by theory and kinetic isotope effects. *Journal of Organic Chemistry* 64:800–2.

Hunkeler, D., N. Andersen, R. Aravena, S. M. Bernasconi, and B. J. Butler. 2001. Hydrogen and carbon isotope fractionation during aerobic biodegradation of benzene. *Environmental Science & Technology* 35:3462–67.

Hunkeler, D., and R. Aravena. 2000. Evidence of substantial carbon isotope fractionation between substrate, inorganic carbon, and biomass during aerobic mineralization of 1,2-dichloroethene by xanthobacter autotrophicus. *Applied and Environmental Microbiology* 66:4870–76.

Hunkeler, D., R. Aravena, J. A. Cherry, and B. Parker. 2003. Carbon isotope fractionation during chemical oxidation of chlorinated ethenes by permanganate: Laboratory and field studies. *Environmental Science Technology* 37:798–804.

Hunkeler, D., R. Aravena, and E. Cox. 2002. Carbon isotopes as a tool to evaluate the origin and fate of vinyl chloride: Laboratory experiments and modeling of isotope evolution. *Environmental Science & Technology* 36:3378–84.

Hunkeler, D., B. J. Butler, R. Aravena, and J. F. Barker. 2001. Monitoring biodegradation of methyl tert-butyl ether (MTBE) using compound-specific carbon isotope analysis. *Environmental Science & Technology* 35:676–81.

Huskey, W. P. 1991. Origins and interpretations of heavy atom isotope effects. In *Enzyme mechanism from isotope effects,* ed. P. F. Cook, pp. 37–72. Boca Raton: FL: CRC Press.

Jeffrey, A. M., H. J. C. Yeh, D. M. Jerina, T. R. Patel, J. F. Davey, and D. T. Gibson. 1975. Initial reactions in the oxidation of naphthalene by *Pseudomonas putida*. *Biochemistry* 14:575–84.

Jin, Y., and J. D. Lipscomb. 1999. Probing the mechanism of C-H activation: Oxidation of methylcubane by soluble methane monooxygenase from *Methylosinus trochosprium* OB3b. *Biochemistry* 3:6178–86.

Karlsson, A., J. V. Parales, R. E. Parales, D. T. Gibson, H. Eklund, and S. Ramaswamy. 2003. Crystal structure of naphthalene dioxygenase: Side-on binding of dioxygen to iron. *Science* 299:1039–42.

Kinnaman, F., D. L. Valentine, and S. C. Tyler. 2007. Carbon and hydrogen isotope fractionation associated with the aerobic oxidation of methane, ethane, propane and butane. *Geochimica Cosmochimica Acta* 71:271–83.

Kniemeyer, O., F. Musat, S. M. Sievert, K. Knittel, H. Wilkes, M. Blumenberg, W. Michaelis, et al. 2007. Anaerobic oxidation of short-chain hydrocarbons by marine sulphate-reducing bacteria. *Nature* 449:898–902.

Kolhatkar, R. V., T. Kuder, P. Philp, J. Allen, and J. T. Wilson. 2002. Use of compound-specific stable carbon isotope analyses to demonstrate anaerobic biodegradation of MTBE in groundwater at a gasoline release site. *Environmental Science & Technology* 36:5139–46.

Kuder, T., P. Philp, R. V. Kolhatkar, J. T. Wilson, and J. Allen. 2002. Application of stable carbon and hydrogen isotopic techniques for monitoring biodegradation on MTBE in the field. *NGWA/API* 371–81.

Kuder, T., J. T. Wilson, P. Kaiser, R. V. Kolhatkar, P. Philp, and J. Allen. 2005. Enrichment of stable carbon and hydrogen isotopes during anaerobic biodegradation of MTBE: Microcosm and field evidence. *Environmental Science & Technology* 39:213–20.

Lange, S. J., and L. Que. 1998. Oxygen activating nonheme iron enzymes. *Current Opinion in Chemical Biology* 2:159–72.

Lee, P. K. H., M. E. Conrad, and L. Alvarez-Cohen. 2007. Stable carbon isotope fractionation of chloroethenes by dehalorespiring isolates. *Environmental Science & Technology* 41:4277–85.

Mancini, S. A., C. E. Devine, M. Elsner, M. E. Nandi, A. C. Ulrich, E. A. Edwards, and B. Sherwood Lollar. 2008. Isotopic evidence suggests different initial reaction mechanisms for anaerobic benzene biodegradation. *Environmental Science & Technology* 42: 8290–8296.

Mancini, S. A., A. C. Ulrich, G. Lacrampe-Couloume, B. E. Sleep, E. A. Edwards, and B. Sherwood Lollar. 2003. Carbon and hydrogen isotopic fractionation during anaerobic biodegradation of benzene. *Applied and Environmental Microbiology* 69:191–98.

Maymo-Gatell, X., Y.-T. Chien, J. M. Gossett, and S. H. Zinder. 1997. Isolation of a bacterium that reductively dechlorinates tetrachloroethene to ethene. *Science* 276:1568–71.

Mazeas, L., and H. Budzinski. 2002. Stable carbon isotopic study ($^{12}C/^{13}C$) of the fate of petrogenic PAHs (methylphenanthrenes) during an in-situ oil spill simulation experiment. *Organic Geochemistry* 33:1253–58.

Mazeas, L., H. Budzinski, and N. Raymond. 2002. Absence of stable carbon isotope fractionation of saturated and polycyclic aromatic hydrocarbons during aerobic bacterial biodegradation. *Organic Geochemistry* 33:1259–72.

McClay, K., B. G. Fox, and R. J. Steffan. 2000. Toluene monooxygenase-catalyzed epoxidation of alkenes. *Applied and Environmental Microbiology* 66:1877–82.

Meckenstock, R. U., B. Morasch, C. Griebler, and H. H. Richnow. 2004. Stable isotope fractionation analysis as a tool to monitor biodegradation in contaminated aquifers. *Journal of Contaminant Hydrology* 75:215–55.

Meckenstock, R. U., B. Morasch, R. Warthmann, B. Schink, E. Annweiler, W. Michaelis, and H. H. Richnow. 1999. $^{13}C/^{12}C$ isotope fractionation of aromatic hydrocarbons during microbial degradation. *Environmental Microbiology* 1:409–14.

Melander, L., and W. H. Saunders. 1980. *Reaction rates of isotopic molecules.* New York: John Wiley.

Morasch, B., H. H. Richnow, B. Schink, and R. U. Meckenstock. 2001. Stable hydrogen isotope fractionation during microbial toluene degradation: Mechanistic and environmental aspects. *Applied and Environmental Microbiology* 67:4842–49.

Morasch, B., H. H. Richnow, B. Schink, A. Vieth, and R. U. Meckenstock. 2002. Carbon and hydrogen stable isotope fractionation during aerobic bacterial degradation of aromatic hydrocarbons. *Applied and Environmental Microbiology* 68:5191–94.

Morasch, B., H. H. Richnow, A. Vieth, B. Schink, and R. U. Meckenstock. 2004. Stable isotope fractionation caused by glycyl radical enzymes during bacterial degradation of aromatic compounds. *Applied and Environmental Microbiology* 70:2935–40.

Nelson, D. J., and R. L. Henley. 1995. Relative rates of permanganate oxidation of functionalized alkenes and the correlation with the ionization-potentials of those alkenes. *Tetrahedron Letters* 36:6375–78.

Nesheim, J. C., and J. D. Lipscomb. 1996. Large kinetic isotope effects in methane oxidation catalyzed by methane monooxygenase: Evidence for C-H bond cleavage in a reaction cycle intermediate. *Biochemistry* 35:10240–47.

Nijenhuis, I., J. Andert, K. Beck, M. Kästner, G. Diekert, and H. H. Richnow. 2005. Stable isotope fractionation of tetrachloroethene during reductive dechlorination by *Sulfurospirillum multivorans* and *Desulfitobacterium sp.* strain PCE-S and abiotic reactions with cyanocobalamin. *Applied and Environmental Microbiology* 71:3413–19.

Nikolausz, M., I. Nijenhuis, K. Ziller, H. H. Richnow, and M. Kästner. 2006. Stable carbon isotope fractionation during degradation of dichloromethane by methylotrophic bacteria. *Environmental Microbiology* 8:156–64.

Nozaki, M., and O. Hayaishi. 1984. *Dioxygenases and monooxygenases, developments in biochemistry. The biology and chemistry of active oxygen*, pp. 68–104. New York: Elsevier.

Peng, X., G. Zhang, F. Chen, and G. Liu. 2004. Stable carbon and hydrogen isotopic fractionations of alkane compounds and crude oil during aerobically microbial degradation. *Chinese Science Bulletin* 49:2620–26.

Penning, H., C. J. Cramer, and M. Elsner. 2008. Rate-dependent carbon and nitrogen kinetic isotope fractionation in hydrolysis of isoproturon. *Environmental Science & Technology* 42:7764–71.

Poulson, S. R., and H. Naraoka. 2002. Carbon isotope fractionation during permanganate oxidation of chlorinated ethylenes (cDCE, TCE, PCE). *Environmental Science & Technology* 36:3270–74.

Pratt, D. A., and W. A. van der Donk. 2006. On the role of alkylcobalamins in the vitamin B_{12}-catalyzed reductive dehalogenation of perchloroethylene and trichloroethylene. *Chemical Communications* 5:558–60.

Richnow, H. H., E. Annweiler, W. Michaelis, and R. U. Meckenstock. 2003. Microbial in situ degradation of aromatic hydrocarbons in a contaminated aquifer monitored by carbon isotope fractionation. *Journal of Contaminant Hydrology* 65:101–20.

Richnow, H. H., R. U. Meckenstock, L. Ask, A. Baun, A. Ledin, and T. H. Christensen. 2003. In situ biodegradation determined by carbon isotope fractionation of aromatic hydrocarbons in an anaerobic landfill leachate plume (Vejen, Denmark). *Journal of Contaminant Hydrology* 59–72.

Rosell, M., D. Barcelo, T. Rohwerder, U. Breuer, M. Gehre, and H. H. Richnow. 2007. Variations in $^{13}C/^{12}C$ and D/H enrichment factors of aerobic bacterial fuel oxygenate degradation. *Environmental Science & Technology* 41:2036–43.

Safinofski, M., and R. U. Meckenstock. 2006. Methylation is the initial reaction in anaerobic naphthalene degradation by a sulfate-reducing enrichment culture. *Environmental Microbiology* 8:347–52.

Scholz-Muramatsu, H., A. Neumann, M. Meßmer, E. Moore, and G. Diekert. 1995. Isolation and characterization of *Dehalospirillum multivorans* gen. no., sp. nov., a tetrachloroethene-utilizing, strictly anaerobic bacterium. *Archives of Microbiology* 163:48–56.

Schuth, C., M. Bill, J. A. C. Barth, G. F. Slater, and R. A. Kalin. 2003. Carbon isotope fractionation during reductive dechlorination of TCE in batch experiments with iron samples from reactive barriers. *Journal of Contaminant Hydrology* 66:25–37.

Selmer, T., A. J. Pierik, and J. Heider. 2005. New glycyl radical enzymes catalysing key metabolic steps in anaerobic bacteria. *Biological Chemistry* 386:981–88.

Sherwood Lollar, B., G. F. Slater, J. Ahad, B. Sleep, J. Spivack, M. Brennan, and P. MacKenzie. 1999. Contrasting carbon isotope fractionation during biodegradation of trichloroethylene and toluene: Implications for intrinsic bioremediation. *Organic Geochemistry* 30:813–20.

Shiner, V. J., and F. P. Wilgis. 1992. Heavy atom isotope rate effects in solvolytic nucleophilic reactions at saturated carbons. In *Isotopes in organic chemistry*. Vol. 8 of *Heavy isotope effects*, eds. E. Buncel and W. H. Saunders. New York: Elsevier.

Singleton, D. A., S. R. Merrigan, J. Liu, and K. N. Houk. 1997. Experimental geometry of the epoxidation transition state. *Journal of the American Chemical Society* 119:3385–86.

Slater, G. F., B. Sherwood Lollar, B. Sleep, and E. Edwards. 2001. Variability in carbon isotopic fractionation during biodegradation of chlorinated ethenes: Implications for field applications. *Environmental Science & Technology* 35:901–7.

Smidt, H., and W. M. De Vos. 2004. Anaerobic microbial dehalogenation. *Annual Review of Microbiology* 58:43–73.

Somsamak, P., H. H. Richnow, and M. M. Häggblom. 2005. Carbon isotopic fractionation during anaerobic biotransformation of methyl bert-butyl ether and tert-amyl methyl ether. *Environmental Science & Technology* 39:103–9.

Somsamak, P., H. H. Richnow, and M. M. Häggblom. 2006. Carbon isotope fractionation during anaerobic degradation of methyl tert-butyl ether under sulfate-reducing and methanogenic conditions. *Applied and Environmental Microbiology* 72:1157–63.

Spormann, A. M., and F. Widdel. 2000. Metabolism of alkylbenzenes, alkanes, and other hydrocarbons in anaerobic bacteria. *Biodegradation* 11:85–105.

Squillace, P. J., M. J. Moran, W. W. Lapham, R. M. Clawges, and J. S. Zogorski. 1999. Volatile organic compounds in untreated ambient groundwater of the United States, 1985–1995. *Environmental Science & Technology* 33:4176–87.

Squillace, P. J., M. J. Moran, and C. V. Price. 2004. VOCs in shallow groundwater in new residential/commercial areas of the United States. *Environmental Science & Technology* 38:5327–38.

Stehmeier, L. G., M. M. Francis, T. R. Jack, E. Diegor, L. Winsor, and T. A. Abrajano. 1999. Field and in vitro evidence for in-situ bioremediation using compound-specific $^{13}C/^{12}C$ ratio monitoring. *Organic Geochemistry* 30:821–33.

Steinbach, A., R. Seifert, E. Annweiler, and W. Michaelis. 2004. Hydrogen and carbon isotope fractionation during anaerobic biodegradation of aromatic hydrocarbons a field study. *Environmental Science & Technology* 38:609–16.

Tiehm, A., K. R. Schmidt, B. Pfeifer, M. Heidinger, and S. Ertl. 2008. Growth kinetics and stable carbon isotope fractionation during aerobic degradation of cis-1,2-dichloroethene and vinyl chloride. *Water Resources* 42:2431–38.

Tobler, N., T. Hofstetter, and R. P. Schwarzenbach. 2008. Carbon and hydrogen isotope fractionation during anaerobic toluene oxidation by *Geobacter metallireducens* with different Fe(III) phases as terminal electron acceptors. *Environmental Science & Technology* 42:7786–92.

Trust Hammer, B., J. G. Mueller, R. B. Coffin, and L. A. Cifuentes. 1995. The biodegradation of fluoranthene as monitored using stable carbon isotopes. In *Monitoring and verification of bioremediation*, eds. R. R. Hinchee, G. S. Douglas, and S. K. Ong, pp. 233–39. Columbus, OH: Battelle Press.

Ulrich, A. C., H. R. Beller, and E. A. Edwards. 2005. Metabolites detected during biodegradation of $^{13}C_6$-benzene in nitrate-reducing and methanogenic enrichment cultures. *Environmental Science & Technology* 39:6681–91.

US-EPA (United States—Environmental Protection Agency). 1999. Use of monitored natural attenuation at superfund, RCRA corrective action, and underground storage tank sites, OSWER directive 9200, pp. 4-17P.

van Beilen, J. B. and E. G. Funhoff. 2005. Expanding the alkane oxygenase toolbox: New enzymes and applications. *Current Opinion in Biotechnology* 16:308–14.

van der Meer, J. R., A. R. W. van Neerven, E. J. De Vries, W. M. De Vos, and A. J. Zehnder. 1991. Cloning and characterization of plasmid-encoded genes for the degradation of 1,2-dichloro-, 1,4-dichloro-, and 1,2,4-trichlorobenzene of *Pseudomonas* sp. strain P51. *Journal of Bacteriology* 173:6–15.

VanStone, N. A., R. Focht, S. A. Mabury, and B. Sherwood Lollar. 2004. Effect of iron type on kinetics and carbon isotopic enrichment of chlorinated ethylenes during abiotic reduction on Fe(0). *Ground Water* 42:268–76.

Vieth, A., and H. Wilkes. 2006. Deciphering biodegradation effects on light hydrocarbons in crude oils using their stable carbon isotopic composition: A case study from the Fullfaks oil field, offshore Norway. *Geochimica Cosmochimica Acta* 70:651–65.

Vogt, C., E. Cyrus, I. Herklotz, D. Schlosser, A. Bahr, S. Herrmann, H. H. Richnow, and A. Fischer. 2008. Evaluation of toluene degradation pathways by two-dimensional stable isotope fractionation. *Environmental Science & Technology* 42:7793–800.

Wackett, L. P., M. S. P. Logan, F. A. Blocki, and C. Bao-Li. 1992. A mechanistic perspective on bacterial metabolism of chlorinated methanes. *Biodegradation* 3:19–36.

Ward, J. A. M., J. M. E. Ahad, G. Lacrampe-Couloume, G. F. Slater, E. A. Edwards, and B. Sherwood Lollar. 2000. Hydrogen isotope fractionation during methanogenic degradation of toluene: Potential for direct verification of bioremediation. *Environmental Science & Technology* 34:4577–81.

Whited, G. M., and D. T. Gibson. 1991. Toluene-4-monooxygenase, a three-component enzyme system that catalyzes the oxidation of toluene to *p*-cresol in *Pseudomonas mendocina* KR1. *Journal Bacteriology* 173:3010–16.

Widdel, F., and R. Rabus. 2001. Anaerobic biodegradation of saturated and aromatic hydrocarbons. *Current Opinion in Biotechnology* 12:259–76.

Wilkes, H., A. Vieth, and R. Elias. 2008. Constraints on the quantitative assessment of in-reservoir biodegradation using compound-specific stable carbon isotopes. *Organic Geochemistry* 39:1215–1221.

Wilkes, H., C. Boreham, G. Harms, K. Zengler, and R. Rabus. 2000. Anaerobic degradation and carbon isotopic fractionation of alkylbenzenes in crude oil by sulphate-reducing bacteria. *Organic Geochemistry* 31:101–15.

Yeh, W. K., D. T. Gibson, and T.-N. Liu. 1977. Toluene dioxygenase: A multicomponent enzyme system. *Biochemical and Biophysical Research Communications* 78:401–10.

Zwank, L., M. Berg, M. Elsner, R. P. Schwarzenbach, and S. B. Haderlein. 2005. New evaluation scheme for two-dimensional isotope analysis to decipher biodegradation processes: Application to groundwater contamination by MTBE. *Environmental Science & Technology* 39:1018–29.

Zwank, L., M. Elsner, A. Aeberhard, R. P. Schwarzenbach, and S. B. Haderlein. 2005. Carbon isotope fractionation in the reductive dehalogenation of carbon tetrachloride at iron (hydr) oxide and iron sulfide minerals. *Environmental Science & Technology* 31:5634–41.

Section II

Isotopes and Microbial
Processes

5 Isotopes and Aerobic Degradation

C. Marjorie Aelion and Silvia A. Mancini

CONTENTS

5.1 DEFINING APPLICATIONS OF ISOTOPES IN AEROBIC ENVIRONMENTS

Aerobic biological processes occur in many compartments of the biosphere and are facilitated by diverse organisms ranging from fungi and bacteria to terrestrial and marine plants. Understanding the effects of these aerobic processes on isotopic fractionation has tremendous practical applications for many scientists including ecologists, microbiologists, hydrogeologists, and geochemists. This chapter will focus on carbon isotopic fractionation but will also explore the applications of oxygen, hydrogen, and chlorine isotope measurements in aerobic environments.

Isotopes of elements such as carbon (^{12}C and ^{13}C) and hydrogen (^{1}H and ^{2}H) react at slightly dissimilar rates during mass differentiating reactions, such as volatilization, sorption, and biodegradation. During biological processes, isotopic fractionation of the natural organic matter (OM) and organic contaminant occurs because it is easier for microorganisms metabolizing these compounds to break bonds containing the light isotope (i.e., ^{12}C-H) than those containing the heavy isotope (i.e., ^{13}C-H). Fractionation causes the remaining natural OM or contaminant in the contaminant pool to become enriched in the heavier isotopes compared to the original isotopic value of the parent compound. A large isotopic fractionation effect (primary kinetic isotope effect (KIE)) can be observed if a bond containing the element of interest is broken or formed in the rate-limiting step of the biodegradation pathway (see Chapter 3). Secondary kinetic isotope effects occur when the substitution is not involved in the bond that is breaking or forming (see Chapter 3). The magnitude of the secondary KIE is normally smaller than that of the primary KIE and more difficult to measure for elements of large mass. For both primary and secondary KIE, changes in ^{1}H and ^{2}H are particularly easy to distinguish because of the large relative differences in mass between the two isotopes. Typically isotopic fractionation effects during biodegradation of organic contaminants are quantified using a Rayleigh model and are described by fractionation factors (α) or enrichment factors (ε; see Chapter 3 for definitions).

The carbon cycle is the foundation for carbon flow in the biosphere. Carbon isotopic compositions of natural organic and inorganic carbon in the Earth's atmosphere and fauna and flora can vary significantly (> 120 per mil) due to chemical and biological fractionation processes. Photosynthesis and aerobic biological processes, such as microbial respiration of OM, play a key role in the evolution of carbon cycling in terrestrial, marine, and subsurface environments and contribute to the observed variation of carbon isotopes in nature. Bacteria play a unique role in the carbon cycle particularly in groundwater environments where other significant biological (i.e., plant root respiration) and chemical (i.e., photooxidation) processes are limited. Many uncontaminated groundwaters are aerobic and natural OM is a primary source of carbon and energy.

In cases where groundwater is contaminated with pollutants such as petroleum hydrocarbons that are a relatively biodegradable carbon source, aerobic microbial degradation of these contaminants can have considerable economic and environmental application. In general, microbial degradation of groundwater contaminants is termed *bioremediation* and is frequently implemented as a cost efficient and less invasive clean-up technique for contaminated groundwater. Typically these groundwaters become anaerobic during early stages of biodegradation and bioremediation is often limited by dissolved oxygen. However, the fringes of the contaminant plume generally remain aerobic and play a significant role in limiting contaminant transport in both the unsaturated and saturated zones in the subsurface.

Because of a large overall range in carbon isotopic compositions, isotope measurements can provide insights required for an understanding of different biological processes, identifying sources of OM, identifying the historical evolution of vegetation changes over time, and even predicting future impacts of global change to terrestrial environments using below ground soil processes (Staddon 2004). Provided that aerobic biodegradation is accompanied by a strong and resolvable isotopic fractionation, carbon, hydrogen, and sometimes chlorine isotope ratio measurements can be used to track the impact of aerobic biodegradation and the fate of groundwater contaminants in the subsurface. In some cases, carbon and hydrogen isotope analysis can provide insight into different aerobic microbial degradation pathways facilitated by specific bacteria such as those associated with methyl tertiary butyl ether (MTBE) degradation in groundwater (Zwank et al. 2005; Elsner et al. 2007).

5.2 PRINCIPLES OF AEROBIC RESPIRATION

Microbial growth is dependent on a number of chemical reactions that take place within a cell for biosynthesis (termed anabolism) and energy production (termed catabolism). Respiration is a catabolic reaction in which organic or inorganic compounds undergo oxidation–reduction reactions to produce energy for cell synthesis, motility, reproduction, and nutrient retrieval. The oxidation–reduction reactions involve transferring electrons from an electron donor compound to one that receives electrons (electron acceptor). Autotrophic respiration does not require organic compounds for growth or reproduction but instead uses inorganic C; energy is supplied from light or the oxidation of inorganic compounds. All heterotrophic respiration processes require an electron acceptor, such as oxygen, nitrate, sulfate, and so on and use OM as an electron donor in the series of oxidation–reduction reactions for catabolism. Aerobic respiration technically must use molecular oxygen as the terminal electron acceptor, but some aerobic microorganisms can also be facultative, meaning they are capable of growing in the absence of oxygen.

The general aerobic microbial respiration reaction is

$$CH_2O + O_2 \rightarrow CO_2 + H_2O \tag{5.1}$$

in which CH_2O represents a generic carbon source as the electron donor or energy source in the reaction. Although many molecules can act as electron acceptors during the oxidation of OM, oxygen is the electron acceptor with the most energy releasing potential for cellular functions and can yield more cellular biomass than other electron acceptors (Madigan, Martinko, and Parker 2000).

Aerobic respiration by microorganisms is responsible for degrading a wide range of organic contaminants present in impacted soils and groundwater. Many indigenous microorganisms are capable of using organic contaminants as the electron donor and carbon source for catabolism and simultaneously degrading harmful environmental contaminants to produce typically benign end products such as carbon dioxide. Many different biodegradation pathways of organic contaminants exist in nature and vary depending on the structure of the organic contaminant and the genetic capability of the degrading organisms present.

In addition to bacteria, fungi successfully degrade organic contaminants in the environment. Fungi are multicellular, nonphotosynthetic heterotrophic protists, and many are obligate aerobes, while bacteria are single-celled protists and are active in a wide range of redox conditions. Fungi may use different enzymatic degradation pathways than those used by bacteria to degrade organic contaminants. Both fungi and bacteria are capable of degrading contaminants by a process called cometabolism. During cometabolism, the degradation of the contaminant of interest depends on the presence and simultaneous degradation of other substrates that serve as the carbon and energy source for the organism. For example during cometabolic methane degradation, some bacteria degrade chlorinated solvents that they would otherwise be unable to degrade.

Enzymes are important components of living cells that catalyze the reactions necessary for cellular respiration. Microbial ecologists and biologists have examined the role of specific enzymes in degradation pathways for some organic chemicals. With this information in hand, specialized bacterial strains that express certain enzymes can be examined during the formation of intermediate and end products under controlled conditions. Information that can be derived from such studies includes the rate at which the degradation reactions occur, the rate at which daughter products are formed, and the rate at which bacteria reproduce using a specific contaminant as a carbon source. For example, one bacterium (*Pseudomonas putida*) that contains a specific enzyme (monooxygenase) can be used in experiments with a specific contaminant to identify the impact of that enzyme on the isotopic signature of the reaction products and remaining substrate. This information can be used to predict the relative contributions of that bacterium and specific enzymatic pathway to contaminant biodegradation in the field, although extrapolating this information to contaminated field sites becomes more difficult as mixtures of both microorganisms and contaminants are present.

Typical oxidative enzymes expressed by bacteria include oxygenases, synthases, and oxidases. Oxidoreductases, one of the six major classes of enzymes, catalyze oxidation–reduction reactions involving the loss or gain of electrons. Examples include dehydrogenases and oxidases. Bacteria commonly using oxidases include *Pseudomonas, Alcaligenes,* and *Flavobacterium.* These bacteria as well as *Acinetobacter* are aerobic gram-negative rods and are commonly represented in soils and sediments. Landfill gases, a second major class of enzymes, catalyze bond formation and include synthases that combine two groups to subsequently form polymers, and carboxylases that form C–C bonds. Most enzymes require either an organic (coenzyme) or metallic cofactor at the active site in order to function (Ucko 1986).

5.3 ISOTOPIC COMPOSITION OF ORGANIC COMPOUNDS AND OTHER CARBON POOLS

In order to assess the impact of aerobic degradation on isotopic signatures, the background signature of the atmosphere and other carbon sources, such as natural OM and environmental contaminants, must be known. Figure 5.1 illustrates the variation in carbon isotopic signatures expressed in δ notation in units of per mil (‰) relative to the Vienna Pee Dee Belemnite (VPDB) standard (see Chapter 1) of some natural carbon substances in the environment.

5.3.1 ATMOSPHERIC CO_2

CO_2 resides in and exchanges between large reservoirs in the atmosphere, oceans, and terrestrial systems as biological, chemical, and physical processes occur. The stable carbon isotope ratio of atmospheric CO_2 has changed over time, decreasing approximately 1.5‰ in the last 200 years due to land use changes such as deforestation and agriculture, and fossil fuel burning that add ^{13}C-depleted C to the atmosphere (Ghosh

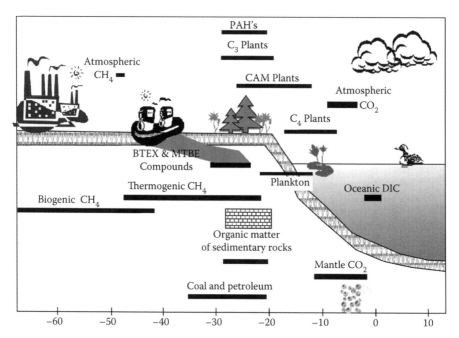

FIGURE 5.1 Range of carbon isotopic signatures of different carbon sources in nature in per mil (‰) versus Vienna Pee Dee Belemnite (VPDB).

and Brand 2003). The isotopic signal of CO_2 exchanged between the atmosphere and ecosystems allows global-scale studies of land–ocean interactions and carbon exchanges.

Isotopic measurements can be used to assess seasonal and annual variability of the terrestrial CO_2 uptake and release. Fluxes of CO_2 from terrestrial sources to the atmosphere occur via respiration that releases CO_2 to the atmosphere and tends to deplete the atmosphere in ^{18}O and ^{13}C. Photosynthesis takes up light CO_2 that tends to enrich the atmosphere in ^{13}C (Yakir and Sternberg 2000). Local conditions will have a significant impact on local fluxes of CO_2 between terrestrial sources and the atmosphere. Water vapor is released during plant transpiration and soil evaporation. Changes in soil moisture and temperature can affect the isotopic signature of atmospheric CO_2. Atmospheric CO_2 has stable carbon isotope values of −7.4 to −12‰, shifting diurnally and depending on the amount of air pollution (Boutton 1991). A mass balance approach can be used to account for C, O, and H inputs and outputs by measuring the fluxes of these gases and liquids in combination with their isotopic signatures. By measuring all environmental media and their respective C, O, and H isotopes including CO_2, OM, liquid water, water vapor, soil, and plant atmospheres, the contribution of different components to the ecosystem exchange can be quantified (Yakir and Sternberg 2000). Figure 5.2 provides some $\delta^{13}C$ values for a variety of these pools and some important biological processes in an uncontaminated environment. In addition to terrestrial sources, the oceans play a major role as a net CO_2 sink from the atmosphere.

5.3.2 Natural Organic Matter (OM)

In addition to the distinct $\delta^{13}C$ values associated with the pools described in the previous section, the C of the biota has a distinct isotopic signature. The carbon isotopic composition of natural OM is primarily determined by a strong isotopic fractionation of CO_2 during photosynthesis. Figure 5.3 provides a detailed illustration of the $\delta^{13}C$ values of the different biota including bacteria, algae, and general plant groups as exists in nature. Most plants, such as those in temperate areas, use the Calvin (C_3) cycle of photosynthesis. C_3 plants initially fix CO_2 into a 3-C compound (3-phosphoglyceric acid) during photosynthesis. This reaction is catalyzed by the enzyme RuBisCo (Staddon 2004). Plants that use the Hatch-Slack (C_4) photosynthetic pathway generally inhabit deserts, tropical, or subtropical grasslands. C_4 plants initially fix CO_2 into a 4-C compound (either malic or aspartic acid) catalyzed by the enzyme phosphoenolpyruvate (PEP) carboxylase. C_4 plants represent < 1% of terrestrial plant species number but are an important ecological group, including grasses, sedges, and important agricultural crop species such as maize and sugarcane.

Photosynthetic pathways of C_3 plants generally result in a net ^{13}C depletion of 16 to 18‰ of the assimilated carbon relative to the atmospheric carbon from which the plant material was derived (Farquhar, O'Leary, and Berry 1982). C_4 plants generally result in a net ^{13}C depletion of about –4‰ (Ghosh and Brand 2003). C_3 plants produce root respired CO_2 values and OM with $\delta^{13}C$ values within the range –25 ± 5‰ (Gleason, Friedman, and Hanshaw 1969; Rightmire and Hanshaw 1973; Busby, Lee, and Hanshaw 1983). C_4 plants yield root respired CO_2 with $\delta^{13}C$

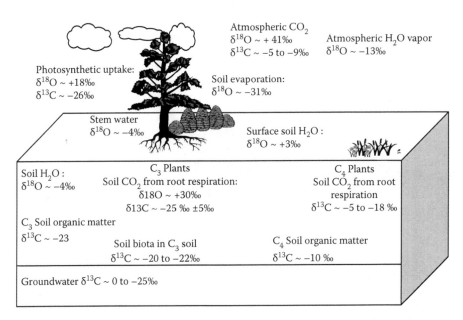

FIGURE 5.2 Carbon and oxygen isotopic values for CO_2, H_2O, and natural organic matter in natural systems.

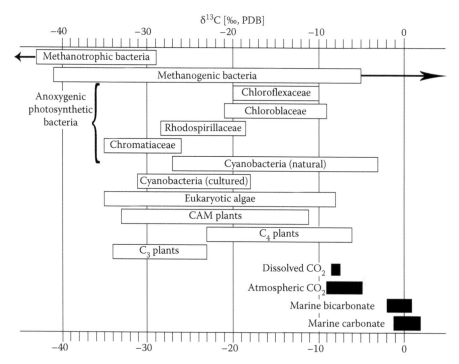

FIGURE 5.3 $\delta^{13}C$ isotopic values in biota including plants, bacteria, and algae. (Modified from Schidlowski, M., *Precambrian Research*, 106, 117–34, 2001. With permission.)

values within the range −5.6 to −18.6‰ (Smith and Epstein 1971). C_3 plant vegetation ranges from approximately −20 to −40‰ and C_4 plant vegetation ranges from approximately −9 to −17‰ $\delta^{13}C$ (Staddon 2004). Figures 5.2 and 5.3 illustrate the ranges of $\delta^{13}C$ values for C_3- and C_4-dominated plant environments and their associated soils, as well as the ranges of $\delta^{13}C$ values for the atmospheric and groundwater reservoirs.

These values represent $\delta^{13}C$ averages in bulk plant materials (Figure 5.3). However, specific components of plants, such as leaves, can be investigated for their isotopic composition in different growth and degradation stages, for example, fresh growth (green), senescent leaves (yellow), and litter (brown). In natural systems *n*-alkanes have been extracted from leaves and their isotopic composition measured. Individual *n*-alkanes (C_{22}–C_{27}) extracted from fresh leaves of *Ginkgo biloba* were depleted in ^{13}C by 2.3 to 5.3‰ compared to bulk fresh leaves that were combusted and for which the CO_2 was purified and the $\delta^{13}C$ values were measured (Nguyen Tu et al. 2004). Average fresh leaves C_{22}–C_{27} *n*-alkanes ranged from $\delta^{13}C$ values of −33.0 to −30‰ and were more ^{13}C-depleted with increasing chain length. Bulk fresh leaves were ~−28‰. Bulk leaf isotopic composition was unchanged through senescence and litter formation, with a $\delta^{13}C$ of ~−27.7‰. The *n*-alkanes associated with these bulk leaves during senescence and litter formation generally were enriched in $\delta^{13}C$ in all *n*-alkanes compared to those of fresh leaves. In addition the values of individual n-alkanes from C_{22} to C_{27} had much smaller ranges in $\delta^{13}C$ values in these leaves compared to those extracted from fresh leaves. Nguyen Tu et al. (2004) postulated that the bulk materials decomposed to litter may be subsequently available for microbial degradation and represent a stable initial $\delta^{13}C$ of natural OM.

$\delta^{13}C$ has been examined in natural environments to identify the sources of OM. OM originates from the decomposition of plants and other living materials including bacteria. Dissolved organic matter (DOM) is generally of interest due to its organic carbon (OC) content and importance in C cycling. Dissolved organic carbon (DOC), which includes molecules that are ≤ 0.45 μm, is chemically complex, and a large fraction of all aquatic DOC (40 to 99%; Morel 1983) includes ill-defined chemical structures, termed humic compounds. Dissolved OC also includes small reactive molecules such as free amino acids, fatty acids, free and combined sugars, and carbohydrates.

$\delta^{13}C$ has been used extensively to identify DOC in estuarine systems. Estuarine OM may originate from upland river sources (primarily C_3 plants), salt marsh OM derived primarily from *Spartina* sp., and microalgae. Humic substances similarly may originate from vascular plants, microalgae, and microbial degradation. Otero et al. (2003) measured $\delta^{13}C$ of bulk DOC and humic substances along a salinity gradient from the freshwater river to the saline estuary in estuarine water in the southeastern United States. Depleted $\delta^{13}C$ DOC suggested terrestrial inputs, and estuarine-marine-derived DOC had an average enrichment of $\delta^{13}C$ DOC from 1 to 2‰. The bulk DOC $\delta^{13}C$ values suggested a nonuniform addition of DOC species to the estuary but also substantial estuarine-derived OM inputs. When examined more closely by specific contribution to the bulk DOC, humic substances $\delta^{13}C$ was more depleted than that of the DOC and showed an increased proportion of terrestrial upland carbon inputs to the estuary. The nonhumic component of DOC was less depleted in $\delta^{13}C$ compared to the humic substances and was assumed to originate from either the decomposition of detrital material from salt marshes or microalgae of estuarine or marine origin. This research is an example of using isotopic signatures of bulk materials whose specific chemical identity is unknown, to identify their possible origins, and assess their contribution to carbon sources in aquatic environments.

Wang, Chen, and Berry (2003) used stable isotopes to examine the sources of OM in marsh sediments by measuring the isotopic signature of long-chain n-alkanes, C_{21} to C_{33} from marsh vascular plants and sediments. It was expected that C_3 plants would have a distinct $\delta^{13}C$ range of -20‰ to ~-34‰ and C_4 plants would have a $\delta^{13}C$ range of -9‰ to -17‰. The $\delta^{13}C$ values measured for *Spartina alterniflora* (a C_4 marsh grass) and *Typha latifolia* (a C_3 marsh grass) ranged from -13.3‰ to -13.9‰ and -24.8‰ and -26.4‰, respectively. Sediments from mudflat sites with no vegetation had a $\delta^{13}C$ value of -19‰, whereas sediments from *Spartina* sites ranged from -17.7 to -20.6‰ and sediment from *Typha* sites ranged from -22.5‰ to -23.4‰. Based on these isotopic signatures, the marsh plants, particularly *Typha*, appeared to contribute significant amounts of OM to the salt marsh sediments.

5.3.3 Environmental Organic Contaminants

Polycyclic aromatic hydrocarbons (PAHs), polychlorinated biphenyls (PCBs), chlorinated solvents, and petroleum hydrocarbons, such as benzene, toluene, ethylbenzene, and xylenes (BTEX) sometimes including fuel oxygenates like MTBE are among the most commonly observed contaminants in the environment and represent the more harmful contaminants known to affect both human and ecological health. Although some of these contaminants may have natural sources in the environment, anthropogenic inputs can be more significant. For example, PAHs are formed naturally from thermal alteration of buried OM, however, anthropogenic inputs, through incomplete combustion of fossil fuels, make PAHs widespread contaminants. In the case of PCBs, their wide range of industrial applications, such as dielectric fluids, flame retardants, pesticides, and their chemical inertness have resulted in their persistence in the environment. These industrial and other commercial contaminants, such as chlorinated solvents and petroleum products, enter the environment during their production, transport, storage, use, and disposal. After entering the environment they partition into various compartments depending on their physical properties such as solubility, octanol-water partition coefficient, and so on. For example, PAHs generated through incomplete combustion are present not only in the atmosphere but also in aquatic sediments as a result of wet and dry deposition.

The carbon isotopic composition of these contaminants varies and ranges are reported in the literature. Various PAHs, obtained from both primary and secondary sources and environmental samples, exhibit $\delta^{13}C$ values of –22.6 to –28.9‰ (O'Malley, Abrajano, and Hellou 1994). PCBs obtained from different chemical manufacturers exhibit a relatively narrow range of $\delta^{13}C$ values of –25.7 to –26.6‰ (Drenzek et al. 2002), although a larger range of values of –18.7 to –28.4‰ is reported for individual congeners (any one particular member of a class of chemical compounds; a specific congener is denoted by a unique chemical structure) between Polychlorinated biphenyl (PCB) mixtures (Jarman et al. 1998). The $\delta^{13}C$ values of pure phase BTEX compounds from at least six different manufacturers range from –23.9 to –29.1‰ (Dempster, Sherwood Lollar, and Feenstra 1997; Harrington et al. 1999; Hunkeler et al. 2001). Petroleum $\delta^{13}C$ values range from –18 to –34‰ (Stahl 1980). MTBE exhibits $\delta^{13}C$ values ranging from –28.8 to –30.7‰ (Smallwood et al. 2001).

The range in the carbon isotopic compositions for a given chemical or group of chemicals reflects the isotopic compositions of the source or raw materials, chemical reactions during the synthesis or natural alteration processes, and operating conditions per chemical manufacturer if chemically derived (van Warmerdam et al. 1995; Beneteau, Aravena, and Frape 1999). In cases where the chemical synthesis or operating conditions do not involve the breakage or formation of a bond containing carbon, the isotopic composition will primarily reflect that of the raw material. For a given manufacturer then, the carbon isotopic composition of a particular chemical may change over the course of the manufacturing history depending on the isotopic composition of the raw material. The raw material for the synthesis of many environmental contaminants is primarily petrochemical feedstock but can sometimes be components of plants (Drenzek et al. 2002). The $\delta^{13}C$ values observed for these contaminants will reflect that of petroleum or plants unless an isotope effect is introduced by chemical reactions during the synthesis of the chemical (Jarman 1998).

5.4   IDENTIFYING CONTAMINANT SOURCES

Distinguishing between different sources of contamination is extremely valuable in environmental law cases such as those associated with the U.S. Environmental Protection Agency's Comprehensive Environmental Response, Compensation, and Liability Act (CERCLA), more commonly known as Superfund. Using environmental forensics, such as stable isotope analysis, the potentially responsible party (PRP) can be identified in order to assign liability to the appropriate polluter to whom clean-up costs are subsequently assessed. Stable isotope analysis can be used to identify and distinguish between different sources of contamination primarily in cases where the carbon isotopic signature of a particular contaminant is conserved (i.e., not affected by biodegradation or physical processes) and provided that isotope ratios from different sources show variations larger than any analytical uncertainty incorporating precision and accuracy (Figure 5.4; Slater 2003; Sherwood Lollar et al. 2007).

A more conclusive approach of tracking the sources of various contaminants on a regional or even global scale is the combination of multiple isotope analyses, such as carbon, hydrogen, and chlorine isotopes, or isotope coupled with other fingerprinting tool such as those used for trace metals. For chlorinated hydrocarbons, carbon and chlorine isotopes are particularly useful. $\delta^{37}Cl$ values of chlorinated semivolatile organic compounds (SVOCs) range from -5.10 to $+1.22‰$ and are noted to have a systematic bias among the manufacturers of these compounds representative of the chemical pathway by which they were synthesized (Drenzek et al. 2002). Plotting the respective $\delta^{13}C$ values versus the $\delta^{37}Cl$ values of these chlorinated SVOCs results in distinct groups of specific contaminants representative of their synthesis, indicating the potential to successfully apply this approach for source identification in the field (Figure 5.5; Drenzek et al. 2002). The more recent analytical developments of hydrogen isotope analysis of chlorinated solvents may provide a further line of evidence for source identification in the field (Shouakar-Stash, Frape, and Drimmie 2003; Chartrand et al. 2007). For petroleum hydrocarbons, the comparison of $\delta^{13}C$ and $\delta^{2}H$ values can provide significant information for source identification. For example, benzene exhibits a large range of $\delta^{2}H$ values from -27.9 to $-96.8‰$ that can help constrain different sources of contamination when coupled with carbon isotope analysis (Hunkeler et al. 2001). Successful applications of this approach were demonstrated at two petroleum contaminated sites where different source areas of groundwater contamination were suspected by distinct pairs of $\delta^{13}C$ and $\delta^{2}H$ values for both benzene and ethylbenzene (Mancini et al. 2002; Mancini, Lacrampe-Couloume, and Sherwood Lollar 2008).

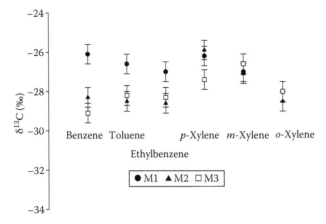

FIGURE 5.4 Demonstration of the potential to use compound specific isotope analysis to distinguish between carbon isotopic signatures of pure product BTEX compounds from three different manufacturers (M1, M2, and M3). Error bars represent the accuracy and reproducibility of ±0.5 per mil (‰) for stable carbon isotope analysis. Differences between carbon isotopic signatures of a chemical from different manufacturers must be greater than the error associated with stable carbon isotope analysis. (Modified from Dempster, H. S., B. Sherwood Lollar, and S. Feenstra, *Environmental Science & Technology,* 31, 3193–97, 1997. With permission.)

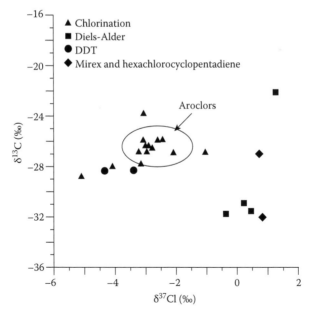

FIGURE 5.5 Identification of chlorinated contaminants using dual isotopic analysis. (Modified from Drenzek, N. J., C. H. Tarr, T. I. Eglinton, L. J. Heraty, N. C. Sturchio, V. J. Shiner, and C. M. Reddy, *Organic Geochemistry,* 33, 437–44, 2002. With permission.)

5.5   ISOTOPIC ANALYSIS OF BIOMARKERS

Biomarkers are compounds specific to a particular organism or group of organisms that can be isolated and measured to allow the study of both microbial processes specific to that organism or group of organisms and the study of larger scales of population and community changes. The microbial formation of biological components from parent carbon can be assessed under aerobic conditions using isotopic fractionation. Bacteria synthesize proteins from amino acids and carbohydrates and lipids that are used in cell wall formation from the building blocks of lipopolysaccharides and peptidoglycans and phospholipids. Phospholipid fatty acids (PLFA) are membrane-bound lipids that can be used as biomarkers because they vary in different organisms. Hopanoids are a diverse group of compounds that are found in the membranes of bacteria and also can be used as biomarkers (Staddon 2004).

Abraham and Hesse (2003) examined carbon fluxes within microbial consortia by measuring the isotopic ratios of individual carbon substrates and comparing them to the taxon-specific biomarkers present in the consortium. They suggested—based on previous literature—that the most useful biomarkers for this approach in bacteria are phospholipids and their fatty acids and outer membrane proteins. For fungi, Abraham and Hesse (2003) measured the isotopic fractionation of carbon from substrates to fatty acids of polar lipids and amino acids. They chose *Rhizopus arrhizus* and *Mortierella isabellina* members of the *Zygomycotina,* and *Fusarium solani* and *Aspergillus niger* from the *Ascomycotina.* The formation of biomarkers impacts the degree of isotopic fractionation between substrate and biomarker. Fatty acids are biosynthesized by multienzyme complex fatty acid synthase and biosynthesis results in palmitic acid (C16:0; Rawn 1989; Abraham and Hesse 2003). Palmitic acid is then the starting material for dehydro-fatty acids and longer chain fatty acids and for these reasons was used as a standard for the fungi in the study. The authors found a ^{13}C enrichment of less than + 1.0‰ of palmitic acid compared to the substrate and a ^{13}C enrichment of + 1.1‰ compared to biomass. However the formation of steric acid (C18:0) from palmitic acid had a fractionation of −4.0‰ compared to the substrate and −4.7‰ compared to palmitic acid. The amino acids alanine, glycine, praline, valine, aspartate, and glutamate were enriched in ^{13}C in all cases; threonine and serine were enriched in most cases; isoleucine and lysine were isotopically similar; and phenylalanine and leucine (−7 to −8‰) were depleted in ^{13}C relative to their substrates from which they were derived.

Not only are specific biomarkers impacted differently compared to the substrates from which they are derived, but the four different fungi also responded differently to various substrates (Abraham and Hesse 2003). Fungi were incubated with either glucose or mannose as a C source and the biomass was subsequently freeze-dried, and a portion of the dried biomass was combusted and analyzed for $\delta^{13}C$. The largest isotopic fractionation was measured when glucose was the substrate, in which case the fungi belonging to the *Zygomycotina* were more depleted in ^{13}C (*Rhizopus* was depleted by 2.44 ± 0.18‰) than the fungi belonging to the *Ascomycotina* (*Fusarium* was depleted by 0.39 ± 0.05‰), possibly due to different ratios of prolongation of the C16:0 precursor to the C18:0 and de novo synthesis of C18:0. When mannose ($\delta^{13}C$ −15.55‰) was the substrate, the fungi were all enriched in $\delta^{13}C$, by 0.88 to 2.43 (average 1.46 ± 0.40‰). Finally when a depleted source of mannose ($\delta^{13}C$ −23.71‰)

was used as the C substrate, the isotopic ratio of the fungi was not different from that of the source. However in this case, *M. isabellina* was depleted and the other three species were enriched. One can quickly see the increase in complexity and also of the information derived, using the isotopic approach as one goes from bulk systems such as plant leaves and fungal biomass, to microbial outer cell membrane proteins and specific amino acids, and effects of various sources of C from which the organisms derive their biomass.

Isotopic fractionation is caused by the selectivity of the enzymes involved in the biosynthesis of the compounds. Carbon isotope analysis may aid in elucidating carbon fluxes in symbioses by determining the isotopic ratios of the individual amino acids of the host and the fungal symbiont to quantify the amino acid sharing between different partners. A significantly large isotopic fractionation is necessary to distinguish differences in these type studies. The isotopic fractionation of the amino acids is greater than that for fatty acids, but correction factors and varying results make the application of isotopic analysis to amino acids more complicated. Because the isotopic fractionation of the amino acids is individual–amino acid dependent, the isotopic fractionation that occurs has to be corrected for fractionation of the individual amino acid that is involved during its biosynthesis. Fatty acids derived from phospholipids are easier to predict but the isotopic fractionations are smaller (averaging 2.5‰) than those of amino acids.

Van der Meer et al. (2003) used bacterial biomarkers in microbial mats in hot springs to determine the dominant microbial metabolic pathway for growth by *Chloroflexus* type bacteria that can grow by photoheterotrophy (using light to incorporate reduced organic compounds), heterotrophy (using aerobic respiration to incorporate reduced organic compounds), and photoautotropy (using light to fix inorganic carbon). Stable carbon isotope fractionation was measured between dissolved inorganic carbon and *Chloroflexus* lipid biomarkers including n-C_{31} through n-C_{40} wax esters, and compared to the cyanobacterial biomarker C_{17} n-alkane, bicarbonate and bulk biomass. Cyanobacteria are thought to be the primary producers in the microbial mats. All *Chloroflexus* wax esters were enriched relative to all cyanobacteria C_{17} n-alkanes (enriched by approximately 6–16‰) and depleted relative to the bulk biomass. Both heterotrophic and autotrophic carbon metabolism were thought to occur in the multiple hot springs bacteria examined, and the cyanobacteria supported photoheterotrophy of *Chloroflexus* in the microbial mats.

5.6   ISOTOPIC COMPOSITION OF CO_2 AND BIOMASS FROM CONTAMINANT BIODEGRADATION

Many isotopic studies used to assess aerobic biodegradation of environmental contaminants measure biomarkers or intermediate and end products (primarily carbon dioxide) generated during degradation of the parent contaminants. Most of these studies are conducted in the laboratory using mainly pure cultures of contaminant degrading bacteria to enhance the rate and extent of biodegradation.

Hall et al. (1999) examined phenol and benzoate fractionation during biodegradation by two different bacteria, *Pseudomonas putida* and *Rhodococcus* sp. and measured isotopic ratios of dissolved inorganic carbon and biomass over time during biodegradation. *P. putida*, a gram negative bacterium, metabolizes phenol using a meta pathway forming pyruvate and acetaldehyde. Benzoate is metabolized by *P. putida* using the ortho pathway with an initial decarboxylation step to catechol and eventually to succinate and acetyl CoA. The metabolic pathways of phenol degradation by *Rhodococcus* sp., a gram positive bacterium, have not been completely elucidated. For both species, biomass was enriched in ^{13}C relative to CO_2 produced (Figure 5.6). In incubation of *P. putida,* which biodegraded phenol and benzoate, microbial biomass was enriched compared to the CO_2 produced by approximately 5‰ after 3 hours. Likewise, for *Rhodococcus* sp., the biomass $\delta^{13}C$ shifted to –17.5‰ at 14 hours of incubation and the difference between CO_2 and biomass produced stabilized to approximately –8‰, with biomass being more enriched than CO_2. Hall et al. (1999) found that aerobic biodegradation produced CO_2 that was slightly depleted in ^{13}C relative to the growth substrate, approximately 3.7‰ by *P. putida* cultured on phenol and benzoate and 5.6‰ by *Rhodococcus* cultured on phenol. Differences between the two bacteria were attributed to bacterial

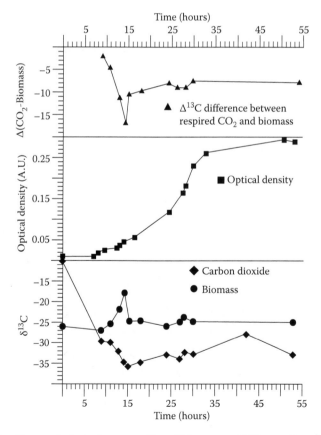

FIGURE 5.6   Changes in isotopic composition of biomass and CO_2 over time during degradation of phenol by *Rhodococcus* sp. (Modified from Hall, J. A., R. M. Kalin, M. J. Larkin, C. C. R. Allen, and D. B. Harper, *Organic Geochemistry,* 30, 801–11, 1999. With permission.)

species specific growth rates and substrate utilization including species specific degradative metabolic pathways that impact fractionation during aerobic biodegradation. Also differences in cell wall structure and subsequent mechanisms of substrate uptake were suspected to cause KIEs. Although in some cases aerobic degradation does not fractionate $\delta^{13}C$ to as large an extent as anaerobic processes, Hall et al. (1999) found significantly depleted $\delta^{13}C$ CO_2 compared to the growth substrate, and differences in depletion occurring throughout the different growth phases of the bacteria from the beginning of the logarithmic growth phase to the end of the logarithmic growth phase.

Some researchers have examined the isotopic signature of CO_2 and dissolved inorganic carbon from aerobic biodegradation of petroleum hydrocarbons (Aggarwal and Hinchee 1991; Landmeyer, Vroblesky, and Chapelle 1996; Aelion, Kirtland, and Stone 1997; Conrad et al. 1999; Kirtland, Aelion, and Stone 2000; Bugna et al. 2004) and chlorinated solvents (Suchomel, Kreamer, and Long 1990; Kirtland et al. 2003; Kirtland, Aelion, and Stone 2005) from contaminated field sites. Results reflect the difficulty in definitively attributing CO_2 produced from biodegradation of contaminants versus that from possible interferences of background CO_2 under field conditions. Other, natural sources of CO_2 exist (plant root respiration in shallow sediments and carbonate dissolution in subsurface sediments) and must be accounted for in field experiments (Figures 5.1 and 5.2).

One example of a successful field application of isotopic assessment of biodegradation was in an aviation gasoline contaminated aquifer where biodegradation of the gasoline constituents produced significantly enriched CO_2 (Conrad et al. 1999). Interestingly, the average $\delta^{13}C$-CO_2 and DIC in the heart of the plume were enriched (average $\delta^{13}C$-CO_2 from soil gas ranging from –6.9 to –14.4‰) suggesting that a large fraction of the CO_2 was produced from acetate fermentation, also resulting in CH_4 production that represented > 20% of the soil gas. In contrast, in the aerobic fringes, the $\delta^{13}C$-CO_2 values were much lower (average $\delta^{13}C$-CO_2 from soil gas ranging from –22.0 to –26.3‰) indicating that aerobic degradation was the main source of this CO_2. Soil OM for the site ranged from –13.5‰ to –26.1‰ and the aviation gasoline was approximately –25‰. Groundwater DIC was always approximately 7‰ heavier than the CO_2, which was attributed to the equilibrium carbon isotopic fractionation that was expected at pH 7, 20°C (Wigley, Plummer, and Pearson 1978). Conrad et al. (1999) concluded that while others reported that the stable isotopic ratios of the substrates and products are not strongly affected by aerobic microbial metabolic processes, in their study the aviation gasoline was –25‰ and the expected $\delta^{13}C$-CO_2 measured at the edges of the plume was less than –25‰ therefore the difference was discernible. However these values were only a few per mil less than the $\delta^{13}C$-CO_2 for plant root respiration and microbial degradation of endogenous plant material.

In an earlier study, Aelion, Kirtland, and Stone (1997) carried out a field study at a gasoline contaminated groundwater site, and measured aerobic degradation and $\delta^{13}C$-CO_2 from soil vapor probes and soil vapor extraction (SVE) wells and $\delta^{13}C$-DIC of groundwater from monitoring wells. $\delta^{13}C$ values of CO_2 and DIC from background samples collected from an uncontaminated forested area approximately 36 km from the contaminated site were –23.9 and –22.9‰, respectively. In the contaminated area, the average vadose zone soil gas $\delta^{13}C$-CO_2 from the vapor probes and SVE wells was −25.8‰ and the average $\delta^{13}C$-DIC of groundwater was −22.7‰, an enrichment of approximately 3‰. Although one of the $\delta^{13}C$-CO_2 values overlapped with background $\delta^{13}C$-CO_2 from uncontaminated areas, on average the stable isotopic values were distinguishable between the contaminated and background sites. This overlap in $\delta^{13}C$-CO_2 values between contaminated areas and uncontaminated areas suggests that the combined measurement of $\delta^{13}C$-CO_2 and ^{14}C-CO_2 (see Chapter 11) may be needed to confirm that the CO_2 is indeed from the degradation of contaminants and not from plant root respiration or microbial metabolism of natural plant materials. Both Aelion, Kirtland, and Stone (1997) and Conrad et al. (1999) used radiocarbon to provide this essential piece of evidence.

Examining aerobic degradation of chlorinated solvents $\delta^{13}C$ values of CO_2 at field sites is complex because multiple sources of CO_2 may be present from the degradation of various chlorinated compounds, and degradation rates are generally low relative to petroleum hydrocarbons. Kirtland et al. (2003) measured trichloroethylene (TCE) and carbon tetrachloride (CT) biodegradation at a contaminated field site from a U.S. Department of Energy facility. Based on $\delta^{13}C$ values of CO_2, sediment OM, groundwater DOC and the chlorinated daughter products cis-1,2-dichloroethylene (1,2-DCE), complete degradation of the chlorinated solvents to CO_2 was assumed to be occurring. Unlike laboratory studies in which 90 to 100% of the parent may be degraded, at these field sites the majority of the parent TCE and CT was not degraded in the vadose zone. DCEs accounted for 10% of the chlorinated hydrocarbon concentrations and small concentrations of ethane and ethene were measured. The daughter product, DCE (−16 to −19‰) was enriched in ^{13}C relative to the TCE parent (−16 to −29‰). $\delta^{13}C$ values of vadose zone sediments and saturated sediment averaged −24.6‰, $\delta^{13}C$ of groundwater DIC ranged from −20.1 to −22.8‰, and $\delta^{13}C$ of CO_2 in vadose zone of the source wells ranged from −18 to −21‰. Because the TCE and DCE may overlap with plant-derived CO_2 sources, it is more difficult at sites such as this that are heavily vegetated with C_3 plants to assess complete biodegradation using the CO_2 isotopic analysis. However over the 114 days of the study, there was a positive correlation in several wells of $\delta^{13}C$-CO_2 values and CO_2 concentrations over time that gave the best indication of aerobic biodegradation. In a similar study Kirtland et al. (2005) used carbon isotopes to examine aerobic degradation of perchloroethylene at a contaminated industrial field site. The average $\delta^{13}C$-CO_2 in the vadose zone of the source area was significantly depleted (−23.5‰) relative to the minimally contaminated areas (−18.4‰) and background control areas (−16.9‰). Although small concentrations of metabolites were measured (from 0.2 to 18% of the chlorinated hydrocarbon concentrations), and CO_2 derived from chlorinated hydrocarbons was calculated to be as small as 0.02–0.06% (v/v), the isotopic method was more sensitive than traditional gas chromatographic methods used to assess biodegradation based on measured concentrations of CO_2 and contaminants.

5.7   COMPOUND SPECIFIC ISOTOPE ANALYSIS (CSIA)

Compound specific isotope analysis (CSIA) has the potential to directly identify and quantify the extent of biodegradation of organic contaminants in the field (Sherwood

Lollar et al. 1999), provided that a significant isotopic fractionation occurs during biodegradation of the contaminant of interest. This type of information is extremely valuable for scientists, environmental consultants, and engineers as they assess and determine the best strategy and long-term treatment plans for contaminant clean-up at field sites. Proving the occurrence of biodegradation in the field is a challenge using traditional measurements of concentration because physical processes such as volatilization, dispersion, and sorption can also cause contaminant attenuation. However, for petroleum and chlorinated hydrocarbons, carbon isotopic fractionation is not significant during volatilization (Harrington et al. 1999; Huang et al. 1999; Poulson and Drever 1999), dissolution (Dempster, Sherwood Lollar, and Feenstra 1997), or adsorption (Harrington et al. 1999; Slater et al. 1999) under equilibrium conditions. Under nonequilibrium conditions, recent studies indicate that remediation techniques such as air sparging in the saturated zone or bioventing in the unsaturated zone (i.e., progressive vaporization) and natural systems where sorption can play a significant role (i.e., high organic content aquifers) can result in isotopic fractionation in ^{13}C of the residual contaminant greater than that associated with analytical uncertainty $\pm0.5\%$ (Harrington et al. 1999; Kopinke et al. 2005). However, these effects have not been demonstrated in the field and the authors note that the importance of sorption-induced effects at field sites is significant primarily for highly hydrophobic organic compounds such as PAHs. Laboratory experiments examining hydrogen isotopic fractionation during volatilization and dissolution (Ward et al. 2000) of aromatic hydrocarbons showed no significant hydrogen isotope fractionation beyond typical analytical uncertainty for compound specific hydrogen isotope analysis ($< \pm5\%$) under equilibrium conditions. Even during sorption of aromatic hydrocarbons to carbonaceous material, only a small hydrogen isotope fractionation of 8‰ at 95% sorption was observed (Schüth et al. 2003). However, in laboratory experiments where aromatic hydrocarbons were progressively vaporized, a significant hydrogen isotopic depletion in ^2H was observed of up to 45‰ for ethylbenzene compared to the original isotopic value (Wang and Huang 2003). The authors note that such a large isotopic depletion occurs primarily at advanced stages of progressive vaporization and therefore, is insignificant in most natural systems where such extensive mass loss is restricted.

In contrast, biodegradation can cause a significant isotopic enrichment with respect to the initial isotopic signature of some organic contaminants. As such, CSIA can be used to distinguish between contaminant mass loss due to subsurface physical and chemical processes and that due to biodegradation in groundwater systems (Hunkeler et al. 1999). Figure 5.7 illustrates the potential to differentiate between different attenuating processes in the field. This technique relies on the faster rate of breakage of bonds containing the lighter isotopes (^{12}C, ^1H) compared to bonds containing the heavy isotopes (^{13}C, ^2H) during mass differentiating reactions. This can result in a progressive enrichment in the heavy isotopes (^{13}C, ^2H) in the remaining contaminant with respect to the initial isotopic signature. Many isotope geochemists and microbiologists have determined that the extent of isotopic fractionation during biodegradation of organic contaminants varies for different compounds, degradation pathways, microbial consortia, and terminal electron accepting processes. More specifically, as in uncontaminated

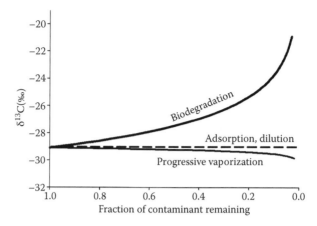

FIGURE 5.7   Illustration of the potential to use compound specific isotope analysis to distinguish between physical processes and biodegradation in the field.

environments discussed in Section 5.3, the specific enzyme involved in the degradation pathway within a bacterial species and the nutrient availability that impacts the efficiency of the enzyme (enzymatic activity) can impact the isotopic fractionation factors. Assuming Michaelis–Menton and steady state enzyme kinetics (Atlas and Bartha 1997), the magnitude of isotopic fractionation depends on the formation-dissociation steps prior to the enzyme–substrate complex and including the irreversible transformation of the substrate. Hence, in a case where the efficiency of the enzymatic reaction is limited by poor binding or low affinity of the substrate to the enzyme (low V_{max}) in the enzyme-complex formation step relative to the chemical transformation of the substrate, the apparent isotopic fractionation can be significant. In contrast, in an extreme case where the enzymatic reaction is completely efficient or has a high affinity (high V_{max}), quantitative conversion of the substrate occurs and no isotope effect will be observed (Northrop 1991; Mancini et al. 2006). While important for understanding the mechanistic relationship between enzymatic efficiency and isotopic fractionation, a recent study suggests that enzyme controlled isotopic fractionation has an insignificant impact in most natural systems but may play a larger role in extreme environments where nutrients may be significantly limited (Mancini et al., 2006).

5.8   AEROBIC BIODEGRADATION OF PETROLEUM HYDROCARBONS

Some researchers have found that little to no carbon isotopic fractionation occurs during aerobic biodegradation of petroleum hydrocarbons. In general, semivolatile organic contaminants including PAHs and PCBs are not fractionated during degradation, that is, their isotopic signatures are conserved. Stahl (1980) observed that the remaining alkanes of petroleum undergoing biodegradation were slightly enriched in ^{13}C but that there were no changes in the isotopic composition of the aromatic components of the petroleum. Stable carbon isotope composition of saturated hydro-carbons and phenanthrene were conserved during marine bacterial degradation of a crude oil. Mazeas, Budzinski, and Raymond (2002) examined n-alkanes (C_{17}-C_{30}) as well as 2-methylphenanthrene as a typical alkylated PAH using pure bacterial strain 2PMII (*Sphingomonas* sp.) isolated from marine sediment. Limited biological fractionation of these compounds in conjunction with limited fractionation by physi-cal processes of evaporation, photooxidation, and OM decomposition (O'Malley, Abrajano, and Hellou 1994; Trust Hammer et al. 1995; Huang 1997; Mazeas, Budzinski, and Raymond 2002), provides an excellent example of how stable isotope analysis has the potential to track the fate of these compounds in the environment (Slater 2003).

Zyakun et al. (2003) measured distinct $\delta^{13}C$ isotopic signatures of n-hexadecane and naphthalene degradation by *Burkholderia* sp. BS3702. They measured *Burkholderia* sp. degradation and growth on n-hexadecane that had an initial $\delta^{13}C$ of −44.6‰. $\delta^{13}C$-CO_2 values averaged −50.2‰, biomass $\delta^{13}C$ values averaged −46.6‰, and unidentified exometabolites values averaged −41.5‰. For *Burkholderia* sp. deg-radation and growth on naphthalene that had an initial $\delta^{13}C$ of −21‰, the $\delta^{13}C$-CO_2 values averaged −24.1‰, biomass $\delta^{13}C$ values averaged −19.2‰, and the unidentified exometabolites averaged −19.1‰. These $\delta^{13}C$ values were significantly different for each category (CO_2, biomass, and unidentified exometabolites) for the two substrates because of the large differences in their starting isotopic signatures. Also, the CO_2 values were depleted and the biomass and unidentified exometabolites were enriched in ^{13}C compared to that of the initial substrate.

For volatile organic petroleum hydrocarbons, such as the BTEX compounds, carbon isotopic fractionation during biodegradation can be more significant than that observed for SVOCs but is controlled primarily by biodegradation pathway and microbial community. For example, Morasch et al. (2002) determined that the first enzyme of attack has important implications for the isotopic fractionation of the remaining contaminant and microbial intermediate and end products formed during aerobic toluene degradation by pure cultures. By using pure bacterial strains known to degrade toluene using different enzymes and initial degradation reac-tions, Morasch et al. (2002) determined that carbon isotopic fractionation varies significantly during aerobic toluene degradation by *Pseudomonas putida* strain mt-2, *Ralstonia pickettii* strain PKO1, and *Pseudomonas putida* strain F1.

Figure 5.8 depicts the three different degradation pathways and initial enzy-matic reactions for the three microorganisms (see Chapter 4, Figure 4.4). The *P. putida* strain mt-2 uses a methyl monooxygenase enzyme to attack the methyl group on the aromatic side-chain (Figure 5.8a); and residual toluene shifted from $\delta^{13}C$ of −29.18‰ to −23.22‰ after 80% of the toluene was degraded with a calculated enrichment factor, ε (see Chapter 4) of −3.3 ± 0.3‰. The *R. pickettii* strain PKO1 uses a ring monooxygenase enzyme for toluene degradation that hydroxylates car-bon three of the aromatic ring (Figure 5.8b). Morasch et al. (2002) assumed that a two-step process occurred in which the electrophilic attack of iron-bound oxygen on the aromatic ring and the formation of a C-O σ-bond occurred. The secondary step in the degradation of toluene occurs as the hydrogen atom bound to this carbon atom is released as a proton and electrons reconstitute the aromaticity of the carbon

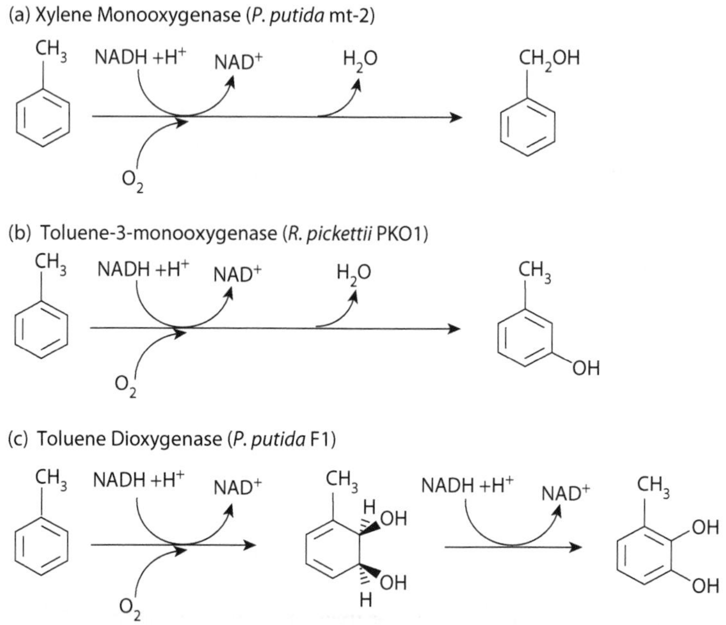

(a) Xylene Monooxygenase (*P. putida* mt-2)

(b) Toluene-3-monooxygenase (*R. pickettii* PKO1)

(c) Toluene Dioxygenase (*P. putida* F1)

FIGURE 5.8 Aerobic biodegradation pathways and initial enzymatic reactions for toluene (also see Chapter 4).

ring. The residual toluene isotope shift was one-third that of the previous *P. putida* strain mt-2 with a calculated enrichment factor of -1.1 ± 0.2‰. Morasch et al. (2002) suggested that the smaller enrichment factor was related to the first step of the reaction; because no distinct bond cleavage occurs in this step, only smaller secondary isotope effects are involved.

The *P. putida* strain F1 uses a ring dioxygenase enzyme for toluene degradation and hydroxylates toluene to the corresponding catechol (Figure 5.8c). Π-electrons of the aromatic substrate are attracted by activated oxygen bound to a catalytic iron center. The first reaction involves a *cis*-dihydrodiol that is dehydrogenated in a subsequent step to form the catechol derivative. No significant isotopic fractionation was observed for this enzyme action with a small calculated enrichment factor of -0.4 ± 0.3.

If methyl monooxygenase, ring oxygenase, and ring dioxygenase enzymes result in distinct isotopic fractionation of the same parent compound during aerobic degradation, the importance of identifying the relative proportion of enzymatic activity that each contributes to contaminant degradation in laboratory and field studies is readily apparent. Once pathways for organisms are known, specific enzyme probes can be useful for quantifying the relative proportion of the enzymatic activity (see Chapter 11; stable isotope probing).

Sherwood Lollar et al. (1999) found no isotopic fractionation during toluene degradation using an undefined mixed consortium of toluene degraders in sand. They showed that the isotopic composition of the residual toluene was identical to the initial isotopic composition of 28‰ even with extensive toluene degradation. These results corroborated previous results based on six sets of experiments on two mixed consortia of toluene degraders. Similarly, Bugna et al. (2004) observed no significant carbon isotopic fractionation during aerobic biodegradation of toluene in an aquifer contaminated with jet fuel. In this study, a mixture of benzene, toluene, naphthalene, xylenes, and decanes was mixed with native soil and buried 4 m below ground in a surficial aquifer. The initial isotopic composition of the organic compounds was compared to the isotopic composition after burial. Initial $\delta^{13}C$ values of the compounds within the mixture ranged from –26.6 to –30.0‰. After 40 and 164 days of incubation, the specific compounds of the hydrocarbon mixture ranged from –26.3 to –30.4‰ and –26.5 to –29.4‰, respectively. Thus chemical migration and partial degradation did not significantly change the isotopic signature of the chemicals from their initial isotopic signature.

During aerobic benzene biodegradation, Stehmeier et al. (1999) measured carbon isotopic shifts of between 0.8 and 2.2‰ in residual benzene when approximately 80 to 90% of the benzene was degraded by a mixed microbial culture. Similarly, Hunkeler et al. (2001) measured a small isotopic fractionation in residual benzene degraded by *Acinetobacter* sp. and *Burkholderia* sp. They measured both the $\delta^{13}C$ and the δ^2H of residual benzene and inorganic carbon produced during biodegradation. The residual benzene $\delta^{13}C$ was enriched by approximately 3‰ (from –27 to –24‰) after seven hours of incubation and biodegradation by *Acinetobacter* while the corresponding δ^2H was enriched by a much greater extent from approximately –27 to –6‰. While a similar pattern held for *Burkholderia,* the differences in fractionation between the $\delta^{13}C$ and δ^2H were not as great as those for *Acinetobacter.* In general, the hydrogen isotope effects for aerobic benzene biodegradation are determined to be smaller than that observed for other biodegradation processes such as the anaerobic degradation of benzene, toluene, or aerobic methane oxidation (Fischer et al. 2008). Benzene oxidation by *Acinetobacter* sp. and *Burkholderia* sp. may have occurred by a monooxygenase or dioxygenase, however the small fractionation suggested relatively little C–H bond breaking in the initial steps of benzene degradation. *Acinetobacter* had distinct isotopic signatures for inorganic carbon and cell biomass suggesting that fractionation can occur at the initial degradation step and also at branch points that continue to deplete the $\delta^{13}C$ of the products and inorganic carbon relative to the remaining substrate. The use of combined C and H isotopes in degradation studies allows comparisons of the δ^2H-$\delta^{13}C$ pairs between species and biodegradation pathways as demonstrated for benzene biodegradation in Mancini et al. (2008) and Fischer et al. (2008).

5.9   AEROBIC BIODEGRADATION OF CHLORINATED HYDROCARBONS

As was observed for toluene degradation, degradation pathways and initial enzymatic reactions are also important considerations for CSIA fractionation studies of chlorinated contaminants. Chlorinated organic contaminants can be degraded cometabolically to carbon dioxide by methane oxidizing bacteria. These bacteria are present in many different aerobic soil and sediment environments (Hanson and Hanson 1996; Brungard et al 2003). Methanotrophs are gram-negative bacteria that are classified based on their assimilation of formaldehyde and physiologic and morphologic features (Hanson 1996). Type I use ribulose monophosphate pathway for formaldehyde assimilation and type II use serine pathway for formaldehyde assimilation (Figure 5.9). Because different enzymes are used by these organisms, differences in isotopic fractionation of substrates may occur during biodegradation. If aerobic conditions exist at contaminated field sites, methanotrophic activity may provide contaminant removal without the formation of toxic intermediate products such as vinyl chloride, which is produced during anaerobic dehalogenation of chlorinated hydrocarbons. In the presence of air and methane, methanotrophs completely degrade chlorinated contaminants using monooxygenase enzymes (Figure 5.10). Methane monooxygenases (MMO) exist in both particulate (pMMO) and dissolved (sMMO) forms. The membrane-bound pMMO is expressed in all methanotrophs. If copper concentrations are limiting, the cytoplasmic sMMO may be expressed. It is thought that pMMO isotopically fractionates methane (discriminates against the ^{13}C) to a greater extent than does sMMO (Jahnke et al. 1999). Thus as degradation proceeds, the relative ^{13}C of the CO_2 product and the methane substrate becomes more and more disparate, particularly if pMMO activity is high. Methane enrichments in ^{13}C have been reported at 30‰ during methanotrophic degradation (Zyakun and Zakharchenko 1998; Brungard et al 2003).

Isotope fractionation of the substrate or specific compound can be impacted not only by the activity of a specific enzyme but also by the type of bond that is broken in a chemical reaction. For example, petroleum hydrocarbons are comprised of

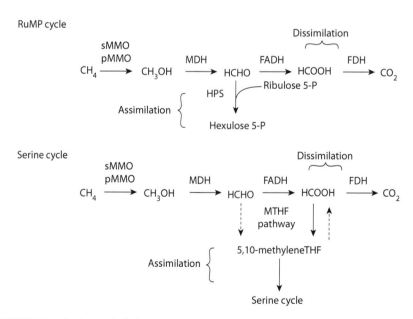

FIGURE 5.9 Serine and ribulose monophosphate pathways used by methanotrophic bacteria. (Modified from Jahnke, L. L., R. E. Summons, J. M. Hope, and D. J. des Marais, *Geochimica Cosmochimica Acta*, 63, 79–93, 1999. With permission.)

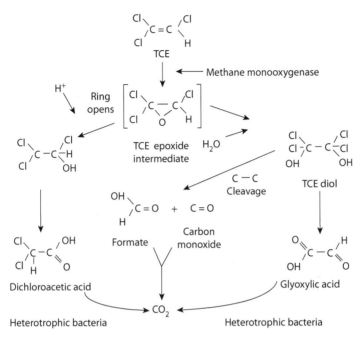

FIGURE 5.10 Proposed pathways of trichloroethylene biodegradation to carbon dioxide by methanotrophic bacteria with the enzyme methane monooxygenase. (Modified from Murray, W. D., and M. Richardson, *Critical Reviews in Environmental Science and Technology*, 23, 195–217, 1993.)

C–C and C–H bonds while chlorinated hydrocarbons have C–C, C–H, and C–Cl bonds. Kinetic isotope theory predicts that large carbon isotopic fractionation can be expected in reactions where heavier atoms are involved in the enzymatic degradation reaction (i.e., C–Cl bonds vs. C–C bonds) and where large changes in vibrational frequencies in the molecule between its ground state and transition state occur. Aerobic vinyl chloride and 1,2-dichloroethane (1,2-DCA) degradation demonstrate this theory well. During aerobic vinyl chloride degradation, a C–C bond breakage is suspected to occur in the initial degradation step and results in a small carbon isotopic fractionation relative to anaerobic biodegradation pathways involving a C–Cl bond breakage (Chu et al. 2004; Chartrand et al. 2005).

Similar results were observed for 1,2-DCA. Aerobic biodegradation of 1,2-DCA may occur by a hydrolytic dehalogenase enzyme in *Xanthobacter autotrophicus* GJ10 and *Ancyclobacter aquaticus* AD20 in which the first step of the reaction requires the cleavage of the C–Cl bond in a biomolecular nucleophilic substitutions (S_N2) reaction, producing 2-chloroethanol (Figure 5.11b). The 2-chloroethanol is considered to be unstable and immediately releases chloride yielding chloroacetaldehyde (Hage and Hartmans 1999). This contrasts with the monooxygenase enzyme used by *Pseudomonas* sp. strain DCA1 in which degradation may occur by the cleavage of the C–H bond producing 1,2-dichloroethanol (Figure 5.11a). The nonlinear relationship between the protein content in the enzyme assay and the

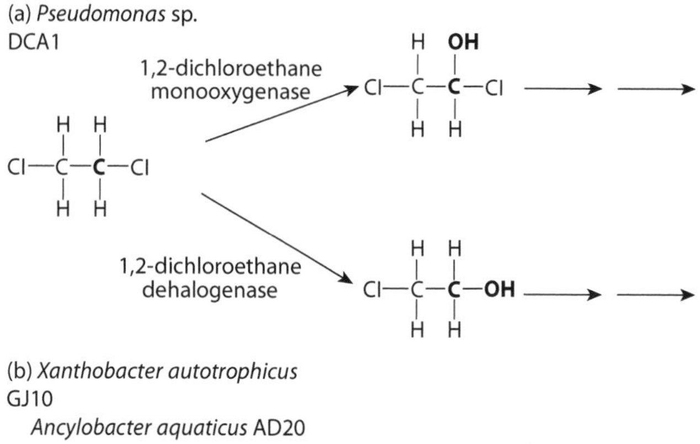

(a) *Pseudomonas* sp.
DCA1

1,2-dichloroethane
monooxygenase

(b) *Xanthobacter autotrophicus*
GJ10
    Ancylobacter aquaticus AD20

1,2-dichloroethane
dehalogenase

FIGURE 5.11   Aerobic biodegradation pathways and initial enzymatic reactions for 1,2-DCA (also see Chapter 4).

specific activity of the monooxygenase could indicate a multicomponent enzyme system and Hage and Hartmans (1999) suggested that the monooxygenase was an inducible enzyme. These characteristics and differences between organisms and their enzyme systems for 1,2-DCA degradation were illustrated by Hirschorn et al. (2004) who observed a bimodal distribution in enrichment factors, one approximately –3.9‰ for *Pseudomonas* and the second –29.2‰ for the other two strains used, *Xanthobacter autotrophicus* GJ10 and *Ancyclobacter aquaticus* AD20 (Figure 5.12). They also concluded that the hydrolytic dehalogenation reactions may have strong isotopic fractionation but would be expected to have weak hydrogen fractionation. The reverse holds true for oxidation reactions in which there may be a weak carbon fractionation but the hydrogen fractionation is expected to be strong. They suggested that measuring both C and H isotopic signatures will help identify the type of degradation pathway during 1,2-DCA degradation by providing corroborative evidence (Chartrand et al. 2007).

Brungard et al. (2003) examined ^{13}C fractionation during degradation of *trans*-1,2-dichloroethylene (*t*-DCE) by type I and type II methanotrophs. Pure cultures of the methanotrophs *Methylomonas methanica* Oak Ridge (type I) and *Methylosinus trichosporium* OB3b (type II) degrading *t*-DCE resulted in an enrichment of the remaining *t*-DCE in solution measured over time. Degradation of *t*-DCE using different enzyme systems was found to generate different isotopic enrichment factors in *t*-DCE under identical growth conditions. The type II methanotrophs had *t*-DCE isotopic enrichment factors approximately twice (–6.7‰) those of the Type I methanotrophs (–3.5‰). Brungard et al. (2003) concluded that enrichment factors were predominantly due to enzyme-mediated reactions and not due to volatilization. This was based on the fact that $\delta^{13}C$ measurements for *t*-DCE in the headspace of the control incubation vials were similar to the liquid stock solution $\delta^{13}C$ values. Other studies showed that stable carbon isotope fractionation varied extensively, from –10 to 34‰, during the oxidation of methane by type I and II methanotrophic bacteria (Coleman, Risatti, and Schoell 1981; Zyakun and Zakharchenko 1998; Brungard et al. 2003).

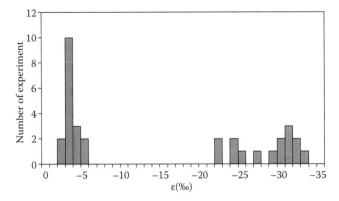

FIGURE 5.12 Bimodal distribution of 1,2-DCA in enrichment cultures and pure cultures. (Modified from Hirschorn, S. K., M. J. Dinglassan, M. Elsner, S. A. Mancini, G. Lacrampe-Couloume, E. A. Edwards, and B. Sherwood Lollar, *Environmental Science Technology,* 38, 4775–81, 2004.)

C_1 compounds contain only one C and therefore no carbon-carbon bonds. C_1 compounds include the aliphatic hydrocarbon methane, chlorinated aliphatics (dichlormethane), the alcohol methanol, formate, and the metal halides methyl chloride (MeCl), methyl bromide (MeBr), and methyl iodide (MeI). Methylotrophs are aerobic organisms that use C_1 compounds. This group includes yeasts such as *Hansenula polymorpha* that use methanol as an energy source, organisms that degrade chlorinated aliphatic hydrocarbons, and bacteria such as *Methylobacterium, Hyphomicrobium,* and *Aminobacter* that degrade methyl halides.

Unlike some organic compounds that are not highly fractionated during chemical reactions, some methyl halides are fractionated during chemical hydrolysis. $\delta^{13}C$ of MeBr increased from $-65‰$ to $+15‰$ and $\delta^{13}C$ of MeI increased from $-64‰$ to $-8‰$ during chemical hydrolysis (Baesman and Miller 2005). Isotope fractionation of C_1 compounds can be used to distinguish between chemical and biological degradation of methyl halides. By comparing chemical degradation and biological degradation of MeBr and MeCl, Miller et al. (2004) showed that chemical hydrolysis dominated the initial removal of MeBr with a KIE of $59‰ \pm 7$ versus the KIE of $69‰ \pm 9$ for the subsequent biological degradation of MeBr. MeCl fractionation only occurred during biological oxidation (KIE $49‰ \pm 3$).

Investigating biological oxidation of C_1 compounds, Miller et al. (2001) measured large isotopic fractionation factors of up to $70‰$ during the oxidation of MeBr and MeCl used as carbon sources by methylotrophs. In their study the $\delta^{13}C$ of the remaining substrate was measured over time during biodegradation. The $\delta^{13}C$ of MeCl increased from -60 to $-25‰$. For irreversible reactions the isotope effect is calculated by regression of the $\delta^{13}C$ values of the residual substrate and the natural log (ln) of the fraction of substrate remaining. The enzymatic pathways are elucidated by isolating the KIEs attributed to the initial enzyme reaction (Miller et al. 2001). In this study the enzymatic pathway was isolated by carrying out the degradation reaction using purified enzymes in the presence of hydrosulfide anion (HS⁻), which mediates the conversion of the MeCl to methane thiol (MeSH). Based on the $\delta^{13}C$ of MeCl measured with the purified cobalamine-dependent enzyme (halomethane:bisulphide/halide ion methyltransferase) over time, a fractionation factor (ϵ) expressed as the deviation from unity × 1,000 was calculated. For MeCl, $\epsilon = 21\%_o$ for degradation was observed using the enzyme system and it was lower than that observed for the MeCl fractionation factor measured for degradation by whole cells $\epsilon > 40\%_o$. The reason for this discrepancy was unknown although one possible explanation was that the fractionation was attributed not only to the initial enzymatic degradation step as was measured in the pure enzyme system, but also in part to enzymes further along the degradation pathway that would have been carried out by the whole cell suspensions subsequently.

Heraty et al. (1999) examined the degradation of dichloromethane (DCM) by pure cultures of a gram negative methylotrophic bacterium. They measured the fractionation of both the $\delta^{13}C$ and the $\delta^{37}Cl$ in the remaining DCM over time during biodegradation. The organisms were shown to fractionate DCM with a fractionation factor (α) for C of 0.9576 ± 0.0015 and for Cl of 0.9962 ± 0.0003. The calculated KIE was greater for C than for Cl, 1.0424 versus 1.0038, respectively. Based on these KIE values, Heraty et al. (1999) concluded that for C, the rate of the initial step of the DCM degradation reaction was 42.4‰ faster for ^{12}C–DCM-bearing molecules than for ^{13}C–DCM-bearing molecules and was 3.8‰ faster when a ^{35}Cl–C bond was broken first versus a ^{37}Cl-C bond. Based on this calculated fractionation and previous literature on possible degradation pathways, Heraty et al. (1999) concluded that the first step in the degradation pathway was a nucleophilic substitution reaction producing inorganic chloride and formaldehyde, although no formaldehyde was measured in the experiments. Based on the graphical representation of the $\delta^{13}C$ and the $\delta^{37}Cl$ data, a linear relation with a slope of 11.2 was observed that Heraty et al. (1999) suggested could be used to estimate the fraction of substrate remaining in other systems.

Griebler et al. (2004) compared aerobic and anaerobic degradation of chlorinated aromatic compounds 1,2,3-trichlorobenzene and 1,2,4-trichlorobenzene. They concluded that for aerobic pure culture, *Pseudomonas* sp., the isotopic signature of the parent compound did not change even though greater than 99% of the added 1,2,4-trichlorobenzene was degraded. This result was in contrast to the anaerobic degradation of 1,2,3-trichlorobenzene by *Dehalococcodies* sp. in which there was an enrichment of the residual compound of approximately 5‰ and the intermediate product formed, 1,3-dichlorobenzene, was depleted relative to the parent. The authors concluded that the chlorodioxygenase enzyme reactions did not cause significant carbon isotope fractionation, similar to results reported for nonchlorinated ring dioxygenases that transform nonchlorinated aromatic compounds. The type of enzyme and its mode of action will determine whether fractionation occurs. Thus, the chlorodioxygenase and nonchlorinated ring dioxygenases act in a similar fashion, by catalyzing the transfer of two activated oxygen atoms to the substrate (Kauppi et al. 1998; Griebler et al. 2004). The oxygen atoms are activated by accepting an electron from the iron at the catalytic site. Reactions involving exclusively π-electrons do not create primary isotope effects.

5.10   AEROBIC BIODEGRADATION OF METHYL TERTIARY BUTYL ETHER (MTBE)

Aerobic biodegradation of MTBE can occur by a few organisms belonging to the genera *Rubrivax, Hydrogenophaga, Methylobacterium, Rhodococcus,* and *Arthrobacter* (see papers in Fayolle, Vandecasteele, and Monot 2001). While these organisms can use MTBE as the sole carbon and energy source for growth, a few organisms have been identified that can cometabolically degrade MTBE with growth on other various hydrocarbons. The metabolic pathway of aerobic MTBE biodegradation is proposed to undergo oxidation in the initial degradation step to form a hemiacetal intermediate, *tert*-butoxymethanol, using a monoxygenase enzyme (Figure 5.13). The subsequent transformation of the *tert*-butoxymethanol intermediate can occur by two different pathways. One involves an enzymatic transformation to form *tert*-butyl formate (TBF) by alcohol dehydrogenase while the other involves an abiotic transformation to form *tert*-butyl alcohol (TBA). All of the pure bacterial strains identified to use MTBE as the carbon and energy source are documented to follow the TBA pathway. Only some cometabolic strains, such as *Mycobacterium* and *Graphium* produce transient TBF (Hardison et al. 1997; Smith, O'Reilly, and Hyman 2003).

MTBE provides an excellent example of the beneficial use of both carbon and hydrogen isotope analysis for the detection and quantification of aerobic MTBE biodegradation in the field. Both primary carbon and hydrogen isotope effects are expected to be significant during aerobic biodegradation of MTBE due to the C–H bond breakage that occurs during the formation of the hemiacetal intermediate, *tert*-butoxymethanol in the initial biodegradation step. Carbon isotopic fractionation during biodegradation of MTBE by pure culture PM1 (Gray et al. 2002) and field microcosms (Hunkeler et al. 2001) was reported to be constant with enrichment factors of –2.0 to –2.4‰ and –1.6 to –2.0‰ respectively (Figure 5.14). Even cometabolic degradation of MTBE in microcosms produced a carbon enrichment factor of –1.5‰ (Hunkeler et al. 2001).

Hydrogen isotopic fractionation was significantly larger in pure culture PM1 and microcosms with enrichment factors of –33 to –37‰ and –29 to –66‰ respectively (Gray et al. 2002; Figure 5.15). Given that hydrogen isotope fractionation is more substantial than carbon isotope fractionation during aerobic biodegradation of MTBE, hydrogen isotope analysis can offer a more sensitive indicator of biodegradation in the field. It should be noted however, that more variability exists for hydrogen isotopic fractionation during biodegradation of MTBE (Gray et al. 2002). Hence, the robust and reproducible nature of carbon isotopic fractionation may provide a more accurate quantitative evaluation of the extent of biodegradation, while the application of both carbon and hydrogen isotope analysis can provide more conclusive evidence and confirmation of biodegradation in the field (Gray et al. 2002).

Combining both carbon and hydrogen isotope analysis at MTBE contaminated field sites has additional benefit. It can provide information on the nature of the biodegradation processes responsible for the observed isotopic fractionation. This approach involves comparing the slope of $\delta^{13}C$ (ordinate) and δ^2H (abscissa) data

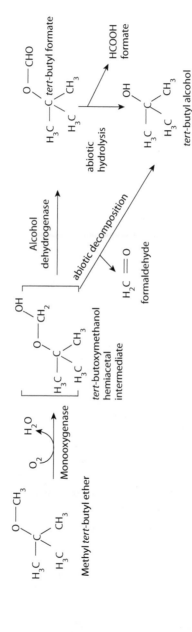

FIGURE 5.13   Metabolic pathway for methyl *tert*-butyl ether degradation using the monoxygenase enzyme.

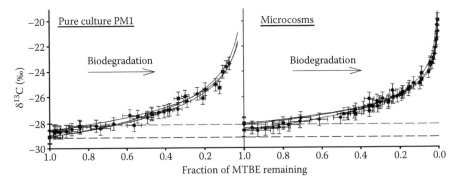

FIGURE 5.14 Carbon isotopic fractionation during aerobic biodegradation of MTBE in pure culture PM1 and in microcosms constructed from aquifer material. Error on carbon isotopic measurements is typically ±0.5 per mil (‰) and includes accuracy and reproducibility associated with stable carbon isotope analysis. The dashed lines represent the error on the initial isotopic value of MTBE. (Modified from Gray, J. R., G. Lacrampe-Couloume, D. Gandhi, K. Scow, R. D. Wilson, D. M. Mackay, and B. Sherwood Lollar, *Environmental Science & Technology,* 36, 1931–38, 2002. With permission.)

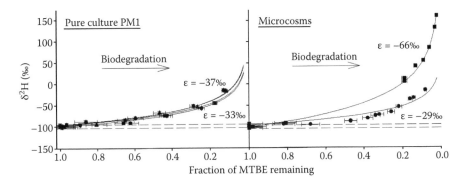

FIGURE 5.15 Hydrogen isotopic fractionation during aerobic biodegradation of MTBE in pure culture PM1 and in microcosms constructed from aquifer material. Error on hydrogen isotopic measurements is typically ±5 per mil (‰) and includes accuracy and reproducibility associated with stable hydrogen isotope analysis. The dashed lines represent the error on the initial isotopic value of MTBE. (Modified from Gray, J. R., G. Lacrampe-Couloume, D. Gandhi, K. Scow, R. D. Wilson, D. M. Mackay, and B. Sherwood Lollar, *Environmental Science & Technology,* 36, 1931–38, 2002. With permission.)

from wells within a groundwater plume versus those characterized in controlled experiments with known degradation processes. Zwank et al. (2005) found that slopes of $\delta^{13}C$ versus δ^2H data from aerobic biodegradation of MTBE in batch experiments reported by Gray et al. (2002) were significantly higher than those reported for anaerobic biodegradation reported by Kuder et al. (2002). This is caused by differences between the aerobic and anaerobic biodegradation pathways. During aerobic biodegradation, MTBE undergoes a C–H bond breakage that results in significant primary carbon and hydrogen isotope effects whereas during anaerobic biodegradation, MTBE undergoes a C–O bond breakage that involves a significant primary carbon isotope effect and only a secondary hydrogen isotope effect. Based on this, Zwank et al. (2005) provide a nice example of the successful application of combining $\delta^{13}C$ versus δ^2H data to determine the nature of the biodegradation process at an MTBE contaminated field site, where the slope of the carbon and hydrogen isotopic field data was consistent with that of anaerobic biodegradation of MTBE reported by Kuder et al. (2002). This type of information can have significant implications for the implementation of successful clean-up strategies at contaminated sites.

5.11 COMBINED ISOTOPIC ANALYSIS OF ^{13}C AND ^{18}O

Combined isotopic analysis may involve several isotope pairs depending on the molecule of interest and the limitations of the analytical methods. ^{13}C, ^2H, ^{37}Cl, ^{18}O, and ^{15}N are present in organic contaminants and in their intermediate and end products. Some isotope ratios can be examined in the same molecule such as the $\delta^{18}O$ and δ^2H of water and $\delta^{13}C$ and $\delta^{18}O$ of CO_2 as was discussed in Section 5.3 (Figure 5.2). Carbon and oxygen represent important elements for the combined isotopic approach for the examination of end products CO_2 and DIC as well as the aerobic terminal electron acceptor oxygen. Aggarwal et al. (1997) measured the $\delta^{13}C$ of CO_2 and the $\delta^{18}O$ of the terminal electron acceptor O_2 during diesel fuel biodegradation. The fuel itself had a $\delta^{13}C$ of $-28.5\%o$. The $\delta^{13}C$ of CO_2 decreased from an initial value of near atmospheric of $-10\%o$ to $-31\%o$ after one day, and then remained relatively unchanged over the remainder of the 42 day incubation, at approximately $-28.5\%o$. The $\delta^{18}O$ of O_2 increased steadily from 22 to $50\%o$ over the first 30 days of incubation, and then decreased suddenly between days 31 to 36 of incubation from $50\%o$ to $25\%o$ (Figure 5.16), while the O_2 concentrations remained relatively unchanged and low. Potential sources of O with low $\delta^{18}O$ were assumed to be water, phosphate, and sulfate. Although not conclusive, the authors suggested that respiration by sulfate reducing bacteria can occur and produce O_2 of low $\delta^{18}O$ following the reaction: $SO_4^{2-} + 2H^+ = H_2S + 2O_2$. This O_2 production can be coupled with reduction to H_2O by aerobic bacteria that would cause no change in O_2 concentration, but would be manifest by depletion in the $\delta^{18}O$ of the O_2. Aggarwal et al. (1997) concluded that isotopic exchange during sulfate reduction was the only plausible explanation for the change in isotopic composition of the O_2. In general, however, sulfate reduction is not considered to occur aerobically and it is possible an alternative explanation exists.

5.12 CONCLUSION

Stable carbon isotopes have been used in many aerobic environmental media, air, water, sediment, and biomass to examine natural carbon cycling as well as contaminant biodegradation. Most nonbiological processes do not fractionate many contaminants significantly relative to biological actions, although some methyl halides are an exception. For some contaminants, aerobic processes do not fractionate C to the

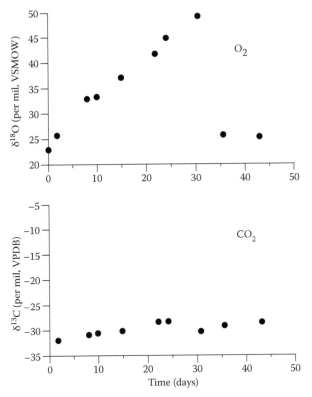

FIGURE 5.16 Changes in isotopic composition of remaining O_2 used as an electron acceptor and CO_2 produced from the degradation of diesel by a mixed bacterial culture. (Modified from Aggarwal, P. K., M. E. Fuller, M. M. Gurgas, J. F. Manning, and M. A. Dillon, *Environmental Science & Technology*, 31, 590–96, 1997. With permission.)

extent that anaerobic processes do. In this case, isotopic fractionation can be used to distinguish aerobic from anaerobic processes and identify sources of contaminants that have different isotopic signatures. For other contaminants, aerobic processes fractionate C and produce changes in the isotopic signatures from the parent to the intermediate or end product. In this case, isotopic signatures can be used to identify aerobic biodegradation if the difference between the remaining parent and the product of interest is greater than the analytical precision of the instrument. As technology continues to be developed with greater precision, the usefulness of information derived from stable isotopes in aerobic degradation studies will increase. Stable isotopes can be used to characterize processes on scales as large as global carbon cycling, to those as small as enzymatic pathways of different species of bacteria, cellular transport of contaminants across cell membranes, and the biosynthesis of specific amino acids. The use of combined isotopic studies (^{13}C, ^{37}Cl, ^{2}H, ^{18}O, and ^{15}N) and combined isotopes of the same molecule are currently being developed and will undoubtedly expand, as will the use of combined $\delta^{13}C$ and naturally occurring radiocarbon (^{14}C) in aerobic studies discussed in Chapter 11.

ACKNOWLEDGMENTS

We thank Melissa Engle for assistance with the preparation of the figures. This work was supported by an ERIG award from the University of South Carolina.

REFERENCES

Abraham, W. R., and C. Hesse. 2003. Isotope fractionations in the biosynthesis of cell components by different fungi: A basis for environmental carbon flux studies. *FEMS Microbiology Ecology* 46:121–28.

Aelion, C. M., B. C. Kirtland, and P. A. Stone. 1997. Radiocarbon assessment of aerobic petroleum bioremediation in the vadose zone and groundwater at an AS/SVE site. *Environmental Science & Technology* 31:3363–70.

Aggarwal, P. K., M. E. Fuller, M. M. Gurgas, J. F. Manning, and M. A. Dillon. 1997. Use of stable oxygen and carbon isotope analyses for monitoring the pathways and rates of intrinsic and enhanced *in situ* biodegradation. *Environmental Science & Technology* 31:590–96.

Aggarwal, P. K., and R. E. Hinchee. 1991. Monitoring in situ biodegradation of hydrocarbons by using stable carbon isotopes. *Environmental Science & Technology* 25:1178–80.

Atlas, R. M., and R. Bartha. 1997. *Microbial ecology: Fundamentals and applications.* 4th ed. Redwood City, CA: Benjamin/Cummings Publishing, Inc.

Baesman, S. M., and L. G. Miller. 2005. Laboratory determination of the caron kinetic isotope effects (KIEs) for reaction of methyl halides with various nucleopiles in solution. *Journal of Atmospheric Chemistry* 52:203–19.

Beneteau, K. M., R. Aravena, and S. K. Frape. 1999. Isotopic characterization of chlorinated solvents: Laboratory and field results. *Organic Geochemistry* 30:739–53.

Boutton, T. W. 1991. Stable carbon isotope ratios of natural materials: I. Sample preparation and mass spectrometric analysis. In *Carbon isotope techniques*, eds. D. C. Coleman and B. Fry, 155–85, San Diego, CA: Academic Press, Inc.

Brungard, K. L., J. Munakata-Marr, C. A. Johnson, and K. W. Mandernack. 2003. Stable carbon isotope fractionation of *trans*-1,2-dichloroethylene during co-metabolic degradation by methanotrophic bacteria. *Chemical Geology* 195:59–67.

Bugna, G. C., J. P. Chanton, C. A. Kelley, T. B. Stauffer, W. G. MacIntyre, and E. L. Libelo. 2004. A field test of $\delta^{13}C$ as tracer of aerobic hydrocarbon degradation. *Organic Geochemistry* 35:123–35.

Busby, J. F., R. W. Lee, and B. B. Hanshaw. 1983. *Major geochemical processes related to the hydrology of the Madison aquifer system and associated rocks in parts of Montana, South Dakota, and Wyoming.* Denver, CO: U.S. Geological Survey, Water Resources Division, Water Resources Investigations 83-4093.

Chartrand, M. M. G., A. Waller, T. E. Mattes, M. Elsner, G. Lacrampe-Couloume, J. M. Gossett, E. A. Edwards, and B. Sherwood Lollar. 2005. Carbon isotopic fractionation during aerobic vinyl chloride degradation. *Environmental Science & Technology* 39:1064–70.

Chartrand, M. M. G., S. K. Hirschorn, G. Lacrampe-Couloume, and B. Sherwood Lollar. 2007. Compound-specific hydrogen isotope analysis of 1,2-dichloroethane: Potential for delineating source and fate of chlorinated hydrocarbon contaminants in groundwater. *Rapid Communications in Mass Spectrometry* 21:184147.

Chu, K., S. Mahendra, D. L. Song, M. E. Conrad, and L. Alvarez-Cohen. 2004. Stable carbon isotope fractionation during aerobic biodegradation of chlorinated ethenes. *Environmental Science & Technology* 38:3126–30.

Coleman, D. D., J. B. Risatti, and M. Schoell. 1981. Fractionation of carbon and hydrogen isotopes by methane-oxidizing bacteria. *Geochimica Cosmochimica Acta.* 45:1033–37.

Conrad, M. E., A. S. Templeton, P. F. Daley, and L. Alvarez-Cohen. 1999. Isotopic evidence for biological controls on migration of petroleum hydrocarbons. *Organic Geochemistry* 30:843–59.

Dempster, H. S., B. Sherwood Lollar, and S. Feenstra. 1997. Tracing organic contaminants in groundwater: A new methodology using compound specific isotopic analysis. *Environmental Science & Technology* 31:3193–97.

Drenzek, N. J., C. H. Tarr, T. I. Eglinton, L. J. Heraty, N. C. Sturchio, V. J. Shiner, and C. M. Reddy. 2002. Stable chlorine and carbon isotopic compositions of selected semivolatile organochlorine compounds. *Organic Geochemistry* 33:437–44.

Elsner, M., J. Mckelvie, G. L. Couloume, B. Sherwood Lollar. 2007. Insight into methyl tert-butyl ether (MTBE) stable isotope fractionation from abiotic reference experiments. *Environmental Science & Technology* 41:5693–700.

Farquhar, G. D., M. H. O'Leary, and J. A. Berry. 1982. On the relationship between carbon isotope discrimination and the inter-cellular carbon-dioxide concentration in leaves. *Australian Journal of Plant Physiology* 9:121–37.

Fayolle, F., J. P. Vandecasteele, and F. Monot. 2001. Microbial degradation and fate in the environment of methyl tert-butyl ether and related fuel oxygenates. *Applied Microbiology and Biotechnology* 56:339–49.

Fischer, A. H. I., S. Herrmann, M. Thullner, A. J. M. Stams, H. H. Richnow, and C. Vogt. 2008. Combined carbon and hydrogen isotope fractionation investigations for elucidating benzene biodegradation pathways. *Environmental Science & Technology* 42:4356–63.

Ghosh, P., and W. A. Brand. 2003. Stable isotope ratio mass spectrometry in global climate change research. *International Journal of Mass Spectrometry* 228:1–33.

Gleason, J. D., I. Friedman, and B. B. Hanshaw. 1969. Extraction of dissolved carbonate species from natural water for carbon-isotope analysis, U.S. Geological Survey Professional Paper 650-D:D248-D250.

Gray, J. R., G. Lacrampe-Couloume, D. Gandhi, K. Scow, R. D. Wilson, D. M. Mackay, and B. Sherwood Lollar. 2002. Carbon and hydrogen isotopic fractionation during biodegradation of methyl *tert*-butyl ether. *Environmental Science & Technology* 36:1931–38.

Griebler, C., L. Adrian, R. U. Meckenstock, and H. H. Richnow. 2004. Stable carbon isotope fractionation during aerobic and anaerobic transformation of trichlorobenzene. *FEMS Microbiology Ecology* 48:313–21.

Hage, J. C., and S. Hartmans. 1999. Monooxygenase-mediated 1,2-dichloroethane degradation by *pseudomonas sp.* strain DCA1. *Applied Environmental Microbiology* 65:2466–70.

Hall, J. A., R. M. Kalin, M. J. Larkin, C. C. R. Allen, and D. B. Harper. 1999. Variation in stable carbon isotope fractionation during aerobic degradation of phenol and benzoate by contaminant degrading bacteria. *Organic Geochemistry* 30:801–11.

Hanson, R. S., and T. E. Hanson. 1996. Methanotrophic Bacteria. *Microbiology Review* 60:439–71.

Hardison, L. K., S. S. Curry, L. M. Ciuffetti, and M. R. Hyman. 1997. Metabolism of diethyl ether and cometabolism of methyl *tert*-butyl ether by a filamentous fungus, a *graphium* sp. *Applied Environmental Microbiology* 63:3059–67.

Harrington, R. R., S. R. Poulson, J. I. Drever, P. J. S. Colberg, and E. F. Kelly. 1999. Carbon isotope systematics of monoaromatic hydrocarbons: Vaporization and adsorption experiments. *Organic Geochemistry* 30:765–75.

Heraty, L. J., M. E. Fuller, L. Huang, T. Abrajano Jr., and N. C. Sturchio. 1999. Isotopic fractionation of carbon and chlorine by microbial degradation of dichloromethane. *Organic Geochemistry* 30:793–99.

Hirschorn, S. K., M. J. Dinglassan, M. Elsner, S. A. Mancini, G. Lacrampe-Couloume, E. A. Edwards, and B. Sherwood Lollar. 2004. Pathway dependent isotopic fractionation during aerobic biodegradation of 1,2-dichloroethane. *Environmental Science Technology* 38:4775–81.

Huang, Y., G. Eglinton, P. Ineson, P. M. Latter, R. Bol, and D. D. Harkness. 1997. Absence of carbon isotope fractionation of individual n-alkanes in a 23-year field decomposition experiment with Calluna vulgaris. *Organic Geochemistry* 23:497–501.

Huang, L., N. C. Sturchio, T. Abrajano Jr., L. J. Heraty, and B. D. Holt. 1999. Carbon and chlorine isotope fractionation of chlorinated aliphatic hydrocarbons by evaporation. *Organic Geochemistry* 30:777–85.

Hunkeler, D., N. Andersen, R. Aravena, S. M. Bernasconi, and B. J. Butler. 2001. Hydrogen and carbon isotope fractionation during aerobic biodegradation of benzene. *Environmental Science & Technology* 35:3462–67.

Hunkeler, D., P. Hohener, S. Bernasconi, and J. Zeyer. 1999. Engineered in situ bioremediation of a petroleum hydrocarbon-contaminated aquifer: Assessment of mineralization based on alkalinity, inorganic carbon and stable carbon isotope balances. *Journal of Contaminant Hydrology* 37:201–23.

Jahnke, L. L., R. E. Summons, J. M. Hope, and D. J. des Marais. 1999. Carbon isotopic fractionation in lipids from methanotrophic bacteria II: The effects of physiology and environmental parameters on the biosynthesis and isotopic signatures of biomarkers. *Geochimica Cosmochimica Acta* 63:79–93.

Jarman, W. M., A. Hilkert, C. E. Bacon, J. W. Collister, K. Ballschmiter, and R. W. Risebrough. 1998. Compound-specific carbon isotopic analysis of aroclors, clophens, kaneclors, and phenoclors. *Environmental Science & Technology* 32:833–36.

Kauppi, B., L. Kyoung E. Carredano, R. E. Parales, D. T. Gibson, H. Eklund, and S. Ramaswamy. 1998. Structure of an aromatic ring-hydroxylating dioxygenase-naphthalene 1,2-dioxygenase. *Structure* 6:571–86.

Kirtland, B. C., C. M. Aelion, and P. A. Stone. 2000. Monitoring anaerobic natural attenuation of petroleum using a novel in situ respiration method in low-permeability sediment. *Bioremediation Journal* 4:187–210.

Kirtland, B. C, C. M. Aelion, and P. A. Stone. 2005. Assessing in situ mineralization of recalcitrant organic compounds in vadose zone sediments using $\delta^{13}C$ and ^{14}C measurements. *Journal of Contaminant Hydrology* 76:1–18.

Kirtland, B. C., C. M. Aelion, P. A. Stone, and D. Hunkeler. 2003. Isotopic and geochemical assessment of in situ biodegradation of chlorinated hydrocarbons. *Environmental Science & Technology* 37:4205–12.

Kopinke, F. D., A. Georgi, M. Voskamp, and H. H. Richnow. 2005. Carbon isotope fractionation of organic contaminants due to retardation on humic substances: Implications for natural attenuation studies in aquifers. *Environmental Science & Technology* 39:6052–62.

Kuder, T., R. P. Philp, R. Kolhatkar, J. T. Wilson, and J. Allen. 2002. Application of stable carbon and hydrogen isotopic techniques for monitoring biodegradation of MTBE in the field. In *National Ground Water Association/American Petroleum Institute petroleum hydrocarbons and organic chemicals in ground water*, pp. 167–177, Washington, DC: American Petroleum Institute.

Landmeyer, J. E., D. A. Vroblesky, and F. H. Chapelle. 1996. Stable carbon isotope evidence of biodegradation zonation in a shallow jet-fuel contaminated aquifer. *Environmental Science & Technology* 30:1120–28.

Madigan, M. T., J. M. Martinko, and J. Parker. 2000. *Brock biology of microorganisms*. 9th ed. Upper Saddle River, NJ: Prentice Hall.

Mancini, S. A., C. E. Devine, M. E. Elsner, M. Nandi, A. C. Ulrich, E. A. Edwards, and B. Sherwood Lollar. 2008. Isotopic evidence shows different reaction mechanisms for anaerobic benzene biodegradation. *Environmental Science & Technology* 42:8290–96.

Mancini, S. A., S. K. Hirschorn, M. E. Elsner, g. Lacrampe-Couloume, B. E. Sleep, E. A. Edwards, and B. Sherwood Lollar. 2006. Effects of trace element concentration on enzyme controlled stable isotopic fractionation during aerobic biodegradation of toluene. *Environmental Science & Technology* 40:7675–81.

Mancini, S. A., G. Lacrampe-Couloume, H. Jonker, B. M. van Breukelen, J. Groen, F. Volkering, and B. Sherwood Lollar. 2002. Hydrogen isotope enrichment: An indicator of biodegradation at a petroleum hydrocarbon contaminated field site. *Environmental Science & Technology* 36:2464–70.

Mancini, S. A., G. Lacrampe-Couloume, and B. Sherwood Lollar. 2008. Source differentiation for benzene and chlorobenzene groundwater contamination: A field application for stable carbon and hydrogen isotope analyses. *Environmental Forensics* 9:177–86.

Mazeas, L., H. Budzinski, and N. Raymond. 2002. Absence of stable carbon isotope fractionation of saturated and polycyclic aromatic hydrocarbons during aerobic bacterial biodegradation, *Organic Geochemistry* 33:1259–72.

Miller, L. G., R. M. Kalin, S. E. McCauley, J. T. G. Hamilton, D. B. Harper, D. B. Millet, R. S. Oremland, and A. H. Goldstein. 2001. Large carbon isotope fractionation associated with oxidation of methyl halides by methylotrophic bacteria. *Proceedings of the National Academy of Sciences* 98:5833–37.

Miller, L. G., K. L. Warner, S. M. Baesman, R. S. Oremland, I. R. McDonald, S. Radajewski, and C. Murrell. 2004. Degradation of methyl bromide and methyl chloride in soil microcosms: Use of stable C isotope fractionation and stable isotope probing to identify reactions and the responsible microorganisms. *Geochimica Cosmochimica Acta* 68 (15):3271–83.

Morasch, B., H. H. Richnow, B. Schink, A. Vieth, and R. U. Meckenstock. 2002. Carbon and hydrogen stable isotope fractionation during aerobic bacterial degradation of aromatic hydrocarbons. *Applied Environmental Microbiology* 68:5191–94.

Morel, F. M. M. 1983. *Principles of aquatic chemistry.* New York: John Wiley & Sons.

Murray, W. D., and M. Richardson. 1993. Progress toward the biological treatment of C(1) and C(2) halogenated hydrocarbons. *Critical Reviews in Environmental Science and Technology* 23:195–217.

Nguyen Tu, T. T., S. Derenne, C. Largeau, G. Bardoux, and A. Mariotti. 2004. Diagenesis effects on specific carbon isotope composition of plant *n*-alkanes. *Organic Geochemistry* 35:317–29.

Northrop, D. B. 1991. Intrinsic isotope effects in enzyme catalyzed reactions. In *Enzyme mechanisms from isotope effects,* ed. P. F. Cook, 181–202. Boca Raton, FL: CRC Press.

O'Malley, V. P., T. A. Abrajano, and J. Hellou. 1994. Determination of the $^{13}C/^{12}C$ ratios of individual PAH from environmental samples: Can PAH sources be apportioned? *Organic Geochemistry* 21:809–22.

Otero, E., R. Culp, J. Noakes, and R. E. Hodson. 2003. The distribution and $\delta^{13}C$ of dissolved organic carbon and its humic fraction in estuaries of southeastern USA. *Estuarine Coastal and Shelf Science* 56:1187–94.

Poulson, S. R., and J. I. Drever. 1999. Stable isotope (C, Cl, and H) fractionation during vaporization of trichloroethylene. *Environmental Science & Technology* 33:3689–94.

Rawn, J. D. 1989. *Biochemistry.* Intl. ed. Burlington, NC: Neil Patterson.

Rightmire, C. T., and B. B. Hanshaw. 1973. Relationship between the carbon isotope composition of soil CO_2 and dissolved carbonate species in groundwater. *Water Resources Research* 9:958–67.

Schidlowski, M. 2001. Carbon isotopes as biogeochemical recorders of lie over 3.8 Ga of Earth history: Evolution of a concept. *Precambrian Research* 106:117–34.

Schüth, C., H. Taubald, N. Bolaño, and K. Maciejczyk. 2003. Carbon and hydrogen isotope effects during sorption of organic contaminants on carbonaceous materials. *Journal of Contaminant Hydrology* 64:269–81.

Sherwood Lollar, B., S. K. Hirschorn, M. M. G. Chartrand, and G. Lacrampe-Couloume. 2007. An approach for assessing total instrumental uncertainty in compound-specific carbon isotope analysis: Implications for environmental remediation studies. *Analytical Chemistry* 79:3469–75.

Sherwood Lollar, B., G. F. Slater, J. Ahad, B. Sleep, J. Spivack, M. Brennan, and P. MacKenzie. 1999. Contrasting carbon isotope fractionation during biodegradation of trichloroethylene and toluene: Implications for intrinsic bioremediation. *Organic Geochemistry* 30:813–20.

Shouakar-Stash, O., S. K. Frape, and R. J. Drimmie. 2003. Stable carbon, hydrogen and chlorine isotope measurements of selected chlorinated solvents. *Journal of Contaminant Hydrology* 60:211–28.

Slater, G. F. 2003. Stable isotope forensics: When isotopes work. *Environmental Forensics* 4:13–23.

Smallwood, B. J., R. P. Philp, T. W. Burgoyne, and J. D. Allen. 2001. The use of stable isotopes to differentiate specific source markers for MTBE. *Environmental Forensics* 2:215–21.

Smith, B. N., and S. Epstein. 1971. Two categories of $^{13}C/^{12}C$ ratios for higher plants. *Plant Physiology* 47:380–84.

Smith, C. A., K. T. O'Reilly, and M. R. Hyman. 2003. Cometabolism of methyl tertiary butyl ester and gaseous *n*-alkanes by pseudomonas mendocina kr-1 grown on C5 to C8 alkanes. *Applied Environmental Microbiology* 69:7385–94.

Staddon, P. L. 2004. Carbon isotopes in functional soil ecology. *Trends in Ecology and Evolution* 19:148–54.

Stahl, W. J. 1980. Compositional changes and $^{13}C/^{12}C$ fractionations during the degradation of hydrocarbons by bacteria. *Geochimica Cosmochimica Acta* 44:1903–7.

Stehmeier, L. G., M. M. Francis, T. R. Jack, E. Diegor, L. Winsor, and T. A. Abrajano, Jr. 1999. Field and in vitro evidence for in situ bioremediation using compound-specific $^{13}C/^{12}C$ ratio monitoring. *Organic Geochemistry* 30:821–33.

Suchomel, K. H., D. K. Kreamer, and A. Long. 1990. Production and transport of carbon dioxide in a contaminated vadose zone: A stable and radioactive carbon isotope study. *Environmental Science Technology* 24:1824–31.

Trust Hammer, B. A., J. G. Mueller, R. B. Coffin, and L. A. Cifuentes. 1995. The biodegradation of fluroanthene as monitored by stable carbon isotopes. In *Monitoring and verification of bioremediation*, ed. R. E. Hinchee, G. S. Douglas, and S. K. Ong, 233–39. Columbus, OH: Battelle Press.

Ucko, D. A. 1986. *Living chemistry: An introduction to general, organic, and biological chemistry*, 2nd ed. Orlando, FL: Harcourt Brace Javanovich, Inc.

van der Meer, M. T. J., S. Schouten, J. S. Sinninghe Damsté, J. W. de Leeuw, and D. M. Ward. 2003. Compound-specific isotopic fractionation patterns suggest different carbon metabolisms among *chloroflexus*-like bacteria in hot-spring microbial mats. *Applied Environmental Microbiology* 69:6000–6.

van Warmerdam, E. M., S. K. Frape, R. Aravena, R. J. Drimmie, H. Flatt, and J. A. Cherry. 1995. Stable chlorine and carbon isotope measurements of selected chlorinated organic solvents. *Applied Geochemistry* 10:547–52.

Wang, X., R. F. Chen, and A. Berry. 2003. Sources and preservation of organic matter in plum island salt marsh sediments (MA, USA): Long-chain *n*-alkanes and stable carbon isotope compositions. *Estuarine Coastal and Shelf Science* 58:917–28.

Wang, Y., and Y. Huang. 2003. Hydrogen isotopic fractionation of petroleum hydrocarbons during vaporization: Implications for assessing artificial and natural remediation of petroleum contamination. *Organic Geochemistry* 18:1641–51.

Ward, J. A., J. M. E. Ahad, G. Lacrampe-Couloume, G. F. Slater, E. A. Edwards, and B. Sherwood Lollar. 2000. Hydrogen isotope fractionation during methanogenic degradation of toluene: Potential for direct verification of bioremediation. *Environmental Science & Technology* 34:4577–81.

Wigley, T. M. L., L. N. Plummer, and F. J. Pearson, Jr. 1978. Mass transfer and carbon isotope evolution in natural water systems. *Geochimica Cosmochimica Acta* 42:1117–39.

Yakir, D., and L. da S. L. Sternberg. 2000. The use of stable isotopes to study ecosystem gas exchange. *Oecologia* 123:297–311.

Zwank, L., M. Berg, M. Elsner, T. C. Schmidt, R. P. Schwarzenbach, and S. B. Haderlein. 2005. New evaluation scheme for two-dimensional isotope analysis to decipher biodegradation processes: Application to groundwater contamination by MTBE. *Environmental Science Technology* 39:1018–29.

Zyakun, A. M., I. A. Kosheleva, V. N. Zakharchenko, A. I. Kudryavtseva, V. A. Peshenko, A. E. Filonov, and A. M. Boronin. 2003. The use of [C-13]/[C-12] ratio for the assay of the microbial oxidation of hydrocarbons. *Microbiology* 72:592–96.

Zyakun, A. M., and V. N. Zakharchenko. 1998. Carbon isotope discrimination by methanotrophic bacteria: Practical use in biotechnological research (Review). *Applied Biochemistry and Microbiology* 34:207–41.

6 Isotopes and Methane Cycling

Edward R. C. Hornibrook and Ramon Aravena

CONTENTS

6.1 INTRODUCTION

This chapter focuses on how stable (^{13}C, ^{12}C, ^{2}H, and ^{1}H) and radiogenic (^{14}C and ^{3}H) isotopes can be used to investigate the origin, transport, and alteration of methane (CH_4) in terrestrial environments. Methane is the most abundant organic gas in the troposphere and plays an important role in both the energy balance and chemistry of the Earth's atmosphere, participating in a range of chemical reactions that directly or indirectly involve CO, CO_2, H_2O, H_2, CH_2O, NO_x, O_3, Cl, F, and OH* (Cicerone and Oremland 1988). The atmospheric lifetime of CH_4 ranges between eight and 14 years with a mean turnover time of ~10 years (Rasmussen and Khalil 1981; Mayer et al. 1982). Destruction of CH_4 in the troposphere and stratosphere is complex but ultimately results in the end products CO_2 and H_2O. The atmospheric mixing ratio of CH_4 has increased from ~700 parts per billion by volume (ppbv) to 1750 ppbv during the period 1750 to 2000 (Etheridge et al. 1998). Methane directly influences the temperature of the lower troposphere through absorption of infrared radiation (IR) and has been estimated to account for ~15% of enhanced radiative forcing of climate as a result of anthropogenic greenhouse gas emissions (Rodhe 1990). The commonly cited ability of CH_4 to absorb thermal radiation at a rate ~25 times that of CO_2 on a molecule-per-molecule basis is a consequence of the relatively low atmospheric abundance of CH_4, and the lack of overlap of absorption bands with CO_2 and H_2O that absorb IR primarily outside the thermal emission window (Senum and Gaffney 1985).

The total global (Tg) budget of CH_4 flux is presently estimated to range from 500 to 600 Tg yr^{-1} (Mikaloff-Fletcher et al. 2004a, 2004b). Numerous natural systems emit CH_4 to the atmosphere, including oceans, fresh water bodies (lakes, ponds, rivers, etc.), wild ruminants, termites, destabilization of gas hydrates within continental-margin sediments and arctic permafrost, and wetlands. Recently Keppler et al. (2006) reported that terrestrial vegetation produces CH_4 aerobically, globally emitting between 62–236 Tg yr^{-1} from living plants and 1–7 Tg yr^{-1} from plant litter. The CH_4 production mechanism apparently is photochemical rather than microbial but the exact nature of the process has yet to be determined. The original estimates of total annual CH_4 flux from this new source have been downsized by 0–85 Tg yr^{-1} based upon modeling efforts (Kirschbaum et al. 2006; Houweling et al. 2006).

Amongst the better established global sources of CH_4, wetlands typically are regarded as the single largest contributor of emissions to the atmosphere each year. The majority of global wetland cover occurs in northern Canada and Siberia, primarily between 50° and 70°N latitude; however, ~60% (66 Tg yr^{-1}) of the total annual flux from wetlands (109 Tg yr^{-1}) is thought to originate from tropical wetlands situated between 20°N and 30°S latitude (Bartlett and Harriss 1993). Mikaloff-Fletcher et al. (2004a, 2004b) have recently suggested that tropical biomes are an even more significant source of CH_4, emitting an estimated 266 ± 25 Tg yr^{-1} of CH_4 from a combination of wetlands and biomass burning.

Major anthropogenic sources of CH_4 are associated with agriculture, disposal of organic waste, and energy production. Agricultural practices that generate significant CH_4 emissions include rice cultivation, livestock (ruminant) production, and burning of biomass during land clearance (Cicerone and Oremland 1988). Landfills remain major point sources of anthropogenic CH_4 despite greater efforts to collect and utilize landfill CH_4 as an energy resource (Bingemer and Crutzen 1987). Annual emissions of CH_4 from exploitation of fossil fuels are difficult to ascertain, but include losses associated with coal mining as well as activities related to petroleum production, including natural gas venting, drilling losses, and leakage during pipeline transmission (Wahlen 1993).

6.2 ORIGINS OF METHANE (CH$_4$)

Methane is formed in terrestrial and marine environments predominantly through the degradation of organic matter in different temperature regimes (Schoell 1988). It is produced exclusively by microorganisms known as Archaea in shallow subsurface sediments, the rumen of cattle and the gut of termites where temperatures typically are low to moderate (~0 to < 50°C). At elevated temperatures in deeply buried sediments, CH_4 and higher homologues (i.e., ethane, propane, butane, etc.), are formed by thermal decomposition of organic compounds. Welhan (1988) has used the term biogenic in reference to all CH_4 derived from organic substances regardless of temperature of formation, suggesting that both microbial and thermogenic CH_4 are included in this category. This use of nomenclature has not been adopted consistently in the literature. The terms biogenic and microbial are used interchangeably in this chapter in reference to CH_4 formed by methanogenic Archaea. The term thermogenic is applied to fossil CH_4 generated primarily by thermal degradation of organic matter; however, we acknowledge that there are issues with such usage because microorganisms are active in deeply buried sediments. It is likely that some accumulations of thermogenic gas contain microbial CH_4 that has been generated at elevated temperatures and that may differ isotopically from CH_4 produced in shallow, low temperature, sedimentary environments.

6.2.1 ABIOGENIC METHANE (CH_4) PRODUCTION

Methane produced in the absence of organic compounds is termed abiogenic or geological and the number of sources of such CH_4 is limited to environments such as mid-ocean ridge hydrothermal systems (e.g., Welhan 1988; Charlou et al. 1998; Kelley and Fruh-Green 1999), ophiolite seeps (Abrajano et al. 1988, 1990), and deep shield rocks (Lollar, Frape, and Weise et al. 1993a; Lollar et al. 2002). The distinguishing feature in abiogenic systems is that energy required for conversion of CO or CO_2 to CH_4 is obtained from H_2 of a geological origin (e.g., serpentinization reactions; Berndt, Allen, and Seyfried 1996) or radiolysis of water (Lin et al. 2005) rather than from organic matter. Again, the potential role of chemolithotrophic microorganisms at depth cannot be excluded as the upper temperature limit for microbial production of CH_4 and the maximum depth of microbial activity within the Earth's crust continue to be extended (Lollar, Frape, Fritz et al. 1993b; Parkes et al. 1994; Stevens and McKinley 1995). Abiogenic and deep thermogenic production of CH_4 are beyond the scope of this review, which will focus upon the isotope systematics of CH_4 production and cycling of CH_4 in shallow, low temperature, freshwater environments, more specifically aquifers, wetlands, and landfills.

6.2.2 MICROBIAL METHANE (CH_4) PRODUCTION

The production of CH_4 in organic-rich, anoxic, freshwater environments requires a microbial consortia that is composed of at least three interacting metabolic groups (Ferry 1999). The first type, known as fermentative bacteria, hydrolyze large organic polymers to long-chain volatile fatty acids (VFAs), alcohols and H_2, CO_2, formate, and acetate. A second catabolic group of microorganisms (acetogenic bacteria) oxidize the higher VFAs and alcohols to acetate and H_2 or formate. Methanogens comprise the third group and represent the terminal step in the anaerobic chain of decay. The metabolic activities of methanogens can be divided into three main types consisting of the methylotrophic, aceticlastic, and CO_2-reduction pathways (Boone, Whitman, and Rouvière 1993).

Methylotrophs catabolize simple organic compounds that contain methyl groups, such as methanol, dimethyl sulfide, and methylated amines. The methyl group is transferred to a methyl carrier (i.e., a coenzyme) and eventually reduced to CH_4 using electrons obtained from H_2 or from oxidation of a portion of the methyl groups. Aceticlastic methanogenesis is a special case of methylotrophism whereby the acetate molecule is both the electron donor and acceptor. During degradation, acetate is cleaved and electrons obtained by oxidizing the carboxyl moiety to CO_2 are used to reduce the methyl group to CH_4 (Ferry 1993). Although relatively few methanogens have been identified that can utilize acetate, acetotrophism is thought to be the predominant methanogenic pathway in anoxic, freshwater environments (Vogels, Keltjens, and Van der Drift 1988). The proportion of CH_4 formed by the aceticlastic pathway may be increased by the presence of homoacetogenic microorganisms (Dolfing 1988). These anabolic bacteria form acetate from CO_2 using reducing equivalents obtained from the oxidation of H_2 or formate. Since methyl-bearing substrates other than acetate (e.g., methanol, dimethyl sulfide, and methylated amines) tend to be of minor importance as CH_4 precursors in the natural environment, the term acetate fermentation is commonly used to refer collectively to all methanogenesis that involves cleavage of a methyl group from an organic substrate. This usage is most prevalent in the biogeochemical literature.

All but a few obligate methylotrophs and acetotrophs are capable of producing CH_4 from CO_2 and H_2 (Müller, Blaut, and Gottschalk 1993). Methanogens utilizing the CO_2-reduction pathway convert CO_2 or bicarbonate to CH_4 using electrons that have been acquired from the oxidation of H_2 or formate (Ferry 1999). Molecular H_2 is a short-lived, but important, intermediate in the breakdown of organic matter in anoxic environments. It is formed during metabolism of carbohydrates, amino acids, aromatics, and fatty acids by fungi, protozoa, and fermentative species of anaerobic bacteria other than methanogens (Lovley and Goodwin 1988; Zinder 1993). Methane-producing bacteria play an essential role in anaerobic systems by removing H_2 before it accumulates in a quantity sufficient to attenuate the metabolism of heterotrophic acetogenic bacteria higher in the chain of decay (Dolfing 2001). The rapid turnover of H_2 between H_2 producers and consumers is known as interspecies hydrogen transfer (Wolin and Miller 1982).

6.3   STABLE ISOTOPES AS A PROXY FOR METHANOGENIC PATHWAYS

During early investigations of stable isotope variations in nature, it was recognized that large kinetic isotope effects (KIEs) are associated with the metabolism of methanogenic microorganisms (e.g., Rosenfeld and Silverman 1959). Numerous subsequent studies reported highly ^{13}C- and ^2H-depleted CH_4 from a variety of anaerobic environments, including lake and marine sediments (Oana and Deevey 1960; Nissenbaum, Presley, and Kaplan 1972; Deuser et al. 1973; Galimov and Kvenvolden 1983; Claypool et al. 1985), landfills (Games and Hayes 1974), glacial-drift deposits

(Wasserburg, Mazor, and Zartman 1963), sewage sludge reactors (Nissenbaum, Presley, and Kaplan 1972), and wetlands (Voytov et al. 1975). By the 1970s, it was well established that CH_4-producing microorganisms utilize only a limited number of simple one carbon compounds (e.g., Barker 1936; Buswell and Solo 1948; Stadtmann and Barker 1949; Mylroie and Hungate 1955; Pine and Barker 1956; Pine and Vishniac 1957; Nottingham and Hungate 1969; Wolfe 1971) and consequently the potential to employ stable isotope ratios as a proxy for CH_4 production pathways was being pursued (Kaplan 1975; Games and Hayes 1976), including the possibility that different species of CH_4-producing microorganisms might possess distinctive KIEs (Belyaev, Laurinavichus, and Gaytan 1977; Games, Hayes, and Gunsalas 1978; Fuchs et al. 1979). Woltemate, Whiticar, and Schoell (1984) reported the first use of $\delta^{13}C$ and δ^2H values to estimate proportions of CH_4 produced by acetate fermentation and CO_2-reduction in an investigation of the origins of CH_4 in sediment-free gas from a shallow freshwater lake. Their stable isotope results suggested ~75% of CH_4 was generated by acetate fermentation, which was consistent with previous studies that employed ^{14}C-labeled substrates to quantify methanogenic pathways in organic-rich, freshwater environments. Burke and Sackett (1986) used a similar approach to estimate that ~50 to 80% of CH_4 in Tampa Bay, Florida estuary sediments was produced by acetate fermentation, noting also that an inverse correlation between $\delta^{13}C$- and δ^2H-CH_4 values could be used to distinguish mixing of CH_4 formed by acetate fermentation and CO_2 reduction from CH_4 remaining after incomplete bacterial methanotrophy that causes enrichment of both ^{13}C and [2]H (i.e., a positive correlation between $\delta^{13}C$ and δ^2H values).

An important milestone in the development of stable isotopes as a proxy for distinguishing methanogenic pathways was the Whiticar, Faber, and Schoell (1986) empirical model, which involved a more elaborate approach utilizing paired $\delta^{13}C$ and δ^2H values from coexisting CH_4 and CO_2, and CH_4 and H_2O. The model was based upon a collection of > 300 stable isotope analyses from a range of terrestrial and marine environments and relied upon the critical assumption that CH_4 production in freshwater environments occurs predominantly via acetate fermentation while in marine systems it is formed mainly by CO_2-reduction coupled to H_2 oxidation because of the depletion of labile substrate as a result of high SO_4^{2-} levels in seawater. This assumption was well supported by investigations of methanogenesis in marine and freshwater systems that utilized ^{14}C-labeled substrates and the addition of specific substrates. Analysis of the stable isotope database suggested that the two primary methanogenic pathways yielded CH_4 with distinctive ranges of $\delta^{13}C$ and δ^2H values: acetate fermentation, $\delta^{13}C \sim -65$ to $-50‰$ and $\delta^2H \sim -400$ to $-250‰$; CO_2-reduction, $\delta^{13}C \sim -110$ to $-60‰$ and $\delta^2H \sim -250$ to $-170‰$ (Whiticar, Faber, and Schoell 1986).

Subsequently it was shown that acetate fermentation can predominate in marine environments when rates of organic matter deposition are high (e.g., Martens et al. 1986; Boehme et al. 1996) and CO_2 reduction can dominate in freshwater systems where organic matter is recalcitrant and pH is low (e.g., Lansdown, Quay, and King 1992; Sizova et al. 2003; Galand et al. 2005; Metje and Frenzel 2005). However, a critical aspect of the Whiticar, Faber, and Schoell (1986) model remains generally true in natural environments: the magnitude of the kinetic isotope effect (KIE) associated with CO_2 reduction ($\Delta^{13}C_{CH_4-CO_2}$~ 40 to 95‰; Whiticar 1999) is larger than for acetate fermentation ($\Delta^{13}C_{CH_4-CH_3COO-}$~ 20 to 25‰; Krzycki et al. 1987, Gelwicks, Risatti, and Hayes 1994), which in most anoxic environments leads to production of CH_4 having more negative $\delta^{13}C$ values. There is also strong evidence that the kinetic isotope effect (KIE) associated with CO_2/H_2 methanogenesis is influenced by temperature (Botz et al. 1996; Whiticar 1999), methanogen species (e.g., Games and Hayes 1976; Fuchs et al. 1979; Balabane et al. 1987), and growth conditions (Botz et al. 1996). Most recently it has been shown that isotopic fractionation associated with CH_4 production from H_2 and CO_2 also varies as a function of Gibbs free energy change (ΔG) where the magnitude of the KIE reaches a maximum as ΔG approaches the thermodynamic limit of CH_4 formation (Valentine et al. 2004; Penning et al. 2005). Because the partial pressure of H_2 (pH_2) exerts an important control on ΔG of CH_4 formation, it is possible that variations in H_2 abundance may influence the range of $\delta^{13}C$-CH_4 values observed in marine and freshwater environments in addition to the effects of pathway predominance. Under highly negative ΔG conditions, the magnitude of fractionation associated with CO_2 reduction is diminished to the extent that $\delta^{13}C$-CH_4 values produced may be indistinguishable from CH_4 formed via acetate fermentation (Penning et al. 2005). Furthermore, Valentine et al. (2004) have suggested that differences in activation energy of biochemical steps in the reduction of CO_2 to CH_4 coupled with conditions of low pH_2 and steady-state metabolism may result in differential reversibility of some catabolic reactions, leading to a mixture of equilibrium and KIEs. If that is the case, then the large ^{13}C-discrimination typically associated with CO_2/H_2 methanogenesis, and traditionally referred to as a KIE, may result from equilibrium isotope effects.

In addition, it may not be possible to predict accurately the $\delta^{13}C$ value of CH_4 produced by acetate fermentation in natural environments because it is uncertain whether KIEs determined in laboratory incubations in the presence of high concentrations of acetic acid (~20 to 25‰; Krzycki et al. 1987; Gelwicks, Risatti, and Hayes 1994) are expressed in situ under conditions where the pore water pool size of acetic acid is small and rapidly turned over (Blair and Carter 1992). Moreover, Valentine et al. (2004) reported that acetate dissimilation at high temperature (61°C) by *Methanosaeta thermophila* resulted in a very small ^{13}C isotope effect (~7‰), yielding CH_4 that had a $\delta^{13}C$ value largely indistinguishable from thermogenic gas.

Potentially the situation is complicated further by autotrophic acetogenesis (i.e., anabolic formation of CH_3COOH from CO_2 and H_2), estimated by Wood and Ljungdahl (1991) to generate ~10% of pore water acetic acid in anaerobic environments. The proportion likely varies as a function of temperature because autotrophic acetogenesis becomes more energetically favorable versus CO_2/H_2 methanogenesis at low temperatures (e.g., Conrad and Wetter 1990, Kotsyurbenko et al. 2001). Acetate formation from CO_2 and H_2 bears a KIE that is comparable in magnitude to that of CO_2 reduction (~58‰; Gelwicks, Risatti, and Hayes 1989). Regardless of acetate pool size effects limiting KIE expression, use of autotrophically formed acetic acid by methanogens will yield CH_4 having $\delta^{13}C$ values indistinguishable from the CO_2 reduction pathway. The contribution of CH_4 from this source in natural environments has yet to be characterized and, at present, little can be done to incorporate its influence into isotope models of methanogenesis.

Despite these issues, the Whiticar, Faber, and Schoell (1986) empirical model and subsequent refinements (e.g., Whiticar 1999; Conrad 2005) remain the most unobtrusive means to investigate methanogenic pathways in situ in anoxic environments (Oremland, Miller, and Whiticar 1987; Burke, Barber, and Sackett 1988; Faber, Stahl, and Whiticar 1990; Fritz et al. 1992; Martens et al. 1992; Wassmann et al. 1992; Sugimoto and Wada 1993; Katz et al. 1995; Revesz et al. 1995; Hornibrook, Longstaffe, and Fyfe 1997, 2000a, 2000b; Popp et al. 1999; Chasar et al. 2000a, 2000b; Chanton et al. 2004). Stable isotopes have been used as proxies to establish that seasonal (Martens et al. 1986; Burke, Martens, and Sackett 1988b; Burke, Barber, and Sackett 1992; Kelley, Dise, and Martens 1992; Blair, Boehme, and Carter 1993) and possibly diurnal (Jedrysek 1995, 1999) variations occur in terminal carbon mineralization processes in natural systems. A few notable studies have coupled stable isotope proxy measurements with conventional use of ^{14}C-labeled substrate tracers (i.e., $^{14}CH_3COONa$ and $NaH^{14}CO_3$) to determine the $\delta^{13}C$ values of CH_4 produced by the two main methanogenic pathways (Table 6.1). In each case, regardless of environment, CO_2/H_2 methanogenesis produced CH_4 that was ^{13}C-depleted relative to CH_4 generated by the acetate fermentation pathways. Although exceptions exist to the original assumptions made by Whiticar, Faber, and Schoell (1986) regarding the occurrence of acetate fermentation and CO_2 reduction in freshwater and marine environments (e.g., Martens et al. 1986; Lansdown, Quay, and King 1992; Waldron, Hall, and Fallick 1999), the premise regarding the stable carbon isotope composition of CH_4 produced via these pathways appears to be valid under most conditions that occur in natural environments. Variable expression of KIEs in situ as a result of differences in growth phase and ΔG of CH_4 formation are unlikely to occur in most natural settings because methanogenesis typically operates as a steady-state process near its thermodynamic limit. This stability is likely driven by the maintenance of very low pH_2 levels required in order for heterotrophic microorganisms to degrade higher alcohols and VFAs. Exceptions may exist in more dynamic environments, such as shallow layers of sediment or peat that are subjected

TABLE 6.1
$\delta^{13}C$ Values of CH_4 Linked to Methanogenic Pathways using ^{14}C Tracers

Environment	δ13C-CH4 (‰) CO₂-reduction	δ13C-CH4 (‰) Acetate Fermentation	Source
Coastal marine	−62‰	−39 to −37‰	Alperin et al. (1992)
Peatland	−73 ± 4‰	n/a	Lansdown et al. (1992)
Rice paddy	−77 to −60‰	−43 to −30‰	Sugimoto & Wada (1993)
Coastal marine	−62 to −58‰	−43 ± 10‰	Blair, Boehme, & Carter (1993)
Freshwater estuary	−72 ± 2.2‰	n/a	Avery & Martens (1999)
Peatland (May)	−72 ± 1.3‰	−43.8 ± 12‰	Avery et al. (1999)
Peatland (June)	−71 ± 1.3‰	−44.5 ± 5.4‰	Avery et al. (1999)

to large diurnal temperature cycles and rapid short-term inputs of labile organic matter from processes such as root exudation.

6.3.1   C AND H ISOTOPES OF METHANE

Future opportunities to investigate relationships between methanogenic pathways and natural abundance level δ^{13}C- and δ^2H-CH$_4$ values should be facilitated by a new ^{13}C-tracer technique reported by Gray, Matthews, and Head (2006). The method involves titrating anaerobic incubations of organic-rich material with different quantities of ^{13}C-labeled acetic acid and bicarbonate, and then determining the relative proportions of methanogenic pathways based upon differences in the ^{13}C content of terminal carbon mineralization end products. The same mass spectrometry instrumentation is employed for δ^{13}C analysis of both ^{13}C-enriched and natural abundance end products, which eliminates the need for facilities to handle and analyze radiocarbon tracers.

Unravelling the hydrogen isotope systematics of biogenic CH$_4$ production has proven to be more problematic than ^{13}C and ^{12}C partitioning. Whiticar, Faber, and Schoell (1986) suggested that H-isotope fractionation between CH$_4$ and coexisting H$_2$O (Δ^2H$_{CH_4-H_2O}$) exhibits distinctive relationships for the two main methanogenic pathways. During CH$_4$ formation by CO$_2$-reduction coupled to H$_2$ oxidation, the four H-atoms are derived from ambient H$_2$O (Daniels et al. 1980) and δ^2H-CH$_4$ values should vary according to Equation 1 in Table 6.2 (Whiticar, Faber, and Schoell 1986). Schoell (1980) and Woltemate, Whiticar, and Schoell (1984) reported earlier versions of this equation having an intercept of –160‰. The difference of 20‰ was explained by Whiticar, Faber, and Schoell (1986) as a refinement, resulting from the use of a larger data set. Sugimoto and Wada (1995) later suggested that the fractionation of Δ^2H ~ –180‰ between CH$_4$ and H$_2$O may apply only in marine environments and reported a much larger Δ^2H$_{CH_4-H_2O}$ value of ~–317‰ for CO$_2$-reduction in freshwater incubations of a rice paddy soil (Table 6.2, Equation 8; Sugimoto and Wada 1995). However, Balabane et al. (1987) also reported unusually depleted δ^2H-CH$_4$ values produced during anaerobic incubations where H$_2$ partial pressures were several orders of magnitude greater than that typically found in the natural environment. Burke (1993) suggested that the effect may have resulted from isotopic exchange between H$_2$ and H$_2$O, which is catalyzed rapidly (i.e., on the order of seconds) by intracellular hydrogenase enzymes with the rate of H-exchange increasing at higher H$_2$ concentrations (Yagi, Tsuda, and Inokuchi 1973). Direct measurements of microbial δ^2H-H$_2$ values have not been reported to date, but presumably microbial H$_2$ is highly ^{2}H-depleted relative to H$_2$O given the extremely negative δ^2H values associated with abiotic H$_2$ (e.g., Goebel et al. 1984). Consequently, elevated levels of H$_2$ could result in ^{2}H-depletion of microbial intracellular H$_2$O, which subsequently may be used to form CH$_4$ with unusually negative δ^2H values. However, Valentine et al. (2004) reported that δ^2H values of CH$_4$ generated by CO$_2$/H$_2$ methanogenesis at elevated temperature (61°C) were highly variable even when extracellular pH_2 levels were constant and therefore suggested that H-isotope effects associated with growth phase and catabolic rate likely are more important than pH_2 levels. In addition, Valentine et al. (2004) measured Δ^2H$_{CH_4-H_2O}$ values of ~160‰ associated with

TABLE 6.2

Equations Relating δ^2H Values of Coexisting CH_4 and H_2O Shown in Figure 6.1

No.	Equation	Origin	Source
1	$\delta D\text{-}CH_4 = 1.000\ \delta D\text{-}H_2O - 180(\pm10)\%o$	CO_2 reduction in situ	Whiticar, Faber, & Schoell (1986)
2	$\delta D\text{-}CH_4 = 0.250\ \delta D\text{-}H_2O - 321\%o$	acetate fermentation in situ	Whiticar et al. (1986)
3	$\delta D\text{-}CH_4 = 0.19\ \delta D\text{-}H_2O - 259\%o$	*Methanosaeta thermophila*	Valentine et al. (2004)
4	$\delta D\text{-}CH_4 = 1.55(\pm0.46)\ \delta D\text{-}H_2O - 145(\pm30)\%o$	Alaskan peatland transect	Chanton, Fields, & Hines (2006)
5	$\delta D\text{-}CH_4 = 0.675(\pm0.10)\ \delta D\text{-}H_2O - 284(\pm6)\%o$	global freshwater in situ	Waldron et al. (1999)
6	$\delta D\text{-}CH_4 = 0.437(\pm0.05)\ \delta D\text{-}H_2O - 302(\pm15)\%o$	acetate fermentation in vivo	Sugimoto & Wada (1995)
7	$\delta D\text{-}CH_4 = 0.444(\pm0.03)\ \delta D\text{-}H_2O - 321(\pm4)\%o$	mesophilic incubations in vivo	Waldron et al. (1998)
8	$\delta D\text{-}CH_4 = 0.683(\pm0.02)\ \delta D\text{-}H_2O - 317(\pm20)\%o$	CO_2 reduction in vivo	Sugimoto & Wada (1995)

high temperature CO_2/H_2 methanogenesis, which agree with the more positive δ^2H-CH_4 values typically observed for biogenic CH_4 in deep marine sediments.

In systems where CH_4 production occurs predominantly via acetate fermentation, Whiticar, Faber, and Schoell (1986) predicted that δ^2H values of coexisting CH_4 and H_2O are governed by Equation 2 in Table 6.2, which assumes that three H atoms are inherited from the methyl group of acetate and one is derived from coexisting H_2O. However, pairs of δ^2H values for coexisting CH_4 and H_2O from methanogenic environments dominated by fermentation typically exhibit a slope of ~0.4 (e.g., Schoell 1980; Woltemate, Whiticar, and Schoell 1984). The deviation from an ideal slope of 0.25 has been interpreted to result from an ~25% contribution of CH_4 from the CO_2-reduction pathway, which is a reasonable interpretation given that CO_2 and H_2 are ubiquitous in anaerobic environments and that most methanogens are either obligate or facultative CO_2-reducers.

Equations from subsequent studies that investigated relationships between δ^2H values of CH_4 and H_2O in laboratory incubations (Sugimoto and Wada 1995; Waldron et al. 1998; Valentine et al. 2004) and natural environments (Waldron et al. 1999; Chanton, Fields, and Hines 2006) also are listed in Table 6.2 (Equations 3 through 7). All the δ^2H-CH_4/H_2O equations are shown in Figure 6.1 numbered according to the list in Table 6.2. The majority of equations determined from in situ samples or heterogeneous incubations are similar to the curve for acetate fermentation (Equation 2) originally proposed by Whiticar, Faber, and Schoell (1986). The δ^2H values of ambient H_2O exert an important control on δ^2H-CH_4 values across latitudinal gradients (Equation 5; Waldron et al. 1999). The relationship is impacted, though, by differences in the predominate pathways of methanogenesis, in particular, with depth in peatlands (Hornibrook, Longstaffe, and Fyfe 2000a; Chanton, Fields, and Hines 2006). The δ^2H-CH_4 values in Figure 6.1 for Turnagain Bog (open triangles), Sifton Bog (open diamonds), Point Pelee Marsh (open circles), and Ellergower Moss (open squares) all increased markedly with depth by amounts that cannot be explained solely by changes in pore water δ^2H values. Thus, equations proposed to describe $\Delta^2H_{CH_4-H_2O}$ relationships in freshwater anoxic environments along latitudinal gradients can be used to provide a reasonably accurate estimate of shallow δ^2H-CH_4 values based on the δ^2H values of meteoric water; however, the δ^2H composition of catotelm CH_4 is likely to deviate from the relationship.

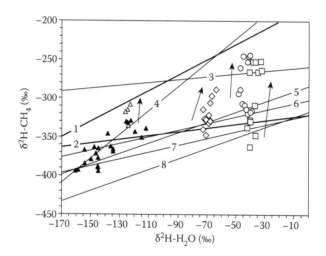

FIGURE 6.1 δ^2H values of coexisting CH_4 and H_2O values from Alaskan peatlands along a N-S transect (▲; Chanton, J. P., D. Fields, and M. E. Hines, *Journal Geophysical Research-Biogeoscience,* 111, G04004, doi:10.1029/2005JG000134, 2006), Turnagain Bog (△; Chanton, J. P., D. Fields, and M. E. Hines, *Journal Geophysical Research-Biogeoscience,* 111, G04004, doi:10.1029/2005JG000134, 2006), Sifton Bog (◇; Hornibrook, E. R. C., F. J. Longstaffe, and W. S. Fyfe, *Isotopes Environmental Health Studies,* 36,151–76, 2000a), Point Pelee Marsh (○; Hornibrook, E. R. C., F. J. Longstaffe, and W. S. Fyfe, *Isotopes Environmental Health Studies,* 36, 151–76, 2000a), and Ellergower Moss (□; Waldron, S., A. J. Hall, and A. E. Fallick. 1999a). The arrows indicate the direction of increasing depth for Turnagain Bog, Sifton Bog, Point Pelee Marsh, and Ellergower Marsh. The two solid lines (labeled 1 and 2) correspond to the first equations describing relationships between δ^2H values of CH_4 and H_2O for CO_2/H_2 and aceticlastic methanogenesis proposed by (Whiticar, M. J., E. Faber, and M. Schoell, *Geochimica Cosmochimica Acta,* 50, 693–709, 1986), that are listed in Table 6.2. The remaining lines are also listed in Table 6.2 and represent subsequently determined equations that attempted to characterize δ^2H values of CH_4 in relation to the H-isotope composition of coexisting water in situ and within laboratory incubations (details provided in Table 6.2).

Within geographically restricted areas, the amount of evaporation can differ significantly between different types of wetlands as demonstrated by the δ^2H-H$_2$O values for the Sifton Bog (*Sphagnum* bog, 43°00'00"N; open diamonds, Figure 6.1) and Point Pelee Marsh (*Typha* marsh, 41°58'00"N; open circles, Figure 6.1) in southwestern Ontario Canada (Hornibrook, Longstaffe, and Fyfe 2000a). Water in the shallow marsh was more ^{2}H-enriched than the bog, and δ^2H-H$_2$O values deviated more strongly from δ^2H values of meteoritic water in the region.

6.4 RADIOCARBON AND METHANE (CH$_4$)

An important constraint in tracing sources of carbon in biogenic and fossil CH$_4$ accumulations has been the use of radiocarbon (expressed in percent Modern Carbon [pMC]). Best estimates suggest that between 15 and 21% of global CH$_4$ emissions are derived from radiocarbon dead sources, such as fossil fuels, organic matter that is > 40 kyr old, and abiogenic (geological) CH$_4$ (Stevens 1988; Wahlen 1994). Measurement of ^{14}CH$_4$ in local environments also provides useful information about carbon sources metabolized by methanogens. For example, degradation of recent photosynthates versus sedimentary or matrix organic matter in peatlands and lakes (Martens et al. 1992; Aravena et al. 1993; Charman, Aravena, and Warner 1994; Chanton et al. 1995; Charman et al. 1999; Dove et al. 1999; Nakagawa, Yoshida, Sugimoto et al. 2002) or methane produced from shallow peat versus deeply buried organic matter, or possibly, deep thermogenic gas in thermokarst areas where significant quantities of CH$_4$ are released from degrading tundra (Zimov et al. 1997; Nakagawa, Yoshida, Nojiri et al. 2002; Walter et al. 2006). The ^{14}C content of CH$_4$ also has been used to evaluate the origin of methane in groundwater (i.e., in situ sources versus migration from bedrock) and in studies dealing with the potential impacts of landfill gases in shallow environments (Coleman, Liu, and Riley 1988; Aravena, Wassenaar, and Barker 1995; Aravena, Wassenaar, and Plummer 1995; Hackley, Liu, and Trainor 1999; Hogg 1997).

6.5 SECONDARY KIEs: METHANE OXIDATION AND TRANSPORT EFFECTS

Primary (i.e., pristine) δ^{13}C- and δ^2H-CH$_4$ values can be altered by secondary processes during transport in groundwater, in particular, at oxic-anoxic or liquid–gas interfaces. Bergamaschi (1997) reported a summary of isotope fractionation factors for partitioning of CH$_4$ isotopologues via bacterial oxidation (Δ^{13}C ~ 3 to 37‰; Δ^2H ~ 24 to 325‰), dissolution in water (Δ^{13}C ~ 0.3 to 0.7‰), and molecular diffusion in air (Δ^{13}C and Δ^2H ~20‰), water (Δ^{13}C ~30‰), and across water–air interfaces (Δ^{13}C ~ 0.5 to 1.3‰). Differences in patterns of covariation of δ^{13}C- and δ^2H-CH$_4$ values provide a means to determine changes in the predominance of methanogenic pathways versus secondary alteration effects. A positive correlation between δ^{13}C- and δ^2H-CH$_4$ values is indicative of secondary alteration because CH$_4$ containing the lighter isotope (^{12}C or ^{1}H) is preferentially consumed in biochemical reactions (e.g., methanotrophy) or transported more quickly by physical processes. A negative correlation between $\delta^{13}C$- and δ^2H-CH_4 values is generally interpreted to result from a change in the relative proportion of CH_4 being produced by acetate fermentation and CO_2 reduction (Schoell 1988).

Use of $\delta^{13}C$ and δ^2H values from residual CH_4 to estimate the degree of oxidation in groundwater systems relies upon an accurate knowledge of the magnitude of isotopic fractionation associated with methanotrophy. Consequently, the large ranges of $\Delta^{13}C$ and Δ^2H values known for this process (e.g., Bergamaschi 1997) results in significant uncertainty in estimates of the extent of CH_4 oxidation in natural systems. Recently, Templeton et al. (2006) suggested that the majority of C-isotope fractionation during methanotrophy is associated with the enzyme methane monooxygenase (MMO) that catalyzes the first oxidation step and that the proportion of CH_4 oxidized is the dominant factor determining the magnitude of ^{13}C fractionation. Using chemostat incubations, they showed that low densities of methanotrophic cells limit the abundance of MMO, which results in a lower proportion of CH_4 being oxidized and consequently larger $\Delta^{13}C$ values. Rayleigh distillation behavior was not observed in their incubations; however, most natural and anthropogenic methanotrophic systems exist in a steady-state condition and thus the densities of methanotroph cells are unlikely to be changing. Rayleigh-type distillation behavior should occur in most cases and use of shifts in $\delta^{13}C$ and δ^2H values to estimate the proportion of CH_4 oxidized remains a valid approach.

6.6   METHANE IN ENVIRONMENTAL SYSTEMS

This section presents a series of case studies from common natural and anthropogenic systems where stable isotopes have been used successfully to study methane dynamics. The list of examples is not exhaustive and is intended to provide an overview of the potential for using $\delta^{13}C$ and δ^2H values of CH_4 to investigate cycling of methane within groundwater, coalbeds, landfills, and wetland systems.

6.6.1   METHANE IN GROUNDWATER

The predominant carbon pools in groundwater are dissolved inorganic carbon (DIC) and dissolved organic carbon (DOC) with CH_4 commonly present as a trace component. However, this simple hydrocarbon can become a major constituent of groundwater and a potentially explosive contaminant when present in high concentrations (Coleman, Liu, and Riley 1988; Hughes, Landon, and Farvolden 1971; Barker and Fritz, 1981a). Methane in groundwater can be associated with in situ production by Archaea or by migration of gases from underlying bedrock reservoirs. Several studies of regional aquifers and aquitards in Canada and the United States have reported significant CH_4 concentrations in these environments associated mainly with microbial degradation of organic matter present in aquifer sediments (Barker and Fritz 1981a; Coleman, Liu, and Riley 1988; Simpkins and Parkin 1993). Methane is also common in aquifers affected by biodegradation of organic contaminants under anaerobic conditions (Herczeg, Richardson, and Dillon 1991).

Isotope approaches to evaluate the origin of CH_4 in regional aquifers typically are based on measurements of ^{13}C, ^{2}H, and ^{14}C content in CH_4, ^{13}C in DIC, and ^{14}C in DOC (Coleman, Liu, and Riley 1988; Aravena, Wassenaar, and Barker 1995; Aravena, Wassenaar, and Plummer 1995; Aravena and Wassenaar 2003). Data reported in these studies clearly showed that CH_4 had an isotope composition typical of biogenic gases and was distinctly different from thermogenic CH_4. For example, Aravena, Wassenaar, and Barker (1995) demonstrated that CH_4 present in a confined sand aquifer was microbial in origin and produced in situ from the degradation of organic matter in the aquifer, thus, excluding a potential thermogenic source present in underlying carbonate rocks (Figure 6.2). In addition, the $\delta^{13}C$- and δ^2H-CH_4 values combined with coexisting δ^2H-H_2O values indicated that CO_2 reduction was the main methanogenic pathway in the aquifers (data not shown). This conclusion also was confirmed by patterns of ^{13}C-enrichment of DIC in both aquifers (Aravena, Wassenaar, and Plummer 1995), a feature that is common in groundwater systems affected by CO_2/H_2 methanogenesis because of preferential uptake of [12]CO_2 by methanogens (Chapelle et al. 1988; Grossman et al. 1989). The ^{14}C content of CH_4 indicated the presence of an in situ carbon source of mid-Wisconsinan age in Alliston (Canada) and Illinois aquifers. The concentration and ^{14}C content of DOC in the Alliston aquifer confirmed the presence of an in situ carbon source (Aravena et al. 1993). These studies demonstrate that the combined use of isotope and hydrogeological tools is a powerful approach for determining the origin of CH_4 in aquifers where gas migration from deep bedrock formations could be occurring.

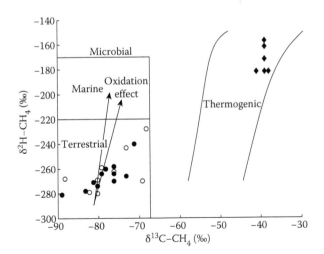

FIGURE 6.2 Relation between δ^2H and $\delta^{13}C$ values in CH_4. (filled circles are overburden wells and the open circle are Bedrock wells) (Modified after Aravena, R., L. Wassenaar, and J. Barker, *Journal of Hydrology,* 173, 51–71, 1995.)

6.6.2 METHANE IN COALBEDS

Most early studies of volatiles in coal basins assumed that coal-seam gases, in particular CH_4, have a thermogenic origin. Thermogenic gases typically are associated with high rank coal whereas biogenic gases are thought to be produced during early stages of the coalification process. Recent studies incorporating isotope measurements and detailed studies of the hydrogeology of coal basins have demonstrated that biogenic methane can be produced in economic quantities at any coal rank (Scott 1993; Scott, Kaiser, and Ayers 1994). Microorganisms are introduced into groundwater systems via recharge pathways and subsequently metabolize C_2 to C_6 (wet) gases, higher alkanes, and other organic components to CH_4 and CO_2.

Several lines of evidence have been used to evaluate the origin of CH_4 in coalbed environments. Thermogenic gases generally contain a relative large amount of C_2 to C_6 hydrocarbons relative to CH_4. The presence of fewer wet gases relative to CH_4 is an indication of a predominance of biogenic CH_4.

Isotope approaches to study the origin of coalbed CH_4, include analysis of CH_4 ($\delta^{13}C$, δ^2H), CO_2 ($\delta^{13}C$), and H_2O (δ^2H). Schoell (1980) proposed that $\delta^{13}C$-CH_4 values less than $-55‰$ indicate a biogenic origin and $\delta^{13}C$ values greater than $-55‰$ are characteristic of thermogenesis. However, $\delta^{13}C$ values greater than $-55‰$ have been reported for biogenic gases (Jenden and Kaplan 1986; Whiticar, Faber, and Schoell 1986), but $\delta^{13}C$ values more negative than $-55‰$ rarely are associated with thermogenic gases (Rice and Claypool 1981; Barker and Fritz 1981a). Rightmire (1984) and Rice (1993) proposed a classification system for evaluation of the origin of coalbed CH_4 similar to the one developed for natural gases (Schoell, 1980). Carbon isotope data collected in several studies in the United States, China, and Canada agree with their approach (Aravena et al. 2003; Minxin et al. 2005; Strąpoć et al. 2007).

The $\delta^{13}C$-CO_2 values resulting from decarboxylation reactions during thermal maturation of coal have been estimated to range from -25 to $-10‰$ (Irwin, Curtis, and Coleman 1977; Chung and Sackett 1979). Therefore, CO_2 released during coalification should be more ^{13}C-depleted than CO_2 associated with methanogenesis. Due to isotope partitioning during the latter process, $\delta^{13}C$ values of CO_2 typically are greater than $0‰$ and attain values as high as $+34‰$ (Scott, Kaiser, and Ayers 1994; Smith and Pallasser 1996; Aravena et al. 2003).

Two recent studies provide good examples of using isotopic approaches to evaluate the origin of coalbed CH_4 in the Elk River Valley, Canada and in the southeastern Illinois Basin in the United States (Aravena et al. 2003; Strąpoć et al. 2007). The Canadian study was part of a general investigation to evaluate the potential environmental impact of coal production in the Elk River Valley watershed (Harrison 1995; Harrison et al. 2000). It reported $\delta^{13}C$-CH_4 values ranging from -65 to $51‰$ with the majority of $\delta^{13}C$ values $< -55‰$. The U.S. study reported $\delta^{13}C$-CH_4 values ranging between -67 to $-57‰$ in Indiana coalbeds and an average $\delta^{13}C$-CH_4 value of $-49 \pm 1.3‰$ for Western Kentucky coalbed gases. The carbon isotope data suggested a predominance of biogenic CH_4 in the Elk River and Indiana coalbeds and thermogenic gases in the Western Kentucky region. A combination of $\delta^{13}C$-CH_4 and δ^2H-CH_4 values clearly separated the two types of gases in the United States and the Canadian studies (Figure 6.3). The δ^2H-CH_4 values measured at the Indiana (-208 to $-195‰$

for CH_4 and -46 to $-39\permil$ for H_2O) and Elk River (-415 to $-288\permil$ for CH_4; -415 to $-288\permil$ for H_2O) sites fit well with the relationship between the δ^2H-H_2O and δ^2H-CH_4 values proposed by Whiticar, Faber, and Schoell (1986) for CO_2 reduction (Figure 6.4), supporting a microbial origin for CH_4 at both sites. The $\delta^{13}C$ values of DIC also support this interpretation because $\delta^{13}C$-CO_2 values at the Indiana site were more enriched (average $+4.3\permil$) than at the Kentucky site (average $-24.3\permil$). The pattern of increasing DIC concentration was accompanied by a trend to more positive $\delta^{13}C$ values (-11.9 to $+34.9\permil$) along the groundwater flow system, which results from preferential uptake of $^{12}CO_2$ by methanogens, and thus further supports a microbial origin for CH_4 in the Elk River Valley. The conclusion based on the isotope data agreed with thermal maturity of the coal (low in Indiana and higher in Kentucky) and gas composition data (dry gases in Indiana and wet gases in Kentucky) in the U.S. study (Strąpoć et al. 2007).

6.6.3 METHANE IN LANDFILLS

Landfilling is the main method for solid waste disposal in North America. Environmental hazards associated with landfills include CH_4 and contaminated liquid production (i.e., leachate) both of which could migrate from the landfill into surrounding environments (Farquhar 1988; Abaci, Edwards, and Whitaker 1993; Robertson, Murphy, and Cherry 1995). Although modern landfills are typically designed with a low permeability liner and leachate collection and gas extraction systems, some degree of contaminant migration from the landfill into nearby soil and groundwater aquifers may still occur. Landfill gas movement through the subsurface and into adjacent buildings can cause serious health and safety hazards from trace levels of carcinogenic gases and from methane's explosion capacity at concentrations of 5–15% vol/vol in air (Metcalfe and Farquhar 1987; Hodgson et al. 1992;

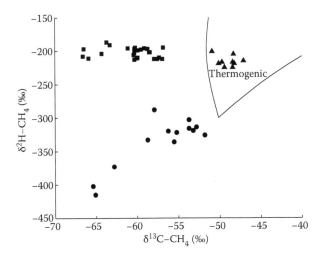

FIGURE 6.3 δ^2H and $\delta^{13}C$ values for coalbed CH_4. (● From Aravena, R., S. M. Harrison, J. F. Barker, H. Abercrombie, and D. Rudolph. 2003; ▲, ■ From Strąpoć, D., M. Mastalerz, C. Eble, and A. Schimmelmann, *Organic Geochemistry*, 38, 267–87, 2007.)

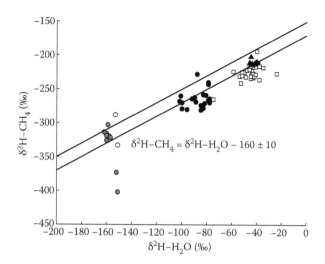

FIGURE 6.4 Relation between δ^2H-CH$_4$ and δ^2H-H$_2$O ($\square$ Strąpoć, D., M. Mastalerz, C. Eble, and A. Schimmelmann, *Organic Geochemistry*, 38, 267–87, 2007; open and filled gray circles; Aravena, R., S. M. Harrison, J. F. Barker, H. Abercrombie, and D. Rudolph, *Chemical Geology*, 195, 219–27, 2003), (•; Aravena, R., L. I. Wassenaar, and J. F. Barker, 173, 51–71, 1995), (filled △, Coleman, D. D., Liu, C. H. L., and Riley, K. M., 71, 23–40, 1988).

Abaci, Edwards, and Whitaker 1993). Although CH$_4$ typically migrates only 10 to 15 meters beyond the landfill periphery before escaping into the atmosphere, it can migrate up to several hundred meters if gas permeability near the ground surface is decreased because of the presence of shallow clay lenses or asphalt surfacing or during periods of frost or meteoric water infiltration (Barber et al. 1990; Williams and Aitkenhead 1991; Hodgson et al. 1992). Also, CH$_4$ in shallow contaminated groundwater plumes extending from landfills can exsolve and accumulate if gas pressure within the unsaturated zone decreases to atmospheric levels.

However, CH$_4$ present in the subsurface near landfills is not necessarily associated with the landfill. Methane can migrate into shallow environments from groundwater where it has been produced in situ by microbial metabolism of organic carbon matter in aquifer sediments (Coleman, Liu, and Riley 1988; Aravena, Wassenaar, and Barker 1995), from bedrock sources probably of thermogenic origin (Barker and Fritz 1981a), or leakage from buried gas pipelines. The last two sources can be identified via analysis of the content of C$_2$ to C$_6$ hydrocarbons. However, analysis of these gases is often insufficient to identify conclusively the CH$_4$ source since the higher homologues can be filtered out or oxidized during migration from depth (Coleman et al. 1978; Kaplan 1994). Thus, isotopes are an important means to access the origin of methane in such environments accurately.

Landfill CH$_4$ is characterized by δ^{13}C and δ^2H values ranging from ~–60 to –48‰ and –350 to –270‰, respectively, which are values typical of CH$_4$ produced by acetate fermentation (Bergmaschi and Harris 1995; Coleman et al. 1995; Hackley, Liu, and Coleman 1996; Desrocher and Sherwood-Lollar 1998; Chanton and Liptay 2000).

In general, landfill CH_4 has an isotopic signature that is distinct from thermogenic and biogenic CH_4 formed via CO_2 reduction in quaternary aquifers ($\delta^{13}C < -60‰$ and $\delta^2H > -250‰$). The other key isotope employed in landfill studies is ^{14}C because municipal waste contains high levels of ^{14}C activity as a result of imprinting from ^{14}C produced by thermonuclear explosions since 1950 (Hackley, Liu, and Coleman 1996; Hackley, Liu, and Trainor 1999). The ^{14}C content of atmospheric CO_2 during 1964–1965 was almost double the pre-1950 level of ~100 pMC (Levin, Munich, and Weiss 1980). Hence CH_4 from domestic landfills is characterized by $\Delta^{14}C$ values > 100 pMC and from industrial landfills by ^{14}C values < 100 pMC, depending upon the proportion of old carbon deposited in the landfill (Figure 6.5). In contrast, thermogenic CH_4 contains a negligible quantity of ^{14}C and microbial CH_4 generated from organic matter deposited in aquifers since the last glaciation, contains significantly less ^{14}C than landfill CH_4 (Coleman, Liu, and Riley 1988; Aravena, Wassenaar, and Barker 1995a; Figure 6.5).

Leachate in modern landfills has been documented to contain tritium (^3H) activity significantly above levels measured in meteoric water (Egboka et al. 1983; Rank et al. 1992; Hackley, Liu, and Coleman 1996). Tritium levels as high as 3900 TU were measured in a domestic landfill in southern Ontario, Canada by Hogg (1997) in contrast to ^3H levels of 20–40 TU in precipitation in the region. The elevated activity of ^3H in landfills is postulated to originate primarily from degradation of discarded consumer products that contain luminescent tritiated paints and gaseous light sources (e.g., watches, instrument dials, etc.). Rank (1993) reported that a single wristwatch containing luminescent paint possesses ~1 billion TU of ^3H. High levels of ^3H in CH_4 can serve as an additional independent tracer for distinguishing gas escaping from landfills when other potential sources of microbial CH_4 are present in an area (Hackley, Liu, and Coleman 1996).

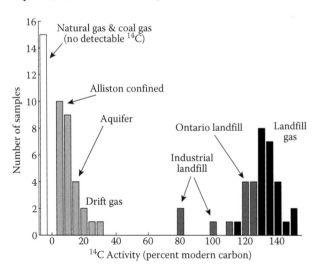

FIGURE 6.5 Activity of ^{14}C in CH_4 from landfills and groundwater. (Modified after Hogg, D, The application of environmental isotopes for evaluating the impact of landfill gas and leachate on shallow groundwater environments. M.Sc. diss., University of Waterloo, Ontario, Canada, 1997.)

Hackley, Liu, and Trainor (1999) conducted a study on the property of a former hazardous waste treatment facility that illustrated the application of ^{13}C, ^{14}C, and ^2H data to determine the source of CH_4 in landfills. The potential sources of CH_4 included a nearby landfill, organic waste from previous impoundments, and natural organic matter in the sediments. The $\delta^{13}C$-CH_4 and δ^2H values ranged, respectively, from −82 to −63‰ and −299 to −272‰, indicating that CH_4 present in deep tills and shallow and intermediate lacustrine deposits was generated from organic matter in the sedimentary deposits. This conclusion was confirmed by ^{14}C data that showed an age trend ranging from 33,000 years BP for the deepest CH_4 to 4,300 years BP for shallow CH_4. No clear evidence of the impact of landfill CH_4 was observed in the study area.

The effect of methanotrophy on the $\delta^{13}C$ and δ^2H values of CH_4 during migration in the unsaturated zone must be considered in landfill studies. Bacterial oxidation of methane enriches residual CH_4 in both ^{13}C and ^2H (Barker and Fritz 1981b; Coleman, Risatti, and Schoell 1981; Whiticar and Faber 1986) with changes in δ^2H values typically being ~8 to 14 times greater than $\delta^{13}C$ values (Coleman, Risatti, and Schoell 1981; Powelson, Charlton, and Abichou 2007). This difference was documented in a study conducted in a landfill constructed in an old quarry located in the Niagara region of Canada (Hogg 1997) where it was necessary to identify the source of CH_4 overburden near the landfill. In addition to the landfill, a second potential source was thermogenesic gas migrating from bedrock. Stable carbon and hydrogen isotope values for CH_4 from the unsaturated zone of the site ranged in composition from that of landfill gas to an apparent thermogenic gas signature (Figure 6.6). The thermogenic endmember composition was obtained from shallow bedrock wells in the area. However, $\Delta^{14}C$ values of > 100 pMC for CH_4 from the unsaturated zone indicated

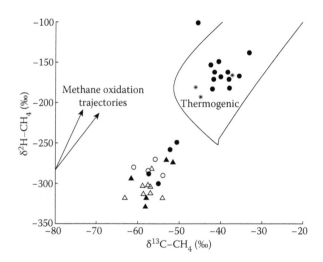

FIGURE 6.6 δ^2H and $\delta^{13}C$ values of CH_4 in landfills and adjacent environments. T bedrock gas; △ landfill unsaturated zone, ▲ landfill saturated zone, ● periphery gas unsaturated zone, ○ periphery gas saturated zone; (Modified after Hogg, D, The application of environmental isotopes for evaluating the impact of landfill gas and leachate on shallow groundwater environments. M.Sc. diss., University of Waterloo, Ontario, Canada, 1997.)

that the ^{13}C- and ^{2}H-enriched compositions, although resembling thermogenic gas, resulted from microbial oxidation of landfill CH_4 during migration (Hogg 1997). Changes in δ^{13}C values caused by CH_4 oxidation also have been applied to quantify the extent of methane oxidation in landfill soil cover (Chanton and Liptay 2000).

6.6.4 METHANE IN WETLANDS

Freshwater wetlands cover ~5.3 million km^2 of the Earth's surface and are the single largest source of CH_4 emissions to the atmosphere each year, contributing ~20% of the total annual global flux or ~140 million metric tons (Mt) of CH_4 per annum (Matthews and Fung 1987). Persistently high water table levels coupled with an abundance of organic matter results in soil anoxia leading to CH_4 production. The stable isotope composition of CH_4 flux from wetlands and other major sources to the atmosphere provides an important constraint in mass balance models of the global CH_4 cycle (Andersen 1996); however, a significant problem remains in the assigning of accurate isotopic values to the different flux sources.

Wetlands display a wide range of δ^{13}C and δ^2H values for both emissions and subsurface CH_4 (Chanton et al. 1989; Kelley, Dise, and Martens 1992; Hornibrook, Longstaffe, and Fyfe 1997, 2000a; Chanton et al. 2004). Considerable effort has been expended in attempting to understand the causes of this variability and, in particular, how systematic patterns in the distribution of δ^{13}C-CH_4 and δ^2H-CH_4 values might be used to understand details of terminal carbon mineralization, gas transport, and carbon recycling in these globally significant short-term carbon pools (e.g., Chanton et al. 1989, 1992; 1995, 2002; Chanton and Dacey 1991; Kelley, Dise, and Martens 1992; Lansdown, Quay, and King 1992; Martens et al. 1992; Chanton and Smith 1993; Aravena et al. 1993; Happell et al. 1993; Happell, Chanton, and Showers 1994; Shannon and White, 1994; Avery et al. 1999; Bellisario et al. 1999; Charman et al. 1999; Hornibrook, Longstaffe, and Fyfe 1997, 2000a, 2000b; Popp et al. 1999; Waldron, Hall, and Fallick 1999a; Chasar et al. 2000a, 2000b; Nakagawa, Yoshida, Sugimoto et al. 2002a; Nakagawa, Yoshida, Nojiri et al. 2002b; Chanton 2005; Sugimoto and Fujita 2006; Bowes and Hornibrook 2006; Hornibrook and Bowes 2007; Prater, Chanton, and Whiting 2007).

Variations in the stable isotope values of primary CH_4 typically reflect differences in methanogenic pathways and the stable isotope composition of substrates (e.g., Chanton et al. 1989, 1995; Hornibrook, Longstaffe, and Fyfe 1997). Variations in the former are most pronounced with depth in peatlands where a gradual transition commonly occurs from ^{13}C-enriched and ^{2}H-depleted CH_4 in shallow peat layers. This is indicative of a predominance of acetate fermentation, to progressively more negative δ^{13}C and more positive δ^2H values in deeper layers, which suggests a shift to CO_2/H_2 methanogenesis (e.g., Hornibrook, Longstaffe, and Fyfe 1997, 2000a, 2000b; Popp et al. 1999; Chasar et al. 2000a, 2000b; Chanton et al. 2004). Changes with depth in δ^{13}C values of coexisting CO_2 and CH_4 are also consistent with such changes in terminal carbon mineralization (Figure 6.7).

Differences exist, though, between wetland types. Minerotrophic wetlands (e.g., fens) commonly exhibit the gradual shift from acetate fermentation to CO_2 reduction. In such wetlands, the δ^{13}C values of CH_4 in the acrotelm tend to be more positive reflecting the δ^{13}C composition of the methyl moiety in heterotrophically produced

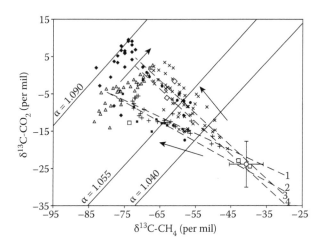

FIGURE 6.7 $\delta^{13}C$ values of dissolved CO_2 and CH_4 in pore waters at Ellergower Moss (♦; Waldron, S., A. J. Hall, and A. E. Fallick, *Global Biogeochemical Cycles*, 13, 93–100, 1999a), Kings Lake Bog (△; Lansdown, J. M., P. D. Quay, and S. L. King, *Geochimica Cosmochimica Acta*, 56, 3493–3503, 1992), Rainy River Peatland (◇; Aravena, R., B. G. Warner, D. J. Charman, L. R. Belyea, S. P. Mathur, and H. Dinel, *Radiocarbon*, 35, 271–76, 1993), Sifton Bog (×; Hornibrook, E. R. C., F. J. Longstaffe, and W. S. Fyfe, *Geochimica Cosmochimica Acta*, 64, 1013–27, 2000b), Point Pelee Marsh (+; Hornibrook, E. R. C., F. J. Longstaffe, and W. S. Fyfe, *Geochimica Cosmochimica Acta*, 64, 1013–27, 2000b) and Minnesota peat bogs 3850 (●) and S4 (■) (Lansdown, J. M., P. D. Quay, and S. L. King, *Geochimica Cosmochimica Acta*, 56, 3493–3503, 1992). Arrows indicate the direction of increasing sample depth. The numbered dashed lines are linear regression curves for the (1) Point Pelee Marsh, (2) bog S4, (3) Sifton Bog, and (4) bog 3850. The open circle (○) with x- and y-error bars is the average value of the intersections of curves 1 to 4. The two open squares (□) are the $\delta^{13}C$ compositions of methyl and carboxyl ends of unaltered acetic acid produced by heterotrophic anaerobic degradation of C_3 vegetation (Sugimoto, A., and E. Wada, *Geochimica Cosmochimica Acta*, 57, 4015–27, 1993). Lines of constant α_C values delineate fields where $\delta^{13}C$-CO_2 and $\delta^{13}C$-CH_4 pairs have been suggested by (Whiticar, M. J., E. Faber, and M. Schoell, *Geochimica Cosmochimica Acta*, 50, 693–709, 1986) to represent CH_4 production by acetate fermentation (~1.040 to 1.055) and CO_2-reduction (~1.055 to 1.090). (Diagram modified after Hornibrook, E. R. C., F. J. Longstaffe, and W. S. Fyfe, *Geochimica Cosmochimica Acta*, 64, 1013–27, 2000b.)

acetic acid (Figure 6.7; Hornibrook, Longstaffe, and Fyfe 2000b). At greater depth, the amount of carbon isotope separation (Δ_C) between CO_2 and CH_4 increases gradually reaching values between 1.060 to 1.080, which is the magnitude of carbon isotope partitioning typically associated with CO_2/H_2 methanogenesis. At shallow depth in minerotrophic wetlands the $\delta^{13}C$ values of CH_4 are largely controlled by the $\delta^{13}C$ composition of organic compounds while at greater depth they reflect KIEs associated with methanogenesis from CO_2. In some ombrogenous (rain fed) mires, acetate fermentation is absent and $\delta^{13}C$-CH_4 values are highly negative in shallow pore water (Hornibrook and Bowes 2007). The trend with depth of $\delta^{13}C$-CH_4 values differs from minerotrophic mires in that it becomes increasingly more positive with an approximately constant Δ_C

value relative to dissolved CO_2 (Figure 6.7), which indicates a predominance of CO_2/H_2 methanogenesis at all depths, something that was confirmed independently using ^{14}C tracers for Kings Lake Bog, Washington State (Lansdown, Quay, and King 1992).

The change in methanogenic pathway with depth appears to result from higher temperatures and a greater abundance of labile substrates in shallow peat layers, the latter most likely derived from root exudates, and a general scarcity of easily degradable organic compounds at depth under cooler conditions. Shifts in $\delta^{13}C$- and δ^2H-CH_4 values related to changes in methanogenic pathway associated with temperature variations or the abundance of labile organic matter also have been reported seasonally (Avery et al. 1999), diurnally (Jedrysek 1995, 1999), and between different wetland types (Nakagawa, Yoshida, Nojiri et al. 2002). Anaerobic decay under substrate limited conditions has been suggested to favor CO_2 reduction in wetlands similar to marine sediments where consumption of labile organic material in the sulfate reduction zone apparently results in a predominance of CH_4 production by CO_2-reduction at depth (Whiticar, Faber, and Schoell 1986).

Miyajima et al. (1997) explained the shift in methanogenic pathways associated with increased recalcitrance of organic matter in a tropical swamp using the metabolic control theory of Westerhoff, Groen, and Wanders (1984). They suggested that a change in the rate limiting step in the anaerobic chain of decay from methanogenic consumption of H_2 under conditions where labile substrates are abundant to one limited by hydrolysis of recalcitrant organic compounds is responsible for the shift from acetate fermentation to CO_2 reduction. A greater prevalence of acetate fermentation was suggested to occur where labile substrates are abundant because of preferential degradation by heterotrophic acetogens of low molecular weight VFAs that produce a higher ratio of acetate/H_2 in order to lower H_2 levels, which, in turn, makes their metabolism less endergonic (Miyajima et al. 1997).

However, organic matter quality does not appear to be the only factor determining methanogenic pathways in wetlands. Acetic acid can be present in abundance at depth in peatlands (e.g., Shannon and White 1994) and, in some cases, for reasons that are unclear, it accumulates during summer months in shallow peat horizons while active methanogenesis is occurring. Hines, Duddleston, and Kiene (2001) and Hines et al. (2008) designated this process acetate-decoupling (i.e., decoupling from terminal carbon mineralization) and it may be responsible for reports of nearly complete predominance of CO_2/H_2 methanogenesis in some peatlands (e.g., Kings Lake Bog; Lansdown, Quay, and King 1992). It may also be responsible for the production of highly ^{13}C-depleted CH_4 in ombrogenous bogs that can result in CH_4 flux having usually negative $\delta^{13}C$ values (Bowes and Hornibrook 2006; Hornibrook and Bowes 2007). The mechanism responsible for the inability of methanogens to utilize acetate in such environments has not been identified; however, Lansdown, Quay, and King (1992) suggested that limited dissociation of acetate under acidic conditions may either inhibit or preclude its use by aceticlastic methanogens as described by Fukuzaki, Nishio, and Nagai (1990). It is also possible that acetate-utilizing methanogens simply do not exist in peatlands. Acid tolerant methanogens isolated from peatlands typically utilize the CO_2/H_2 pathway (e.g., Brauer et al. 2006; Rooney-Varga et al. 2007) and decreasing the pH of anoxic peat incubations has been shown to increase the prevalence of CO_2/H_2 methanogenesis and the production of highly ^{13}C-depleted CH_4 (Kotsyurbenko et al. 2007).

Bellasario et al. (1999) reported that minerotrophic wetlands, which have both higher net ecosystem productivity (NEC) and higher rates of CH_4 flux, on average contain pore water CH_4 that has more positive $\delta^{13}C$ values than ombrogenous bogs. Methane from minerotrophic wetlands generally contained a greater abundance of ^{14}C (i.e., $\Delta^{14}C > 120$ pMC) and CH_4 flux rates correlated well with the quantity of aboveground sedge biomass. These characteristics suggest that vascular vegetation plays a significant role in CH_4 production via exudation of labile photosynthate in the rhizosphere, leading to a greater predominance of aceticlastic methanogenesis. The vascular plants also enhance CH_4 flux rates by providing a conduit to the atmosphere that bypasses the oxic-anoxic interface in pore water (Armstrong 1978).

Changes in the stable carbon isotope composition of methane precursors does not appear to influence $\delta^{13}C$-CH_4 values significantly in northern wetlands; however, C_4 plant species can contribute substantial amounts of biomass in tropical and sub-tropical wetlands. For example, floating mats of C_4 grass (Capims) proliferate in the Amazon Basin floodplain during the wet season, producing copious amounts of ^{13}C-enriched CH_4 (Chanton et al. 1989). A significant proportion of the CH_4 is released to the atmosphere via ebullition resulting in pristine subsurface $\delta^{13}C$-CH_4 values being largely retained in emissions (Wassmann et al. 1992). Stevens and Rust (1982) suggested that tropical papyrus marshes, which are dominated by C_4 sedges, could be an important source of ^{13}C-enriched CH_4 to the atmosphere. However, to date only one study has reported $\delta^{13}C$-CH_4 values from papyrus wetlands (Tyler et al. 1988), and those results were inconclusive because an absence of $\delta^{2}H$-CH_4 data prevented unusually positive $\delta^{13}C$-CH_4 values from being interpreted unambiguously as resulting from the influence of C_4 substrates or the effects of methanotrophy on primary $^{13}C/^{12}C$ ratios. Additional stable isotope data for CH_4 emissions from tropical wetlands are necessary for more rigorous interpretation of $\delta^{13}C$-CH_4 values measured from CH_4 in air trapped in ice cores spanning the last glacial-interglacial transition when low latitude wetlands presumably contributed significantly to the global CH_4 flux signal (Schaefer et al. 2006). Data to date suggest that CH_4 emissions from low latitude wetlands are more ^{13}C-enriched than their northern counterparts (e.g., Chanton et al. 1989; Nakagawa, Yoshida, Sugimoto et al. 2002). In addition to the likely influence of ^{13}C-enriched C_4 biomass, the difference probably also arises from higher NEC and temperature producing an abundance of labile substrates, which contribute to a greater prevalence of aceticlastic methanogenesis.

The isotopic composition of precursor compounds also affects $\delta^{2}H$-CH_4 values via the influence of the $^{2}H/^{1}H$ ratio in H_2O (Waldron et al. 1999). Meteoric H_2O at low latitudes typically is ^{2}H-enriched because of higher temperatures and limited Rayleigh distillation effects on moisture air masses that are derived largely from seawater evaporation in the tropics (Gat 1996). Consequently, $\delta^{2}H$-CH_4 values generally are more positive in tropical wetlands despite the greater prevalence of aceticlastic methanogenesis. The $\delta^{2}H$ values of meteoric H_2O decrease by ~120‰ from the tropics to high latitude regions and this change is reflected in H-isotope composition of CH_4 in far northern wetlands that commonly have $\delta^{2}H$ values $< -400‰$ (Waldron et al. 1999; Chanton, Fields, and Hines 2006).

Secondary KIEs associated with gas transport mechanisms can significantly influence $\delta^{13}C$- and $\delta^{2}H$-CH_4 values in wetlands. Methane is released from wetlands to the atmosphere via three main routes: (i) diffusion across the water–air interface, (ii) plant-mediated transport by passive diffusion and effusion, or active ventilation of submerged tissues, and (iii) ebullition of exsolved gas bubbles (Chanton 2005). Rapid transport mechanisms, such as ebullition, are nonfractionating with respect to stable isotopes (e.g., Chanton and Martens 1988), hence pristine $\delta^{13}C$ values in sedimentary CH_4 typically are preserved in CH_4 flux to the atmosphere. In contrast, slow upward diffusion of CH_4 through wetland soil generally exposes the CH_4 to methanotrophs at the oxic-anoxic interface, which can result in a significant fraction of the CH_4 being consumed before it reaches the atmosphere. Under such conditions the residual CH_4 normally will become both ^{13}C- and ^2H-enriched (e.g., Happell, Chanton, and Showers 1994). The presence of vascular flora usually enhances rates of CH_4 emission from wetland soils but effects on sedimentary $\delta^{13}C$- and δ^2H-CH_4 values will depend upon whether the plants actively ventilate their root systems or conduct CH_4 by passive diffusion through aerenchymatous tissue (Chanton and Dacey 1991; Chanton & Whiting 1996). The latter can result in carbon isotope fractionation as great as ~20‰ in contrast to CH_4 transport by active ventilation that is largely nonfractionating (Chanton 2005). This secondary KIE when coupled with a predominance of CO_2/H_2 methanogenesis in ombrogenous peatlands produces CH_4 flux to the atmosphere having highly negative $\delta^{13}C$ values (< −85‰; Bowes and Hornibrook 2006), which potentially can be used to distinguish isotopically the CH_4 emission signal from northern bogs and fens (Chanton et al. 1989; Kelley, Dise, and Martens 1992; Bellasario et al. 1999; Hornibrook, Longstaffe, and Fyfe 2000b; Hornibrook and Bowes 2007).

6.7 SUMMARY

Stable isotopes provide an unobtrusive means to investigate physical, chemical, and biological processes associated with methane cycling in groundwater environments. Mass dependent fractionation of ^{13}C versus ^{12}C and ^2H versus ^1H at natural abundance levels can yield information about terminal carbon mineralization pathways producing CH_4, the degree to which pristine CH_4 has been oxidized by methanotrophic bacteria, the physical mechanisms by which CH_4 is transported, and the level of mixing, if any, with deeply sourced thermogenic natural gas. Natural abundance level stable isotope data for CH_4 can be interpreted more conclusively in many cases when complemented by ^{14}C data, in particular, instances when the provenance of CH_4 is of interest.

The recent development of new techniques for rapid and accurate measurement of stable isotope ratios in CH_4 at a range of concentrations and sample sizes has made it possible for isotope specialists and nonspecialists to routinely use stable isotopes as a tool to study methane cycling. Such techniques currently enable the investigation of methanogenesis, methanotrophy, and methane transport at scales ranging from microbial niches to basin scale groundwater systems.

ACKNOWLEDGMENT

We thank Jeffrey Chanton (Florida State University) and Keith Hackley (Illinois State Geological Survey) for their insightful reviews that helped to improve this manuscript.

REFERENCES

Abaci, S., J. S. Edwards, and B. N. Whitaker. 1993. Migration of methane and leachate from landfill. In *Environmental management, geo-water & engineering aspects*, eds. R. N. Chowdhury and S. M. Sivakumar, 347–50. Rotterdam: Balkema.

Abrajano, T. A., N. C. Sturchio, J. K. Bohlke, G. L. Lyon, R. J. Poreda, and C. M. Stevens. 1988. Methane-hydrogen gas seeps, Zambales Ophiolite, Philippines: Deep or shallow origin. *Chemical Geology* 71:211–22.

Abrajano, T. A., N. C. Sturchio, B. M. Kennedy, G. L. Lyon, K. Muehlenbachs, and J. K. Bohlke. 1990. Geochemistry of reduced gas related to serpentinization of the Zambales ophiolite, Philippines. *Applied Geochemistry* 5:625–30.

Alperin, M. J., N. E. Blair, D. B. Albert, T. M. Hoehler, and C. S. Martens. 1992. Factors that control the stable carbon isotopic composition of methane produced in an anoxic marine sediment. *Global Biogeochemical Cycles* 6:271–91.

Andersen, B. L. 1996. Modeling isotopic fractionation in systems with multiple sources and sinks with application to atmospheric CH_4. *Global Biogeochemical Cycles* 10:191–96.

Aravena, R., S. M. Harrison, J. F. Barker, H. Abercrombie, and D. Rudolph. 2003. Origin of methane in the Elk Valley coalfield, southeastern British Columbia, Canada. *Chemical Geology* 195:219–27.

Aravena, R., B. G. Warner, D. J. Charman, L. R. Belyea, S. P. Mathur, and H. Dinel. 1993. Carbon isotopic composition of deep carbon gases in an ombrogenous peatland, north-western Ontario, Canada. *Radiocarbon* 35:271–76.

Aravena, R., and L. I. Wassenaar. 2003. Dissolved organic carbon and methane in a regional confined aquifer: Evidence for associated subsurface sources. *Applied Geochemistry* 8:483–93.

Aravena, R., L. Wassenaar, and J. Barker. 1995. Distribution and isotopic characterization of methane in a confined aquifer in southern Ontario, Canada. *Journal of Hydrology* 173:51–71.

Aravena, R., L. Wassenaar, and L. Plummer. 1995. Estimating ^{14}C groundwater ages in a methanogenic aquifer. *Water Resources Research* 31:2307–17.

Armstrong, W. 1978. Root aeration in the wetland condition. In *Plant life in anaerobic environments*, eds. D. D. Hook and R. R. M. Crawford, 269–97. Ann Arbor, MI: Ann Arbor Science Publishers.

Avery, G. B., and C. S. Martens. 1999. Controls on the stable carbon isotopic composition of biogenic methane produced in a tidal freshwater estuarine sediment. *Geochimica Cosmochimica. Acta* 63:1075–82.

Avery, G. B., R. D. Shannon, J. R. White, C. S. Martens, and M. J. Alperin. 1999. Effect of seasonal changes in the pathways of methanogenesis on the $\delta^{13}C$ values of pore water methane in a Michigan peatland. *Global Biogeochemical Cycles* 13:475–84.

Balabane, M., E. Galimov, M. Hermann, and R. Létolle. 1987. Hydrogen and carbon isotope fractionation during experimental production of bacterial methane. *Organic Geochemistry* 11:115–19.

Barber, C., G. B. Davis, D. Briegel, and J. K. Ward. 1990. Factors controlling the concentration of methane and other volatiles in groundwater and soil-gas around a waste site. *Journal of Contaminant Hydrology* 5:155–69.

Barker, H. A. 1936. On the biochemistry of the methane fermentation. *Arch. Mikrobiol.* 7:420–38.

Barker, J. F., and P. Fritz. 1981a. The occurrence and origin of methane in some groundwater flow system. *Canadian Journal of Earth Science* 18:1802–16.

Barker, J. F., and P. Fritz. 1981b. Carbon isotope fractionation during microbial methane oxidation. *Nature* 293:289–91.

Bartlett, K. B., and R. C. Harriss. 1993. Review and assessment of methane emissions from wetlands. *Chemosphere* 26:261–320.

Bellisario, L. M., J. L. Bubier, T. R. Moore, and J. P. Chanton. 1999. Controls on CH_4 emissions from a northern peatland. *Global Biogeochemical Cycles* 13:81–91.

Belyaev, S. S., K. S. Laurinavichus, and V. I. Gaytan. 1977. Modern microbiological formation of methane in the Quaternary and Pliocene rocks of the Caspian Sea. *Geochemistry International* 4:172–76.

Bergamaschi, P. 1997. Seasonal variations of stable hydrogen and carbon isotope ratios in methane from a Chinese rice paddy. *Journal Geophysical Research-Atmospheres* 102:25383–93.

Bergamaschi, P., and G. W. Harris. 1995. Measurements of stable isotope ratios ($^{13}CH_4/^{12}CH_4$; $^{12}CH_3D/^{12}CH_4$) in landfill methane using a tunable diode laser absorption spectrometer. *Global Biogeochemical Cycles* 9:439–47.

Berndt, M. E., D. E. Allen, and W. E. Seyfried. 1996. Reduction of CO_2 during serpentinization of olivine at 300°C and 500 bars. *Geology* 24:351–54.

Bingemer, H. G., and P. J. Crutzen. 1987. The production of methane from solid wastes. *Journal Geophysical Research* 92:2181–87.

Blair, N. E., S. E. Boehme, and W. D. Carter. 1993. The carbon isotope biogeochemistry of methane production in anoxic sediments: 1. Field observations. In *Biogeochemistry of global change: Radiatively active trace gases*, ed. R. S. Oremland, 574–93. New York: Chapman & Hall.

Blair, N. E., and W. D. J. Carter. 1992. The carbon isotope biogeochemistry of acetate from a methanogenic marine sediment. *Geochimica Cosmochimica Acta* 56:1247–58.

Boehme, S. E., N. E. Blair, J. P. Chanton, and C. S. Martens. 1996. A mass balance of ^{13}C and ^{12}C in an organic-rich methane-producing marine sediment. *Geochimica Cosmochimica Acta* 60:3835–48.

Boone, D. R., W. B. Whitman, and P. Rouvière. 1993. Diversity and taxonomy of methanogens. In *Methanogenesis: Ecology, physiology, biochemistry and genetics*, ed. J. G. Ferry, 35–82. New York: Chapman & Hall.

Botz, R., H. D. Pokojski, M. Schmitt, and M. Thomm. 1996. Carbon isotope fractionation during bacterial methanogenesis by CO_2 reduction. *Organic Geochemistry* 25:255–62.

Bowes, H. L., and E. R. C. Hornibrook. 2006. Emission of highly ^{13}C-depleted methane from an upland blanket mire. *Geophysical Research Letters* 33:L04401, doi:10.1029/2005GL025209.

Brauer, S. L., E. Yashiro, N. G. Ueno, J. B. Yavitt, and S. H. Zinder. 2006. Characterization of acid tolerant H_2/CO_2-utilizing methanogenic enrichment cultures from an acidic peat bog in New York State. *FEMS Microbiology Ecology* 57:206–16.

Burke, R. A. J. 1993. Possible influence of hydrogen concentration on microbial methane stable hydrogen isotopic composition. *Chemosphere* 26:55–67.

Burke, R. A. J., T. R. Barber, and W. M. Sackett. 1988. Methane flux and stable hydrogen and carbon isotope composition of sedimentary methane from the Florida Everglades. *Global Biogeochemical Cycles* 2:329–40.

Burke, R. A. J., T. R. Barber, and W. M. Sackett. 1992. Seasonal variations of stable hydrogen and carbon isotope ratios of methane in subtropical freshwater sediments. *Global Biogeochemical Cycles* 6:125–38.

Burke, R. A. J., C. S. Martens, and W. M. Sackett. 1988. Seasonal variations of D/H and $^{13}C/^{12}C$ ratios of microbial methane in surface sediments. *Nature* 332:829–31.

Burke, R. A. J., and W. M. Sackett. 1986. Stable hydrogen and carbon isotopic compositions of biogenic methanes from several shallow aquatic environments. In *Organic marine geochemistry*, ed. M. L. Sohn, 297–313. Washington, DC: American Chemical Society.

Buswell, A. M., and F. W. Solo. 1948. The mechanism of the methane fermentation. *Journal of the American Chemical Society* 70:1179–80.

Chanton, J. P. 2005. The effect of gas transport on the isotope signature of methane in wetlands. *Organic Geochemistry* 36:753–68.

Chanton, J. P., and J. W. H. Dacey. 1991. Effects of vegetation on methane flux, reservoirs, and carbon isotopic composition. In *Trace gas emissions from plants*, eds. T. D. Sharkey, E. A. Holland, and H. A. Mooney, 65–92. San Diego, CA: Academic.

Chanton, J., and K. Liptay. 2000. Seasonal variation in methane oxidation in a landfill cover soil as determined by an in situ stable isotope technique. *Global Biogeochemical Cycles* 14:51–60.

Chanton, J. P., and C. S. Martens. 1988. Seasonal variations in ebullitive flux and carbon isotopic composition of methane in a tidal freshwater estuary. *Global Biogeochemical Cycles* 2:289–98.

Chanton, J. P., and L. K. Smith. 1993. Seasonal variations in the isotopic composition of methane associated with aquatic macrophytes. In *Biogeochemistry of global change: Radiatively active trace gases*, ed. R. S. Oremland, 619–32. New York: Chapman & Hall.

Chanton, J. P., and G. J. Whiting. 1996. Methane stable isotopic distributions as indicators of gas transport mechanisms in emergent aquatic plants. *Aquatic Botany* 54:227–36.

Chanton, J. P., T. J. Arkebauer, H. S. Harden, and S. B. Verma. 2002. Diel variation in lacunal CH_4 and CO_2 concentration and $\delta^{13}C$ in *Phragmites australis*. *Biogeochemistry* 59:287–301.

Chanton, J. P., J. E. Bauer, P. A. Glaser, D. I. Siegel, C. A. Kelley, S. C. Tyler, E. H. Romanowicz, and A. Lazrus. 1995. Radiocarbon evidence for the substrates supporting methane formation within northern Minnesota peatlands. *Geochimica Cosmochimica Acta* 59:3663–68.

Chanton, J. P., L. I. Chasar, P. Glaser, and D. A. Siegel. 2004. Carbon and hydrogen isotope effects in microbial methane from terrestrial environments. In *Stable isotopes and biosphere-atmosphere interactions: Processes and biological controls*, eds. L. B. Flanagan, J. R. Ehleringer, and D. E. Pataki, 85–105. London: Elsevier Academic Press.

Chanton, J. P., P. M. Crill, K. B. Bartlett, and C. S. Martens. 1989. Amazon Capims (floating grassmats): A source of ^{13}C enriched methane to the troposphere. *Geophysical Research Letters* 16:799–802.

Chanton, J. P., D. Fields, and M. E. Hines. 2006. Controls on the hydrogen isotopic composition of biogenic methane from high-latitude terrestrial wetlands. *Journal Geophysical Research-Biogeoscience* 111:G04004, doi:10.1029/2005JG000134.

Chanton, J. P., G. J. Whiting, W. J. Showers, and P. M. Crill. 1992. Methane flux from *Peltandra virginica*: Stable isotope tracing and chamber effects. *Global Biogeochemical Cycles* 6:15–31.

Chapelle, F. H., J. T. Morris, P. B. McMahon, and J. L. Zelibor, Jr. 1988. Bacterial metabolism and the $\delta^{13}C$ composition of ground water, Floridian aquifer system, South Carolina. *Geology* 16:117–21.

Charlou, J. L., Y. Fouquet, H. Bougault, J. P. Donval, J. Etoubleau, P. Jean-Baptiste, A. Dapoigny, P. Appriou, and P. A. Rona. 1998. Intense CH_4 plumes generated by serpentinization of ultramafic rocks at the intersection of the 15°20′N fracture zone and the Mid-Atlantic Ridge. *Geochimica Cosmochimica Acta* 62:2323–33.

Charman, D. J., R. Aravena, C. L. Bryant, and D. D. Harkness. 1999. Carbon isotopes in peat, DOC, CO_2, and CH_4 in a holocene peatland on Dartmoor, Southwest England. *Geology* 27:539–42.

Charman, D. J., R. Aravena, and B. G. Warner. 1994. Carbon dynamics in a forested peatland in north-eastern Ontario, Canada. *Journal of Ecology* 82:55–62.

Chasar, L. S., J. P. Chanton, P. H. Glaser, and D. I. Siegel. 2000a. Methane concentration and stable isotope distribution as evidence of rhizospheric processes: Comparison of a fen and bog in the Glacial Lake Agassiz Peatland complex. *Annals of Botany-London* 86:655–63.

Chasar, L. S., J. P. Chanton, P. H. Glaser, D. I. Siegel, and J. S. Rivers. 2000b. Radiocarbon and stable carbon isotopic evidence for transport and transformation of dissolved organic carbon, dissolved inorganic carbon, and CH_4 in a northern Minnesota peatland. *Global Biogeochemistry Cycles* 14:1095–108.

Chung, H. M., and W. M. Sackett. 1979. Use of stable carbon isotope compositions of pyrolytically derived methane as maturity indices for carbonaceous materials. *Geochimica Cosmochimica Acta* 43:1979–88.

Cicerone, R. J., and R. S. Oremland. 1988. Biogeochemical aspects of atmospheric methane. *Global Biogeochemical Cycles* 2:299–327.

Claypool, G. E., C. N. Threlkeld, P. N. Mankiewicz, M. A. Arthur, and T. F. Anderson. 1985. Isotopic composition of interstitial fluids and origin of methane in slope sediment of the middle American trench, deep sea drilling project leg 84. In *Initial reports of the deep sea drilling project,* ed. Susan O, 683–691. Washington, DC: U.S. Government Printing Office.

Coleman, D. D., C. L. Liu, K. C. Hackley, and S. R. Pelphrey. 1995. Isotopic identification of landfill methane. *Environmental Geoscience* 2:95–103.

Coleman, D. D., C. L. Liu, and K. M. Riley. 1988. Microbial methane in the shallow Paleozoic sediments and glacial deposits of Illinois, U.S.A. *Chemical Geology* 71:23–40.

Coleman, D. D., W. F. Meents, C. L. Liu, and R. A. Keogh. 1978. Isotopic identification of leakage gas from underground storage reservoirs: A progress report. *Illinois State Geological Survey, Illinois Petroleum* 111:1–10.

Coleman, D. D., J. B. Risatti, and M. Schoell. 1981. Fractionation of carbon and hydrogen isotopes by methane-oxidizing bacteria. *Geochimica Cosmochimica Acta* 45:1033–37.

Conrad, R. 2005. Quantification of methanogenic pathways using stable carbon isotopic signatures: A review and a proposal. *Organic Geochemistry* 36:739–52.

Conrad, R., and B. Wetter. 1990. Influence of temperature on energetics of hydrogen metabolism in homoacetogenic, methanogenic, and other anaerobic bacteria. *Archives of Microbiology* 155:94–98.

Daniels, L., G. Fulton, R. W. Spencer, and W. H. Ormejohnson. 1980. Origin of hydrogen in methane produced by *Methanobacterium thermoautotrophicum. Journal Bacteriology* 141:694–98.

Desrocher, S., and B. Sherwood-Lollar. 1998. Isotopic constraints on off-site migration of landfill CH_4. *Ground Water* 36:801–9.

Deuser, W. G., E. T. Degens, G. R. Harvey, and M. Rubin. 1973. Methane in Lake Kivu: New data on its origin. *Science* 181:51–54.

Dolfing, J. 1988. Acetogenesis. In *Biology of anaerobic microorganisms,* ed. A. J. B. Zehnder, 417–68. New York: John Wiley & Sons.

Dolfing, J. 2001. The microbial logic behind the prevalence of incomplete oxidation of organic compounds by acetogenic bacteria in methanogenic environments. *Microbial Ecology* 41:83–89.

Dove, A., N. Roulet, P. Crill, J. Chanton, and R. Bourbonniere. 1999. Methane dynamics of a northern boreal beaver pond. *Ecoscience* 6:577–86.

Egboka, B. C. E., J. A. Cherry, R. N. Farvolden, and E. O. Frind. 1983. Migration of contaminants in groundwater at a landfill: A case study. 3. Tritium as an indicator of dispersion and recharge. *Journal of Hydrology* 63:51–80.

Etheridge, D. M., L. P. Steele, R. J. Francey, and R. L. Langenfelds. 1998. Atmospheric methane between 1000 AD and present: Evidence of anthropogenic emissions and climatic variability. *Journal Geophysical Research-Atmosphere* 103:15979–93.

Faber, E., W. J. Stahl, and M. J. Whiticar. 1990. Carbon and hydrogen isotope variations in marine sediment gases. In *Facets of modern biogeochemistry,* eds. V. Ittekkot, S. Kempe, W. Michaelis, and A. Spitzy, 205–13. New York: Springer-Verlag.

Farquhar, G. J. 1988. Leachate: Production and characterization. *Canadian Journal of Civil Engineers* 16:317–25.

Ferry, J. G. 1993. Fermentation of acetate. In *Methanogenesis: Ecology, physiology, biochemistry and genetics*, ed. J. G. Ferry, 304–34. New York: Chapman & Hall.

Ferry , J. G. 1999. Enzymology of one-carbon metabolism in methanogenic pathways. *FEMS Microbiology Review* 23:13–38.

Fritz, P., I. D. Clark, J. C. Fontes, M. J. Whiticar, and E. Faber. 1992. Deuterium and ^{13}C evidence for low temperature production of hydrogen and methane in highly alkaline groundwater environment in Oman. In *Water rock interaction*, Vol. 1 of *Low temperature environments*, eds. Y. K. Kharaka and A. S. Maest, 793–96. Rotterdam: Balkema.

Fuchs, G. D., R. Thauer, H. Ziegler, and W. Stichler. 1979. Carbon isotope fractionation by *Methanobacterium thermoautotrophicum*. *Archives Microbiology* 120:135–39.

Fukuzaki, S., N. Nishio, and S. Nagai. 1990. Kinetics of the methanogenic fermentation of acetate. *Applied Environmental Microbiology* 56:3158–63.

Galand, P. E., H. Fritze, R. Conrad, and K. Yrjala. 2005. Pathways for methanogenesis and diversity of methanogenic archaea in three boreal peatland ecosystems. *Applied Environmental Microbiology* 71:2195–98.

Galimov, E. M., and K. A. Kvenvolden. 1983. Concentrations and carbon isotopic compositions of CH_4 and CO_2 in gas from sediments of the Blake outer ridge, deep sea drilling project leg 76. In *Initial reports of the deep sea drilling project*, 76, eds. R. E. Sheridan, F. M. Gradstein et. al., 403–7. Washington, DC: U.S. Government Printing Office.

Games, L. M., and J. M. Hayes. 1974. Carbon in ground water at the Columbus, Indiana landfill. In *Solid waste disposal by land burial in southern Indiana*, eds. D. B. Waldrip and R. V. Ruhe, 81–110. Bloomington: Indiana University Water Resources Research Center.

Games, L. M., and J. M. Hayes. 1976. On the mechanisms of CO_2 and CH_4 production in natural anaerobic environments. In *Environmental biogeochemistry* Vol. 1 of *Carbon, nitrogen, phosphorus, sulfur and selenium cycles*, ed. J. O. Nriagu, 51–73. Ann Arbor, MI: Ann Arbor Science.

Games, L. M., J. M. Hayes, and R. D. Gunsalas. 1978. Methane producing bacteria: Natural fractionations of the stable carbon isotopes. *Geochimica Cosmochimica Acta* 42:1295–97.

Gat, J. R. 1996. Oxygen and hydrogen isotopes in the hydrologic cycle. *Annual Review of Earth and Planetary Sciences* 24:225–62.

Gelwicks, J., J. B. Risatti, and J. M. Hayes. 1989. Carbon isotope effects associated with autotrophic acetogenesis. *Organic Geochemistry* 14:441–46.

Gelwicks, J. T., J. B. Risatti, and J. M. Hayes. 1994. Carbon isotope effects associated with aceticlastic methanogenesis. *Applied Environmental Microbiology* 60:467–72.

Goebel, E. D., R. M. Coveney, E. E. Angino, E. J. Zeller, and G. A. M. Dreschhoff. 1984. Geology, composition, isotopes of naturally occurring H_2-N_2 rich gas from wells near Junction City, Kansas. *Oil Gas Journal* 82:215–22.

Gray, N. D., J. N. S. Matthews, and I. M. Head. 2006. A stable isotope titration method to determine the contribution of acetate disproportionation and carbon dioxide reduction to methanogenesis. *Journal of Microbiological Methods* 65:180–86.

Grossman, E. L., B. K. Coffman, S. J. Fritz, and H. Wada. 1989. Bacterial production of methane and its influence on ground-water chemistry in east-central Texas aquifers. *Geology* 17:495–99.

Hackley, K. C., C. L. Liu, and D. D. Coleman. 1996. Environmental isotope characteristics of landfill leachates and gases. *Ground Water* 34:827–36.

Hackley, K. C., C. L. Liu, and D. Trainor. 1999. Isotopic identification of the source of methane in subsurface sediments of an area surrounded by waste disposal facilities. *Applied Geochemistry* 14:119–31.

Happell, J. D., J. P. Chanton, and W. S. Showers. 1994. The influence of methane oxidation on the stable isotopic composition of methane emitted from Florida swamp forests. *Geochimica Cosmochimica Acta* 58:4377–88.

Happell, J. D., J. P. Chanton, G. J. Whiting, and W. J. Showers. 1993. Stable isotopes as tracers of methane dynamics in Everglades marshes with and without active populations of methane oxidizing bacteria. *Journal Geophysical Research-Atmospheres* 98:14771–82.

Harrison, S. M. 1995. The hydrogeology and hydrochemistry of a potential coalbed methane exploration area, Elk River Valley, southeastern British Columbia. M.Sc. diss., University of Waterloo, Ontario, Canada.

Harrison, S. M., J. W. Molson, H. J. Abercrombie, J. F. Barker, D. Rudolph, and R. Aravena. 2000. Hydrogeology of a coal-seam gas exploration area, southeastern British Columbia, Canada: Part 1. Groundwater flow systems. *Hydrogeology Journal* 8:608–22.

Herczeg, A. L., S. B. Richardson, and P. J. Dillon. 1991. Importance of methanogenesis for organic carbon mineralisation in groundwater contaminated by liquid effluent, South Australia. *Applied Geochemistry* 6:533–42.

Hines, M. E., K. N. Duddleston, and R. P. Kiene. 2001. Carbon flow to acetate and C-1 compounds in northern wetlands. *Geophysical Research Letters* 28:4251–54.

Hines, M. E., K. N. Duddleston, J. N. Rooney-Varga, D. Fields, and J. P. Chanton. 2008. Uncoupling of acetate degradation from methane formation in Alaskan wetlands: Connections to vegetation distribution. *Global Biogeochemical Cycles* 22:GB2017, doi:10.1029/2006GB002903.

Hodgson, A. T., K. Garbesi, R. G. Sextro, and J. M. Daisey. 1992. Soil-gas contamination and entry of volatile organic compounds into a house near a landfill. *Journal Air Waste Management* 42:277–83.

Hogg, D. 1997. The application of environmental isotopes for evaluating the impact of landfill gas and leachate on shallow groundwater environments. M.Sc. diss., University of Waterloo, Ontario, Canada.

Hornibrook, E. R. C., and H. L. Bowes. 2007. Trophic status impacts both the magnitude and stable carbon isotope composition of methane flux from peatlands. *Geophysical Research Letters* 34:doi:10.1029/2007GL031231.

Hornibrook, E. R. C., F. J. Longstaffe, and W. S. Fyfe. 1997. Spatial distribution of microbial methane production pathways in temperate zone wetland soils: Stable carbon and hydrogen isotope evidence. *Geochimica Cosmochimica Acta* 61:745–53.

Hornibrook, E. R. C., F. J. Longstaffe, and W. S. Fyfe. 2000a. Factors influencing stable-isotope ratios in CH_4 and CO_2 within subenvironments of freshwater wetlands: Implications for δ-signatures of emissions. *Isotopes Environmental Health Studies* 36:151–76.

Hornibrook, E. R. C., F. J. Longstaffe, and W. S. Fyfe. 2000b. Evolution of stable carbon isotope compositions for methane and carbon dioxide in freshwater wetlands and other anaerobic environments. *Geochimica Cosmochimica Acta* 64:1013–27.

Houweling, S., T. Rockmann, I. Aben, F. Keppler, M. Krol, J. F. Meirink, E. J. Dlugokencky, and C. Frankenberg. 2006. Atmospheric constraints on global emissions of methane from plants. *Geophysical Research Letters* 33:L15821, doi:10.1029/2006GL026162.

Hughes, G. M., R. A. Landon, and R. N. Farvolden. 1971. *Hydrogeology of solid waste disposal sites in Northern Illinois,* Report No. SW-12D. Washington DC: US Environmental Protection Agency.

Irwin, H., C. Curtis, and M. L. Coleman. 1977. Isotopic evidence for the source of diagenetic carbonates formed during burial of organic-rich sediments. *Nature* 269:209–13.

Jedrysek, M. O. 1995. Carbon isotope evidence for diurnal variations in methanogenesis in freshwater lake sediments. *Geochimica Cosmochimica Acta* 59:557–61.

Jedrysek, M. O. 1999. Spatial and temporal patterns in diurnal variations of carbon isotope ratios of early-diagenetic methane from freshwater sediments. *Chemical Geology* 159:241–62.

Jenden, P. D., and I. R. Kaplan. 1986. Comparison of microbial gases from the middle American trench and Scripps submarine canyon: Implications for the origin of natural gas. *Applied Geochemistry* 1:631–46.

Kaplan, I. R. 1975. Stable isotopes as a guide to biogeochemical processes. *Proceedings of the Royal Society of London B-Biological Sciences* 189:183–211.

Kaplan, I. R. 1994, March. Formation mechanisms and differentiation of methane and hydrogen sulfide in various environments. In *Proceedings of the SWANA 17th Annual Landfill Gas Symposium*, Long Beach, CA.

Katz, B. G., L. N. Plummer, E. Busenberg, K. M. Revesz, B. F. Jones, and T. M. Lee. 1995. Chemical evolution of groundwater near a sinkhole lake, Northern Florida: 2. Chemical patterns, mass transfer modeling, and rates of mass transfer reactions. *Water Resources Research* 31:1565–84.

Kelley, C. A., N. B. Dise, and C. S. Martens. 1992. Temporal variations in the stable carbon isotopic composition of methane emitted from Minnesota peatlands. *Global Biogeochemical Cycles* 6:263–69.

Kelley, D. S., and G. L. Fruh-Green. 1999. Abiogenic methane in deep-seated mid-ocean ridge environments: Insights from stable isotope analyses. *Journal Geophysical Research-Solid Earth* 104:10439–60.

Keppler, F., J. T. G. Hamilton, M. Brass, and T. Rockmann. 2006. Methane emissions from terrestrial plants under aerobic conditions. *Nature* 439:187–91.

Kirschbaum, M. U. F., D. Bruhn, D. M. Etheridge, J. R. Evans, G. D. Farquhar, R. M. Gifford, K. I. Paul, and A. J. Winters. 2006. A comment on the quantitative significance of aerobic methane release by plants. *Functional Plant Biology* 33:521–30.

Kotsyurbenko, O. R., A. W. Friedrich, M. V. Simankova, A. N. Nozhevnikova, P. N. Golyshin, K. N. Timmis, and R. Conrad. 2007. Shift from acetoclastic to H_2-dependent methanogenes is in a West Siberian peat bog at low pH values and isolation of an acidophilic *Methanobacterium* strain. *Applied Environmental Microbiology* 73:2344–48.

Kotsyurbenko, O. R., M. V. Glagolev, A. N. Nozhevnikova, and R. Conrad. 2001. Competition between homoacetogenic bacteria and methanogenic archaea for hydrogen at low temperature. *FEMS Microbiology Ecology* 38:153–59.

Krzycki, J. A., W. R. Kenealy, M. J. DeNiro, and J. G. Zeikus. 1987. Stable carbon isotope fractionation by *Methanosarcina barkeri* during methanogenesis from acetate, methanol, or carbon dioxide-hydrogen. *Applied Environmental Microbiology* 53:2597–99.

Lansdown, J. M., 1992. The carbon and hydrogen stable isotope composition of methane released from natural wetlands and ruminants. PhD diss., University of Washington.

Lansdown, J. M., P. D. Quay, and S. L. King. 1992. CH_4 production via CO_2 reduction in a temperate bog: A source of ^{13}C-depleted CH_4. *Geochimica Cosmochimica Acta* 56:3493–3503.

Levin, I., K. O. Munich, and W. Weiss. 1980. The effect of anthropogenic CO_2 and ^{14}C sources on the distribution of ^{14}C in the atmosphere. *Radiocarbon* 22:379–91.

Lin, L. H., G. F. Slater, B. S. Lollar, G. Lacrampe-Couloume, and T. C. Onstott. 2005. The yield and isotopic composition of radiolytic H_2, a potential energy source for the deep subsurface biosphere. *Geochimica Cosmochimica Acta* 69:893–903.

Lollar, B. S., S. K. Frape, P. Fritz, S. A. Macko, J. A. Welhan, R. Blomqvist, and P. W. Lahermo. 1993. Evidence for bacterially generated hydrocarbon gas in Canadian shield and Fennoscandian shield rocks. *Geochimica Cosmochimica Acta* 57:5073–85.

Lollar, B. S., S. K. Frape, S. M. Weise, P. Fritz, S. A. Macko, and J. A. Welhan. 1993. Abiogenic methanogenesis in crystalline rocks. *Geochimica Cosmochimica Acta* 57:5087–97.

Lollar, B. S., J. Ward, G. Slater, G. Lacrampe-Couloume, J. Hall, L. H. Lin, D. Moser, and T. C. Onstott. 2002. Hydrogen and hydrocarbon gases in crystalline rock: Implications for the deep biosphere. *Geochimica Cosmochimica Acta* 66:A706.

Lovley, D. R., and S. Goodwin. 1988. Hydrogen concentrations as an indicator of the predominant terminal electron-accepting reactions in aquatic sediments. *Geochimica Cosmochimica Acta* 52:2993–3003.

Martens, C. S., N. E. Blair, C. D. Green, and D. J. Des Marais. 1986. Seasonal variation in the stable carbon isotope signature of biogenic methane in a coastal sediment. *Science* 233:1300–303.

Martens, C. S., C. A. Kelley, J. P. Chanton, and W. J. Showers. 1992. Carbon and hydrogen isotopic characterization of methane from wetlands and lakes of the Yukon-Kuskokwim Delta, Western Alaska. *Journal Geophysical Research-Atmospheres* 97:16689–701.

Matthews, E., and I. Fung. 1987. Methane emission from natural wetlands: Global distribution, area, and environmental characteristics of sources. *Global Biogeochemical Cycles.* 1:61–86.

Mayer, E. W., D. R. Blake, S. C. Tyler, T. Makide, D. C. Montague, and F. S. Rowland. 1982. Methane interhemispheric concentration gradient and atmospheric residence time. *Proceedings National Academy Science* 79:1366–70.

Metcalfe, D. E., and G. J. Farquhar. 1987. Modeling gas migration through unsaturated soils from waste disposal sites. *Water Air Soil Pollution* 32:247–59.

Metje, M., and P. Frenzel. 2005. Effect of temperature on anaerobic ethanol oxidation and methanogenesis in acidic peat from a northern wetland. *Applied Environmental Microbiology* 71:8191–200.

Mikaloff Fletcher, S. E., P. P. Tans, L. M. Bruhwiler, J. B. Miller, and M. Heimann. 2004a. CH_4 sources estimated from atmospheric observations of CH_4 and its $^{13}C/^{12}C$ isotopic ratios: 1. Inverse modeling of source processes. *Global Biogeochemical Cycles* 18:GB4004, doi:10.1029/2004GB002223.

Mikaloff Fletcher, S. E., P. P. Tans, L. M. Bruhwiler, J. B. Miller, and M. Heimann. 2004b. CH_4 sources estimated from atmospheric observations of CH_4 and its $^{13}C/^{12}C$ isotopic ratios: 2. Inverse modeling of CH_4 fluxes from geographical regions. *Global Biogeochemical Cycles* 18:GB4005, doi:10.1029/2004GB002224.

Minxin, T., W. Wanchun, X. Guangxin, L. Jinying, W. Yanlong, Z. Xiaojun, Z. Hong, S. Baoguang, and G. Bo. 2005. Secondary biogenic coalbed gas in some coal fields of China. *Chinese Science Bulletin* 50:24–29.

Miyajima, T., E. Wada, Y. T. Hanba, and P. Vijarnsorn. 1997. Anaerobic mineralization of indigenous organic matters and methanogenesis in tropical wetland soils. *Geochimica Cosmochimica Acta* 61:3739–51.

Müller, V., M. Blaut, and G. Gottschalk. 1993. Bioenergetics of methanogenesis. In *Methanogenesis: Ecology, physiology, biochemistry and genetics*, ed. J. G. Ferry, 360–406. New York: Chapman & Hall.

Mylroie, R. L., and R. E. Hungate. 1955. Experiments on the methane bacteria in sludge. *Canadian Journal Microbiology* 1:55–64.

Nakagawa, F., N. Yoshida, Y. Nojiri, and V. N. Makarov. 2002. Production of methane from alasses in Eastern Siberia: Implications from its ^{14}C and stable isotopic compositions. *Global Biogeochemical Cycles* 16:1041, doi:10.1029/2000GB001384.

Nakagawa, F., N. Yoshida, A. Sugimoto, E. Wada, T. Yoshioka, S. Ueda, and P. Vijarnsorn. 2002. Stable isotope and radiocarbon compositions of methane emitted from tropical rice paddies and swamps in Southern Thailand. *Biogeochemistry* 61:1–19.

Nissenbaum, A., B. J. Presley, and I. R. Kaplan. 1972. Early diagenesis in a reducing fjord, Saanich-Inlet, British Columbia I. Chemical and isotopic changes in major components of interstitial water. *Geochimica Cosmochimica Acta* 36:1007–27.

Nottingham, P. M., and R. E. Hungate. 1969. Methanogenic fermentation of benzoate. *Journal Bacteriology* 98:1170–72.

Oana, S., and E. S. Deevey. 1960. Carbon 13 in lake waters, and its possible bearing on paleolimnology. *American Journal Science* 258A:253–72.

Oremland, R. S., L. J. Miller, and M. J. Whiticar. 1987. Sources and flux of natural gas from Mono Lake, California. *Geochimica Cosmochimica Acta* 51:2915–29.

Parkes, R. J., B. A. Cragg, S. J Bale, J. M. Getliff, K. Goodman, P. A. Rochelle, J. C. Fry, A. J. Weightman, and S. M. Harvey. 1994. Deep bacterial biosphere in Pacific Ocean sediments. *Nature* 371:410–13.

Penning, H., C. M. Plugge, P. E. Galand, and R. Conrad. 2005. Variation of carbon isotope fractionation in hydrogenotrophic methanogenic microbial cultures and environmental samples at different energy status. *Global Change Biology* 11:2103–113.

Pine, M. J., and H. A. Barker. 1956. Studies on methane fermentation XII. The pathway of hydrogen in the acetate fermentation. *Journal Bacteriology* 71:644–48.

Pine, M. J., and W. Vishniac. 1957. The methane fermentation of acetate and methanol. *Journal Bacteriology* 73:736–42.

Popp, T. J., J. P. Chanton, G. J. Whiting, and N. Grant. 1999. Methane stable isotope distribution at a Carex dominated fen in north central Alberta. *Global Biogeochemical Cycles* 13:1063–77.

Powelson, D. K., J. P. Charlton, and T. Abichou. 2007. Methane oxidation in biofilters measured by mass-balance and stable isotope methods. *Environmental Science & Technology* 41:620–25.

Prater, J. L., J. P. Chanton, and G. J. Whiting. 2007. Variation in methane production pathways associated with permafrost decomposition in collapse scar bogs of Alberta, Canada. *Global Biogeochemical Cycles* 21:GB4004, doi:10.1029/2006GB002866.

Rank, D. 1993. Environmental tritium in hydrology: present state. In *Advances in liquid scintillation spectrometry 1992*, eds. J. E. Noakes, F. Schonjofer, and H. A. Polach, 327–34, Tucson: Radiocarbon Publishers, University of Arizona.

Rank, D., W. Papesch, V. Rajner, and G. Riehl-Herwirsch. 1992. Environmental isotopes study at the Breitenau experimental landfill (Lower Austria). In *Tracer hydrology*, eds. H. Hötzl and A. Werner, 173–77. Rotterdam: Balkema.

Rasmussen, R. A., and M. A. K. Khalil. 1981. Atmospheric methane: Trends and seasonal cycles. *Journal Geophysical Research* 86:9826–32.

Revesz, K., T. B. Coplen, M. J. Baedecker, P. D. Glynn, and M. Hult. 1995. Methane production and consumption monitored by stable H and C isotope ratios at a crude oil spill site, Bemidji, Minnesota. *Applied Geochemistry* 10:505–16.

Rice, D. D. 1993. Composition and Origins of Coalbed Gas. In *Hydrocarbons from coal: AAPG Studies in Geology Series #38*, eds. B. E. Law and D. D. Rice, 159–84. Tulsa, OK: American Association of Petroleum Geologists.

Rice, D. D., and G. E. Claypool. 1981. Generation, accumulation and resource potential of biogenic gas. *AAPG Bulletin* 65:5–25.

Rightmire, C. T. 1984. Coalbed methane resource. In *Coalbed methane resources of the United States: AAPG Studies in Geology #17*, eds. C. T. Rightmire, G. E. Eddy, J. N. Kirr, 1–13. Tulsa, OK: American Association of Petroleum Geologists.

Robertson, W. D., A. M. Murphy, and J. A. Cherry. 1995. Low-cost on-site treatment of landfill leachate using infiltration beds: Preliminary field trial. *Ground Water Monitoring Remediation* 15:107–15.

Rodhe, H. 1990. A comparison of the contribution of various gases to the greenhouse effect. *Science* 248:1217–19.

Rooney-Varga, J. N., M. W. Giewat, K. N. Duddleston, J. P. Chanton, and M. E. Hines. 2007. Links between archaeal community structure, vegetation type and methanogenic pathway in Alaskan peatlands. *FEMS Microbiology Ecology* 60:240–51.

Rosenfeld, W. D., and S. R. Silvermann. 1959. Carbon isotope fractionation in bacterial production of methane. *Science* 130:1658–59.

Schaefer, H., M. J. Whiticar, E. J. Brook, V. V. Petrenko, D. F. Ferretti, and J. P. Severinghaus. 2006. Ice record of $\delta^{13}C$ for atmospheric CH_4 across the Younger Dryas-Preboreal transition. *Science* 313:1109–12.

Schoell, M. 1980. The hydrogen and carbon isotopic composition of methane from natural gases of various origins. *Geochimica Cosmochimica Acta* 44:649–61.

Schoell, M. 1988. Multiple origins of methane in the Earth. *Chemical Geology* 71:1–10.

Scott, A. R. 1993. Composition and origin of coalbed gases from selected basins in the United States. *Proceeding of the 1993 international methane symposium, Birmingham, Alabama*, 1:207–22.

Scott, A. R., W. R. Kaiser, and W. B. Ayers. 1994. Thermogenic and secondary biogenic gases, San-Juan Basin, Colorado and New-Mexico: Implications for coalbed gas producibility. *AAPG Bulletin* 78:1186–1209.

Senum, G. I., and J. S. Gaffney. 1985. A re-examination of the tropospheric methane cycle: Geophysical implications. In *The carbon cycle and atmospheric CO_2: natural variations archean to present*, eds. E. T. Sundquist and W. S. Broecker, 61–9. Washington, DC: AGU Geophysical Monograph Series.

Shannon, R. D., and J. R. White. 1994. A three-year study of controls on methane emissions from two Michigan peatlands. *Biogeochemistry* 27:35–60.

Simpkins, W. W., and T. B. Parkin. 1993. Hydrogeology and redox geochemistry of CH_4 in a late Wisconsinan till and loess sequence in central Iowa. *Water Resources Research* 29:3643–57.

Sizova, M. V., N. S. Panikov, T. P. Tourova, and P. W. Flanagan. 2003. Isolation and characterization of oligotrophic acido-tolerant methanogenic consortia from a Sphagnum peat bog. *FEMS Microbiology Ecology* 45:301–15.

Smith, J. W., and R. J. Pallasser. 1996. Microbial origin of Australian coalbed methane. *AAPG Bulletin* 80:891–97.

Stadtmann, T. C., and H. A. Barker. 1949. Studies on the methane fermentation, VI. Tracer experiments of the mechanism of methane formation. *Archives Biochemistry and Biophysics* 21:256–64.

Stevens, C. M. 1988. Atmospheric methane. *Chemical Geology* 71:11–21.

Stevens, T. O., and J. P. McKinley. 1995. Lithoautotrophic microbial ecosystems in deep basalt aquifers. *Science* 270:450–54.

Stevens, C. M., and F. E. Rust. 1982. The carbon isotopic composition of atmospheric methane. *Journal Geophysical Research* 87:4879–82.

Stevens, T. O., and J. P. McKinley. 1995. Lithoautotrophic microbial ecosystems in deep basalt aquifers. *Science* 270:450–54.

Strąpoć, D., M. Mastalerz, C. Eble, and A. Schimmelmann. 2007. Characterization of the origin of coalbed gases in southeastern Illinois basin by compound-specific carbon and hydrogen stable isotope ratios. *Organic Geochemistry* 38:267–87.

Sugimoto, A., and N. Fujita. 2006. Hydrogen concentration and stable isotopic composition of methane in bubble gas observed in a natural wetland. *Biogeochemistry* 81:33–44.

Sugimoto, A., and E. Wada. 1993. Carbon isotopic composition of bacterial methane in a soil incubation experiment: Contributions of acetate and CO_2/H_2. *Geochimica Cosmochimica Acta* 57:4015–27.

Sugimoto, A., and E. Wada. 1995. Hydrogen isotopic composition of bacterial methane: CO_2/H_2 reduction and acetate fermentation. *Geochimica Cosmochimica Acta* 59:1329–37.

Templeton, A. S., K. H. Chu, L. Alvarez-Cohen, and M. E. Conrad. 2006. Variable carbon isotope fractionation expressed by aerobic CH_4-oxidizing bacteria. *Geochimica Cosmochimica Acta* 70:1739–52.

Tyler, S. C., P. R. Zimmerman, C. Cumberbatch, J. P. Greenberg, C. Westberg, and J. P. E. C. Darlington. 1988. Measurement and interpretation of δ¹³C of methane from termites, rice paddies, and wetlands in Kenya. *Global Biogeochemical Cycles* 2:341–55.

Valentine, D. L., A. Chidthaisong, A. Rice, W. S. Reeburgh, and S. C. Tyler. 2004. Carbon and hydrogen isotope fractionation by moderately thermophilic methanogens. *Geochimica Cosmochimica Acta* 68:1571–90.

Vogels, G. D., J. T. Keltjens, and C. Van der Drift. 1988. Biochemistry of methane production. In *Biology of anaerobic microorganisms*, ed. A. J. B. Zehnder, 707–70. New York: John Wiley & Sons.

Voytov, G. I., V. S. Lebeder, G. G. Blokhina, and L. F. Cherevichnaya. 1975. Chemical and isotopic composition of subsoil and marsh gases of the Kola Tundras. *Doklady Akademii Nauk SSSR* 224:186–88.

Wahlen, M. 1993. The global methane cycle. *Annual Review Earth Planetary Sciences* 21:407–26.

Wahlen, M. 1994. Carbon dioxide, carbon monoxide and methane in the atmosphere: Abundance and isotopic composition. In *Stable isotopes in ecology and environmental science*, eds. K. Lajtha and R. H. Michener, 93–113. London: Blackwell Scientific.

Waldron, S., A. J. Hall, and A. E. Fallick. 1999a. Enigmatic stable isotope dynamics of deep peat methane. *Global Biogeochemical Cycles* 13:93–100.

Waldron, S., J. M. Lansdown, E. M. Scott, A. E. Fallick, and A. J. Hall. 1999b. The global influence of the hydrogen isotope composition of water on that of bacteriogenic methane from shallow freshwater environments. *Geochimica Cosmochimica Acta* 63:2237–45.

Waldron, S., I. A. Watson-Craik, A. J. Hall, and A. E. Fallick. 1998. The carbon and hydrogen stable isotope composition of bacteriogenic methane: A laboratory study using a landfill inoculum. *Geomicrobiology Journal* 15:157–69.

Walter, K. M., S. A. Zimov, J. P. Chanton, D. Verbyla, and F. S. Chapin. 2006. Methane bubbling from Siberian thaw lakes as a positive feedback to climate warming. *Nature* 443:71–5.

Wasserburg, G. S., E. Mazor, and R. E. Zartman. 1963. Isotopic and chemical composition of some terrestrial natural gases. In *Earth science and meteoritics*, eds. J. Geiss and E. D. Goldberg, 219–40. Amsterdam: North-Holland.

Wassmann, R., U. G. Thein, M. J. Whiticar, H. Rennenberg, W. Seiler, and W. J. Junk. 1992. Methane emissions from the Amazon floodplain: Characterization of production and transport. *Global Biogeochemical Cycles* 6:3–13.

Welhan, J. A. 1988. Origins of methane in hydrothermal systems. *Chemical Geology* 71:183–98.

Westerhoff, H. V., A. K. Groen, and J. A. Wanders. 1984. Modern theories of metabolic control and their applications. *Bioscience Reports* 4:1–22.

Whiticar, M. J. 1999. Carbon and hydrogen isotope systematics of bacterial formation and oxidation of methane. *Chemical Geology* 161:291–314.

Whiticar, M. J., and E. Faber. 1986. Methane oxidation in sediment and water column environments: Isotope evidence. *Organic Geochemistry* 10:759–68.

Whiticar, M. J., E. Faber, and M. Schoell. 1986. Biogenic methane formation in marine and freshwater environments: CO_2 reduction vs. acetate fermentation: Isotope evidence. *Geochimica Cosmochimica Acta* 50:693–709.

Williams, G. M., and N. Aitkenhead. 1991. Lessons from Loscoe: The uncontrolled migration of landfill gas. *Quarterly Journal Engineering Geology Hydrogeology* 24:191–207.

Wolfe, R. S. 1971. Microbial formation of methane. *Advances Microbiology Physiology* 6:107–46.

Wolin, M. J., and T. L. Miller. 1982. Interspecies hydrogen transfer: 15 years later. *ASM News* 48:561–65.

Woltemate, I., M. J. Whiticar, and M. Schoell. 1984. Carbon and hydrogen isotopic composition of bacterial methane in a shallow freshwater lake. *Limnology Oceanography* 29:985–92.

Wood, H. G., and Ljungdahl, L. G. 1991. Autotrophic character of the acetogenic bacteria. In *Variations in autotrophic life*, eds. J. M. Shively and L. L. Barton, 201–50. San Diego, CA: Academic Press.

Yagi, T., M. Tsuda, and H. Inokuchi. 1973. Kinetic studies on hydrogenase parahydrogen-orthohydrogen conversion and hydrogen deuterium exchange reactions. *Journal Biochemistry* 73:1069–81.

Zimov, S. A., Y. V. Voropaev, I. P. Semiletov, S. P. Davidov, S. F. Prosiannikov, F. S. Chapin, M. C. Chapin, S. Trumbore, and S. Tyler. 1997. North Siberian lakes: A methane source fueled by pleistocene carbon. *Science* 277:800–02.

Zinder, S. H. 1993. Physiological ecology of methanogens. In *Methanogenesis: Ecology, physiology, biochemistry and genetic*, ed. J. G. Ferry, 128–206. New York: Chapman & Hall.

7 Isotopes and Processes in the Nitrogen and Sulfur Cycles

Ramon Aravena and Bernhard Mayer

CONTENTS

7.1 INTRODUCTION

This chapter describes the basic principles that govern the application of environmental isotopes in nitrogen and sulfur cycling studies in groundwater. The most relevant questions concerning evaluation of sources (fingerprinting) and processes that control nitrate and sulfate in the hydrosphere are addressed. The first part of this chapter will focus on nitrate studies and the second part will address sulfur cycling.

Nitrate is one of the most common contaminants found in groundwater and surface water. Elevated nitrate concentrations in groundwater are a significant threat facing water resources worldwide and are becoming critical in rural areas (Spalding and Exner 1993; Nolan, Hitt, and Ruddy 2002; Tredoux, Engelbrecht, and Talma 2005). In agricultural areas of North America, for example, 5–46% of the domestic water wells exceeded the drinking water quality standard (Hamilton and Helsel 1995; Goss, Barry, and Rudolph 1998). The rising trend of nitrate is mainly related to historical agricultural practices that started during the mid-twentieth century. The potential human health risk associated with the consumption of nitrate-contaminated drinking water is related to the occurrence of methaemoglobinaemia or baby blue syndrome (Hesseling, Toens, and Visser 1991; Fan and Steinberg 1996). High concentrations of nitrate in drinking water have been linked to livestock poisoning (Tredoux, Engelbrecht, and Talma 2005) but no clear evidence has been found for a link between nitrate and cancer in the human population (World Health Organization [WHO] 2004). Beside potential health effects, nitrate contamination is also causing adverse economic effects (O'Neil and Raucher 1990). Based on the potential human health risk related to nitrate, the WHO recommended 11.5 mg NO_3-N/L as a permissible level for nitrate in potable water. The drinking water limit in Canada, the United States, and Australia is 10 mg NO_3-N/L. High levels of nitrate also deteriorate the health of surface water through eutrophication (Mason 2002). Nutrient rich water tends to favor excessive growth of aquatic plants like algae that can deplete dissolved oxygen levels by algae respiration or decay. Low dissolved oxygen concentrations cause adverse effects in freshwater invertebrates and fish and also pose problems for the use of water for activities such as livestock watering, water sports, and navigation.

In general, nitrate contamination of groundwater is attributed to long-term point sources (manure and sewage lagoons, septic systems) and nonpoint or diffuse sources (fertilizers and manure spreading) associated with agricultural practices. Inputs of sewage through leaks in sewage networks are also causing significant nitrate con-tamination in urban environments (Whitehead, Hiscock, and Dennis 1999; Lerner et al. 1999; Tellam and Thomas 2002; Iriarte 2004).

During the last decades, a significant amount of research has been conducted to evaluate origin, fate, and transport of nitrate in groundwater (e.g., Gillham and Cherry 1978; Ronen, Kanfi, and Magaritz 1983; Andersen and Kristiansen 1984; Postma et al. 1991; Wassenaar 1995). The long-term history of nitrate input in recharge areas, the groundwater flow system, and nitrate attenuation processes control the distribution of nitrate in groundwater. Identifying processes that affect nitrate can be problematic in regional cases due to variability in the physical and geochemical settings present in the aquifer. Relationships between the distribution of agricultural contaminants in surficial aquifers and corresponding stream responses to contamination are somewhat complicated where regional changes in agricultural practices and groundwater residence times are on the same time scale (Böhlke and Denver 1995). The implementation of better groundwater management practices in aquifers contaminated with nitrate also requires a good understanding of the sources and processes that control the concentration of nitrate in the contaminated groundwater. This information is also essential for evaluating the efficiency of ben-eficial management practices (BMPs) implemented in some countries during the last decade in order to reduce nitrate contamination in aquifers (Wassenaar, Hendry, and Harrington 2006).

This chapter describes the use of naturally occurring ^{15}N and ^{18}O stable isotopes that form part of the nitrate molecule as tools to evaluate nitrate sources and pro-cesses that control nitrate concentrations in recharge areas and along the groundwa-ter flow systems of contaminated aquifers. The use of ^{15}N and ^{18}O to evaluate the role of riparian zone attenuation will be also covered in this chapter.

7.2   NITROGEN ISOTOPE FUNDAMENTALS

The abundance of nitrogen in the whole earth is 0.003%, with most of the nitrogen, 97.76% located in rocks, 2.01% in the atmosphere, and the rest is in the hydro-sphere and biosphere. Nitrogen has two stable isotopes, ^{14}N being the most abun-dant and ^{15}N. Atmospheric nitrogen is composed of 99.64% ^{14}N and 0.36% ^{15}N. Oxygen has three stable isotopes ^{16}O, ^{17}O, and ^{18}O and their natural abundances are 99.763%, 0.0375%, and 0.1995%, respectively (Chapter 1). Stable isotopes of nitrogen and oxygen are naturally incorporated into the major nitrogen contain-ing compounds that can be present in groundwater (e.g., NH_4^+, NO_3^-, NO_2^-, N_2O, and N_2).

Nitrogen ($^{15}N/^{14}N$) and oxygen ($^{18}O/^{16}O$) isotope ratios are reported in delta (δ) per mil (‰) units relative to the international reference materials atmospheric nitro-gen and V-SMOW, respectively (see Chapter 1). Detailed information about current laboratory protocols for ^{15}N and ^{18}O analyses can be found in Silva et al. (2000), Sigman et al. (2001), Casciotti et al. (2002), and Coplen, Böhlke, and Casciotti (2004). The analytical precision for nitrate is in the order of 0.1 to 0.5‰ for both isotopes, respectively.

7.3   NITROGEN CYCLE

The isotope composition of the main nitrogen pools involved in nitrogen cycling in groundwater (soil organic nitrogen, ammonium, and nitrate) is imprinted by the isotope composition of the sources (internal and external) and processes that affect these pools in the unsaturated and saturated zones. The external sources are fertilizers and nitrogen derived from animal and domestic waste. In some specific areas of the world, soil nitrogen (internal source) has been shown to be an important source of high nitrate concentration in groundwater (Heaton, Talma, and Vogel 1983; Edmunds and Gaye 1997).

The main biological reactions that control nitrogen cycling in soil and groundwater are assimilation, nitrification, and denitrification.

7.3.1   ASSIMILATION

It seems that nitrogen assimilation by plants is not a highly fractionating process. In the case of N_2 fixing plants, their $\delta^{15}N$ values are very close to that of atmospheric nitrogen (Kohl and Shearer 1980; Mariotti et al. 1980). A similar pattern is observed in mature nonfixing plants that showed $\delta^{15}N$ values very similar to those of the nitrate source (Mariotti et al. 1980). Therefore nitrogen assimilation by plants usually does not significantly change the ^{15}N content of the nitrogen source in soils (soil organic nitrogen and fertilizers).

7.3.2   NITRIFICATION

Nitrification involves a series of bacteria mediated reactions that convert organic nitrogen and ammonium to nitrate. It is a redox reaction, converting a reduced N species to a more oxidized form. Nitrification is also an exothermic reaction and more energy for biosynthesis is obtained from the oxidation of NH_4^+ to NO_2^- (–84 kcal/mole) than from the subsequent oxidation of NO_2^- to NO_3^- (–18 kcal/mole). The degree of nitrification is controlled by oxygen and the actual limit depends more on the rate of O_2 diffusion into the soil than the O_2 level. Nitrification can occur rapidly, even down to the concentration of about 0.3 ppm dissolved O_2. Other factors that control nitrification rates include pH, $[NH_4^+]$, $[NO_3^-]$ and soil moisture level.

$$Organic - N \rightarrow NH_4^+ \rightarrow NO_2^- \rightarrow NO_3^- \tag{7.1}$$

The organic nitrogen to ammonium step involves very little isotope fractionation. However, the conversion of ammonium to nitrate is a highly fractionating process. As a result of nitrification, the residual ammonium pool becomes enriched in ^{15}N, while the nitrate produced is depleted in ^{15}N relative to the ammonium source. The kinetic isotope fractionation factor for the conversion of ammonium to nitrate ranges between –5 and –35‰ (Delwiche and Steyn 1970; Freyer and Aly 1975; Mariotti et al. 1981). However, field evidence showed that ammonium is often rapidly and completely converted to nitrate and in that case the isotope composition of the nitrate tends to reflect the isotopic composition of the ammonium source. In case a significant mass of ammonium is present in the unsaturated zone and ammonium is partially nitrified, the first nitrate fraction will be significantly more ^{15}N depleted than the ammonium source, and subsequently the nitrate will tend toward the isotopic composition of the ammonium source as nitrification progresses toward completion (Feigin et al. 1974).

The oxygen isotope composition of nitrate produced by nitrification is mainly controlled by the isotope composition of the oxygen sources involved in the process. The conversion of ammonium to nitrate includes the incorporation of three oxygen molecules, two are coming from the water and one comes from dissolved atmospheric oxygen (Anderson and Hooper 1983; Hollocher 1984). Therefore the $\delta^{18}O$ of the nitrate produced from nitrification can be calculated using the following expression

$$\delta^{18}O = 2/3\ \delta^{18}O\text{-}H_2O + 1/3\ \delta^{18}O\text{-}O_2 \tag{7.2}$$

Atmospheric oxygen has a $\delta^{18}O$ value of $+23.5‰$ (Kroopnick and Craig 1972). For $\delta^{18}O\text{-}H_2O$ values around $-10‰$, the $\delta^{18}O$ of the nitrate produced from nitrification should be around $-1‰$. Slightly higher $\delta^{18}O\text{-}NO_3^-$ values compared to these theoretical predictions have been observed at field sites and have been attributed to ^{18}O enrichment of the water due to evaporation, ^{18}O enrichment of O_2 due to respiration, or unknown isotope effects occurring during nitrification (Wassenaar 1995; Kendall and McDonnell 1998).

7.3.3 DENITRIFICATION

Denitrification is a microbially mediated redox process, which involves the conversion of nitrate via a multistep process, to dinitrogen gas.

$$NO_3^- \rightarrow NO_2^- \rightarrow NO \rightarrow N_2O \rightarrow N_2. \tag{7.3}$$

Denitrifiers are ubiquitous in soils, surface water, and groundwater and can be found in shallow and deep aquifers. Conditions that favor denitrification include anaerobic environments and an ample supply of nitrate, however it seems that some organisms can denitrify even in the presence of some oxygen (Robertson and Kuenen 1991). Nitrate reduction by heterotrophic bacteria, which get their energy and carbon from the oxidation of organic compounds, seems to be the main mechanism for nitrate attenuation in soils and aquifers. This reaction can be expressed by

$$NO_3^- + 5C + 2H_2O \rightarrow 1/2N_2 + 4HCO_3^- + CO_2 \tag{7.4}$$

However, there is evidence that some types of denitrifiers (autotrophs) can also obtain their energy from the oxidation of inorganic compounds such as sulfide minerals. This reaction is assisted by *Thiobacillus denitrificans* and can be expressed by

$$14NO_3^- + 5FeS_2 + 4H^+ \rightarrow 7N_2 + 10\ SO_4^- + 5\ Fe + 2H_2O. \tag{7.5}$$

Furthermore, denitrification can also occur through oxidation of ferrous iron (Postma 1990)

$$NO_3^- + 5Fe^{2+} + 6H^+ \rightarrow 1/2N_2 + 5\ Fe^{3+} + 3H_2O. \tag{7.6}$$

During denitrification, due to a small difference in reaction rates for nitrate molecules with the lighter isotopes (^{14}N, ^{16}O) compared to nitrate molecules with the heavier isotopes (^{15}N, ^{18}O), the isotopically light nitrate molecules tend to be reduced faster than the isotopically heavy nitrate molecules. Consequently, the residual nitrate becomes enriched in the heavy isotopes ^{15}N and ^{18}O (Mariotti 1986; Mariotti, Landreau, and Simon 1988; Aravena and Robertson 1998; Devito et al. 2000). This pattern can be described as a Rayleigh distillation process: $\delta X_{res} = \delta X_o + \varepsilon \ln (C_{res}/Co)$, where o and res represent the initial and residual nitrate, respectively: δX is the $\delta^{15}N$ or the $\delta^{18}O$ value of NO_3^-, C is the nitrate concentration and ε is the enrichment factor for ^{15}N or ^{18}O and related to the fractionation factor α by $\varepsilon = (\alpha - 1) \times 1'000$.

The degree of isotope fractionation observed for a particular reaction is often strongly influenced by reaction rate. Therefore, the more important factors that influence the magnitude of denitrification fractionation factors are those that exert strong influence on denitrification rates including temperature and the availability of O_2, NO_3^-, and organic carbon. For example, nitrification rates influence substrate pools while soil water content influences the transfer of substrates and O_2 to denitrification sites (Cornwell, Kemp, and Kana 1999; Herbert 1999). Furthermore, the in-situ denitrification rate in aquifers is controlled by a complex interaction between these factors. This explains the wide range of enrichment factors for ^{15}N (–11 to –33) reported for denitrification in different studies. Mariotti, Germon, and Leclerc (1982) found an inverse relation between reaction rate and isotopic enrichment and the greatest enrichments were obtained for low reduction rates.

Several denitrification studies in groundwater have found a linear relationship between $\delta^{15}N-NO_3^-$ and the $\delta^{18}O-NO_3^-$ with a slope ranging between 0.48 and 0.67, implying that the isotopic composition of both nitrogen and oxygen increases during denitrification in a ratio of about 2:1 (Böttcher et al. 1990; Aravena and Robertson 1998; Mengis et al. 1999; Cey et al. 1999; Devito et al. 2000; Fukada et al. 2004). Mathematical modeling of denitrification using parallel, first-order kinetics, and reaction rate constants produced a theoretical value of 0.51 for the $\delta^{18}O$ versus $\delta^{15}N$ slope, which is in agreement with the field data (Chen and MacQuarrie 2004). Therefore, the combination of ^{15}N and ^{18}O analysis of nitrate is a powerful tool to demonstrate unequivocally the occurrence of denitrification in groundwater.

7.3.4 DISSIMILATORY NITRATE REDUCTION TO NH_4^+ AMMONIUM (DNRA)

Dissimilatory nitrate reduction to NH_4^+ ammonium (DNRA) is another anaerobic reduction process that can be used by fermentative bacteria to transform nitrate to ammonium (Korom 1992). The reaction can be expressed by

$$2H^+ + NO_3^- + 2CH_2O \rightarrow NH_4^+ + 2\ CO_2 + H_2O \tag{7.7}$$

However, there is not much evidence to support the occurrence of this process in groundwater. One important difference between DNRA and denitrification is that fermentative bacteria that carry out DNRA are obligate anaerobes (Hill 1996). Hence they have less chance to occupy the niches that denitrifiers (facultative anaerobic heterotrophs) can, particularly in the soil and the unsaturated zone.

7.3.5 NITROUS OXIDE

Nitrous oxide (N_2O) is an important atmospheric trace greenhouse gas contributing to global warming and destruction of the ozone layer (Cicerone 1989). N_2O can be produced during nitrification and can also be produced and consumed during denitrification (Kuenen and Robertson 1988), and is therefore an integral part of the nitrogen cycle in groundwater. The reactions involving N_2O can be expressed by

$$\rightarrow N_2O$$

$$NH_4^+ \rightarrow NO_2^- \rightarrow NO_3^-\ \text{Nitrification} \tag{7.8}$$

$$NO_3^- \rightarrow NO_2^- \rightarrow N_2O \rightarrow N_2\ \text{Denitrification.} \tag{7.9}$$

Few studies have been done to evaluate the processes that control N_2O concentrations in groundwater (Ronen, Magaritz, and Almon 1988; Ueda, Ogura, and Wada 1991; Spalding and Parrot 1994; Hiscock et al. 2003). The most common approach in these studies is to use the concentration patterns of O_2, NO_3^- and N_2O to evaluate sources of N_2O. For example, based on a positive correlation between concentrations of N_2O and NO_3^-, a study in regional aquifers in the United Kingdom inferred that nitrification was the principal source of N_2O in the groundwater (Hiscock et al. 2003).

Since soil N cycling is one of the main sources of N_2O to the atmosphere, several studies have been conducted to evaluate the application of ^{15}N and [18]O to determine N_2O production pathways in soils (Webster and Hopkins 1996; Perez et al. 2000, 2001; Bol et al. 2003; Rock, Ellert, and Mayer 2007). These studies have shown that the stable isotope composition of N_2O produced during nitrification is very different and more ^{15}N and [18]O depleted than the isotope composition of N_2O produced by denitrification (Bol et al. 2003). Therefore, the relative importance of these two processes could be determined using ^{15}N and [18]O analyses. These results open a new possibility for utilizing ^{15}N and [18]O in N_2O to investigate the importance of nitrification and denitrification in nitrate cycling in groundwater.

7.4   ISOTOPE FINGERPRINT OF NITRATE SOURCES

7.4.1   SOILS

In general, soil N is usually more abundant in ^{15}N than atmospheric N. A survey that included 124 surface soils, representing crops (57), pasture (15), native herbs (19), and forests (13) from 20 U.S. states showed a range of δ^{15}N values between 5.1 and 12.3‰ with a mean value of 9.2‰ (Shearer, Kohl, and Chien 1978). A Canadian study showed no significant differences of δ^{15}N values collected for surface soils from cultivated (38 samples) and native pasture (11 samples) areas in central Saskatchewan, Canada. The mean value for both types of soils was 8.6 ± 1.9‰ (Karamanos, Voroney, and Rennie 1981). It seems a combination of factors and processes (including nitrification, denitrification, and volatilization among others) control the δ^{15}N values of total N in soils. The study by Shearer, Kohl, and Chien (1978) showed that less than half of the isotopic variation obtained in their study could be accounted for by differences in environmental variables (mean annual temperature and precipitation) or soil characteristic (C and N content, clay and sand content, pH). The range for δ^{15}N values representing soil N compiled from different studies is reported in Figure 7.1. The δ^{18}O values for nitrate originated by nitrification of soil N should range between 0 and +6‰ Figure 7.2).

7.4.2   FERTILIZERS

The nitrogen source for production of synthetic fertilizers is atmospheric nitrogen, which is characterized by a δ^{15}N of 0‰. The total nitrogen in fertilizers is characterized by δ^{15}N values ranging between –3.3 and +3.9‰ with an average value of 0.2‰ (Vitòria et al. 2004). Nitrate fertilizers showed a δ^{15}N range between

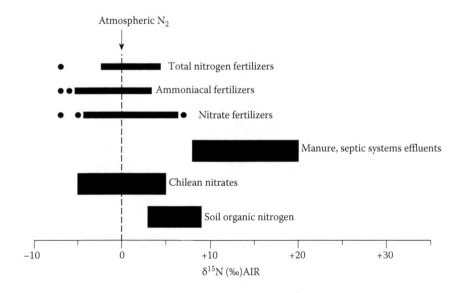

FIGURE 7.1   Isotopic composition of nitrogen compounds.

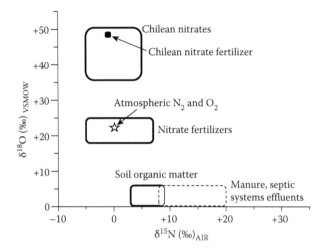

FIGURE 7.2   Relationship between $\delta^{18}O$ and $\delta^{15}N$ values for nitrate of different origin.

–7.5 and +6.6‰ with an average value of 1.8‰ and ammonium fertilizers varies between –7.4 and +3.6‰ with average value of –0.6‰ (Figure 7.1). The nitrate fertilizers are also characterized by $\delta^{18}O$ values ranging between +17.4 and +25.0‰ (Figure 7.2) with an average value of +22.1‰, which is very similar to the $\delta^{18}O$ value of the atmospheric oxygen (+23.5‰), the oxygen source for NO_3^- fertilizer production. One minor group of fertilizers, which are characterized by very high $\delta^{18}O$ values ranging between +36 and +50‰, are natural nitrate deposits formed in the Atacama Desert in Chile (Figure 7.2). The source of the Chilean nitrates is mainly atmospheric deposition and the extremely high $\delta^{18}O$-NO_3^- values are related to chemical reactions in the upper atmosphere (Böhlke, Ericksen, and Revesz 1997).

7.4.3   ANIMAL AND DOMESTIC WASTE

The $\delta^{15}N$ values of fresh animal waste range between +2 and +8‰ and tend to be more ^{15}N enriched than the average isotope composition of industrial fertilizers (Gormly and Spalding 1979; Wassenaar 1995). The isotopic difference between these two nitrogen pools increases significantly due to isotopic fractionation associated with the volatilization of ammonia, which is derived from the hydrolysis of urea. This process (Equation 7.9) favors off-gassing of the $^{14}NH_3$, leaving the residual NH_3 enriched in ^{15}N relative to the manure source material. The reported values for the isotope equilibrium fractionation for NH_3 volatilization is about 1.034 (Urey 1947; Kirshenbaum et al. 1947), meaning that at equilibrium the $\delta^{15}N$ value of gaseous ammonia should be 34‰ lower than that of the initial aqueous ammonia. Volatilization can occur during manure storage in lagoons and also in stockpiles or in the unsaturated zone. The effect of volatilization is clearly observed on data collected from a swine lagoon, which showed $\delta^{15}N$-NH_4 values ranging from 9.8 to 18.4‰ and a clear pattern of summertime ^{15}N enrichment (due to enhanced volatilization) and winter depletion (Karr et al. 2001). Karr, Showers, and Jennings (2003) observed a similar pattern in a dairy lagoon. $\delta^{15}N$ values collected during a study in British Columbia (Wassenaar 1995) showed $\delta^{15}N$ values ranging between 7.9 and 8.6‰ for total organic nitrogen in poultry manure. Soil nitrate collected beneath the poultry manure stockpile showed significant ^{15}N enrichment, with $\delta^{15}N$ values as high as +16‰ associated with ammonia volatilization in the stockpile or the unsaturated zone.

$$NH_3$$
$$\uparrow$$

$$CO\,(NH_2)_2 + H_2O \rightarrow CO_2 + 2NH_3 \leftrightarrow NH_4^+. \tag{7.10}$$

The nitrogen isotope ratios reported for nitrate of domestic waste origin also showed the effect of ammonia volatilization (Kreitler and Browning 1983; Mariotti, Landreau, and Simon 1988; Smith, Howes, and Duff 1991; Aravena, Evans, and Cherry 1993). Nitrate in groundwater impacted by sewage showed a $\delta^{15}N$ range between 8.1 and 24‰, which is higher than the $\delta^{15}N$ range of 5–8‰ reported for fresh sewage (Rivers et al. 1996; O'Connell, Neal, and Hedges). An overall $\delta^{15}N$ range for nitrate of manure and sewage origin is reported in Figure 7.1. The effect of volatilization is also observed in cultivated basic soils that are heavily fertilized with ammonium-based fertilizers. These soils tend to produce nitrate with higher $\delta^{15}N$ values than ammonium fertilizers and nitrate from cultivated nonfertilized soil (Kreitler 1979; Vitòria et al. 2005). Depending on the $\delta^{18}O$ value of the water, nitrate of manure and sewage origin tend to be in the $\delta^{18}O$ range of 0–6‰ (Figure 7.2).

7.5   NITROGEN CASE STUDIES IN GROUNDWATER

The main issues that need to be addressed in aquifers contaminated with nitrate are: (1) sources of the contamination, and (2) processes that control the nitrate concentration in the aquifer. The potential impact of nitrate-contaminated groundwater discharging into surface water bodies is also an important issue that needs to be investigated especially in agriculture landscapes (e.g., Rock and Mayer 2004). A significant number of studies applying stable isotopes analyses have been done in agriculture and urban landscape to address the issues mentioned above. This section will illustrate the application of ^{15}N and ^{18}O using selected case studies that cover the following topics: (a) evaluation of sources of nitrate in groundwater, (b) evaluation of processes that attenuate nitrate in groundwater, and (c) evaluation of the role of riparian zones in the attenuation of nitrate at the groundwater-surface water interface. Reviews on this topic can be found in Heaton (1986) and Kendall and Aravena (2000). A comprehensive description of the use of ^{15}N and ^{18}O in nitrogen studies in surface water is presented in Kendall and McDonnell (1998).

7.5.1   SOURCES OF NITRATE IN GROUNDWATER

Development of short- and long-term groundwater management strategies for aquifers contaminated or at risk of being contaminated with nitrate requires a good understanding of the sources of nitrate and the processes that affect nitrate concentrations from recharge areas toward the production wells. It is common to have multiple sources of nitrate, including inorganic and organic fertilizers and domestic waste (septic system and lagoon) in areas affected by intense agriculture activities. Aquifers in urban and rural environments can also be impacted by nitrate inputs from the leakage of sewage networks and septic systems. A significant number of studies have applied stable nitrogen isotopes to differentiate inorganic fertilizers versus animal waste, fertilizers versus septic system, and natural soil N versus fertilizers and animal or domestic waste (Kreitler and Jones 1975; Kreitler 1979; Kreitler and Browning 1983; Wassenaar 1995; Panno et al. 2001).

7.5.1.1 Industrial and Organic Fertilizers

Most of the nitrate contamination documented in groundwater in regional aquifers is attributed to excessive use of fertilizers (inorganic and animal waste) in agricultural areas. Sometimes nitrate contamination in production wells can be associated to old agricultural practices in recharge areas. Since agricultural practices could have changed from an intense use of industrial fertilizers to a more intensive use of organic fertilizers (manure) or vice versa, it is quite difficult, based only on nitrate concentration data, to identify the sources of nitrate in contaminated aquifers. Several studies have used ^{15}N analysis to evaluate sources of nitrate in areas where industrial fertilizers and animal waste (manure and sewage) were or are potential sources of nitrate (Kreitler and Jones 1975; Kreitler 1979; Kreitler and Browning 1983; Wassenaar 1995; Panno et al. 2001; Wassenaar, Hendry, and Harrington 2006).

One of the key aspects of nitrate studies in groundwater is the understanding of the role of macropore flow on contaminant transport. Due to the distinct isotopic compositions of the main nitrate sources in groundwater, it is possible to use ^{15}N as a tracer to evaluate fast transport via macropores of nitrate that has not been affected by nitrogen cycling in the unsaturated zone. This approach was well documented in a mantled karst aquifer in southern Indiana (Wells and Krothe 1989). The mantle is a clayey soil with an average thickness of 10 m and covered by less than 2 m of loess. In this study 20 wells were sampled before and after inorganic fertilizers had been applied in the area. The data before fertilization showed $\delta^{15}N$-NO_3^- values ranging from 6.8 to 17.6‰, with a mean value of 10.3‰. These data showed that most of the nitrate was of human or animal origin (septic tanks and manure). After fertilization, the $\delta^{15}N$-NO_3^- values ranged from 3.6 to 14.7‰ with a mean value of 8.3‰. Nineteen of the wells showed a shift toward lower $\delta^{15}N$-NO_3^- values and the nitrate concentrations also increased in 14 of the 20 wells after fertilization, clearly showing the input and fast transport of fertilizers by macropore flow into the underlying karst aquifer. A later study in the same area showed no systematic changes in nitrate concentration before and after fertilization and no major shift toward lower $\delta^{15}N$-NO_3^- values were observed in this study (Iqbal, Krothe, Spalding 1997). The different response in the aquifer was attributed to differences in recharge conditions. The study of Wells and Krothe (1989) was done during a wet year and the study of Iqbal, Krothe, Spalding (1997) was carried out during a dry year, which prevented significant leaching and infiltration. However, the $\delta^{15}N$-NO_3^- values collected in the unsaturated zone showed a more complex response in nitrate transport linked to differences in recharge mechanism, climate, and land use (Iqbal, Krothe, Spalding 1997).

7.5.1.2 Soil Nitrogen

In most cases, a high level of nitrate associated with soil nitrogen has been caused by increasing mineralization resulting from the increased cultivation of soils in agriculturally productive countries (e.g., Kreitler and Jones 1975). However as discussed below, natural soil nitrogen in uncultivated areas can also be a significant source of nitrate to groundwater (Heaton, Talma, and Vogel 1983; Edmunds and Gaye 1997).

One excellent study carried out by Kreitler and Jones (1975) in Runnels County, Texas, showed the importance of soil nitrogen as a source of nitrate and the effect of land use on the $\delta^{15}N-NO_3^-$ values of soil nitrate and groundwater nitrate. Nitrate collected in soil profiles beneath cultivated soils without any influence of animal waste showed $\delta^{15}N-NO_3$ values ranging between 3 and 7‰. Similar data collected from soils in a barnyard setting showed much higher $\delta^{15}N-NO_3^-$ values ranging between 10 and 20‰. These values are within the $\delta^{15}N-NO_3^-$ range reported for nitrate under soils affected by animal manure (Jones 1973; Wassenaar 1995). The $\delta^{15}N-NO_3^-$ values higher than 10‰ were also observed for nitrate collected under septic tank drain fields, which are within the range for nitrate of septic system origin reported in other more recent studies (e.g., Aravena et al. 1993). The influence of grazing cattle (animal manure) in a cultivated field with corn is clearly observed in nitrate depth profiles that show an isotopic shift from values around 7‰ in the deepest part of the soil profile toward elevated $\delta^{15}N-NO_3^-$ values as high as 18‰ in the shallow part of the soil profile, clearly indicating the more recent impact by cattle grazing in the shallow soil. Groundwater samples showed a wide range of $\delta^{15}N-NO_3$ values (1 to 15‰), with most data <10‰, clearly demonstrating that soil organic nitrogen was the main source of nitrate in Runnels County, Texas.

Several studies have shown that natural soil organic nitrogen can be an important source of nitrate in semiarid and arid zones (Heaton, Talma, and Vogel 1983; Heaton 1984; Edmunds and Gaye 1997). Nitrate concentrations between 20 to 100 mg-N/L were found in both shallow groundwater and pore water in the unsaturated zone in uncultivated and rain-fed agricultural areas of northern Senegal (Edmunds and Gaye 1997). High concentrations of NO_3-N, up to 37 mg/L were also documented in groundwater in the western part of the Kalahari and the origin of the nitrate was indeed confirmed by ^{15}N isotope analysis (Heaton, Talma, and Vogel 1983). The source of the natural nitrate was nitrogen-fixing vegetation that had accumulated soil organic nitrogen during the last 30,000 years, as demonstrated in a groundwater study in the Sahara and Sahel regions (Edmunds 1999). The role of nitrogen-fixing vegetation in nitrate contamination in arid regions was also demonstrated in a recent study that showed that soil nitrate concentrations were much higher under nitrogen-fixing vegetation than under nonnitrogen-fixing vegetation (Deans et al. 2005).

A recent paleohydrogeological study carried out in the regional High Plains aquifer in the western United States showed the importance of natural nitrogen in groundwater, and was able to link the nitrate isotope fingerprint found in the groundwater to changes in recharge conditions along a north-south transect in the High Plains (McMahon and Böhlke 2006). Groundwater representing recharge over the last 10,000 years showed a $\delta^{15}N-NO_3$ range from 1.3 to 12.3‰ and increased systematically from the northern to the southern High Plains. This large range complicated the use of ^{15}N to evaluate the impact of fertilizers and manure in groundwater from recent agriculture practices. However, the recent groundwater in at least two of the regions is characterized by much higher nitrate concentrations and different $\delta^{15}N$ values than the old groundwater. Based on an inverse relation between the $\delta^{15}N\text{-}NO_3$ and the $NO_3{}^-/Cl$ ratios, it was postulated that the $\delta^{15}N\text{-}NO_3$ range was consistent with isotopically selective losses of N (NH_3 volatilization and denitrification) increasing from the wetter north to the drier south in the High Plains. This study was able to show that ^{15}N of natural nitrate at a regional scale seems to be controlled by climate and possible local variations in soil, vegetation, topography, and moisture conditions.

7.5.1.3 Sewage

The impact of sewage on groundwater can be due to direct input from septic systems and cesspools, use of treated wastewater for irrigation, and due to leakage of the sewage networks in urban areas. It is well documented that septic systems and cesspools are one of the most numerous point sources of nitrate to shallow groundwater systems (Viraraghavan and Warnock 1975; Starr and Sawhney 1980; Robertson, Cherry, and Sudicky 1991; Wilhelm, Schiff, and Robertson 1994). Due to the overlap between the $\delta^{15}N\text{-}NO_3$ values of soils (high end) and those of animal waste (lower end) with $\delta^{15}N\text{-}NO_3$ of septic systems, only a few studies have been able to use ^{15}N to clearly demonstrate the impact of septic systems and cesspools on nitrate contamination in groundwater. Based on $\delta^{15}N\text{-}NO_3$ values ranging between 14 and 24‰ obtained in a freshwater aquifer on Grand Cayman Island, which is populated by small communities and isolated houses, Kreitler and Browning (1983) showed that septic tanks and cesspools are the main sources of nitrate contamination of the fresh groundwater supply on the island. In the case of using wastewater for irrigation, ^{15}N has proven to be a sensitive tracer for evaluating the impact of wastewater on groundwater resources (Smith, Howes, and Duff 1991; Spalding et al. 1993; Moore et al. 2006). For example, a groundwater study carried out in the Livermore Valley, which lies 60 km east of San Francisco, California, showed that based on $\delta^{15}N\text{-}NO_3$ values ranging between 13.0‰ and 29.2‰, the shallow aquifer in the northern part of the valley had been impacted by nitrate of wastewater origin as a result of irrigation practices. No evidence of this impact was found in the deep aquifer since the $\delta^{15}N\text{-}NO_3$ values in the deep wells ranged between 6.7 to 8.8‰, implying the existence of a different nitrate source (Moore et al. 2006).

The increasing population in urban centers has significantly increased the demand for drinking water and has subsequently focused attention on utilization of groundwater from urban aquifers. Several studies have shown nitrate contamination in urban environments (Hiscock et al. 1997; Whitehead, Hiscock, and Dennis 1999; Lerner et al. 1999; Tellam and Thomas 2002) and a few studies have applied isotope techniques to evaluate the sources of nitrate contamination in these environments (Rivers et al. 1996; Aravena et al. 1999; Silva et al. 2002, Fukada, Hiscock, and Dennis 2004; Iriarte 2004). One of these studies was carried out in the city of Nottingham, UK (Fukada, Hiscock, and Dennis 2004). Two representative zones in the city, Old Banford (residential and industrial) and the Meadows (residential) were selected for the study and instrumented with one multilevel piezometer nest at each site. The sandstone unconfined aquifer in the Old Banford area contained elevated nitrate concentrations ranging from 58.5 to 66.7 mg/L, $\delta^{15}N\text{-}NO_3{}^-$ values between 9.2 and 11‰ and no differences were observed between the shallow and deeper part of the aquifer. The $\delta^{15}N\text{-}NO_3^-$ values were very similar to those of raw sewage (Fairbairn 1996) obtained in one of the city sewage treatment plants, supporting the argument presented by Fukada, Hiscock, and Dennis (2004) that the high nitrate concentration found in the aquifer was related to leakage of the sewage distribution system. $\delta^{18}O\text{-}NO_3^-$ values, although slightly more ^{18}O enriched than the theoretical value based on Equation 7.2, also indicated that the nitrate originated from nitrification of sewage NH_4^+. The data collected in the Meadows area is discussed in the nitrate attenuation section.

A recent study in an alluvial unconfined aquifer located in the city of Santiago, Chile, showed an increase in nitrate concentration along the groundwater flow path and the highest nitrate concentrations were observed underneath one of the oldest urban areas of the city (Iriarte 2004). A pattern of increasing nitrate concentration associated with a trend toward elevated $\delta^{15}N$ values in nitrate clearly showed that the source of the nitrate was leakage of the sewage distribution system. The high nitrate groundwater was also characterized by distinct isotopic compositions of H_2O ($\delta^{18}O$, δ^2H) and sulfate ($\delta^{34}S$, $\delta^{18}O$) similar to the source of the potable water for the city of Santiago, the Maipo River. These additional isotopic similarities lend support to the interpretations based on the ^{15}N data.

7.5.1.4   Industrial Sources

Industrial processes such as the manufacturing of ammunition and the production of coke from coal can be point sources of contamination that can generate very high nitrate concentrations in groundwater (DiGnazio et al. 1998; Beller et al. 2004; Talma and Meyer 2005). One of the most comprehensive studies was done at the site of a coking plant of Mittal Steel at Valderbijlpark, South Africa (Talma and Meyer 2005). The coking process converts coal to coke (essentially pure carbon) under high-temperature, anoxic conditions and releases the other constituents of coal in gaseous form. The hot gas is a mixture of ammonia and a range of hydrocarbons (from methane to heavy tars and creosotes). Water is progressively sprayed into the gas lines to cool the hot gases. After most of the organic compounds have been removed, the ammonia solution is collected and transferred to an ammonia sulfate plant. The effluent from the ammonia sulfate plant is collected in evaporation ponds. Talma and Meyer (2005) showed that the liquid in the wastewater evaporation ponds contained ammonia at levels of 41–472 mg/L, nitrate between 1 and 11 mg/L, and $\delta^{15}N$ values between 25 and 59‰. These nitrogen isotope ratios, which are the result of isotope fractionation effects between gas and solvents within the ammonia extraction plant, are more ^{15}N enriched than the most common nitrate sources in groundwater. It was found that these high nitrogen isotope ratios persisted throughout the groundwater system enabling a distinction between the industrial pollutant N and other N sources (Talma and Meyer 2005).

7.5.2   Nitrate Attenuation in Groundwater

Nitrate attenuation in groundwater can be the result of dilution due to mixing of high and low nitrate concentration groundwater or denitrification. Due to the complexity of groundwater flow systems and sometimes a lack of detailed instrumentation, it is quite difficult, based on hydrogeological and chemical data alone, to conclusively document the occurrence of mixing (dilution) and denitrification in contaminated groundwater. Several studies have shown that ^{15}N and ^{18}O are very efficient tracers to evaluate the role of denitrification in groundwater (Böttcher et al. 1990; Smith, Howes, and Duff 1991; Aravena and Robertson 1998; Fukada, Hiscock, and Dennis 2004).

A groundwater study carried out in the Coet-Dan catchment located approximately 70 km southwest of Rennes in western France, offers a good example of the application of multiple isotope ($\delta^{18}O$, δ^2H in water, $\delta^{34}S$ and $\delta^{18}O$ in sulfate, and $\delta^{18}N$ and $\delta^{18}O$ in nitrate) and chemical (anions and cations) parameters to evaluate the main processes involved in nitrate attenuation on an aquifer contaminated with nitrate from agriculture activities (Pauwels, Foucher, and Kloppmann 2000). The aquifer is composed of an upper compartment that acts as storage unit and is comprised of weathered facies with interstitial porosity, as well as the colluvial, loamy, or sedimentary cover. The lower compartment that serves as a transmissive role is comprised of schist with fissure and fracture porosity. The connection between these two compartments seems to be poor but piezometric and chemical data showed significant downward, as well as upward, flow (Pauwels, Foucher, and Kloppmann 2000). The shallow groundwater is characterized by nitrate concentrations up to 200 mg/L and $\delta^{15}N$ values around 10‰ associated with input from manure sources. Fertilizer input during some parts of the year was also inferred from $\delta^{15}N$ data with values less than 10‰. A pattern of ^{15}N enrichment accompanied by decreasing nitrate concentrations was explained as a mixing of low nitrate strongly denitrified groundwater with nitrate-rich groundwater. The role of denitrification at this site was also supported by N_2 data that showed an excess of N_2 in comparison with air and data collected in a previous tracer experiment (Pauwels et al. 1998).

In the case of an aquifer impacted by urban activities in the city of Nottingham (Fukada, Hiscock, and Dennis 2004), the data collected in the Old Banford area contained elevated nitrate concentrations ranging from 58.5 to 66.7 mg/L, the $\delta^{15}N$ values were between 9.2 and 11‰ and no differences were observed between the shallow and deeper part of the aquifer. A significantly different pattern was observed in the Meadows area. The nitrate concentrations were much lower ranging between 7.6 and 31.3 mg/L and the $\delta^{15}N$ values were much higher ranging between 24 and 42‰ than at the old Banford area Figure 7.3). The $\delta^{18}O$ values of nitrate at the Meadows were also elevated varying between +20.5 to 29.4‰. The linear correlation observed between the isotope composition and the natural log of the nitrate concentration Figure 7.4, the ^{18}O data showed a similar pattern) provides clear evidence that denitrification controlled nitrate concentrations in the Meadows area. The enrichment factors calculated in this study, −13.7 for ^{15}N and −6.9 for ^{18}O, are within the range of values reported for denitrification in water (Böttcher et al. 1990; Kendall 1998; Fukada et al. 2003). The linear correlation between $\delta^{15}N$ and $\delta^{18}O$ values (Figure 7.4), with ^{15}N enrichment twice the ^{18}O enrichment, also supports the occurrence of denitrification (Aravena and Robertson 1998; Cey et al. 1999; Fukada et al. 2003). It seems that the presence of overlying fill material (reclaimed swampland) and longer groundwater residence times controlled by geological features promoted denitrification in the Meadows area.

It is well documented that nitrate plumes generated by septic systems in aerobic groundwater can travel for long distances without any apparent attenuation

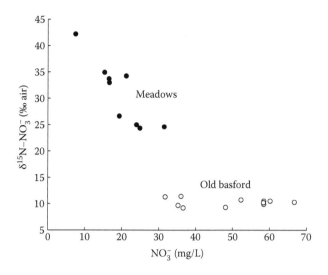

FIGURE 7.3   Relationship between $\delta^{15}N$ values and concentration of nitrate in urban settings (From Fukada, T., K. M. Hiscock, and P. F. Dennis, *Applied Geochemistry,* 19, 709–19, 2004).

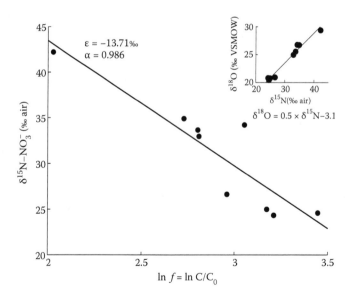

FIGURE 7.4   Cross plot of the $\delta^{15}N$ of nitrate versus $\ln f$, where f is the ratio of the nitrate concentration C/C_0; Insert: relationship between $\delta^{15}N$ and $\delta^{18}O$ for nitrate in an urban setting (From Fukada, T., K. M. Hiscock, and P. F. Dennis, *Applied Geochemistry,* 19, 709–19, 2004).

(Aravena, Evans, and Cherry 1993). However if the redox conditions are conducive for denitrification (anaerobic and plenty of electron donors), these plumes tend to travel only short distances thereby limiting their impact on nearby surface water bodies. This is the case for a nitrate plume generated by a large septic system that served a campground (200 persons in the summer) in southern Ontario (Aravena and Robertson 1998). The plume showed a nitrate concentration around 65 mg-N/L at the water table and a trend of decreasing nitrate concentrations with depth, reaching values as low as 0.3 mg-N/L at five meters depth. This decreasing nitrate concentration trend was accompanied by a ^{15}N enrichment trend starting with $\delta^{15}N$-NO_3^- values around 6–10‰ at the water table and reaching $\delta^{15}N$-NO_3^- values as high as +58.3‰ in the deepest piezometers (Figure 7.5). This pattern and the linear relationship observed between $\delta^{15}N$-NO_3^- and $\delta^{18}O$-NO_3^- values clearly demonstrated that the attenuation of nitrate was controlled mainly by denitrification. This study showed that organic carbon was the main electron donor and that about 20% of the nitrate was reduced by iron sulfide. The latter process was documented by a trend of increasing sulfate concentrations accompanied with a change in the isotopic composition of sulfate, with $\delta^{34}S$ values trending toward those of pyrite with increasing depth. The optimum conditions for denitrification at this site were triggered by the presence of pyrite in the sand aquifer and high concentrations of Dissolved organic Carbon (DOC), which consumed oxygen in the unsaturated zone and acted as electron donors in the saturated zone.

7.5.3 Groundwater–Surface Water Interaction

The fate of nitrate in groundwater is intimately linked to redox conditions. Organic carbon and sulfide minerals acting as electron donors play a significant role in the

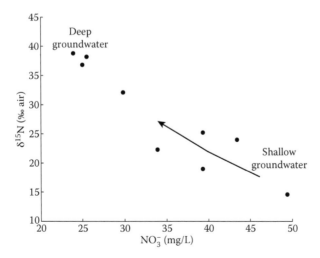

FIGURE 7.5 Isotope and concentration patterns for nitrate in a septic system plume (From Aravena, R., and W. Robertson, *Ground Water*, 36, 975–82, 1998).

process of denitrification (Korom 1992; Hedin et al. 1998). Sediments rich in organic carbon characterize riparian areas (wetlands and buffer strips) near rivers and lakes. Therefore, these areas have the potential to create the optimal conditions for denitrification, which are anaerobic conditions and an abundant carbon source for the denitrifying bacteria. During the last decades, a significant amount of work has been conducted to understand the role of riparian zones as buffer zones to reduce nitrate loading from terrestrial to aquatic ecosystems (Peterjohn and Correll 1984; Lowrance 1992; Haycock and Pinay 1993; Pinay, Rogues, and Fabre 1993; Jordan, Correll, and Weller 1993; Gilliam 1994; Hill 1996; Jacinthe et al. 1998; Hill et al. 2000). Most of the studies have been done in riparian zones characterized by shallow aquitards and a few studies included a detailed evaluation of the nitrate plume and the groundwater flow system. In general, several lines of evidence, including redox conditions (dissolved oxygen and dissolved organic carbon), in situ acetylene block technique and aquifer characterization (stratigraphy and flow system), have been used to evaluate the role of denitrification in the attenuation of nitrate in riparian zones. During the last decade, a number of studies have added ^{15}N and ^{18}O in nitrate as new tools to evaluate denitrification in riparian zones (Cey et al. 1999; Mengis et al. 1999; Devito et al. 2000; Fukada, Hiscock, and Dennis 2003; Vidon and Hill 2004a,b). One comprehensive hydrogeological and geochemical study that illustrated very well the use of ^{15}N and ^{18}O in riparian zones is the study carried out in a riparian wetland (Boyne River) in southern Ontario reported by Devito et al. (2000) and Hill et al. (2000). The riparian wetland had been impacted by a high concentration of nitrate groundwater from up-gradient agricultural activities. A 150 m nitrate plume showed a pattern of high nitrate concentration ranging from 20 to 30 mg-N/L in the up-gradient areas and a trend toward lower concentration along the groundwater flow system reaching values <5 mg-N/L in the areas near the river. The decreasing nitrate concentrations were accompanied by an increase in DOC concentrations. A later study (Hill et al. 2000) showed that the groundwater is aerobic in the up-gradient areas and became anaerobic in the areas of low nitrate concentration. All the evidence indicated that denitrification was the main process responsible for the attenuation of nitrate along the groundwater flow system. This conclusion was confirmed by isotope data that showed a ^{15}N enrichment trend starting with $\delta^{15}N\text{-}NO_3^-$ value of ~6‰ in the up-gradient areas (high nitrate concentration) shifting toward values as high as 27‰ in the down-gradient areas (low nitrate concentration). This pattern and the linear relationship observed between $\delta^{15}N\text{-}NO_3^-$ and $\delta^{18}O\text{-}NO_3^-$ clearly demonstrated that denitrification was the key process responsible for nitrate attenuation at the Boyne River wetland (Devito et al. 2000).

Based on a ^{15}N enrichment pattern associated with a trend toward lower DO concentration plus other geochemical evidence, Vidon and Hill (2004b) were able to demonstrate that denitrification was playing a significant role on nitrate attenuation in eight riparian wetlands in Ontario, Canada. The isotope and geochemical research on riparian zones has been further advanced by the use of numerical simulations. Chen and MacQuarrie (2004) compared numerical simulation using a reactive transport model to field data collected in the riparian wetland discussed above (Devito et al. 2000) and were able to reproduce the essential reactive transport processes for the major reactive species, NO_3^- and DOC, including the ^{15}N pattern observed at the site.

In general, riparian zones are in areas of groundwater discharge. Heavy groundwater extraction, however, can reverse the natural flow directions creating the conditions necessary for river infiltration into the aquifer. This is the case at a site near Torgau in eastern Germany, located in the River Elbe basin where several production wells about 300 m from the river are in operation (Trettin, Johnson, and Todd 1999). This study showed that river bank infiltration affected the quality of the groundwater extracted from production wells in the Torgau aquifer. A later study in the same area provides a good example of the use of ^{15}N and ^{18}O to evaluate the fate of nitrate as it moved from the river bank infiltration areas into the aquifer (Fukada et al. 2003). Two cross sections perpendicular to the river and comprised of several piezometer nests were used in this study. The δ^{15}N-NO$_3^-$ values ranged between 8.9 (River Elbe) to 20.6‰ and the δ^{18}O-NO$_3^-$ values ranged between 8.0 (River Elbe) to 16.8‰. The ^{15}N and ^{18}O enrichment trend seemed to correlate with distance from the river and was accompanied by a decrease in nitrate concentration in one of the cross sections. This pattern was not as clear in the other cross sections and was explained by a difference in nitrate input to the aquifer during different infiltration periods and differences in groundwater residence times in the aquifer (Fukada et al. 2003). Overall, the nitrogen and oxygen isotope data showed a clear trend of isotope enrichment correlated with decreasing nitrate concentrations, identifying denitrification as the main mechanism for nitrate attenuation at the study site. This conclusion was corroborated by the linear relationship observed between the δ^{15}N-NO$_3^-$ and δ^{18}O-NO$_3^-$ data that showed the typical trend attributed to denitrification in other studies (Cey et al. 1999; Mengis et al. 1999; Devito et al. 2000). The slope of the relationship between the δ^{15}N and δ^{18}O values of nitrate was 1.3 and is at the lower end for values reported in other studies (1.5 and 2.1; Cey et al. 1999; Mengis et al. 1999; Devito et al. 2000).

One of the main conclusions of the work in riparian zones is that understanding the groundwater flow system is essential for evaluating the fate of nitrate in these ecosystems. When assessing whether a riparian zone might be effective at attenuating nitrate, it is essential to fully characterize both the horizontal and vertical variation in geological and geochemical conditions within the zone. One study that clearly illustrates this case was carried out in an agricultural watershed in southern Ontario (Cey et al. 1999). Based on a detailed evaluation of the geology and groundwater flow system, this study showed that most of the nitrate-contaminated groundwater from the agricultural field discharged into the regional groundwater flow system and it did not discharge into the stream. The nitrate data showed a decrease in nitrate concentration in the direction of the regional groundwater flow system and near the stream. The combination of the ^{15}N and ^{18}O data confirmed than the trend toward lower nitrate concentrations observed in the regional groundwater flow component was related to denitrification and the low nitrate and chloride observed near the stream was due to the low application rate of manure fertilizers in that area (Cey et al. 1999).

7.6 EVALUATION OF LONG-TERM CHANGES IN MANAGEMENT PRACTICES

Due to an increasing demand for potable water in rural communities and urban centers, water managers and agricultural organizations started to focus their efforts on reducing nitrate contamination in regional aquifers. Since the 1990s, agricultural nutrient beneficial management programs started to be implemented with the aim of reducing the environmental impact of agricultural nutrients on groundwater and surface water quality. Few studies have been performed to evaluate the long-term ability of BMPs to reduce nitrate contamination in groundwater. A recent study illustrates the use of multiple isotopes and geochemical tracers, including ^{15}N and ^{18}O, to evaluate the efficiency of BMPs in the Abbotsford-Sumas regional sand and gravel aquifer, which is used as a water supply for over 100,000 people in southwestern British Columbia, Canada, and Washington State, USA (Wassenaar, Hendry, and Harrington 2006). Several studies have shown widespread nitrate contamination in the aquifer and many wells exceed the drinking water standards for nitrate (Wassenaar 1995, Mitchell et al. 2003). A previous isotope study performed in 1993 showed that the aquifer was impacted by nitrate derived from manure used as a fertilizer and also by leaching from manure stockpiles that were stored on top of the aquifer without any protective lining (Wassenaar 1995). The initiatives promoted by the BMP program in the Abbotsford region included the development of alternative markets for manure utilization and changes in crop practices including diversification to blueberries because of their low nutrient requirements and a shift from manure to inorganic fertilizers. A comparison between nitrate data collected in monitoring and domestic wells in 1993 and 2004 showed no significant changes in nitrate concentration. Furthermore, the comparison of ^{15}N-NO_3^- data collected in 1993 and 2004 showed, in spite of the implementation of BMPs in the region, manure remained the main source of nitrate in the aquifer in part due to comparatively long groundwater mean residence times (Wassenaar, Hendry, and Harrington 2006). However, a more focused evaluation of isotope data collected from groundwater with a residence time <5 years showed a trend toward lower $\delta^{15}N$ values, implying a shift from manure-derived nitrate toward fertilizer-derived nitrate in this groundwater. This information could not be obtained based solely on nitrate concentration patterns and therefore shows the uniqueness and value of the isotope fingerprinting approach to evaluate changes in nitrate sources associated with implementation of BMPs in the agriculture sector.

7.7   SULFUR ISOTOPE FUNDAMENTALS

Sulfate is a major constituent of many surface waters and groundwaters. Identifying potential sources and the biogeochemical history of dissolved sulfate is often of significant interest for water resource managers, since sulfate may cause acidification (Reuss and Johnson 1986), contributes to the hardness of water, and is an important terminal electron acceptor in reductive processes (Appelo and Postma 2005). For the delineation of various sources of sulfate in the hydrosphere, the application of stable isotope techniques has proven most useful (e.g., Hitchon and Krouse 1972; Rightmire et al. 1974; Fritz and Drimmie 1991; Krouse and Mayer 2000; Knoeller, Trettin, and Strauch 2005; Schiff et al. 2005; Bottrell et al. 2008). This approach is particularly successful when both the sulfur and oxygen isotope abundance ratios of sulfate are determined. Furthermore, sulfur transformation processes can be identified by analyzing the isotopic composition of sulfate and other sulfur species, particularly if the observed trends in the isotope ratios are interpreted in concert with hydrological and chemical parameters. This section describes the use of sulfur and oxygen isotope ratios as tools to evaluate sulfate sources and microbially mediated processes that control sulfate concentrations in the hydrosphere and in sedimentary systems.

Sulfur, element 16, has four stable isotopes ^{32}S, ^{33}S, ^{34}S, and ^{36}S. The abundance ratio of the two most abundant stable sulfur isotopes $(^{34}S/^{32}S)$ of a sample is measured relative to that of a reference material and expressed as $\delta^{34}S$ (Chapter 1). The historical reference for the $\delta^{34}S$ scale was CDT troilite (FeS) from the Cañon Diablo meteorite. However, Beaudoin et al. (1994) concluded that variations in $\delta^{34}S$ values of CDT due to inhomogeneity were at least 0.4‰. Consequently, an advisory committee of the International Atomic Energy Agency (IAEA, Vienna, Austria) established a V-CDT scale in 1993, on which an Ag_2S reference material IAEA-S-1 was defined as having a $\delta^{34}S$ value of –0.30‰ (Coplen and Krouse 1998). Further reference materials such as IAEA-S-2 (+22.67‰) and IAEA-S-3 (–32.55‰) with widely differing $\delta^{34}S$ values (Coplen et al. 2002) are available from the IAEA for calibration and normalization purposes.

Of oxygen's three stable isotopes, ^{16}O and ^{18}O are usually measured on a $\delta^{18}O$ scale (Chapter 1). For oxygen in both SO_4^{2-} and H_2O, the reference is V-SMOW (Vienna Standard Mean Ocean Water), which is the recommended replacement for SMOW referred to in earlier literature. V-SMOW by definition has a $\delta^{18}O$ value of 0‰. The oxygen isotope composition of sulfate is influenced by the $\delta^{18}O$ value of atmospheric O_2 and that of water in which it forms. Atmospheric O_2 has a $\delta^{18}O$ value of +23.5‰ (Kroopnick and Craig 1972; Horibe, Shigehara, and Takakuwa 1973). Because of differences in the vapor pressure of isotopic species of H_2O, the $\delta^{18}O$ value of precipitation ranges from slightly positive to as low as –55 ‰ dependent upon temperature, location, storm path, and other factors. The oxygen isotope composition of groundwater is usually similar to that of mean annual precipitation in the recharge area. Reviews on natural oxygen isotope abundance variations in water include Epstein and Mayeda (1953), Craig (1961), Dansgaard (1964), Moser and Rauert (1980), Clark and Fritz (1997), and Kendall and McDonnell (1998) among others.

7.8 THE SULFUR CYCLE

The variability in terrestrial sulfur isotope compositions is quite large due to the wide range of valence states (+6 to –2) for the element and its many functional groups. Thode, Macnamara, and Collins (1949) effectively established the general pattern of sulfur isotope abundance variations in nature. Troilite in meteorites is markedly consistent isotopically within 2‰ of the average $\delta^{34}S$ value of the Earth's crust. As theoretically predicted by statistical thermodynamics, sulfur in higher valence states tends to be enriched in the heavier isotopes. However, the sulfur cycle is far from isotopic equilibrium and unidirectional processes are primarily responsible for the isotope abundance variations. Therefore, natural sulfur compounds can be found with $\delta^{34}S$ values covering a wide range from lower than –50‰ to higher than +50‰. Reviews on natural variations in $\delta^{34}S$ values include Nielsen (1979), Krouse (1980), Krouse and Tabatabai (1986), Canfield (2001a), Krouse and Grinenko (1991), and Hoefs (1997) among others.

7.8.1 Isotopic Fingerprint of Sulfate Sources

Numerous case studies investigating the isotopic composition of sulfate in precipitation (e.g., Östlund 1957; Jensen and Nakai 1961; Grey and Jensen 1972), river water (e.g., Longinelli and Cortecci 1970; Hitchon and Krouse 1972; Chukhrov et al. 1975), lake water (e.g., Deevey, Nakai, and Stuiver 1963; Matrosov et al. 1975), and groundwater (e.g., Müller, Nielsen, and Ricke 1966; Rightmire et al. 1974; Zuppi, Fontes, and Letolle 1974; Fontes and Zuppi 1976) have been conducted throughout the last five decades with the objective to determine the sources of sulfate. Generally, dissolved sulfate in the hydrosphere can be of atmospheric, pedospheric, or lithospheric origin. In addition, anthropogenic sulfate sources with varying isotopic compositions are ubiquitous in many watersheds worldwide.

Typical isotopic compositions of sulfate derived from different sources are summarized in Figure 7.6. In industrialized countries, the majority of sulfur in atmospheric sulfate deposition is of anthropogenic origin (e.g., Benkovitz et al. 1996). The $\delta^{34}S$ values between -1 and $+6‰$ were found to be characteristic for atmospheric sulfate deposition in these regions (Mizutani and Rafter 1969b; Cortecci and Longinelli 1970; Wadleigh, Schwarcz, and Kramer 1994; Herut et al. 1995; Moerth and Torssander 1995; Wakshal and Nielsen 1982; Alewell et al. 2000). Elsewhere, natural sulfur compounds may contribute significantly to atmospheric sulfate deposition. Sulfur from volcanic emissions often has $\delta^{34}S$ values near $+5‰$ (Lein 1991) and may be isotopically indistinguishable from atmospheric sulfur of anthropogenic origin. Emissions of reduced biogenic sulfur gases from wetlands may result in negative

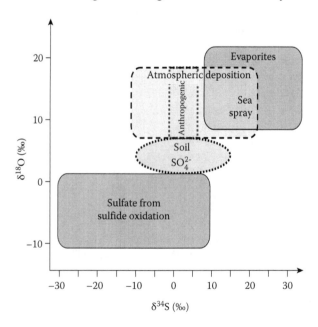

FIGURE 7.6 Typical ranges for $\delta^{34}S$ and $\delta^{18}O$ values of sulfate from (a) evaporites, (b) atmospheric deposition, (c) soil solutions, and (d) sulfide oxidation (Modified from Mayer, B., *Isotopes in the water cycle: Past present and future of a developing science,* Springer, Dordrecht, 2005).

$\delta^{34}S$ values of atmospheric sulfate deposition (Nakai and Jensen 1967; Nriagu and Coker 1978). In coastal regions, sea spray is often the dominant source of atmospheric sulfate and $\delta^{34}S$ values as high as +21‰ have been observed (Nielsen 1974; Chivas et al. 1991; Wadleigh, Schwarcz, and Kramer 1994; Wadleigh, Schwarcz, and Kramer 1996). The oxygen isotope composition of atmospheric sulfate is dependent on the oxidation conditions in the atmosphere and the $\delta^{18}O$ value of the moisture involved in the oxidation of precursor compounds such as SO_2. Comprehensive reviews on the relevant oxidation mechanisms and the associated isotope effects have been provided by Holt and Kumar (1991) and Van Stempvoort and Krouse (1994). In temperate regions, $\delta^{18}O$ values between +5 and +17‰ are typical for sulfate in atmospheric deposition (e.g., Rafter and Mizutani 1967; Cortecci and Longinelli 1970), with the lower values observed in winter and the high values associated with summer precipitation.

Lithospheric sulfur occurs predominantly in sedimentary rocks, either as sulfate minerals (gypsum, anhydrite, etc.) in evaporite deposits or as reduced inorganic sulfur in minerals such as pyrite, which are common in shale and many other rock types. The isotopic composition of evaporite sulfate deposited throughout Earth's history has been summarized by Claypool et al. (1980) and more recently by Strauss (1997). The $\delta^{34}S$ values range between 35‰ (late Precambrian and Cambrian) and 8‰ (Permian). The respective $\delta^{18}O$ values for evaporite sulfate vary between +7 and +20‰ dependent upon geological age (Claypool et al. 1980). Current ocean water sulfate has a $\delta^{34}S$ value of 21‰ (e.g., Rees, Jenkins, and Monster 1978) and a $\delta^{18}O$ value near 9.5‰ (Lloyd 1967). Reduced inorganic sedimentary sulfur compounds often (but not always) have negative $\delta^{34}S$ values (e.g., Migdisov, Ronov, and Grinenko 1983; Strauss 1999). The $\delta^{34}S$ value of sulfate generated via oxidation of reduced inorganic sulfur is usually similar to that of the precursor minerals, since there tends to be only minor sulfur isotope fractionation during oxidation (Price and Shieh 1979). The oxygen isotope ratios of sulfate generated by sulfide oxidation are typically lower than those of sulfate from evaporitic or atmospheric sources (Figure 7.6). During the oxidation of sulfide minerals to SO_4^{2-}, four oxygen atoms are incorporated into the newly formed sulfate, which are either derived from H_2O or atmospheric O_2. The proportion of water and atmospheric oxygen incorporated into the sulfate molecule is dependent on whether the reaction occurs under reducing or oxidizing conditions (e.g., Taylor, Wheeler, and Nordstrom 1984; Taylor and Wheeler 1994; Balci et al. 2007). Regardless of the exact stoichiometry, sulfate derived from oxidation of sulfide minerals often has a distinct isotopic composition with comparatively low $\delta^{34}S$ and $\delta^{18}O$ values.

Additionally, anthropogenic sources may contribute sulfate with characteristic, but widely differing isotope compositions depending on location, to groundwater and surface water. For example, municipalities using surface water for their drinking water supply often use aluminium sulfate as an additive for settling particulate matter during water processing. Soaps and detergents contain considerable amounts of sulfate ($\delta^{34}S$ values often near +1‰ (Mayer et al. 2006), which is released into rivers via waste water effluents. Drywalls in leaky landfills or by-products of the sour gas industry are other examples for anthropogenic sulfate point sources. Ammonium sulfate is a widely used fertilizer in agriculture and on golf courses. The fertilizer-derived sulfate may infiltrate into groundwater or surface water bodies since it is usually applied in rates far exceeding the sulfur requirements of the biosphere.

7.9 ISOTOPE EFFECTS DURING MICROBIAL SULFUR TRANSFORMATION PROCESSES

There are five types of microbial conversions that may occur in water-unsaturated or water-saturated environments (Figure 7.7)

1. Immobilization; conversion of simple inorganic-S compounds, usually sulfates, into organic sulfur compounds, for example, by microorganisms or macrophytes. This process is also known as assimilatory bacterial sulfate reduction.
2. Mineralization; cleaving of large organic-S molecules into smaller entities and eventual conversion to SO_4^{2-}.
3. Bacterial dissimilatory sulfate reduction; SO_4^{2-} reduced to intermediate valence state S-containing anions with the final product being sulfide.
4. Oxidation; conversion of S, thiosulfates, polythionates, sulfides, and so on to SO_4^{2-}.
5. And finally, disproportionation of sulfur intermediate compounds.

Microorganisms have long been known to impart isotope fractionation during their metabolism (Harrison and Thode 1957; Harrison and Thode 1958; Kaplan and Rittenberg 1964; Chambers and Trudinger 1979). Therefore, sulfur isotope studies

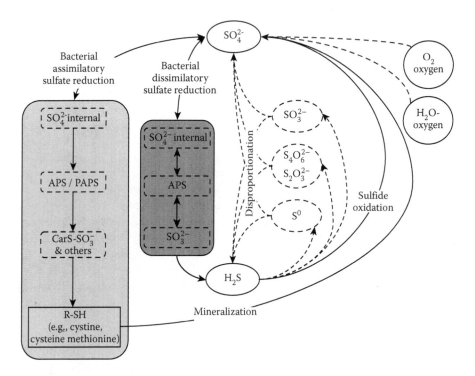

FIGURE 7.7 Schematic summary of microbially mediated sulfur transformation.

can play a key role in identifying the occurrence and extent of microbial processes in aquatic and sedimentary systems.

7.9.1   IMMOBILIZATION

The formation of carbon-bonded organic sulfur via assimilatory sulfate reduction (Figure 7.8) is a cell-internal multistep conversion of sulfate via adenosine-5′-phosphpsulfate (APS) and sulfite to compounds such as cysteine, cystine, and methionine (Brunold 1990; Cram 1990). In contrast to bacterial dissimilatory sulfate reduction (Section 7.5.3), sulfur isotope selectivity during bacterial assimilatory sulfate reduction is small, typically of the order of <2‰ (Kaplan and Rittenberg 1964; Trust and Fry 1992). Consequently, the $\delta^{34}S$ values of the formed carbon-bonded sulfur compounds are usually within 1 to 2‰ of that of dissolved sulfate in the system.

7.9.2   MINERALIZATION

The reverse process to immobilization is mineralization of organic sulfur compounds to SO_4^{2-} by microorganisms (Figure 7.7). The lack of direct analytical techniques to determine sulfur isotope ratios of organic sulfur compounds (Mayer and Krouse 2004) makes an assessment of the extent of isotope effects during mineralization challenging. There is some evidence from studies in forest soils that ^{32}S is favored slightly during the mineralization of organic sulfur compounds, leading to an enrichment of ^{34}S in the remaining organic soil sulfur (Mayer, Fritz, and Krouse 1992; Gebauer et al. 1994; Novak, Buzek, and Adamova 1999; Norman et al. 2002). This effect is likely in the range of <2‰.

It is important to realize that during mineralization of carbon-bonded sulfur compounds four new oxygen atoms are incorporated into the newly formed SO_4^{2-}. The majority of this oxygen is derived from water with negative $\delta^{18}O$ values depending on location, while a smaller proportion of the oxygen may be derived from atmospheric O_2 (Mayer et al. 1995a). As a consequence, sulfate derived from mineralization often has $\delta^{18}O$ values between 0 and +6‰ (Figure 7.6).

7.9.3   BACTERIAL DISSIMILATORY SULFATE REDUCTION

Dissimilatory sulfate reduction under environmental conditions is conducted by a specialized group of prokaryotic bacteria, who gain energy for their growth by catalyzing chemical reactions in which organic carbon (or H_2) is oxidized, while sulfate is reduced (Canfield 2001a)

$$2 \, CH_2O + SO_4^{2-} \rightarrow H_2S + 2 \, HCO_3^{-}. \qquad (7.11)$$

It has long been known that the light isotope ^{32}S is preferentially metabolized during bacterial dissimilatory sulfate reduction (e.g., Harrison and Thode 1958; Mizutani and Rafter 1969a). Consequently, the remaining sulfate becomes progressively enriched in ^{34}S as sulfate concentrations decrease. The standard Rees (1973) model comprised of uptake of sulfate into the cell (step 1: $\varepsilon \sim -3\text{‰}$), the activation of sulfate as APS (step 2: $\varepsilon \sim 0\text{‰}$), the reduction to sulfite (step 3: $\varepsilon \sim 22\text{‰}$), and the subsequent conversion to H_2S (step 4: $\varepsilon \sim 24\text{‰}$) is often used to explain sulfur isotope fractionation during bacterial dissimilatory sulfate reduction (with ε representing the enrichment factors for the individual steps). The overall sulfur isotope fractionation during BSR is dependent upon the rate-limiting step and can vary from less than 10‰ to more than 45‰. Factors influencing the extent of sulfur isotope fractionation include the sulfate concentration in the system (Harrison and Thode 1957; Kaplan and Rittenberg 1964), the availability and type of carbon source (Canfield 2001b; Detmers et al. 2001), temperature (Canfield, Olesen, and Cox 2006), and cell specific reduction rates (Habicht and Canfield 1997). Although sulfur isotope enrichment factors during bacterial dissimilatory sulfate reduction can be quite variable dependent upon environmental conditions (0 to 46‰), they often range between 10 and 20‰ in hydrological settings (e.g., Strebel, Boettcher, and Fritz 1990).

Historically, oxygen isotope effects during bacterial dissimilatory sulfate reduction have been studied in less detail than sulfur isotope fractionation. Early work by Mizutani and Rafter (1973) suggested that the $\delta^{34}S$ and $\delta^{18}O$ values in the remaining sulfate increase at a ratio of 4:1. More recent work has shown that this is generally not the case and that oxygen isotope ratios may approach a constant value while $\delta^{34}S$ values continue to increase (Mizutani and Rafter 1969a; Fritz et al. 1989; Aharon and Fu 2000; Boettcher, Thamdrup, and Vennemann 2001; Aharon and Fu 2003). It has been suggested and demonstrated that this is caused by equilibrium oxygen isotope exchange between water and the intermediate sulfite followed by back-reactions influencing the $\delta^{18}O$ value of sulfate in solution (Mizutani and Rafter 1973; Fritz et al. 1989; Brunner et al. 2005; Knoeller et al. 2006; Mangalo et al. 2007). Spence et al. (2001) demonstrated for a phenol-contaminated aquifer that bacterial (dissimilatory) sulfate reduction proceeded with significant sulfur isotope selectivity but no concurrent change in the oxygen isotope ratios of the remaining sulfate.

7.9.4 SULFIDE OXIDATION

The pathways of sulfide oxidation in nature are varied, including inorganic oxidation of sulfide to sulfate and other intermediate sulfur compounds and biologically mediated oxidation of sulfide (Figure 7.7), and are often rather poorly understood (Canfield 2001a). In most cases oxidation of sulfur in lower valence states is accompanied by relatively small sulfur isotope fractionation (Kaplan and Rittenberg 1964; Fry, Gest, and Hayes 1984; Fry et al. 1986; Fry et al. 1988; Canfield 2001a; Balci et al. 2007), one reason being that reactants such as metal sulfides are often in the solid phase.

The oxygen isotope behavior during chemical and bacterial oxidation of sulfur compounds is rather complex since there are two sources of oxygen, O_2 and H_2O, involved. The topic has been thoroughly reviewed with experimental data for various environments examined by Van Stempvoort and Krouse (1994), Taylor and Wheeler (1994), and Balci et al. (2007). During aerobic oxidation, typically more than two-thirds of the oxygen in SO_4^{2-} is derived from H_2O accompanied by a small oxygen isotope enrichment factor of about 3‰ and a larger enrichment factor of circa 10‰

for O_2 incorporation. During oxidation of sulfide under anaerobic conditions, for example, by using dissolved Fe(III), all of the sulfate-oxygen is derived from water (Balci et al. 2007). Independent upon whether the oxidation of sulfides occurs abiotic or mediated by bacteria, the newly formed sulfate will have $\delta^{18}O$ values below 5‰ and as low as –20‰ dependent on the environmental conditions. It is also important to note that oxygen isotope exchange between water and sulfate is extremely slow at near neutral pH values and environmental temperatures (Lloyd 1967; Chiba and Sakai 1985) Therefore, once sulfate has formed via oxidation, its oxygen isotope ratio is usually preserved.

7.9.5 DISPROPORTIONATION OF SULFUR INTERMEDIATE COMPOUNDS

It has also been shown that some microorganisms are capable of disproportionating sulfur intermediate compounds such as sulfite (SO_3^{2-}), thiosulfate ($S_2O_3^{2-}$), and elemental sulfur to sulfide and sulfate (Bak and Pfennig 1987; Habicht, Canfield, and Rethmeier 1998; Thamdrup et al. 1993; Figure 7.8). Sulfur isotope fractionation associated with the disproportionation of elemental sulfur has been shown to rather constantly deplete sulfide in ^{34}S by circa 6‰, whereas the sulfate is enriched in ^{34}S by usually circa 18‰ (Canfield 1998; Canfield and Thamdrup 1994). During disproportionation of sulfite, larger depletions of sulfide in ^{34}S by 21–37‰ have been observed, whereas sulfate was found to be enriched in ^{34}S by 9–12‰ (Habicht, Canfield, and Rethmeier 1998). Whether this process is of relevance in aquifers featuring reducing conditions is currently unclear.

7.10 CASE STUDIES

The above described principles of isotope distribution in the sulfur cycle have been utilized in numerous case studies throughout the last five decades to determine sources and the biogeochemical history of sulfur compounds in terrestrial and aquatic ecosystems (e.g., Krouse and Grinenko 1991). The following sections discuss selected examples with specific relevance to microbial conversions in the hydrosphere.

7.10.1 TRACING SULFATE DURING RECHARGE

Atmospheric deposition contains sulfate typically in low concentrations (usually <10 mg/L) that is transported into groundwater and surface water during recharge and runoff. It is interesting to note that sulfate in the water-unsaturated soil zone often has $\delta^{34}S$ values similar to those of atmospherically derived sulfate, but its oxygen isotope ratios are significantly depleted in ^{18}O. This indicates that sulfate does not behave conservatively in the water-unsaturated zone during groundwater recharge (Caron et al. 1986; Mayer et al. 1995b). Particularly in forest soils, SO_4^{2-} may be converted to carbon-bonded S compounds (immobilization). During the subsequent reoxidation of these organic S compounds (mineralization), new oxygen atoms are incorporated into the generated sulfate molecule. These microbially mediated reactions explain why soil sulfate, and consequently sulfate in freshly recharged groundwater, may have lower $\delta^{18}O$ values than sulfate in atmospheric deposition (Figure 7.6).

7.10.2 ABIOTIC AND MICROBIAL SULFIDE OXIDATION

When sulfide-containing sediments are exposed to oxidizing conditions, the sulfide may oxidize to sulfate. This results in increasing sulfate concentrations in the aquifer accompanied by decreasing $\delta^{34}S$ values (Hendry, Krouse, and Shakur 1989; Fennell and Bentley 1998) in case that the $\delta^{34}S$ values of the sulfide minerals are significantly lower than those of the background sulfate in the groundwater. While this process may occur naturally (e.g., due to water table fluctuations), it is often enhanced by excessive pumping of groundwater (Krouse and Mayer 2000). Figure 7.8 illustrates data from a case study in southern Ontario (Canada), where sulfate concentrations as high as 2300 mg/L were observed in shallow depths in a low permeability till in a highly industrialized area (Abbot 1987). Sulfur associated with ashes generated by hydroelectric carbon plants were suggested as a potential sulfate source and its $\delta^{34}S$ values were found to range between 3 and 8‰. Dissolved sulfate in shallow groundwater, however, was found to have much lower $\delta^{34}S$ values around −15‰ (Figure 7.8), revealing that ash-derived sulfate was not the source of groundwater sulfate. Also, sulfate from the evaporitic Silurian Salina Formation with $\delta^{34}S$ values between 24 and 32‰ was excluded as a potential sulfate source due to its high sulfur isotope ratios being inconsistent with those of groundwater sulfate. Due to the negative $\delta^{34}S$ values of groundwater sulfate, it was concluded that the source of the sulfate was oxidation of pyrite present in the tills. This process occurred during low water table conditions in dry periods.

It has been proposed in earlier studies that the proportions of water-oxygen and atmospheric oxygen incorporated into the sulfate molecule during sulfide oxidation are not only dependent on whether the reaction occurs under reducing or oxidizing

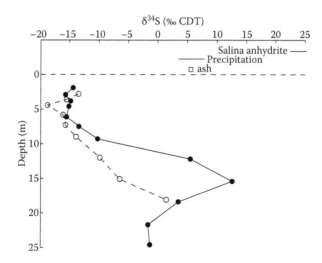

FIGURE 7.8 $\delta^{34}S$ values for sources of sulfate and depth profile of $\delta^{34}S$ values in sulfate in tills sediments (From Abbott, D. E., The origin of sulphur in sulphate-rich shallow groundwater of the St. Clair Plain, southwestern Ontario. MSc thesis, Earth Sciences, University of Waterloo, Ontario, Canada, 1987).

conditions, but also on whether the reactions proceed abiotically or microbially medi-ated (e.g., Taylor et al. 1984; Taylor and Wheeler 1994). Therefore, it was hypoth-esized that oxygen isotope ratios of sulfate may reveal whether sulfide oxidation occurred chemically or biologically. Recent pyrite oxidation experiments by Balci et al. (2007) confirmed that sulfide oxidation under reducing conditions using dis-solved $Fe(III)_{aq}$ yielded lower $\delta^{18}O$ values for sulfate than pyrite oxidation under aer-obic conditions using O_2 as an oxidizing agent. These authors, however, also found that the obtained $\delta^{18}O$ values of sulfate did not vary dependent on the presence or absence of bacteria under their experimental conditions. Based on the similarity of the obtained oxygen isotope ratios of sulfate from biologically mediated and abiotic experiments, Balci et al. (2007) suggested that $\delta^{18}O$ values of sulfate cannot be used to distinguish biological and abiotic mechanisms of pyrite oxidation.

7.10.3 IN-SITU BIOREMEDIATION INVOLVING BACTERIAL SULFATE REDUCTION

Bacterial dissimilatory sulfate reduction is a common process in anoxic aquifers that results in characteristic patters of decreasing sulfate concentrations accompanied by increasing $\delta^{34}S$ values of the remaining sulfate (e.g., Robertson, Cherry, and Schiff 1989; Strebel, Boettcher, and Fritz 1990). The produced H_2S gives a foul smell and is highly toxic and therefore is undesirable in drinking water. Often H_2S reacts with Fe to form Fe-sulfide minerals, potentially clogging the well screens (Van Beek and Van der Kooij 1982). Figure 7.8, displaying data from the case study in Southern Ontario, reveals that at depths of more than 10 meters $\delta^{34}S$ values increased by more than 15‰ with increasing depth. The increasing $\delta^{34}S$ values were accompanied by decreasing sulfate concentrations (Figure 7.8), suggesting that bacterial dissimila-

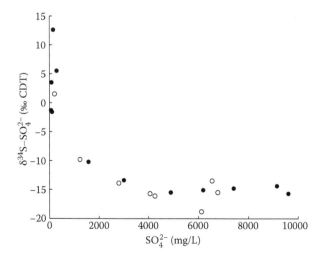

FIGURE 7.9 $\delta^{34}S$ values in sulfate versus sulfate concentration in tills sediments (From Abbott, D. E., The origin of sulphur in sulphate-rich shallow groundwater of the St. Clair Plain, south-western Ontario. MSc thesis, Earth Sciences, University of Waterloo, Ontario, Canada, 1987).

tory sulfate reduction was at least partially responsible for sulfate attenuation in the deeper part of the till aquifer.

Bacterial dissimilatory sulfate reduction plays an important role in natural attenuation, an effective low-cost method for clean up of chemical and petroleum contamination in aquifers. Under anaerobic conditions, the reduction of nitrate, iron, manganese, and sulfate are important terminal electron accepting processes. Since in most groundwater systems sulfate is more abundant than more energy-favored electron acceptors such as oxygen and nitrate, bacterial dissimilatory sulfate reduction is often the dominant terminal electron accepting process in the natural attenuation (e.g., of hydrocarbons; Chapelle et al. 1996; Wiedemeier et al. 1999). Several studies have demonstrated that the biodegradation of hydrocarbons and BTEX in groundwater is linked to bacterial dissimilatory reduction of sulfate (Reinhard et al. 1997; Anderson and Lovley 2000; Van Stempvoort et al. 2004; Knoeller et al. 2006). Isotope analyses constitute an elegant tool to differentiate between sulfate concentration decreases caused by dilution versus bacterial (dissimilatory) sulfate reduction (Knoeller et al. 2006). This is a prerequisite for determining sulfate reduction rates (McGuire et al. 2002; Van Stempvoort, Armstrong, and Mayer 2007a) and for assessing the affectivity of monitored natural attenuation. In situations where sulfate appears to limit bacterial dissimilatory sulfate reduction, sulfate additions have proven to be a viable approach for enhancing in-situ bioremediation (Weiner, Lauck, and Lovley 1998). Sulfur isotope techniques can play an important role in tracing the fate of the added sulfate in the aquifer (Van Stempvoort, Armstrong, and Mayer 2007a, b).

7.10.4 Denitrification Coupled to Pyrite Oxidation

This chapter has first described sources and processes that affect N-containing compounds and their isotopic compositions followed by a summary of reactions that influence the isotopic composition of S-bearing compounds in aqueous and sedimentary systems. The reduction of nitrate by pyrite is a naturally occurring process that has been widely reported in the literature (Koelle et al. 1983; Robertson, Russel, and Cherry 1996), in which the N and S isotopic systems are intimately linked with each other. The process involves the oxidation of sulfide coupled with denitrification

$$5FeS_2 + 14\ NO_3^- + 4H^+ \rightarrow 7N_2 + 5Fe^{2+} + 10SO_4^{2-} + H_2O \qquad (7.12)$$

often followed by oxidation of Fe^{2+} to Fe^{3+} (Appelo and Postma 2005). This process results in characteristic patterns of decreasing nitrate concentrations and increasing $\delta^{15}N$ and $\delta^{18}O$ values in the remaining nitrate coupled with increasing sulfate (and bicarbonate) concentrations. The sulfur isotope composition of the groundwater sulfate will tend toward the usually negative $\delta^{34}S$ values of the sedimentary pyrite in the aquifer as sulfate concentrations increase. These patterns were observed in a nitrate plume generated by a large communal septic system and based on chemical mass balances it was postulated that 20% of the nitrate reduction was mediated by pyrite and 80% by organic carbon (Aravena and Robertson 1998).

7.10.5 SOURING CONTROL

The generation of H_2S via bacterial dissimilatory sulfate reduction, often referred to as souring, is a major problem in various industries including oil production and processing. Nitrate and nitrite additions are an effective tool to reduce or eliminate souring (Hubert and Voordouw 2007). However, it is often not clear whether this is achieved by stimulating nitrate reducing bacteria that out-compete sulfate reducers for electron donors, a process referred to as biocompetitive exclusion. Alternately, souring may be controlled by lithotrophic nitrate and nitrite reducing sulfide oxidizing bacteria that eliminate H_2S via oxidative processes. Hubert, Voordouw, and Mayer (2009) have recently shown that isotope analyses on sulfate and sulfide are an effective tool to distinguish between these two processes. In particular, oxygen isotope analyses on SO_4^{2-} yield much lower $\delta^{18}O$ values in cases that lithotrophic nitrate and nitrite reducing sulfide oxidizing bacteria eliminate H_2S, as compared to $\delta^{18}O$ values of sulfate resulting from biocompetitive exclusion. This is another example where stable isotope analyses can provide effective insights into complex microbial processes in subsurface environments.

7.11 SUMMARY

This chapter described the rationale behind the application of nitrogen, sulfur, and oxygen isotopes for fingerprinting sources of nitrate and sulfate and processes that affect these compounds in groundwater. A wide range of applications in local and regional groundwater flow systems have been presented, highlighting the most common research questions concerning nitrate and sulfate that can be addressed using environmental isotopes in groundwater studies. The case studies have focused mainly on the application of environmental isotopes, however in all of these studies, a detailed evaluation of the groundwater geochemistry and groundwater flow system provided the framework for a successful application of the isotope tool. If applied in concert with hydrodynamic and hydrochemical techniques, stable isotope techniques provide an additional tool that can provide unique insights into the nitrogen and sulfur cycle in groundwater studies.

ACKNOWLEDGMENT

We would like to thanks Leonard Wassenaar (National Hydrology Research Center, Environment Canada) for his insightful review that helped to improve this manuscript.

REFERENCES

Abbott, D. E., 1987. The origin of sulphur in sulphate-rich shallow groundwater of the St. Clair Plain, southwestern Ontario. MSc thesis, Earth Sciences, University of Waterloo, Ontario, Canada.

Aharon, P., and B. Fu. 2000. Microbial sulfate reduction rates and sulfur and oxygen isotope fractionations at oil and gas seeps in deepwater Gulf of Mexico. *Geochimica Cosmochimica Acta* 64:233–46.

Aharon, P., and B. Fu. 2003. Sulfur and oxygen isotopes of coeval sulfate-sulfide in pore fluids of cold seep sediments with sharp redox gradients. *Chemical Geology* 195:201–8.

Alewell, C., M. J. Mitchell, G. E. Likens, and H. R. Krouse. 2000. Assessing the origin of sulfate deposition at the Hubbard Brook experimental forest. *Journal of Environmental Quality* 29:759–67.

Andersen, L. J., and H. Kristiansen. 1984. Nitrate in groundwater and surface water related to land use in the Karup basin, Denmark. *Environmental. Geology* 5:207–12.

Anderson, K. K., and A. B. Hooper. 1983. O_2 and H_2O are each the source of one O in NO_2-produced from NH_3 by Nitrosomas-^{15}N-NMR evidence. *FEBS Letters* 64:236–40.

Anderson, R. T., and D. R. Lovley. 2000. Anaerobic bioremediation of benzene under sulfate-reducing conditions in a petroleum-contaminated aquifer. *Environmental Science & Technology* 34:2261–66.

Appelo, C. A. J., and D. Postma. 2005. *Geochemistry, groundwater and pollution.* Leiden, The Netherlands: A A Balkema Publishers.

Aravena, R., M. Auge, N. Bucich, and M. I. Nagy. 1999. Evaluation of the origin of ground-water nitrate in the city of La Plata, Argentina, using isotope techniques. In *Proceeding XXIX IAH congress, hydrogeology and land use management,* Bratislava, September 6–10.

Aravena, R., M. L. Evans, and J. A. Cherry. 1993. Stable isotopes of oxygen and nitrogen in source identification of nitrate from septic tanks. *Ground Water* 31:180–86.

Aravena, R., and W. Robertson. 1998. The use of multiple isotope tracers to evaluate deni-trification in groundwater: A case study in a large septic system plume. *Ground Water* 36:975–82.

Bak, F., and N. Pfennig. 1987. Chemolithotropic growth of *Desulfovibrio sulfodismutans* sp. nov. by disproportionation of inorganic sulfur compounds. *Archives of Microbiology* 147:184–89.

Balci, N., W. C. Shanks III, B. Mayer, and K. W. Mandernack. 2007. Oxygen and sulfur isotope systematics of sulfate produced by bacterial and abiotic oxidation of pyrite. *Geochimica Cosmochimica Acta* 71:3796–811.

Beaudoin, G., B. E. Taylor, D. Rumble III, and M. Thiemens. 1994. Variations in the sul-fur isotope composition of troilite from Canyon Diablo iron meteorite. *Geochimica Cosmochimica Acta* 58:4253–255.

Beller, H. R., V. Madrid, G. Bryant Hudson, W. W. McNab, and T. Carlsen,. 2004. Biogeochemistry and natural attenuation of nitrate in groundwater at an explosives test facility. *Applied Geochemistry* 19:1483–494.

Benkovitz, C., M. Scholtz, J. Pacyna, L. Tarrasón, J. Dignon, E. Voldner, P. Spiro, J. Logan, and T. Graedel. 1996. Global gridded inventories of anthropogenic emissions of sulfur and nitrogen. *Journal of Geophysical Research* 101:29239–53.

Boettcher, M. E., B. Thamdrup, and T. W. Vennemann. 2001. Oxygen and sulfur isotope frac-tionation associated with bacterial disproportionation of elemental sulfur. *Geochimica Cosmochimica Acta* 65:1601–9.

Böhlke, J. K., and J. M. Denver. 1995. Combined use of groundwater dating, chemical, and isotopic analyses to resolve the history and fate of nitrate contamination in two agricultural watersheds, Atlantic coastal plain, Maryland. *Water Resources Research* 31:2319–39.

Böhlke , J. K., G. E. Ericksen, and K. Revesz. 1997. Stable isotope evidence for an atmo-spheric origin of desert nitrate deposits in northern Chile and southern California, U.S.A. *Chemical Geology* 136:135.

Bol, R., S. Toyoda, S. Yamulki, J. M. B. Hawkins, L. M. Cardenas, and N. Yoshida. 2003. Dual isotope and isotopomer ratios of N_2O emitted from a temperate grassland soil after ferti-liser application. *Rapid Communications in Mass Spectrometry* 17:2550–56.

Böttcher, J., O. Strebel, S. Voerkelius, and H. L. Schmidt. 1990. Using isotope fractionation of nitrate-nitrogen and nitrate-oxygen for evaluation of microbial denitrification in a sandy aquifer. *Journal of Hydrology* 114:413.

Bottrell, S. H., J. Tellam, R. Barlett, and A. Hughes. 2008. Isotopic composition of sulfate as tracer of natural and anthropogenic influences on groundwater geochemistry in an urban sandstone aquifer, Birmingham, UK. *Applied Geochemistry* 23:2381–94.

Brunner, B., S. M. Bernasconi, J. Kleikemper, and H. Schroth. 2005. A model for oxygen and sulfur isotope fractionation in sulfate during bacterial sulfate reduction processes. *Geochimica Cosmochimica Acta* 69:4773–85.

Brunold, C. 1990. Reduction of sulfate to sulfide. In *Sulfur nutrition and sulfur assimilation in higherplants*, eds. H. Rennenberg, C. Brunold, L. J. DeKok, and I. Stulen, 11–31. The Hague: SPB Academic Publishing.

Canfield, D. E. 1998. Isotope fractionation and sulfur metabolism by pure and enriched cultures of elemental sulfur disproportionating bacteria. *Limnology and Oceanography* 43:253–64.

Canfield, D. E. 2001a. Biogeochemistry of sulfur isotopes. In *Stable isotope geochemistry. Reviews in mineralogy and geochemistry,* eds. J. W. Valley and D. R. Cole, 607–36. Washington, DC: The Mineralogical Society of America.

Canfield, D. E. 2001b. Isotope fractionation by natural populations of sulfate-reducing bacteria. *Geochimica Cosmochimica Acta* 65:1117–124.

Canfield, D. E., C. A. Olesen, and R. P. Cox. 2006. Temperature and its control of isotope fractionation by a sulfate-reducing bacterium. *Geochimica Cosmochimica Acta* 70:548–61.

Canfield, D. E., and B. Thamdrup. 1994. The production of ^{34}S depleted sulfide during bacterial disproportionation of elemental sulfur. *Science* 266:1973–75.

Caron, F., A. Tessier, J. R. Kramer, H. P. Schwarcz, and C. E. Rees. 1986. Sulfur and oxygen isotopes of sulfur in precipitation and lakewater, Quebec, Canada. *Applied Geochemistry* 1:601–6.

Casciotti, K. L., D. M. Sigman, M. Hastings, J. K. Böhlke, and A. Hilkert. 2002. Measurement of the oxygen isotopic composition of nitrate in seawater and freshwater using the denitrifier method. *Analytical Chemistry* 74:4905.

Cey E. E., D. L. Rudolph R. Aravena, and G. Parkin. 1999. Role of the riparian zone in controlling the distribution and fate of agricultural nitrogen near a small stream in southern Ontario. *J. Contaminant Hydrology* 37:45–67.

Chambers, L.A., and P. A. Trudinger. 1979. Microbiological fractionation of stable sulfur isotopes: A review and critique. *Geomicrobiology. Journal* 1:249–93.

Chapelle, F. H., P. M. Bradley, D. R. Lovely, and D. A. Vroblesky. 1996. Measuring the rates of biodegradation in a contaminated aquifer using field and laboratory methods. *Ground Water* 34:691–98.

Chen, D. J. Z., and K. T. B. MacQuarrie. 2004. Numerical simulation of organic carbon, nitrate, and nitrogen isotope behavior during denitrification in a riparian zone. *Journal of Hydrology* 293:235–54.

Chiba, H., and H. Sakai. 1985. Oxygen isotope exchange rate between dissolved sulfate and water at hydrothermal temperatures. *Geochimica Cosmochimica Acta* 49:993–1000.

Chivas, A. R., A. S. Andrew, W. B. Lyons, M. I. Bird, and T. H. Donelly. 1991. Isotopic constraints on the origin of salts in Australian playas. 1. Sulphur. *Palaeogeography Palaeoclimatology Palaeoecology* 84:309–32.

Chukhrov, F. V., V. S. Churikov, L. P. Yermilova, and L. P. Nosik. 1975. On the variation of sulfur isotopic composition in some natural waters. *Geochemistry International* 12:20–33.

Cicerone, R. J. 1989. Analysis of sources and sinks of atmospheric nitrous oxide (N_2O). *Journal of Geophysical Research* 94:18265–71.

Clark, I., and P. Fritz. 1997. *Environmental isotopes in hydrogeology*. Boca Raton, New York: Lewis Publishers.

Claypool, G. E., W. T. Holser, I. R. Kaplan, H. Sakai, and I. Zak. 1980. The age curves of sulfur and oxygen isotopes in marine sulfate and their mutual interpretation. *Chemical Geology* 28:199–260.

Coplen, T. B., J. K. Böhlke, and K. L. Casciotti. 2004. Using dual bacterial denitrification to improve ^{15}N determinations of nitrates containing mass independent [17]O. *Rapid Communications in Mass Spectrometry* 18:245.

Coplen, T. B., J. A. Hopple, J. K. Böhlke, et al. 2002. Compilation of minimum and maximum isotope ratios of selected elements in naturally occurring terrestrial materials and reagents. 01-4222, *USGS Water-Resources Investigations Report* 01-4222, Reston, Virginia.

Coplen, T., and H. R. Krouse. 1998. Sulphur isotope data consistency improved. *Nature* 392:32.

Cornwell J. C., W. M. Kemp, and T. M. Kana. 1999 Denitrification in coastal ecosystems: Methods, environmental controls, and ecosystem level controls, a review. *Aquatic Ecology* 33:41–54.

Cortecci, G., and A. Longinelli. 1970. Isotopic composition of sulfate in rain water, Pisa, Italy. *Earth and Planetary Science Letters* 8:36–40.

Craig, H. 1961. Isotopic variations in meteoric waters. *Science* 133:1833–34.

Cram, W. J. 1990. Uptake and transport of sulfate. In *Sulfur nutrition and sulfur assimilation in higher plants*, eds. H. Rennenberg, C. Brunold, L. J. DeKok, and I. Stulen, 3–11. The Hague: SPB Academic Publishing.

Dansgaard, W. 1964. Stable isotopes in precipitation. *Tellus* 16:436–68.

Deans, J. D., W. M. Edmunds, D. K. Lindley, C. B. Gaye, B. Dreyfus, J. J. Nizinski, M. Neyra, R. C. Munro, and K. Ingleby. 2005. Nitrogen in interstitial waters in the Sahel: Natural baseline, pollutant or resource? *Plant Soil* 271:47–62.

Deevey, E. S., N. Nakai, and M. Stuiver. 1963. Fractionation of sulfur and carbon isotopes in a meromictic lake. *Science* 139:407–8.

Delwiche, C. C., and P. L. Steyn. 1970. Nitrogen isotope fractionation in soils and microbial reactions. *Environmental Science & Technology* 4:929–35.

Detmers, J., V. Bruchert, K. S. Habicht, and J. Kuever. 2001. Diversity of sulfur isotope fractionations by sulfate-reducing prokaryotes. *Applied Environmental Microbiology* 67:888–94.

Devito, K. J., D. Fitzgerald, A. R. Hill, and R. Aravena. 2000. Nitrate dynamics in relation to lithology and hydrologic flow path in a river riparian zone. *Journal of Environmental Quality* 29:1075–84.

DiGnazio, F. J., N. C. Krothe, S. J. Baedke, and R. F. Spalding. 1998. Nitrate derived from explosive sources in a karst aquifer beneath the Ammunition Burning Ground, Crane Naval Surface Warfare Center, Indiana, USA. *Journal of Hydrology* 206:164–75.

Edmunds, W. M. 1999. Groundwater nitrate as a palaeoenvironmental indicator. In *Geochemistry of the Earth's surface*, ed. Armnannsson, 35–39. Rotterdam: Balkema.

Edmunds W. M., and C. B. Gaye. 1997. Naturally high nitrate concentrations in groundwaters from the Sahel. *Journal of Environmental Quality* 26:1231–39.

Epstein, S. and T. Mayeda. 1953. Variation of O-18 content of waters from natural sources. *Geochimica Cosmochimica Acta* 4:213–24.

Fairbairn, J. D. 1996. Application of nitrogen isotope techniques to assess the impact of sewer leakage on shallow urban groundwater quality in the Sherwood sandstone aquifer of Nottingham. MSc thesis, University of East Anglia, Norwich.

Fan, A. M., and V. E. Steinberg. 1996. Health implications of nitrate and nitrite in drinking water: An update on methemoglobinemia occurrence and reproductive and developmental toxicity. *Regulatory Toxicology and Pharmacology* 23:35–43.

Feigin, A., D. H. Kohl,, G. Shearer, and B. Commoner. 1974. Variation in the natural abundance in nitrate mineralised during the incubation of several Illinois soils. *Soil Science Society of America Proceedings* 38:465–71.

Fennell, J., and L. R. Bentley. 1998. Distribution of sulfate and organic carbon in a prairie till setting: Natural versus industrial sources. *Water Resources Research* 34:1781–94.

Fontes, J. C., and G. M. Zuppi. 1976. Isotopes and water chemistry in sulphide-bearing springs of central Italy. In *Interpretation of environmental isotope and hydrochemical data in ground water hydology*, ed. IAEA, 143–58. Vienna, Austria.

Freyer, H. D., and A. I. M. Aly. 1975. Nitrogen-15 studies on identifying fertilizer excess in environmental systems. In *Isotope ratios as pollutant source and behaviour indicators*, ed. IAEA, 21–32. Vienna, Austria.

Fritz, P., G. M. Basharmal, R. J. Drimmie, J. Ibsen, and R. M. Qureshi. 1989. Oxygen isotope exchange between sulphate and water during bacterial reduction of sulphate. *Chemical Geology* 79:99–105.

Fritz, P., and R. J. Drimmie. 1991. Ground water. In *Stable isotope: Natural and anthropogenic sulphur in the environment*, SCOPE 43, eds. H. R. Krouse and V. A. Grinenko, 228–41. Chichester: Wiley.

Fry, B., J. Cox, H. Gest, and J. M. Hayes. 1986. Discrimination between ^{34}S and ^{32}S during bacterial metabolism of inorganic sulfur compounds. *Journal of Bacteriology* 165:328–30.

Fry, B., H. Gest, and J. M. Hayes. 1984. Isotope effects associated with the anaerobic oxidation of sulfide by the purple photosynthetic bacterium *Chromatium vinosum*. FEMS *Microbiology Letters* 22:283–87.

Fry, B., W. Ruf, H. Gest, and J. M. Hayes. 1988. Sulfur isotope effects associated with oxidation of sulfide by O_2 in aqueous solution. *Chemical Geology* 73:205–10.

Fukada, T., K. M. Hiscock, and P. F. Dennis. 2004. A dual-isotope approach of the nitrogen hydrochemistry of an urban aquifer. *Applied Geochemistry* 19:709–19.

Fukada, T., K. M. Hiscock, P. F. Dennis, and T. Grischek. 2003. A dual isotope approach to identify denitrification in groundwater at a river bank infiltration site. *Water Research* 37:3070–78.

Gebauer, G., A. H. Giesemann, E. D. Schulze, and H. J. Jager. 1994. Isotope ratios and concentrations of sulfur and nitrogen in needles and soils of *Picea abies* stands as influenced by atmospheric deposition of sulfur and nitrogen compounds. *Plant Soil* 164:267–81.

Gillham, R.W., and J. A. Cherry. 1978. Field evidence of denitrification in shallow ground water flow systems. *Water Pollution Research Journal of Canada* 13:53–71.

Gilliam, J. W. 1994. Riparian wetlands and water quality. *Journal of Environmental Quality* 23:896–900.

Gormly, J. R., and R. F. Spalding. 1979. Sources and concentrations of nitrate–nitrogen in ground water of the central Platte Region, Nebraska. *Ground Water* 17:291–301.

Goss, M.J., D. A. J. Barry, and D. L. Rudolph. 1998. Contamination in Ontario farmstead domestic wells and its association with agriculture: 1. Results from drinking water wells. *Journal of Contaminant Hydrology* 32:267–93.

Grey, D. C., and M. L. Jensen. 1972. Bacteriogenic sulfur in air pollution. *Science*. 177:1099–100.

Habicht, K. S., and D. E. Canfield. 1997. Sulfur isotope fractionation during bacterial sulfate reduction in organic-rich sediments. *Geochimica Cosmochimica Acta* 61:5351–361.

Habicht, K. S., D. E. Canfield, and J. Rethmeier. 1998. Sulfur isotope fractionation during bacterial reduction and disproportionation of thiosulfate and sulfite. *Geochimica Cosmochimica Acta* 62:2585–595.

Hamilton, P. A., and D. R. Helsel. 1995. Effects of agriculture on groundwater quality in 5 regions of the United States. *Ground Water* 33:217–26.

Harrison, A. G., and H. G. Thode. 1957. The kinetic isotope effect in the chemical reduction of sulphate. *Transactions of the Faraday Society* 53:1648–51.

Harrison, A. G., and H. G. Thode. 1958. Mechanisms of bacterial reduction of sulphate from isotope fractionation studies. *Transactions of the Faraday Society* 54:84–92.

Haycock, N. E., and G. Pinay. 1993. Groundwater nitrate dynamics in grass and poplar vegetated riparian buffer strips during the winter. *Journal of Environmental Quality* 22:273–78.

Heaton, T. H. E. 1984. Sources of the nitrate in phreatic groundwater in the western Kalahari. *Journal of Hydrology* 67:249–59.

Heaton, T. H. E. 1986. Isotopic studies of nitrogen pollution in the hydrosphere and atmosphere: A review. *Chemical Geology* 59: 87–102.

Heaton, T. H. E., A. S. Talma, and J. C. Vogel. 1983. Origin and history of nitrate in confined groundwater in the western Kalahari. *Journal of Hydrology* 62:243–62.

Hedin, L. O., J. C. Von Fischer, N. E. Ostrom, B. P. Kennedy, M. G. Brown, and G. P. Robertson. 1998. Thermodynamic constraints on nitrogen transformations and other biogeochemical processes at soil-stream interfaces. *Ecology* 79:684–703.

Hendry, M. J., H. R. Krouse, and M. A. Shakur. 1989. Interpretation of oxygen and sulfur isotopes from dissolved sulfates in tills of southern Alberta, Canada. *Water Resources Research* 25:567–72.

Herbert, R. A. 1999. Nitrogen cycling in coastal marine ecosystems. *FEMS Microbiology Reviews.* 23:563–90.

Herut, B., B. Spiro, A. Starinsky, and A. Katz. 1995. Sources of sulfur in rainwater as indicated by isotopic $\delta^{34}S$ data and chemical composition, Israel. *Atmospheric Environment* 29:851–57.

Hesseling, P. B., P. D. Toens, and H. Visser. 1991. An epidemiological survey to assess the effect of well-water nitrates on infant health at Rietfontein in the northern Cape Province, South Africa. S.A. *Journal of Science* 87:300–4.

Hill, A. R. 1996. Nitrate removal in stream riparian zones. *Journal of Environmental Quality* 25:743–55.

Hill, A. R., K. Devito, S. Campagnolo, and K. Sanmugadas. 2000. Subsurface denitrification in a forested riparian zone; interactions between hydrology and supplies of nitrate and organic carbon. *Biogeochemistry* 51:193–223.

Hiscock, K. M., A. S. Bateman, I. H. Muhlherr, T. Fukada, and P. F. Dennis. 2003. Indirect emissions of nitrous oxide from regional aquifers in the United Kingdom. *Environmental Science & Technology* 37:3507–12.

Hiscock, K. M., P. F. Dennis, N. A. Feast, and J. D. Fairbairn. 1997. Experience in the use of stable nitrogen isotopes to distinguish groundwater contamination from leaking sewers in urban areas. In *Groundwater in the urban environment: Problems, processes and management,* ed. J. Chilton, 427–32. Rotterdam: Balkema.

Hitchon, B., and H. R. Krouse. 1972. Hydrogeochemistry of the surface waters of the Mackenzie River drainage basin, Canada - III. Stable isotopes of oxygen, carbon and sulphur. *Geochimica Cosmochimica Acta* 36:1337–57.

Hoefs, J. 1997. *Stable Isotope Geochemistry*, Berlin: Springer Verlag.

Hollocher, T. C. 1984. Source of oxygen atoms in nitrate in the oxidation of nitrite by *Nitrobacter agilis* and evidence against a P-O-N anhydride mechanism in oxidative phosphorylation. *Archives of Biochemistry and Biophysics* 233:721–27.

Holt, B. D., and R. Kumar. 1991. Oxygen isotope fractionation for understanding the sulfur cycle. In *Stable isotopes:Natural and anthropogenic sulphur in the environment*, SCOPE 43, eds. H. R. Krouse and V. A. Grinenko, 55–64. Chichester: Wiley.

Horibe, Y., K. Shigehara, and Y. Takakuwa. 1973. Isotopic separation factors of carbon-dioxide-water system and isotopic composition of atmospheric oxygen. *Journal of Geophysical Research* 78:2625–29.

Hubert, C., and G. Voordouw. 2007. Oil field souring control by nitrate-reducing *Sulfurospirillum* spp. that outcompete sulfate-reducing bacteria for organic electron donors. *Applied Environmental Microbiology* 73:2644–52.

Hubert, C., G. Voordouw, and B. Mayer. 2009. Elucidating microbial processes in nitrate- and sulfate-reducing systems using sulfur and oxygen isotope ratios: The example of oil reservoir souring control. *Geochimica Cosmochimica Acta.* In press.

Iqbal, M. Z., N. C. Krothe, R. F. Spalding. 1997. Nitrogen isotope indicators of seasonal source variability to groundwater. *Environmental Geology* 32:210–18.

Iriarte, S. 2004. Impact of urban recharge on long-term management of Santiago Norte aquifer, Santiago-Chile. MSc thesis, Earth Sciences, University of Waterloo,Ontario, Canada.

Jacinthe, P. A., P. M. Groffman, A. J. Gold, and A. Mosier. 1998. Patchiness in microbial nitrogen transformation in groundwater in a riparian zone. *Journal of Environmental Quality* 27:156–64.

Jensen, M. L., and N. Nakai. 1961. Sources and isotopic composition of atmospheric sulfur. *Science* 134:2102–4.

Jones, M. J. 1973. Time of application of nitrogen fertilizer to maize at Samaru, Nigeria. *Experimental Agriculture* 9:113–20.

Jordan, T. E., D. L. Correll, and D. E. Weller. 1993. Nutrient interception by a riparian forest receiving inputs from adjacent cropland, *Journal of Environmental Quality* 22:467–73.

Kaplan, I. R., and S. C. Rittenberg. 1964. Microbiological fractionation of sulphur isotopes. *Journal of General Microbiology* 34:195–212.

Karamanos, R. E., R. P. Voroney, and D. A. Rennie. 1981. Variation in natural N-15 abundance of central Saskatchewan soils. *Soil Science Society of America Journal* 45:826–28.

Karr, J. D., W. J. Showers, J. W. Gilliam, and A. S. Andres. 2001. Tracing nitrate transport and environmental impact from intensive swine farming using delta N-15. *Journal of Environmental Quality* 30:1163–75.

Karr, J. D., W. J. Showers, and J. D. Jennings. 2003. Low-level nitrate export from confined dairy farming detected in North Carolina streams using $\delta^{15}N$. *Agriculture, Ecosystems and Environment* 95:103–10.

Kendall, C., and R. Aravena. 2000. Nitrate isotopes in groundwater systems. In: *Environmental tracers in subsurface hydrology.* eds. P. G. Cook and A. L. Herczeg, 261–97. Norwell, MA: Kluwer Academic Publishers.

Kendall, C., and J. J. McDonnell. 1998. *Isotope tracers in catchment hydrology.* Amsterdam: Elsevier.

Kirshenbaum, I., J. S. Smith, T. Crow, J. Graff, and R. Mckee. 1947. Separation of the nitrogen isotopes by the exchange reaction between ammonia and solutions of ammonium nitrate. *Journal of Chemical Physics* 15:440–46.

Knoeller, K., R. Trettin, and G. Strauch. 2005. Sulphur cycling in the drinking water catchment area of Torgau-Mockritz (Germany): Insights from hydrochemical and stable isotope investigations. *Hydrological Processes* 19:3445–65.

Knoeller, K., C. Vogt, H. Richnow, and S. Weise. 2006. Sulfur and oxygen isotope fractionation during benzene, toluene, ethyl benzene and xylene degradation by sulfate-reducing bacteria. *Environmental Science & Technology* 40:3879–85.

Koelle, W., P. Werner, O. Strebel, and J. Boettcher. 1983. Denitrifikation in einem reduzierenden grundwasserleiter. *Vom Wasser* 61:125–47.

Kohl, D. H., and G. Shearer. 1980. Isotopic fractionation associated with symbiotic N_2 fixation and uptake of NO_3^- by plants. *Plant Physiology* 66:51–56.

Korom, S. F. 1992. Natural denitrification in the saturated zone: A review. *Water Resources Research* 28:1657–68.

Kreitler, C. W. 1979. Nitrogen-isotope ratio studies of soils and groundwater nitrate from alluvial fan aquifers in Texas. *Journal of Hydrology* 42:147.

Kreitler, C. W., and L. A. Browning. 1983. Nitrogen isotope analysis of groundwater nitrate in carbonate aquifers: Natural sources versus human pollution. *Journal of Hydrology* 61:285–301.

Kreitler, C. W., and D. C. Jones. 1975. Natural soil nitrate: The cause of the nitrate contamination of groundwater in Runnels County, Texas. *Ground Water* 13:53–62.

Kroopnick, P. M., and H. Craig. 1972. Atmospheric oxygen: Isotopic composition and solubility fractionation. *Science* 175:54–55.

Krouse, H. R. 1980. Sulphur isotopes in our environment. In *Handbook of environmental isotope geochemistry*, eds. P. Fritz and J. C. Fontes, 435–71. Amsterdam: Elsevier.

Krouse, H. R., and V. A. Grinenko. 1991. *Stable isotopes: Natural and anthropogenic sulphur in the environment*, SCOPE 43. Chichester: John Wiley.

Krouse, H. R., and B. Mayer. 2000. Sulphur and oxygen isotopes in sulphate. In *Environmental tracers in subsurface hydrology*, eds. P. G. Cook and A. L. Herczeg, 195–231. Boston: Kluwer.

Krouse, H. R., and M. A. Tabatabai. 1986. Stable sulfur isotopes. In *Sulfur in agriculture*, ed. M. A. Tabatabai, 169–205. Madison, WI: *American Society of Agronomy, Crop Science Society of America, Soil Science Society of America.*

Kuenen, J. G., and L. A. Robertson. 1988. Ecology of nitrification and denitrification. In *The nitrogen and sulphur cycles. Forty-second symposium of the society for general microbiology*, eds. J. A. Cole and S. J. Ferguson, 161–221. Cambridge: Cambridge University Press.

Lein, A. Y. 1991. Flux of volcanic sulphur to the atmosphere and isotopic composition of total sulphur. In *Stable isotopes: Natural and anthropogenic sulfur in the environment*, SCOPE 43, eds. H. R. Krouse and V. A. Grinenko, 116–25. Chichester: Wiley.

Lerner, D. N., Y. Yang, M. H. Barrett, and J. H. Tellam. 1999. Loading of non-agricultural nitrogen in urban groundwater. In *Impacts of urban growth on surface and groundwater quality*. Proceedings of IUGG 99 Symposium HS5, Birmingham, July 1999. IAHS Publication No. 259, 117–23.

Lloyd, R. M. 1967. Oxygen-18 composition of oceanic sulfate. *Science* 156:1228–31.

Longinelli, A., and G. Cortecci. 1970. Isotopic abundance of oxygen and sulfur in sulfate ions from river water. *Earth and Planetary Science Letters* 7:376–80.

Lowrance, R. 1992. Groundwater nitrate and denitrification in a coastal riparian forest. *Journal of Environmental Quality* 21:401–5.

Mangalo, M., R. U. Meckenstock, W. Stichler, and F. Einsiedl. 2007. Stable isotope fractionation during bacterial sulfate reduction is controlled by reoxidation of intermediates. *Geochimica Cosmochimica Acta* 71:4161–71.

Mariotti, A. 1986. Denitrification in groundwaters, principles and methods for its identification: A review. *Journal of Hydrology* 88:1–23.

Mariotti, A., J. C. Germon, P. Hubert, P. Kaiser, R. Letolle, A. Tardieux, and P. Tardieux. 1981. Experimental determination of nitrogen kinetic isotope fractionation of nitrogen isotopes during denitrification. *Plant Soil* 62:413–30.

Mariotti, A., J. C. Germon, and A. Leclerc. 1982. Nitrogen isotope fractionation associated with the $NO_2^- $-$N_2O$ step of denitrification in soils. *Canadian Journal of Soil Science* 62:227–41.

Mariotti, A., A. Landreau, and B. Simon. 1988. [15]N isotope biogeochemistry and natural denitrification process in groundwater: Application to the Chalk aquifer of northern France. *Geochimica Cosmochimica Acta* 52:1869–78.

Mariotti, A., D. Pierre, J. C. Vedy, S. Bruckert, and J. Guillemot. 1980. The abundance of natural nitrogen 15 in the organic matter of soils along an altitudinal gradient (Chablais, Haute Savoie, France). *Catena* 7:293–300.

Mason, C. F. 2002. *Biology of freshwater pollution*, 4th ed. Harlow, Essex: Prentice Hall.

Matrosov, A. G., Y. N. Chebotarev, A. J. Kudryavtseva, A. M. Zyukun, and M. V. Ivanov. 1975. Sulfur isotope composition in freshwater lakes containing H_2S. *Geochemistry International* 12:217–21.

Mayer, B. 2005. Assesing sources and transformations of sulphate and nitrate in the hydrosphere using isotope techniques. In: *Isotopes in the water cycle: Past present and future of a developing science,* eds. P. K. Aggarwal, J. R. Gat, and F. O. Froehlich, 67–89. Dordrecht: Springer.

Mayer, B., K. H. Feger, A. Giesemann, and H. J. Jaeger. 1995b. Interpretation of sulfur cycling in two catchments in the Black Forest (Germany) using stable sulfur and oxygen isotope data. *Biogeochemistry* 30:31–58.

Mayer, B., P. Fritz, and H. R. Krouse. 1992. Sulphur isotope discrimination during sulphur transformations in aerated forest soils. In *Sulphur transformations in soil ecosystems,* eds. M. J. Hendry and H. R. Krouse, 161–71. Saskatoon, Canada: National Hydrology Research Institute.

Mayer, B., P. Fritz, J. Prietzel, and H. R. Krouse. 1995a. The use of stable sulfur and oxygen isotope ratios for interpreting the mobility of sulfate in aerobic forest soils. *Applied Geochemistry* 10:161–73.

Mayer, B., and H. R. Krouse. 2004. Procedures for sulfur isotope abundance studies. In *Handbook of stable isotope analytical techniques*, ed. P. De Groot, 538–96. Amsterdam: Elsevier.

Mayer, B., L. Rock, L. Hogberg, et al. 2006. Delineating sources of sulfate and nitrate in rivers and streams by combining hydrological, chemical and isotopic techniques In *Proceedings of international conference on isotopes in environmental studies: Aquatic forum 2004,* ed. IAEA, 148–49. Monaco: IAEA.

McGuire, J. T., D. T. Long, M. J. Klug, S. K. Haack, and D. W. Hyndman. 2002. Evaluating the behavior of oxygen, nitrate, and sulfate during recharge and quantifying reduction rates in a contaminated aquifer. *Environmental Science & Technology* 36:2693–700.

McMahon, P. B., and J. K. Böhlke. 2006. Regional patterns in the isotopic composition of natural and anthropogenic nitrate, in groundwater, High Plains, U.S.A. *Environmental Science & Technology* 40:2965–70.

Mengis, M., S. L. Schiff, M. Harris, M. C. English, R. Aravena, R. J. Elgood, and A. MacLean. 1999. Multiple geochemical and isotopic approaches for assessing ground water NO3 elimination in a riparian zone. *Ground Water* 37:448–57.

Migdisov, A. A., A. B. Ronov, and V. A. Grinenko. 1983. The sulfur cycle in the lithosphere. In *The global biogeochemical sulphur cycle*, SCOPE 19, eds. M. V. Ivanov and J. R. Freney, 25–95. Chichester: John Wiley.

Mitchell, R. J., R. S. Babcock, S. Gelinas, L. Nanus, and D. E. Stasney. 2003. Nitrate distributions and source identification in the Abbotsford-Sumas aquifer, northwestern Washington State. *Journal of Environmental Quality* 32:789–800.

Mizutani, Y., and T. A. Rafter. 1969a. Oxygen isotopic composition of sulphates: part 4. bacterial fractionation of oxygen isotopes in reduction of sulphate and in the oxidation of sulphur. *New Zealand Journal of Science* 12:60–68.

Mizutani, Y., and T. S. Rafter. 1969b. Isotopic composition of sulphate in rain water, Gracefield, New Zealand. *New Zealand Journal of Science* 12:69–80.

Mizutani, Y., and T. A. Rafter. 1973. Isotopic behaviour of sulphate oxygen in the bacterial reduction of sulphate. *Geochemistry Journal* 6:183–91.

Moerth, C. M., and P. Torssander. 1995. Sulfur and oxygen isotope ratios in sulfate during an acidification reversal study at Lake Garsjon, western Sweden. *Water Air Soil Pollution* 79:261–78.

Moore, K. B., B. Ekwurzel, B. K. Esser, G. B. Hudson, J. E. Moran. 2006. Sources of groundwater nitrate revealed using residence time and isotope methods. *Applied Geochemistry* 21:1016–29.

Moser, H., and W. Rauert. 1980. *Isotopenmethoden in der hydrologie. Lehrbuch der hydrogeologie, Band 8*. Berlin, Stuttgart: Gebrueder Borntraeger.

Müller, G., H. Nielsen, and W. Ricke. 1966. Sulphur isotope abundance in formation waters and evaporites from northern and southern Germany. *Chemical Geology* 1:211–20.

Nakai, N., and Jensen, M. L. 1967. Sources of atmospheric sulfur compounds. *Geochemistry Journal* 1:199–210.

Nielsen, H. 1974. Isotopic composition of the major contributors to the atmospheric sulfur. *Tellus* 26:213–21.

Nielsen, H. 1979. Sulfur isotopes. In *Lectures in isotope geology*, eds. E. Jager and J. Hunziker, 283–312. Berlin: Springer-Verlag.

Nolan, B. T., K. J. Hitt, and B. C. Ruddy. 2002. Probability of nitrate contamination of recently recharged groundwaters in the conterminous United States. *Environmental Science & Technology* 36:2138–45.

Norman, A. L., A. H. Giesemann, H. R. Krouse, and H. J. Jaeger. 2002. Sulphur isotope fractionation during sulphur mineralization: Results of an incubation-extraction experiment with a Black Forest soil. *Soil Biology & Biochemistry* 34:1425–38.

Novak, M., F. Buzek, and M. Adamova. 1999. Vertical trends in $\delta^{13}C$, $\delta^{15}N$, and $\delta^{34}S$ ratios in bulk sphagnum peat. *Soil Biology & Biochemistry* 31:1343–46.

Nriagu, J. O., and R. D. Coker. 1978. Isotopic composition of sulfur in precipitation within the Great Lakes Basin. *Tellus* 30:365–75.

O'Connell, T. C., K. Neal, and R. E. M. Hedges. 2001. Isolation of urinary urea for natural abundance nitrogen isotopic analysis. *Stable isotope mass spectrometry users group meeting*, SUERC, Glasgow.

O'Neil, W. B., and R. S. Raucher. 1990. The costs of groundwater contamination. *Journal of Soil and Water Conservation* 45:180–83.

Östlund, G. 1957. Isotopic composition of sulfur in precipitation and sea-water. *Tellus* 11:478–80.

Panno, S. V., K. C. Hackley, H. H. Hwang, and W. R. Kelly. 2001. Determination of the sources of nitrate contamination in Karst springs using isotopic and chemical indicators. *Chemical Geology* 179:113–28.

Pauwels, H., Foucher, J. C., and W. Kloppmann. 2000. Denitrification and mixing in a schist aquifer: Influence on water chemistry and isotopes. *Chemical Geology* 168:307–24.

Pauwels, H., W. Kloppmann, J. C. Foucher, A. Martelat, and V. Fritsche. 1998. Field tracer test for denitrification in a pyrite-bearing schist aquifer. *Applied Geochemistry* 13:767–78.

Pérez, T., S. E. Trumbore, S. C. Tyler, E. A. Davidson, M. Keller, and P. B. de Camargo. 2000. Isotopic variability of N_2O emissions from tropical forest soils. *Global Biogeochemical Cycles* 14:525–35.

Pérez, T., S. E. Trumbore, S. C. Tyler, P. A. Matson, I. Ortiz- Monasterio, T. Rahn, and D. W. T. Griffith. 2001. Identifying the agricultural imprint on the global N_2O budget using stable isotopes. *Journal of Geophysical Research* 106:9869–78.

Peterjohn, W. T., and T. Correll. 1984. Nutrient dynamics in an agricultural watershed observations on the role of a riparian forest. *Ecology* 65:1466–75.

Pinay, G., L. Rogues, and A. Fabre. 1993. Spatial and temporal patterns of denitrification in a riparian forest. *Journal of Applied Ecology* 30:581–91.

Postma, D. 1990. Kinetics of nitrate reduction in a sandy aquifer. *Geochimica Cosmochimica Acta* 54:903–8.

Postma, D., C. Boesen, H. Kristiansen, and F. Larsen. 1991. Nitrate reduction in an unconfined sandy aquifer: Water chemistry, reduction processes, and geochemical modelling. *Water Resources Research* 2:2027–45.

Price, F. T., and Y. N. Shieh. 1979. The distribution and isotopic composition of sulfur in coals from the Illinois Basin. *Economic Geology* 74:1445–61.

Rafter, T. A., and Y. Mizutani. 1967. Oxygen isotopic composition of sulphate, 2. Preliminary results on oxygen isotopic variation in sulphates and the relationship to their environment and their $\delta^{34}S$ values. *New Zealand Journal of Science* 10:816–40.

Rees, C. E. 1973. A steady-state model for sulphur isotope fractionation in bacterial reduction processes. *Geochimica Cosmochimica Acta* 37:1141–62.

Rees, C. E., W. J. Jenkins, and J. Monster. 1978. The sulphur isotopic composition of ocean water sulphate. *Geochimica Cosmochimica Acta* 42:377–81.

Reinhard, M., S. Shang, P. K. Kitanidis, E. Orwin, G. D. Hopkins, and C. A. LeBron. 1997. *In situ* BTEX biotransformation under enhanced nitrate- and sulfate-reducing conditions. *Environmental Science & Technology* 31:28–36.

Reuss, J. O., and D. W. Johnson. 1986. Acid deposition and the acidification of soils and water. *Ecological Studies, 59.* New York: Springer Verlag.

Rightmire, C. T., F. J. Pearson, W. Back, R. O. Rye, and B. B. Hanshaw. 1974. Distribution of sulfur isotopes of sulfates in ground waters from the principal artesian aquifer of Florida and the Edwards aquifer of Texas, United States of America. In *Proceedings of an symposium on isotope techniques in ground water hydrology,* ed. IAEA, 191–207. Vienna.

Rivers, C. N., M. H. Barrett, K. M. Hiscock, P. F. Dennis, N. A. Feast, and D. N. Lerner. 1996. Use of nitrogen isotopes to identify nitrogen contamination of the Sherwood sandstone aquifer beneath the city of Nottingham, United Kingdom. *Hydrogeology Journal* 4:90–102.

Robertson, L. A., and J. G. Kuenen. 1991. Physiology of nitrifying and denitrifying bacteria. In *Microbial production and consumption of greenhouse gases: Methane nitrogen oxides and halomethanes,* eds. J. E. Rogers and W. B. Whitman, 189–99. Washington, DC: American Society for Microbiology.

Robertson, W. D., J. A. Cherry, and S. L. Schiff. 1989. Atmospheric sulfur deposition 1950–1985 inferred from sulfate in groundwater. *Water Resources Research* 25:1111–23.

Robertson, W. D., J. A. Cherry, and E. A. Sudicky. 1991. Groundwater contamination from two small septic systems on sand aquifers. *Ground Water* 29:82–92.

Robertson, W. D., B. M. Russel, and J. A. Cherry. 1996. Attenuation of nitrate in aquitrad sediments of southern Ontario. *Journal of Hydrology* 180:267–81.

Rock, L., B. H. Ellert, and B. Mayer. 2007. Isotopic composition of tropospheric and soil N_2O from various depths of five agricultural plots with contrasting crops and nitrogen amendments. *Journal of Geophysical Research* 112:D18303. Doi:10.1029/2006JD008330.

Rock, L., and B. Mayer. 2004. Isotopic assessment of sources of surface water nitrate within the Oldman River basin, southern Alberta, Canada. *Water Air Soil Pollution Focus* 4:545–62.

Ronen, D., Y. Kanfi, and M. Magaritz. 1983. Sources of nitrates in groundwater of the coastal plain of Israel (evolution of ideas). *Water Resources Research* 17:1499–503.

Ronen, D., M. Magaritz, and E. Almon. 1988. Contaminated aquifers are a forgotten component of the global N_2O budget. *Nature* 335:57–59.

Schiff, S. L., J. Spoelstra, R. G. Semkin, and D. S. Jeffries. 2005. Drought induced pulses of SO_4^{2-} from a Canadian shield wetland: Use of $\delta^{34}S$ and $\delta^{18}O$ in SO_4^{2-} to determine sources of sulfate. *Applied Geochemistry* 20:691–700.

Shearer, G., D. H. Kohl, and S. H. Chien. 1978. The nitrogen-15 abundance in a wide variery of soils. *Soil Science Society of America Journal* 42: 899–902.

Sigman, D. M., K. L. Casciotti, M. Andreani, C. Barford, M. Galanter, and J. K. Böhlke. 2001. A bacterial method for the nitrogen isotopic analysis of nitrate in seawater and freshwater. *Analytical Chemistry* 73:4145.

Silva, S. R., P. B. Ging, R. W. Lee, J. C. Ebbert, A. J. Tesorierio, and L. E. Inkpen. 2002. Forensic application of nitrogen and oxygen isotopes in tracing nitrate sources in urban environments. *Environmental Forensics* 3:125–30.

Silva, S. R., C. Kendall, D. H. Wilkison, A. C. Ziegler, C. C. Y. Chang, and R. J. Avanzino. 2000. A new method for collection of nitrate from fresh water and analysis of the nitrogen and oxygen isotope ratios. *Journal of Hydrology* 228:22–36.

Smith, R. L., B. L. Howes, and J. H. Duff. 1991. Denitrification in nitrate-contaminated groundwater: Occurrence in steep vertical geochemical gradients. *Geochimica Cosmochimica Acta* 55:1815–25.

Spalding, R. F., and M. E. Exner. 1993. Occurrence of nitrate in groundwater-A review. *Journal of Environmental Quality* 22:392–402.

Spalding, R. F., M. E. Exner, G. E. Martin, and D. D. Snow. 1993. Effects of sludge disposal on groundwater nitrate concentrations. *Journal of Hydrology* 142:213–28.

Spalding, R., and J. Parrott. 1994. Shallow groundwater denitrification. *Science of the Total Environment* 141:17–25.

Spence, M. J., S. H. Bottrell, S. F. Thornton, and D. N. Lerner. 2001. Isotopic modelling of the significance of bacterial sulphate reduction for phenol attenuation in a contaminated aquifer. *Journal of Contaminant Hydrology* 53(3–4):285–304.

Starr, F. L., and B. L. Sawhney. 1980. Movement of nitrogen and carbon from a septic system drainfield. *Water Air and Soil Pollution* 13:113–28.

Strauss, H. 1997. The isotopic composition of sedimentary sulfur through time. *Palaeogeography Palaeoclimatology Palaeoecology* 132:97–118.

Strauss, H. 1999. Geological evolution from isotope proxy signals - sulfur. *Chemical Geology* 161:89–101.

Strebel, O., J. Boettcher, and P. Fritz. 1990. Use of isotope fractionation of sulfate-sulfur and sulfate-oxygen to assess bacterial desulfurication in a sandy aquifer. *Journal of Hydrology* 121:155–72.

Talma, A. S., and R. Meyer. 2005. The application of 'anomalous' nitrogen isotopic tracers in industrial groundwater pollution situations. Biennial conference: Ground water division of GSSA, Pretoria, 7–9 March 2005 (on CD).

Taylor, B. E., and M. C. Wheeler. 1994. Sulfur- and oxygen-isotope geochemistry of acid mine drainage in the western United States: Field and experimental studies revisited. In *Environmental geochemistry of sulfide oxidation*, ACS Symposium Series, eds. C. N. Alpers and D. W. Blowes, 481–514. Washington, DC: American Chemical Society.

Taylor, B. E., M. C. Wheeler, and D. K. Nordstrom. 1984. Isotope composition of sulphate in acid mine drainage as measure of bacterial oxidation. *Nature* 308:538–41.

Tellam, J. H., and A. Thomas. 2002. Well water quality and pollutant source distributions in an urban aquifer. In *Current problems in hydrogeology in urban areas, urban agglomerates and industrial centres.* eds. W. F. Howard and R. G. Israfilov, 139–58. NATO Science Series, IV. Earth and Environmental Sciences 8.

Thamdrup, B., K. Finster, J. W. Hansen, and F. Bak. 1993. Bacterial disproportionation of elemental sulfur coupled to chemical reduction of iron or manganese. *Applied Environmental Microbiology* 59:101–8.

Thode, H., J. Macnamara, and C. B. Collins. 1949. Natural variations in the isotopic content of sulphur and their significance. *Canadian Journal of. Research* 27:361–73.

Tredoux, G., J. F. P. Engelbrecht, and A. S. Talma. 2005. Groundwater nitrate in arid and semi-arid parts of southern Africa. In *Environmental geology in semi-arid environments*, eds. H. Vogel and C. Chilume, 121–33. Lobatse, Botswana: DGS.

Trettin, C. C., D. W. Johnson, and D. E. Todd. 1999. Forest nutrient and carbon pools at Walker Branch Watershed: Changes during a 21-year period. *Soil Science Society of America Journal* 63:1436–8.

Trust, B. A., and B. Fry. 1992. Stable sulphur isotopes in plants: A review. *Plant Cell and Environment* 15:1105–10.

Ueda, S., N. Ogura, and E. Wada. 1991. Nitrogen stable isotope ratio of groundwater N_2O. *Geophysical Research Letters* 18:1449–52.

Urey, H. C. 1947. The thermodynamic properties of isotopic substances. *Journal of chemical Society*, 1947: 562–81.

Van Beek, C. G. E. M., and D. Van der Kooij. 1982. Sulfate reducing bacteria in groundwater from clogging and nonclogging shallow wells in the Netherlands river region. *Ground Water* 20:298–302.

Van Stempvoort, D. R., J. Armstrong, and B. Mayer. 2007a. Microbial reduction of sulfate injected to gas condensate plumes in cold water. *Journal of Contaminant Hydrology* 92:184–207.

Van Stempvoort, D. R., J. Armstrong, and B. Mayer. 2007b. Seasonal reacharge and replenishment of sulfate associated with biodegradation of a hydrocarbon plume. *Ground Water Monitoring Remediation* 27:110–21.

Van Stempvoort, D. R., and H. R. Krouse. 1994. Controls of $\delta^{18}O$ in sulfate: Review of experimental data and application to specific environments. In *Environmental geochemistry of sulfide oxidation*, ACS Symposium Series, eds. C. N. Alpers and D. W. Blowes, 446–80. Washington, DC: *American Chemical Society*.

Van Stempvoort, D. R., H. Maathius, E. Jaworski, B. Mayer, and K. Rich. 2004. Oxidation of fugitive methane in ground water linked to bacterial sulfate reduction. *Ground Water* 43:187–99.

Vidon, P., and A. R. Hill. 2004a. Landscape controls on the hydrology of stream riparian zones. *Journal of Hydrology* 292:210–28.

Vidon, P., and A. R. Hill. 2004b. Landscape controls on nitrate removal in stream riparian zones. *Water Resources Research* 40:W03201, Doi: 10.1029=2003WR002473.

Viraraghavan, T., and R. G. Warnock. 1975. Current Canadian practice in using septic tank systems. *Canadian Journal of Public Health* 66:213–20.

Vitòria, L., N. Otero, A. Soler, and A. Canals. 2004. Fertilizer characterization: isotopic data (N, S, O, C, and Sr). *Environmental Science & Technology* 38:3254–62.

Vitòria, L., A. Soler, R. Aravena, and A. Canals. Multi-isotopic approach (^{15}N, ^{13}C, ^{34}S, ^{18}O and D) for tracing agriculture contamination in groundwater (Maresme, NE Spain). 2005. In *Environmental chemistry: Green chemistry and pollutants in ecosystems.* Vol 1, No. XXVI, eds. E. Lichtfouse, S. Dudd, and D. Robert, 137–47. Springer-Verlag.

Wadleigh, M. A., H. P. Schwarcz, and J. R. Kramer. 1994. Sulphur isotope tests of seasalt correction factors in precipitation: Nova Scotia, Canada. *Water Air and Soil Pollution* 77:1–16.

Wadleigh, M. A., H. P. Schwarcz, and J. R. Kramer. 1996. Isotopic evidence for the origin of sulphate in coastal rain. *Tellus* 48B:44–59.

Wakshal, E., and H. Nielsen. 1982. Variations of $\delta^{34}S(SO_4)$, $\delta^{18}O(H_2O)$ and Cl/SO_4 ratio in rainwater over northern Israel, from the Mediterranean coast to Jordan Rift Valley and Golan Heights. *Earth and Planetary Science Letters* 61:272–82.

Wassenaar, L. I. 1995. Evaluation of the origin and fate of nitrate in the Abbotsford aquifer using the isotopes of ^{15}N and ^{18}O in NO_3. *Applied Geochemistry* 10:391–405.

Wassenaar, L. I., M. J. Hendry, and N. Harrington. 2006. Decadal geochemical and isotopic trends for nitrate in a transboundary aquifer and implications for agricultural beneficial management practices. *Environmental Science & Technology* 40:4626–32.

Webster, E. A., and D. W. Hopkins. 1996. Nitrogen and oxygen isotope ratios of nitrous oxide emitted from soils and produced by nitrifying and denitrifying bacteria. *Biology and Fertility of Soils* 22:326–30.

Wells, E. R., and N. C. Krothe. 1989. Seasonal fluctuation in $\delta^{15}N$ of groundwater nitrate in a mantled karst aquifer due to macropore transport of fertilizer-derived nitrate. *Journal of Hydrology* 112:91–20.

Weiner, J. M., T. S. Lauck, and D. R. Lovley. 1998. Enhanced anaerobic benzene degradation with the addition of sulfate. *Bioremediation Journal* 2:159–73.

Whitehead, E., K. M. Hiscock, and P. F. Dennis. 1999. Evidence for sewage contamination of the Sherwood Sandstone aquifer beneath Liverpool, UK. In *Impacts of urban growth on surface water and groundwater quality* (Proc. IUGG 99 Symp. HS5, Birmingham, July 1999). IAHS Publ. no. 259, 179–85.

Wiedemeier, T. H., H. S. Rifai, C. J. Newell, and J. T. Wilson. 1999. *Natural attenuation of fuels and chlorinated solvents in the subsurface.* New York: Wiley.

Wilhelm, S. R., S. L. Schiff, and W. D. Robertson. 1994. Chemical fate and transport in a domestic septic system: Unsaturated and saturated zone geochemistry. *Environmental. Toxicology and Chemistry* 13:193–203.

World Health Organization. 2004. *Guidelines for drinking water quality*, 3rd ed. Geneva: WHO.

Zuppi, G. M., J. C. Fontes, and R. Letolle. 1974. Isotopes du milieu et circulations d'eaux sulfurees dans le Latinum. In *Isotope techniques in groundwater hydrology*, ed. IAEA, 341–61. Vienna, Austria: IAEA.

Section III

Isotopes in Field Applications

8 Investigating the Origin and Fate of Organic Contaminants in Groundwater Using Stable Isotope Analysis

Daniel Hunkeler and Ramon Aravena

CONTENTS

8.1   INTRODUCTION

This chapter discusses how isotope data can be used to demonstrate, and in some cases, quantify the extent of contaminant transformation at the field scale in the context of natural attenuation or engineered remediation. Furthermore, it discusses how isotope data can help to differentiate between different sources of the same contaminant.

When implementing natural attenuation or engineered remediation to remediate contaminated aquifers, there is usually a need to demonstrate occurrence of contaminant degradation. Here contaminant degradation refers to any biotic or abiotic process that leads to a transformation or complete mineralization of a compound. Concentration measurements alone often do not unequivocally demonstrate contaminant degradation since concentrations also vary due to physical processes (dilution, sorption) in the aquifer, mixing of different waters during sampling, or due to injection of treatment solution during engineered remediation. Therefore, there is a need for additional evidence for contaminant degradation, especially if the contaminants are degraded to end products that are abundant in groundwater such as CO_2 or if some of the end product is already present at the source (e.g., TBA at MTBE sites). Compound-specific isotope analysis can provide compelling evidence for contaminant degradation provided that the expected process is associated with sufficient isotope fractionation, which is the case for many common groundwater contaminants (see Chapter 4). The basic principle of the method is that the degrading compound becomes increasingly enriched in the heavy isotope as degradation proceeds. The isotope ratios of the contaminant provide a qualitative and sometimes also quantitative indication for the degree of degradation. The quantitative evaluation of field isotope data can be based on the Rayleigh equation (Section 8.4) or by incorporating the isotope fractionation into a reactive transport model (Section 8.5). The strategies to evaluate field isotope data will be illustrated using selected case studies. A more extensive overview of case studies from the scientific literature can be found in Table 8.1.

8.2   ISOTOPE TRENDS DURING CONTAMINANT DEGRADATION

In the following, the expected isotope evolution in a plume undergoing contaminant degradation is illustrated in a simplified way. For the case of constant groundwater flow velocity and a constant first-order degradation rate, we can derive an equation that describes the isotope evolution by combining the simplified Rayleigh equation

$$\delta^{13}C = \delta^{13}C_0 + \varepsilon_{bulk}(\text{‰}) \times \ln f_{DEG} \tag{8.1}$$

with the first-order degradation law

$$f_{DEG} = \exp(-k_t \times t) \tag{8.2}$$

which leads to

$$\delta^{13}C = \delta^{13}C_0 - \varepsilon_{bulk}(\text{‰}) \times k_t \times t \tag{8.3}$$

TABLE 8.1
Overview of Selected Field Studies on the use of Isotopes to Assess Reactive Processes

Compounds Chlorinated Hydrocarbons[a]	Source	Site – Aquifer Type – Flow Rate – Installation – Study Scale	Remediation	Data Evaluation	Reference
PCE, TCE, DCE, VC, Ethene	Transfer facility	– Sandy aquifer – 0.02–2 m/year – Short screen (1.4 to 2.5 m) monitoring wells – 30 m	– Natural attenuation		Hunkeler, Aravena, and Butler 1999
PCE, TCE	Air Force Base	– Sand and gravel aquifer – 14.4 m/year – Monitoring wells – 1400 m	– Natural attenuation	– B/f calc.	Sherwood Lollar et al. 2001
PCE, TCE, cDCE, VC	Air Force Base	– Clayey gravel – 0.9 m/d – Monitoring wells – 10 m	– Bio-augmentation	– Rate calculation	Morrill et al. 2005

(Continued)

TABLE 8.1 (Continued)

Compounds Chlorinated Hydrocarbons[a]	Source	Site – Aquifer Type – Flow Rate – Installation – Study Scale	Remediation	Data Evaluation	Reference
PCE, TCE, cDCE, VC, Ethene	Liquid waste injection site	– Basalt – n.d. – Monitoring wells	– Enhanced bioremediation (Lactate inj.)		Song et al. 2002
PCE, TCE, DCE, VC, CT, CF, DCM, 12 DCA, 11 DCA, 112 TCA, 1122 PCA	Waste lagoons	– Sandy aquifer – 0.3–4.8 m/d – Monitoring wells – 450 m	– Source excavation	– B/f calc. – Enrichment factor	Hunkeler, Aravena, and Butler 1999
TCE	Industrial site	– Sandy aquifer – Enclosed source – Multilevel wells – 15 m	Permanganate injection	–	Hunkeler et al. 2003
PCE, TCE, DCE, VC	Industrial site	– Sand and gravel – n.d. – Monitoring wells – 1300 m	– Natural attenuation		Imfeld et al. 2008
Petroleum Hydrocarbons[b]					
B, E	Industrial site	– Sandy aquifer – 5 m/year – Multilevel wells – 160 m	– Natural attenuation	– B/f calc.	Mancini et al. 2002

T, E, X, TMB, NAP	Landfill	– Sandy aquifer – n.d. – Piezometer (10cm) – 47m	– Natural attenuation	– B/f calc.	Richnow, Meckenstock et al. 2003
T, E, NAP	Industrial site	– Sand and gravel aquifer – 0.2–0.4m/d – Fully screened monitoring wells (7–15 m) – 700 m	– Natural attenuation	– B/f calc.	Richnow, Annweiler et al. 2003
B, MTBE	Retail filling station	– Chalk aquifer – n.d. – Multilevel wells – 125 m	– Natural attenuation	– Rayleigh plot	Spence et al. 2005
B, T, E, X, NAP, mNAP	Gaswork	– Gravel and sand aquifer – n.d. – Fully screened monitoring wells – 100 m	– Natural attenuation	– Rayleigh plot – B/f calc.	Griebler et al. 2004
T, X, mNAP	Gaswork	– Sand – Gravel and sand – 2 to 4.2 m/d – 130 m	– Natural attenuation	– Rayleigh plot	Steinbach et al. 2004

(Continued)

TABLE 8.1　(Continued)

Gasoline Additives[c]	Source	Site –Aquifer Type –Flow Rate –Installation –Study Scale	Remediation	Data Evaluation	Reference
Gasoline Additives[c]					
MTBE	Gasoline station	– Sandy aquifer – 3 to 26 m/year – Monitoring wells (4m) – 25 m	– Natural attenuation	– Rayleigh plot – Enrichment factor	Kolhatkar et al. 2002
MTBE	Industrial landfill	– n.d. – n.d. – Monitoring wells – 460 m	– Pump and treat	– Dual isotope	Zwank et al. 2005
MTBE	Gasoline station	– Sandy aquifer – 0.55 m/d – Monitoring wells – 28 m	– Natural attenuation	– Rate calculation	McKelvie et al. 2007

[a] PCE: Tetrachloroethene; TCE: Trichloroethene; DCE: Dichloroethene; VC: Vinylchloride; CT: Carbon tetrachloride; CF: Chloroform; DCM: Dichloromethane; 12 DCA: 1,2-Dichloroethane; 11 DCA: 1,1-Dichloroethane; 112 TCA: 1,1,2-Trichloroethane; 1122 PCE: 1,1,2,2-Tetrachloroethane.

[b] B: Benzene; T: Toluene; E: Ethylbenzene; X: Xylene; NAP: Naphtalene; mNAP: methyl-Naphthalene.

[c] MTBE: Methyl-*tert*-butyl-ether.

[d] B/F Calc: Calculation of Degree of Biodegradation or Fraction Remaining; Rate Calculation: Calculation of First-Order Rate Constant; Enrichment Factor: Calculation of Field-Based Isotope Enrichment Factor; Dual Isotope: Measurement of Isotopes of Two Elements in Compound to Identify Reaction Mechanism

By replacing the travel time with

$$t = \frac{x}{v_{cont}} \qquad (8.4)$$

we obtain

$$\delta^{13}C = \delta^{13}C_0 - \varepsilon_{bulk}(‰) \times k_t \times \frac{x}{v_{cont}} \qquad (8.5)$$

With:

$\delta^{13}C_0$	Isotope ratio at the source
$\delta^{13}C$	Isotope ratio at a downgradient well
$\varepsilon_{bulk}(‰)$	Isotope enrichment factor in per mill
x	Travel distance
v_{cont}	Contaminant migration velocity
k_t	First-order rate constant
f_{DEG}	Fraction of compound that is degraded (for closed system $f_{DEG} = C/C_0$).

According to Equation 8.5, the isotope ratio is expected to increase linearly with distance (Figure 8.1a). How rapidly the isotope ratio increases depends on the isotope enrichment factor, the degradation rate constant, and the contaminant migration velocity. In reality, the isotope evolution will often be more complex because the velocity and degradation rate often vary in space. For example in the case of reductive dechlorination driven by dissolved electron donor from the source zone, the electron donor may become depleted and degradation slows down or stops downgradient of the source (Figure 8.1b); or the transformation rate may be controlled by mixing of contaminated groundwater with dissolved electron acceptors (Figure 8.1c). In both cases the isotope evolution will be far from linear with distance.

8.3 SAMPLING STRATEGIES

Usually isotope analysis is carried out in a second phase after a detailed site characterization including concentration analysis and characterization of prevailing geochemical conditions. Based on this data, it may be possible to postulate possible degradation processes at a site but it may remain uncertain if they actually occur. Isotope data can serve as a tool to validate a conceptual model of reactive processes or to distinguish between two competing conceptual models (e.g., dilution only versus reactive process). For this purpose usually a subset of the available monitoring wells is sampled for isotope analysis. The sampling strategy will depend on the question that is addressed, the conceptual model of transformation processes that is tested, and on the available monitoring network. Frequently monitoring wells are aligned along transects to locate the plume core and delineate the lateral extent of the plume. A common strategy is to sample along the core of the plume to evaluate if this high concentration zone is subject to transformation. However, transformation may mainly occur at the plume fringe and samples need to be taken there as well

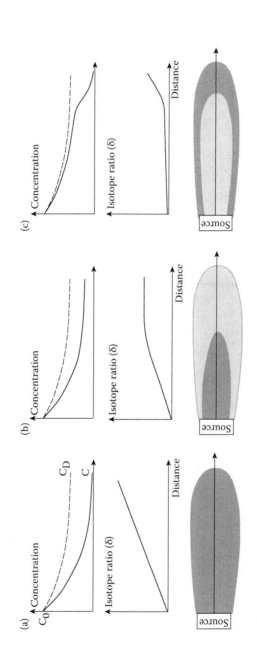

FIGURE 8.1 Concentration and isotope evolution along plume center line and schematic illustration of contaminant plume with zone where degradation takes place (dark gray). (a) Constant degradation rate throughout plume, (b) Degradation limited by availability of dissolved electron donor from source zone, and (c) Degradation limited by availability of dissolved electron donor.

(Figure 8.1c). This can be challenging because the reactive zone may be of limited extent and concentrations low. In addition to sampling the plume, it is important to know the starting isotope signature either by sampling some source material (e.g., NAPL) or, since this is often not possible, by sampling high concentration areas in the source zone. Especially for degradation processes that show relatively little isotope fractionation, it is important that the concentration contrast between source zone and plume wells is sufficiently large for isotope fractionation to become detectable. Using carbon isotopes as an example, the minimal necessary decrease in concentration can be calculated based on the Rayleigh equation

$$\frac{C}{C_0} = \exp\left(\frac{\Delta \delta^{13}C}{\varepsilon_{bulk}(\%_0)}\right) \qquad (8.6)$$

Although the uncertainty of the analytical method is typically around 0.3–0.5‰ for carbon, the shift in isotope ratios should typically be at least 2‰ since small isotope fractionation associated with various processes adds additional variability. According to the equation, the minimal concentration decrease is larger the smaller the isotope enrichment factor is. Note that this equation assumes that the concentration decreases solely due to transformation while in reality, dilution contributes as well. Thus, if possible, the actual concentration decrease should be substantially larger than the calculated one.

8.4 EVALUATING ISOTOPE DATA BASED ON THE RAYLEIGH EQUATION

Before illustrating different strategies to evaluate isotope data based on the Rayleigh equation, some key mathematical expressions will be defined. The simplified form of the Rayleigh equation is given by

$$\delta^{13}C = \delta^{13}C_0 + \varepsilon_{bulk}(\%_0) \times \ln f_{DEG}$$

In the Rayleigh equation f_{DEG} corresponds to the fractionation of compound remaining (i.e., the fraction of compound that has not been subject to degradation yet). As illustrated in Figure 8.2a, in a closed laboratory microcosm, the concentration decrease is usually solely due to degradation and hence f_{DEG} corresponds to

$$f_{DEG} = \frac{C}{C_0} \qquad (8.7)$$

where C is the concentration at any moment in time and C_0 the initial concentration.
However, at a field site, the contaminant concentration usually decreased due to dilution in addition to degradation (Figure 8.2b). In this case, f_{DEG} can be considered

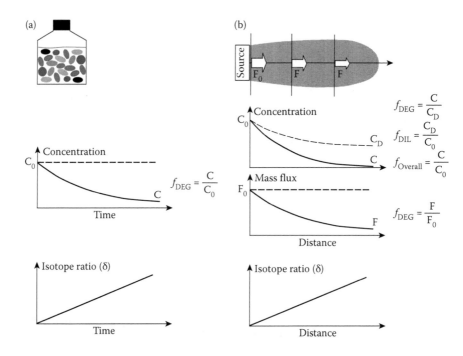

FIGURE 8.2 Comparison of concentration and isotope evolution in (a) laboratory microcosm and (b) field study. f_{DEG}: Concentration decrease due to degradation; f_{DIL}: Concentration decrease due to dilution; $f_{Overall}$: Total concentration decrease.

to correspond to the concentration decrease normalized to the concentration of a (hypothetical) conservative tracer (Abe and Hunkeler 2006)

$$f_{DEG} = \frac{C}{C_D} \qquad (8.8)$$

where C is the measured concentration and C_D is the concentration of a hypothetical tracer leaving the source at the same concentration as the contaminant. Alternatively, f_{DEG} can be considered to correspond to the mass flux reduction. At steady state, the mass flux remains constant with distance if dilution only occurs, while the concentration decreases.

The amount of dilution that a hypothetical tracer would experience is given by

$$f_{DIL} = \frac{C_D}{C_0} \qquad (8.9)$$

where C_0 is the initial concentration at the source.

The overall concentration decrease is given by

$$f_{Overall} = \frac{C}{C_0} \qquad (8.10)$$

The product of the f_{DEG} and f_{DIL} corresponds to the overall concentration decrease

$$f_{DEG} \times f_{DIL} = \frac{C}{C_D} \times \frac{C_D}{C_0} = \frac{C}{C_0} = f_{Overall} \qquad (8.11)$$

As illustrated by Equation 8.1, the isotope enrichment factor $\varepsilon_{bulk}(\permil)$ is a key parameter that controls the isotope evolution and its value needs to be known for a quantitative evaluation of isotope data. The factors that influence the magnitude of isotope enrichment factors are discussed in detail in Chapter 4. A key element is usually the mechanism of the initial transformation step. Hence, it is important to know the degradation mechanism before proceeding to calculations. Since most compounds (e.g., BTEX, chlorinated ethenes) are degraded by a different mechanism under oxic and anoxic conditions, the redox conditions usually need to be known in order to choose the appropriate isotope enrichment factor. However, for some compounds, even under the same redox conditions, transformation can proceed by different mechanisms leading to a different magnitude of isotope fractionation (e.g., aerobic degradation of 1,2-DCA). Furthermore, it is sometimes difficult to unambiguously identify the prevailing redox conditions. As will be shown in Section 8.4.5, if isotope ratios of two elements are measured it is sometimes possible to identify the reaction mechanism using field isotope data and then chose the appropriate isotope enrichment factors for calculations.

In the following, the strategies for quantitative evaluation of isotope ratios are discussed based on data from a gasoline contaminated site (Figure 8.3). An overview of other published field studies can be found in Table 8.1. For simplicity, only the monitoring wells with the highest concentration in each transect are considered, which represent the core of the plume using o-xylene as an example. According to Equation 8.6, the minimal concentration decrease (C/C_0) for a shift of $2\permil$ corresponds to 0.43 assuming an average isotope enrichment factor of $-2.4\permil$ for anaerobic o-xylene degradation. The actual concentration decrease between the source and MW-III is much larger $(C/C_0 = 0.01)$. Hence, a potentially significant change in $\delta^{13}C$ should be detectable at the site if biodegradation is an important process contributing to the concentration reduction.

8.4.1 RAYLEIGH PLOT

A first approach for evaluating isotope data is to plot $\delta^{13}C$ versus $\ln(C/C_0)$, or $\ln(C)$, according to the simplified Rayleigh equation (Figure 8.4) or as $1000 \cdot \ln(R/R_0)$ versus $\ln(C/C_0)$ according to the full Rayleigh equation. The resulting plots are frequently denoted as Rayleigh plots. The two lines in Figure 8.4 correspond to the expected isotope evolution if concentrations were affected by biodegradation only (full line) and dilution only (dashed line). The line for biodegradation only corresponds to the trend expected during biodegradation in a laboratory microcosm and is defined by the simplified Rayleigh equation. Hence, its slope corresponds to the isotope enrichment factor. The measured points are located between the two lines with one exception indicting that biodegradation as well as dilution contribute to the reduction of the contaminant concentration.

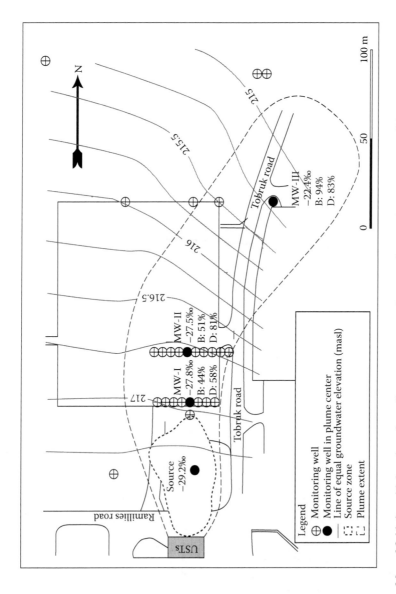

FIGURE 8.3 Map of field site with location of gasoline source zone and extent of contaminant plume. The $\delta^{13}C$ and calculated extent of biodegradation B and dilution D of o-xylene is illustrated for the sampling point with the highest concentration in each transect. D: Concentration decrease due to dilution; B: Amount of biodegradation (see text for details). Note that both B and D can reach a maximum of 100%.

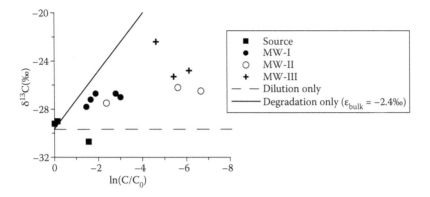

FIGURE 8.4 Rayleigh plot of field isotope data. The full line corresponds to the expected isotope evolution if concentrations were affected by biodegradation only, similarly to a closed laboratory microcosms, the dashed line if they were affected by dilution only.

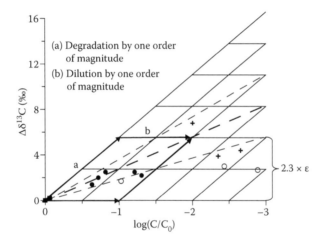

FIGURE 8.5 Effect of biodegradation and dilution on the location of points in the Rayleigh plot.

The relative contribution of biodegradation and dilution to the concentration reduction can easily be illustrated in quantitative terms in a Rayleigh plot as illustrated in Figure 8.5 where the concentration decrease is given on a log instead of an ln scale. According to the simplified Rayleigh equation, the shift in $\Delta\delta^{13}C$ due to degradation only is given by

$$\Delta\delta^{13}C = \delta^{13}C - \delta^{13}C_0 = \varepsilon_{bulk}(‰) \times \ln f_{DEG} = 2.3 \times \varepsilon_{bulk}(‰) \times \log f_{DEG} \quad (8.12)$$

For each order of magnitude of reduction of concentration decrease due to degradation, the point on the Rayleigh plot shifts by 1 on the $\log(C/C_0)$ axis and by 2.3 ε on the $\delta^{13}C$ axis (arrow "a" in Figure 8.5). In contrast, if only dilution takes place, the point shifts by 1 on the $\log(C/C_0)$ axis without any change in isotope ratio (arrow "b" in Figure 8.5). If both processes take place, the final location is independent of the order of the process (i.e., independent of whether shift a takes place before or after shift b). In reality, biodegradation and dilution take place at the same time and the position of the point reflects the overall contribution of biodegradation and dilution to the concentration reduction. Lines that illustrate the relative contribution of the process can also be included in the Rayleigh plot. For example, the bold dashed line in Figure 8.5 corresponds to the expected trend if degradation and dilution equally contribute to the concentration reduction. The thinner dashed lines correspond to the trends if degradation proceeds at twice the rate of dilution (upper line) or half its rate (lower line). Hence in the example, degradation and dilution approximately equally contribute to concentration reduction until transect 1 while in transect 2 and further downgradient dilution generally dominates over degradation.

8.4.2   QUANTIFYING DEGRADATION AND DILUTION

At field sites, we are particularly interested in the extent of degradation. We can calculate the f_{DEG} based on the Rayleigh equation

$$f_{DEG} = \exp\left(\frac{\Delta\delta^{13}C}{\varepsilon_{bulk}(\%o)} \right) \qquad (8.13)$$

Since there is usually no conservative tracer present at a field site, we generally do not know f_{DIL}, but we do know the overall concentration decrease $f_{Overall}$ based on concentrations at the source (C_0) and at downgradient locations (C). We can estimate f_{DIL} by rearrangement Equation 8.11

$$f_{DIL} = \frac{f_{Overall}}{f_{DEG}} = \frac{C/C_0}{f_{DEG}} \qquad (8.14)$$

Hence, we can estimate f_{DEG} and f_{DIL} from concentration and isotope data, provided that we know the isotope enrichment factor associated with the process. To illustrate the relative importance of degradation and dilution, we can plot the corresponding fractions on a X-Y plot (Van Breukelen 2007). Note that Figure 8.6 is related to Figure 8.5 with the difference that the grid is de-skewed and the $\delta^{13}C$ axis was replaced by the degree of biodegradation according to Equation 8.13.

Instead of expressing the amount of degradation and dilution as fractions, we can also express it as percentage degradation, B (Meckenstock et al. 2004) and percentage dilution D (Van Breukelen 2007), respectively

$$B = (1 - f_{DEG}) \cdot 100$$

$$D = (1 - f_{DIL}) \, 100$$

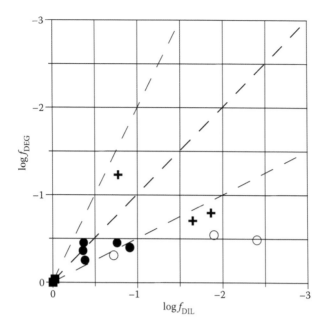

FIGURE 8.6 Contribution of degradation and dilution to concentration reduction. f_{DEG}: Concentration decrease due to degradation; f_{DIL}: Concentration decrease due to dilution. The total concentration decrease is given by $f_{overall} = f_{DEG} f_{DIL}$.

Both the percentage of degradation and dilution varies between 0 and 100%. D corresponds to the concentration reduction in percentage with respect to the initial concentration, B corresponds to the concentration reduction in percentage with respect to the expected concentration if dilution only occurred. The calculated percentages are particularly useful for illustrating the spatial variability of biodegradation and dilution (Figure 8.3).

8.4.3 ESTIMATING DEGRADATION RATES

Instead of estimating the fraction remaining or the percentage of biodegradation, it is also possible to estimate first-order degradation rate constants Equation 8.3 or Equation 8.5

$$\Delta\delta^{13}C = -\varepsilon_{bulk}(‰) \times k_t \times t = -\varepsilon_{bulk}(‰) \times k_t \times \frac{x}{v_{cont}}$$

The degradation rate constant can be calculated based on the change in isotope ratio, $\Delta\delta^{13}C$, between the source and a downgradient well separated by a distance x or contaminant travel time t

$$k_t = -\frac{\Delta\delta^{13}C \times v_{cont}}{\varepsilon_{bulk}(‰) \times x} = -\frac{\Delta\delta^{13}C}{\varepsilon_{bulk}(‰) \times t} \qquad (8.15)$$

Alternatively, the degradation rate can be estimated by plotting $\Delta\delta^{13}C$ versus x or t for several points at different distances from the source followed by linear regression. The slope corresponds to $-\varepsilon_{bulk}(\text{‰})\times k_t/v_{cont}$ or $-\varepsilon_{bulk}(\text{‰})\times k_t$.

The obtained degradation rate constant is directly proportional to the velocity of contaminant migration that can be difficult to determine accurately. It not only depends on the groundwater flow velocity but also on the magnitude of retardation. The simplest approach to obtain the contaminant velocity is to estimate the groundwater flow velocity using Darcy's Law and the retardation factor based on the organic carbon content

$$v_{cont} = \frac{K \times i}{n \times R} \tag{8.16}$$

where K is the hydraulic conductivity, i the hydraulic gradient, n the porosity, and R the retardation factor.

Alternatively, a first-order degradation rate constant with respect to distance instead of time can be defined as follows

$$k_x = \frac{k_t}{v_{cont}} = -\frac{\Delta\delta^{13}C}{\varepsilon_{bulk}(\text{‰}) \times x} \tag{8.17}$$

The isotope evolution as a function of distance is then given by

$$\Delta\delta^{13}C = -\varepsilon_{bulk}(\text{‰}) \times k_x \times x \tag{8.18}$$

The quantification of k_x does not require information about the groundwater flow velocity. Similarly as we can define the half-life time of a substance, we can also define the half-life distance based on Equation 8.17

$$X_{1/2} = \frac{\ln 2}{k_x} = -\frac{\varepsilon_{bulk}(\text{‰}) \times x \times \ln 2}{\Delta\delta^{13}C} \tag{8.19}$$

For the case study, linear regression based on Equation 8.18 was used to estimate the first-order degradation rate. The isotope evolution can be described by two lines in a $\delta^{13}C$ versus x plot (Figure 8.7): One following the points corresponding to the center of the plume (in vertical direction), and another, with a steeper slope, corresponding to points at the plume fringe. According to Equation 8.18, the slope corresponds to $-\varepsilon_{bulk}\times x$. Thus if the isotope enrichment factor is known, the degradation rate constant can be estimated from the slope. For the plume center a smaller degradation rate constant was obtained than for the plume fringe probably

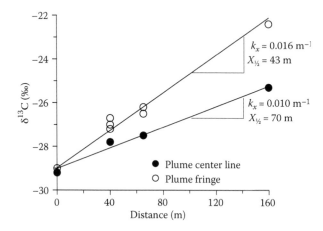

FIGURE 8.7 Estimation of the first-order rate constant and half-life from isotope data. The rate constants (k_x) are expressed in m^{-1} and the half-life $(X_{1/2})$ in m flow distance and thus are independent of the groundwater flow velocity.

because stronger oxidants are depleted in the plume core while they are available at the plume fringe (Figure 8.7). The average degradation rate obtained for the plume fringe does not necessarily reflect the kinetics of microbial degradation but the rate of mixing of water with organic contaminants and with dissolved electron acceptors, respectively.

The obtained degradation rate can be incorporated into a reactive transport model or be used to estimate the remaining travel distance until concentrations have dropped below regulatory limits. In the first case, the degradation rate with respect to time is required, in the second case, the calculation can be directly carried out with the degradation rate with respect to distance eliminating uncertainties associated with quantifying the contaminant migration velocity. The concentration decrease due to biodegradation only is given by

$$\frac{C_x}{C_{x,0}} = \exp(-k_x \times \Delta x) \tag{8.20}$$

Assuming that $C_{x,0}$ corresponds to the concentration measured at a certain downgradient location we can calculate the necessary additional flow distance x until the concentration C_x (e.g., regulatory limit) is reached

$$\Delta x = -\frac{\ln(C_x/C_{x,0})}{k_x} \tag{8.21}$$

Note that the equation yields a conservative estimate as the concentration will also decrease due to dilution.

Using the example again, it is interesting to know what farther flow distance is required until a concentration of <1ug/L is reached corresponding, for example, to the groundwater quality goal in Switzerland. The highest concentration in the most downgradient multilevel well amounts to 23 ug/L. Using the degradation rate for the plume core, $0.010m^{-1}$, we obtain a distance of 313 m based on Equation 8.21.

8.4.4 QUANTIFICATION OF MASS LOSS

For risk assessment, the mass flux of contaminants toward a receptor rather than the local concentration at a specific point is a key piece of information. The mass flux reduction between the source and a control plane perpendicular to groundwater flow can be estimated based on isotope data if a sufficient number of sampling points are available along the control plane. The remaining mass flux of a compound at the control plain F_x can be estimated by multiplying the concentration at each sampling point n with a cross-sectional area A_n for which the concentration is representative and the water flux q_n

$$F_x = \sum_n C_n \times A_n \times q_n = C_o \times \sum_n \frac{C_n}{C_o} \times A_n \times q_n = C_o \times \sum_n f_{\text{Overall},n} \times A_n \times q_n$$

(8.22)

Where C_n is the concentration in well n C_0 is the average concentration at the contaminant source A_n is the cross-sectional area for which well n is representative q_n is the water flux.

The expected mass flux if no degradation had occurred, F_D corresponds to

$$F_D = \sum_n C_{D,n} \times A_n \times q_n = C_o \times \sum_n \frac{C_{D,n}}{C_o} \times A_n \times q_n = C_o \times \sum_n f_{\text{DIL},n} \times A_n \times q_n$$

(8.23)

The relative decrease of the mass flux is given by

$$\frac{\Delta F}{F} = \frac{F_D - F_x}{F_D} = \frac{\sum_n (f_{\text{DIL},n} - f_{\text{Overall},n}) \times A_n \times q_n}{\sum_n f_{\text{DIL},n} \times A_n \times q_n} = \frac{\sum_n f_{\text{Overall},n} \times \left(\frac{1}{f_{\text{DEG},n}} - 1\right) \times A_n \times q_n}{\sum_n \frac{f_{\text{Overall},n}}{f_{\text{DEG},n}} \times A_n \times q_n}$$

(8.24)

Then $f_{\text{Overall},n}$ can be calculated according to Equation 8.11 and $f_{\text{DEG},n}$ according to Equation 8.13. The method requires only one control plane in addition to concentration and isotope data from some source sampling points. In contrast, concentration-based methods require two control planes and the calculated decrease of mass flux may often be within the range of uncertainty, which is usually quite large. The isotope-based mass flux decrease has to be considered as a minimal value because at some wells complete degradation may have occurred and for these wells the original mass flux before degradation cannot be reconstructed (nominator in Equation 8.24).

8.4.5 IDENTIFYING REACTION MECHANISMS USING DUAL ISOTOPE PLOTS

Frequently, the degradation of the same compounds by various mechanisms involves cleavage of different bonds containing other elements. Therefore, if isotope ratios of several elements are measured, the isotope fractionation pattern between different elements varies depending on the degradation mechanism. According to the simplified Rayleigh equation, the shift in isotope ratio for an element E_i is given by

$$\Delta \delta^H E_i = \varepsilon_{\text{bulk},i} \times \ln f_{\text{DEG}} \qquad (8.25)$$

Dividing Equation 8.25 for one element by the same equation for another element leads to

$$\frac{\Delta \delta^H E_1}{\Delta \delta^H E_2} = \frac{\varepsilon_{\text{bulk},1}(\text{‰})}{\varepsilon_{\text{bulk},2}(\text{‰})} \qquad (8.26)$$

The ratio of the shifts in the isotope ratio of two elements corresponds in good approximation to the ratio between the isotope enrichment factors for the two elements. The reaction mechanism can be identified by plotting the isotope ratios for two elements on an X–Y plot and comparing the slope of the obtained relationship with slopes calculated using the enrichment factors for different reactions from laboratory experiments. For example, MTBE degradation under oxic and anoxic conditions leads to different slopes on a dual isotope plot (e.g., Kuder et al. 2005; Zwank et al. 2005). Using this approach, it was demonstrated for a number of sites that biodegradation of MTBE predominantly occurred under anoxic conditions (Figure 8.8). The dual isotope approach also applies for other compounds such as the transformation of chlorinated ethenes by oxidation or reductive dechlorination (Abe, Aravena, Zopfi et al., 2009). The measurement of multiple isotopes can also be useful to distinguish spatical variations in isotope ratio due to transformation processes from spatial variations due to different sources (see Section 8.8).

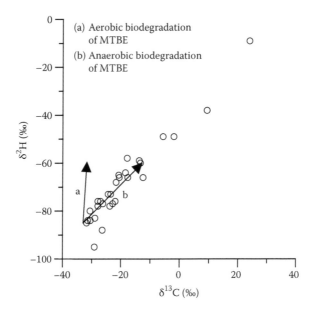

FIGURE 8.8 Dual isotope plot for aerobic and anaerobic biodegradation of MTBE. Comparison of field $\delta^{13}C$ versus δ^2H data with slopes from laboratory experiments (arrows). (Data from Gray, J. R., G. Lacrampe-Couloume, D. Gandhi, K. M. Scow, R. D. Wilson, D. M. Mackay, and B. Sherwood Lollar, *Environmental Science & Technology,* 36, 1931–38, 2002; Kuder, T., J. T. Wilson, P. Kaiser, R. Kolhatkar, P. Philp, and J. Allen, *Environmental Science & Technology,* 39, 213–20, 2005.)

8.5 INCORPORATION OF ISOTOPE FRACTIONATION INTO REACTIVE TRANSPORT MODELS

All calculation procedures introduced so far are based on the Rayleigh equation. While these calculations can be carried out rapidly and provide insight in the magnitude of contaminant degradation, they have their limitations. They only apply to one step degradation processes. Furthermore, degradation has to be the only fractioning process, which is usually the case in groundwater but not in the unsaturated zone where diffusion can fractionate isotopes as well (Bouchard et al. 2008). Finally, the Rayleigh equation relies on the assumption of a homogeneous, well-mixed contaminant reservoir that is not strictly met for groundwater, where transport of molecules may occur at different rates.

A more general approach to simulate the evolution of isotope ratios is to incorporate isotope fractionation into reactive transport models. The basic principle of this approach is to treat the molecules with a different isotopic composition as separate compounds (e.g., molecules with ^{12}C only, molecules with one ^{13}C) that are undergoing the same processes but degrade at slightly different rates. Based on the concentrations of different compounds at a location, the isotope ratio can then be calculated. Calculated isotope ratios can be compared to measured isotope ratio to verify that the reactive processes were correctly quantified (verification). Alternatively, reaction rate constants and other parameters can be adjusted until calculated isotope ratios match measured isotope ratios (calibration procedure). By using reactive transport models to simulate isotope data, different processes that fractionate (e.g., degradation and diffusion) can be treated simultaneously and also isotope fractionation during multistep transformation such as reductive dechlorination of chlorinated ethenes can be simulated. In the following, the calculation procedure is discussed in more detail and the procedures necessary for obtaining the different parameters are outlined.

For each compound with a different isotopic composition, a partial differential equation is formulated describing transport (advection and dispersion) and reaction. For transport and reaction in one dimension in flowing groundwater, the equation corresponds to

$$\frac{\partial C_i}{\partial t} = -v_x \times \frac{\partial C_i}{\partial x} + D_L \times \frac{\partial^2 C_i}{\partial x^2} - r \tag{8.27}$$

where C_i is the concentration of compound i v_x is the groundwater velocity D_L is the dispersion coefficient and r is the reaction rate.

To illustrate different possibilities to subdivide a compound into species with a different isotope distribution, we consider again the degradation of o-xylene and assume that anaerobic degradation dominates, which involves an initial reaction at

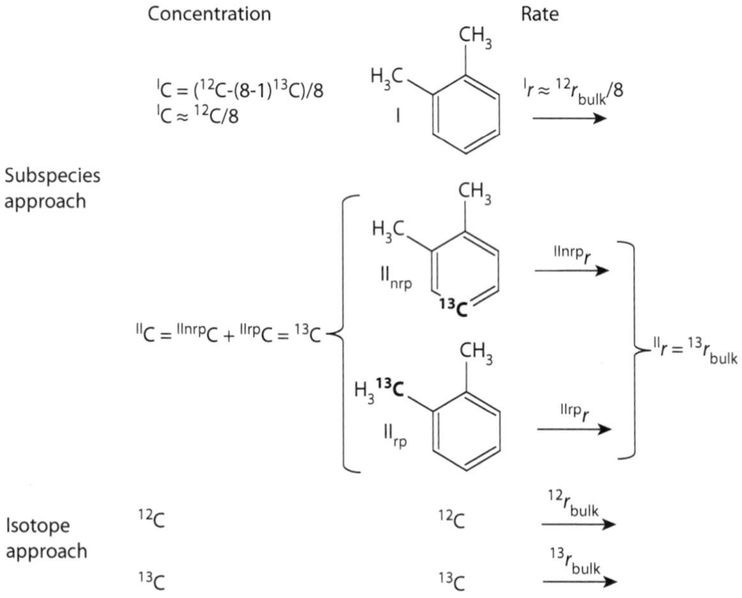

FIGURE 8.9 Subdivision of o-xylene into forms with a different carbon isotope composition (subspecies approach) and notation for the corresponding concentrations and reaction rates.

a methyl group (Figure 8.9). In the most rigorous way, the compound is subdivided into three different forms: molecules with light isotope only (version I), molecules with a heavy isotope at a location that does not participate in the reaction (i.e., the aromatic ring; version IInrp), and molecules with a heavy isotope at any of the reacting positions (i.e., the methyl group; version IIrp). This subdivision is only feasible if we know the exact reaction mechanisms (i.e., how many reacting positions are present in the molecule and where they are located).

A simpler approach is to divide the compounds into two groups, one containing molecules with light isotopes only (group I) and another containing all compounds that have one heavy isotope anywhere (group II consisting of IInrp and IIrp). Note that molecules of group I and II are often denoted as isotopologues (=molecules that differ only in their isotopic compositions) while molecules IInrp and IIrp are isotopomers (=molecules having the same number of isotopic atoms but at different positions). Since the subdivision into the three forms shown in Figure 8.9 includes both isotopologues and isotopmers, a more general term, subspecies, is used here for the different forms of the compound. The two subspecies approach is usually preferable because no assumptions about the number of reacting and nonreacting processes need to be made. Although the two subspecies approach does not fully capture the behavior of all subspecies, it provides results comparable to the three subspecies approach (Bouchard et al. 2008). Therefore, in the following, we only consider the two subspecies approach.

Isotope fractionation takes place strictly between different subspecies. Hence when simulating the isotope evolution, we should simulate the subspecies behavior and then calculate the isotope ratios from the subspecies concentrations. However, when considering the relationship between subspecies and isotope ratio, we realize that the subspecies ratio is in very good approximation always proportional to the isotope ratio for elements where the heavy isotope is present at a low abundance (e.g., C, H, N, O, but not Cl)

$$\frac{^{II}C}{^{I}C} = \frac{^{H}C}{(^{L}C - (n-1) \times {}^{H}C)/n} \approx \frac{^{H}C}{^{L}C/n} = \frac{^{H}C}{^{L}C} \times n \qquad (8.28)$$

where ^{I}C and ^{II}C are the molar concentrations of subspecies I and II, respectively, ^{H}C and ^{L}C the concentrations of heavy and light isotope, respectively, and n the number of atoms of interest in the molecule. Equation 8.28 takes into account that for elements with a low abundance of the heavy isotopes, molecules with more than one heavy isotope are very rare. The concentration of molecules with a heavy isotope is equivalent to the concentration of heavy isotopes. To calculate the concentration of molecules with light isotopes only, the light isotopes present in the molecules with a heavy isotope need to be deduced from the total concentration of light isotopes. Due to the proportionality between subspecies and isotope ratios, and given that we are interested in changes in isotope ratios at the end, we can approximate the behavior of the subspecies I and II by simulating the evolution of the concentration of heavy and light isotopes directly (isotope approach). While the direct simulation of concentrations of isotopes is frequently appropriate, especially for simulating reactive processes in groundwater, the subspecies approach should be used to simulate processes where the molecule mass influences isotope ratios, especially diffusion. It is more appropriate to attribute different diffusion coefficients to different subspecies rather than to individual atoms.

When simulating the isotope or subspecies evolution, we need, in addition to the common transport parameters, the initial concentrations and transformation rates for the two isotopes or subspecies, respectively. The initial concentrations can be calculated from the initial isotope ratio as shown in Table 8.2. As a next step we need to know the reaction rates for the two isotopes or subspecies. In case of the isotope approach, we can use the definition of the isotope fractionation factor to derive an expression that relates the reaction rates for the two isotopes. For a kinetic process, the fractionation factor is defined as follows

$$\alpha_{\text{bulk}} = \frac{d^H P/d^L P}{{}^H C/{}^L C} \tag{8.29}$$

where ${}^H C/{}^L C$ is the isotope ratio of the reacting compounds and $d^H P/d^L P$ the ratio between heavy and light instantaneous product. The amount of instantaneous product that is formed corresponds to inverse of the amount of compound that is removed from the substrate pool (i.e., it corresponds to the reaction rate)

$$\frac{d^H P}{d^L P} = \frac{{}^H r_{\text{bulk}}}{{}^L r_{\text{bulk}}} = \alpha_{\text{bulk}} \times \frac{{}^H C}{{}^L C} = \alpha_{\text{bulk}} \times R \tag{8.30}$$

where R is the isotope ratio of the reacting compounds, ${}^L r_{\text{bulk}}$ and ${}^H r_{\text{bulk}}$ are the reaction rates of light and heavy isotopes, respectively.

The total reaction rate r is given by

$$r = ({}^L r_{\text{bulk}} + {}^H r_{\text{bulk}})/n \tag{8.31}$$

Here total and subspecies rates have the units of "mole molecules volume^{-1} time^{-1}" while isotopes rates have the units of "mole atoms^{-1} time^{-1}".

Combining Equation 8.30 and Equation 8.31, the following equations are obtained to calculate the reaction rate of light and heavy isotopes from the total reaction rate

$$^L r_{\text{bulk}} = \frac{r \times n}{1 + \alpha_{\text{bulk}} \times R} \approx r \times n \tag{8.32}$$

$$^H r_{\text{bulk}} = \frac{r \times n}{1 + 1/(\alpha_{\text{bulk}} \times R)} \approx r \times n \times \alpha_{\text{bulk}} \times R \tag{8.33}$$

The approximations can be made because $\alpha_{\text{bulk}} \times R \ll 1$ and because they still conform to Equation 8.30. Equation 8.32 and Equation 8.33 apply independent of the type

TABLE 8.2

Equations to Calculate Initial Concentrations and First-Order Rates for Isotope and Subspecies Approach. ^{L}C Concentration of ^{12}C, ^{H}C Concentration of ^{13}C, ^{I}C Concentration of Subspecies I, ^{II}C Concentration of Subspecies II, R_0 Initial Isotope Ratio, n Number of Carbon Atoms in Molecule, ^{tot}k Overall Degradation Rate Constant and α_{bulk} Bulk Isotope Fractionation Factor

	Isotope Approach		Two Subspecies Approach	
	^{12}C	^{13}C	I	II
Initial concentration	$^L C_0 = \dfrac{1}{1+R_0} \times {}^{tot}C_0 \times n$	$^H C_0 = \dfrac{R_0}{1+R_0} \times {}^{tot}C_0 \times n$	$^I C_0 = \dfrac{{}^L C_0 - (n-1) \times {}^H C_0}{n} \approx \dfrac{{}^L C_0}{n}$	$^{II} C_0 = \dfrac{R_0}{1+R_0} \times {}^{tot}C_0 \times n = {}^H C_0$
First-order rate expression	$^L r_{bulk} = {}^{tot}k \times {}^L C$	$^H r_{bulk} = {}^{tot}k \times \alpha_{bulk} \times {}^H C$	$^I r = {}^{tot}k \times {}^I C$	$^{II} r = {}^{tot}k \times \alpha_{bulk} \times {}^{II} C$

of rate expression (e.g., zero-order, first-order, Monod) used to describe the reaction rate. In the case of the first-order rate law, the rate expression can be stated directly as

$$^{L}r_{\text{bulk}} = {}^{L}k_{\text{bulk}} \times {}^{L}C \approx {}^{\text{tot}}k \times {}^{L}C \tag{8.34}$$

$$^{H}r_{\text{bulk}} = {}^{H}k_{\text{bulk}} \times {}^{H}C = {}^{L}k_{\text{bulk}} \times \alpha_{\text{bulk}} \times {}^{H}C \approx {}^{\text{tot}}k \times \alpha_{\text{bulk}} \times {}^{H}C \tag{8.35}$$

where $^{\text{tot}}k$ is the overall reaction rate. The approximation can be made because the light isotopes dominate and again the approximate equation still conforms to Equation 8.30.

If we want to simulate the subspecies evolution rather then the isotope evolution, the reaction rates for the subspecies need to be known. Using the two subspecies approach, the reaction rate of group II corresponds exactly to ^{H}k because each heavy isotope corresponds to one molecule with a heavy isotope, hence $^{II}r = {}^{H}r_{\text{bulk}}$. The reaction rate of ^{I}r is approximately n times smaller than the reaction rate of ^{L}r because there are approximately n times more light isotopes than there are molecules with light isotopes. Hence $^{I}r \approx {}^{L}r_{\text{bulk}}/n$. The relationship is valid in good approximation although some of the light isotopes are present in molecules with a heavy isotope. Using Equation 8.32 and Equation 8.33, the following rate expressions for subspecies are obtained

$$^{I}r = \frac{^{L}r_{\text{bulk}}}{n} = \frac{r}{1 + \alpha_{\text{bulk}} \times R_{E}} \approx r \tag{8.36}$$

$$^{II}r = {}^{H}r_{\text{bulk}} = \frac{r \times n}{1 + 1/(\alpha_{\text{bulk}} \times R_{E})} \approx r \times n \times \alpha_{\text{bulk}} \times R_{E} \approx r \times \alpha_{\text{bulk}} \times R_{M} \tag{8.37}$$

where $R_{M} = {}^{II}C/{}^{I}C \approx n \times R$ is the subspecies ratio.

Assuming again first-order kinetics and using Equation 8.34 and Equation 8.35, the following rate expressions are obtained

$$^{I}r = \frac{^{L}r_{\text{bulk}}}{n} = {}^{L}k_{\text{bulk}} \times \frac{^{L}C}{n} \approx {}^{\text{tot}}k \times {}^{I}C \tag{8.38}$$

$$^{II}r = {}^{H}k_{\text{bulk}} \times {}^{H}C = {}^{L}k_{\text{bulk}} \times \alpha_{\text{bulk}} \times {}^{H}C \approx {}^{\text{tot}}k \times \alpha_{\text{bulk}} \times {}^{II}C \tag{8.39}$$

Hence, in good approximation, the subspecies rates can be calculated based on the isotope fractionation factor from laboratory experiments. Equations to calculate the initial concentration and first order rate expressions for different isotopes and subspecies, respectively are summarized in Table 8.2.

In the following, the incorporation of isotope fractionation into a reactive model is illustrated for the gasoline spill case study (Figure 8.3). The aim of the simulations was to reproduce the observed isotope and concentration evolution of o-xylene along the plume center line by adjusting the overall first-order rate constant (^{tot}k) and the source width that is not exactly known. For simplicity, the isotope approach was chosen using a 2D analytical solution to describe reactive transport. Two equations were established describing the evolution of ^{L}C (Figure 8.10a) and ^{H}C (Figure 8.10b), respectively after 20 years, the time elapse between the discovery of the spill and the sampling. All parameters were identical in the two equations except for the degradation rate, which was calculated according to Table 8.2 using an isotope fractionation factor of 0.9976 and by varying ^{tot}k. Initial concentrations were calculated using the equations given in Table 8.2 based on the measured concentration and isotope ratio at the source. A Pe number of 10 and ratio of 10 between longitudinal to transversal dispersivity was assumed.

A good fit between the calculated and measured isotope ratio was obtained for a degradation rate of 0.0016 d^{-1} or 0.017 m^{-1}. The calculated and measured concentration matched best for a source width of 8 m. By fitting analytical solutions simultaneously to isotope and concentration data, not only the biodegradation rate can be

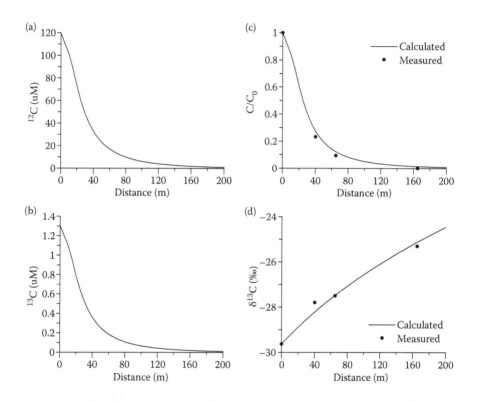

FIGURE 8.10 Measured and calculated concentration and isotope evolution. The (a) ^{12}C and (b) ^{13}C evolution was simulated using a 2D analytical solution to the advection-dispersion equation. The initial concentrations of ^{12}C and ^{13}C were calculated based on a measured initial concentration of 1600 ug/L (=120 uMC) and an initial $\delta^{13}C$ of −29.6‰ as outlined in Table 8.2. The source width and degradation rate constant was adjusted until (c) relative concentration and (d) isotope ratios matched measured values. A Pe number and ratio between longitudinal to transversal dispersivities of 10 was assumed. The flow velocity corresponded to 0.094 m/d.

calculated but it is also possible to estimate or verify the magnitude of dispersion and the source width. These parameters may be used as a starting point for a more complex numerical model.

When comparing the rate constants for the plume cores derived using the Rayleigh approach (0.01 m^{-1}) and by fitting a reactive transport model to the isotope data (0.016 m^{-1}), it becomes apparent that the two rates are different. This is due to the different conceptual model of contaminant transport inherent in the Rayleigh model and the advection-dispersion model. The Rayleigh model relies on the assumption that (i) isotope fractionation takes place out of a contaminant pool of fixed volume (by setting the fraction remaining f as C/C_0), and (ii) the contaminant pool is well mixed as it migrates toward the observation well. As illustrated above, it is usually straightforward to introduce a correction to take into account that molecules migrating away from the source become distributed in an increasingly large volume by defining $f_{DEG}=C/C_{DIL}$. However, it is more difficult to correct for the second factor. The Rayleigh equation would apply exactly if the molecules that leave the source at a certain moment in time take the same time to reach the monitoring well. This corresponds to a situation where only advective transport occurs. However, in reality, the molecules will travel along different paths and at variable velocities due to physical heterogeneity. In order to investigate how variable the travel time is, we could inject a tracer pulse in the source zone and observe the arrival of the tracer at the monitoring well. A wider break through curve indicates a more variable travel time distribution of the molecules. The travel time variability can be characterized using the dispersivity or, if we relate the dispersivity to the travel distance, the Peclet number. A higher Peclet number, corresponds to a more compact breakthrough curve and hence a more uniform travel time distribution. The error introduced by the simplified conceptual model of solute transport inherent in the Rayleigh equation can be estimated using the following approach (Abe and Hunkeler 2006)

1. We simulate the expected isotope and concentration evolution at a site using a reactive transport model that incorporates advective–dispersive transport.
2. We quantify the "true" amount of biodegradation by comparing the simulated concentration data of a degradable compound (C) to that of a conservative tracer (C_D). The obtained values are denoted as $f_{true} = C/C_D$ and B_{true} because they correspond to the actual amount of degradation.
3. We estimate the same parameters ($f_{Rayleigh}$ and $B_{Rayleigh}$) and the first order rate constant $k_{Rayleigh}$, based on the isotope data using the Rayleigh approach (see Section 8.3), which in contrast to the advection-dispersion concept, does not take into account travel time variability.
4. By comparing the values obtained by the two approaches we can gain insight to what extent the simplified conceptual model of transport affects the quantification of the parameter f, B and k.

This procedure was applied for variable Peclet numbers and for a different degree of degradation expressed by the Damköhler number (Da). Da numbers of 1, 3, and 5 correspond to 63, 95, and 99%, respectively, of degradation.

As illustrated in Figure 8.11, by using the Rayleigh equation, we generally overestimate the fraction remaining f (i.e., we underestimate the amount of degradation;

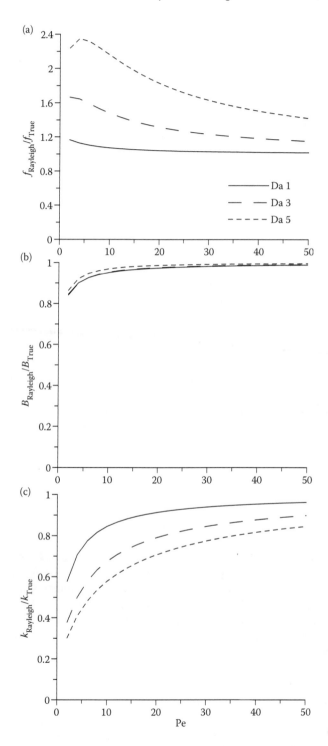

Abe and Hunkeler 2006). The overestimation islarger for smaller Peclet numbers (i.e., a more variable travel time distribution). Furthermore, is larger for an increasing degree of transformation (larger Da). An analogous conclusion is reached when we consider B rather than f. If transport velocities vary (small Pe), molecules that have experienced a lot of degradation and hence show strong isotope enrichment will be mixed with molecules with little degradation and a small shift in isotope ratio. This mixing dampens the shift in the isotope ratio and hence the calculated degree of degradation based on the Rayleigh equation is smaller than in reality. For large Pe, advective transport dominates, which corresponds to the conceptual model inherent in the Rayleigh equation. The Rayleigh approach also underestimates the degradation rate constant (Figure 8.11c) as we have already seen by comparing the degradation rate constant calculated based on the Rayleigh approach with the degradation rate constant obtained by fitting an analytical model to the data. For a typical condition at field sites (Pe around 10), the error introduced by applying the Rayleigh equation to field conditions is relatively small. Independent of which calculation is carried out, the Rayleigh estimates tend to underestimate the actual amount of degradation and can be considered as a conservative estimate.

8.6   ISOTOPE EVOLUTION OF THE PRODUCT

Although one could argue that when a clearly identifiable product is accumulating during a reaction, it is not necessary to use isotope analysis since the accumulating product is a clear indicator for transformation. However, concentration data of reactant and product are not always sufficient to clearly demonstrate or quantify a process. The product may already be present at the sources such as tributyl-alcohol (TBA) at an MTBE source or TCE at a PCE source. Furthermore, the same degradation product may originate from several compounds present at the site (e.g., ethene from VC or 1,2-DCA), which complicates the identification of an active degradation pathway. Finally, the degradation product may itself be degraded to a nonunique product such as the oxidation of cDCE or VC to CO_2. In such cases, the coordinated shifts in isotope ratios of primary compound and product can help to substantiate a transformation process.

The expected isotope evolution of the product varies depending on whether or not all the atoms of an element are conserved in the product. If all the atoms are conserved (e.g., carbon during reductive dechlorination of TCE to cis-DCE), the isotope ratio of the product is initially depleted in the heavy isotope and then approaches the initial isotope ratio of the primary compound as the reaction approaches completion and no further transformation takes place (Figure 8.12b). However, during many reactions only part

FIGURE 8.11 (Opposite) Error introduced in (a) calculated fraction remaining f, (b) percentage of biodegradation B and (c) first-order rate constant k due to the simplified conceptual model of solute transport inherent in the Rayleigh model. Comparison of Rayleigh model, which only takes into account advective transport (Rayleigh) with a reference model including advection and dispersion (true). The Peclet number (Pe) characterizes the degree of dispersion. The smaller the number the more dispersion occurs (i.e., the more variable are contaminant transport velocities). The Damköhler number (Da) characterizes the amount of degradation. The larger the number the further degradation has proceeded between the source and a downgradient well.

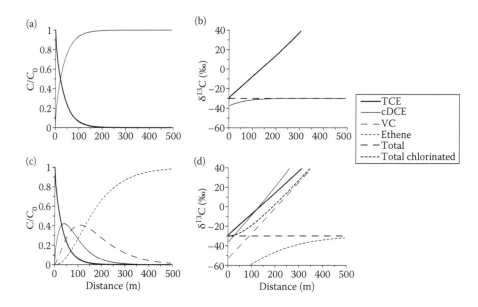

FIGURE 8.12   Concentration and carbon isotope evolution of reactant and product(s) for (a,b) single and (c,d) multiple reductive dechlorination of trichloroethene.

of the atoms of an element are transferred to the product as, for example, for TBA from MTBE, which contains only four of the five carbons of the MTBE while the methyl-ester group is removed. The isotope ratio of the product will depend on the isotope ratio of the group that is removed and on the possible isotope fractionation during removal of the atoms. Finally, the isotope evolution of the product can also be influenced by the presence of the degradation product already at the source. The measured isotope ratio of the product will then reflect the mixing of already present compound with freshly formed product as illustrated for dichloroelimination of 1,1,2–Trichloroethane (1,1,2-TCA) to VC at a field site where VC was already present at the source (Figure 8.13). The freshly produced VC is depleted in ^{13}C compared to the VC at the source and therefore, the $\delta^{13}C$ of VC evolves toward more negative $\delta^{13}C$ values while the parent compound becomes enriched in ^{13}C (Hunkeler, Aravena, and Cox 2002).

8.7   MULTISTEP TRANSFORMATION

For some compounds, the transformation product itself may be subject to transformation in a sequential process as observed for reductive dechlorination of chlorinated ethenes. For carbon in chlorinated ethenes, each product of reductive dechlorination is initially depleted in ^{13}C compared to its precursor and subsequently becomes enriched in ^{13}C (Figure 8.12d). The enrichment is due to two reasons; the product simply follows the isotope evolution of the reactant and in addition becomes enriched if further transformation is associated with isotope fractionation. The product can become more enriched in the heavy isotope than the reactant if its further transformation is rapid or associated with large isotope fractionation as illustrated for cDCE in Figure 8.12d.

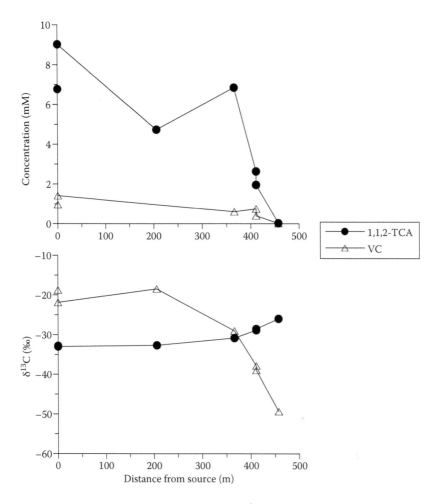

FIGURE 8.13   Concentration and carbon isotope evolution during dichloroelimination of 1,1,2-trichloroethane (1,1,2-TCA) to vinyl chloride (VC). The VC is already present at the source.

The final product in the chain (cDCE in Figure 8.12b and ethene in Figure 8.12d) approaches the initial isotope ratio of the parent compound. Hence, by comparing the isotope ratio of a daughter product to the initial isotope ratio of the parent compound, we can evaluate if substantial transformation beyond a daughter product occurs. In addition to the isotope ratio of individual compounds, the overall isotope ratio of several compounds in a degradation chain can provide insight into transformation processes (Hunkeler, Aravena, and Butler 1999; Hunkeler et al. 2005). The overall isotope ratio can be calculated as follows, which is sometimes denoted as isotope balance

$$\delta^{13}C_{total} = \frac{\sum_{i}^{n} C_i \times \delta^{13}C_i}{\sum_{i}^{n} C_i} \qquad (8.40)$$

where $\delta^{13}C_{total}$ is the average isotope ratio of n compounds, $\delta^{13}C_i$ is the isotope ratio of an individual compound i, and C_i the concentration of compound i on a molar basis. The overall isotope ratio of all chlorinated ethenes and ethene should remain constant if only reductive dechlorination occurs (Figure 8.12d). Furthermore, the overall isotope ratio of all chlorinated ethenes (this time without ethene) is a good indictor on whether or not complete reductive dechlorination of some of the molecules has occurred. If complete dechlorination occurs, the overall isotope ratios start to increase above the initial isotope ratio of the parent compound and approaches the isotope evolution of vinyl chloride (total chlorinated, Figure 8.12d). The interpretation of the overall isotope ratio becomes more complicated if the compounds migrate at different velocities, which can cause changes in the overall isotope ratio along the migration path as well (Van Breukelen et al. 2005).

In the following, the use of carbon isotope ratio to evaluate degradation of chlorinated ethenes is illustrated for a field site where a PCE plume discharges across a sandy-silty streambed (Abe, Aravena, Parker et al., 2009). Within the streambed groundwater discharge rates and redox conditions strongly vary due to physical and geochemical heterogeneity. Groundwater flow is upward toward the streambed surface and at several locations concentration profiles were determined. At a sampling location with intermediate redox conditions, the $\delta^{13}C$ of cDCE corresponds to that of the original parent compound throughout the profile indicating that no further transformation of cDCE occurs (Figure 8.14a). At a second location under sulfate reducing conditions, cDCE and VC are clearly more enriched than the original parent compound and show a trend toward more enriched values (i.e. more positive) in the profile. Ethene approaches the original parent compound ratio indicating that it is the dominant end product at this location (Figure 8.14b). The overall isotope ratio closely follows the initial isotope ratio of the parent compound confirming that reductive dechlorination to ethene is the dominant process. In contrast, at another location, even the ethene becomes enriched compared to the original parent compound demonstrating its further transformation probably to ethane, which is also present in the profile (Figure 8.14c).

8.8   SOURCE FINGERPRINTING

In industrialized and urbanized area, there are often several sources of the same contaminant. For example, there may be multiple spills of chlorinated solvents such as PCE or several gas stations potentially with spills containing MTBE. Due to the complexity of transport processes in the subsurface, it can be difficult to attribute downgradient contamination to a specific source. Industrial solvents from different manufacturers or petroleum hydrocarbons of different origin have frequently different isotopic compositions that opens the possibility of using isotope analysis for source identification. When using this method, it has to be taken into account that isotope ratios can change due to biodegradation and that two sources can accidentally have the same isotope ratio. The combined measurement of isotopes of several elements opens additional possibilities to differentiate between contaminant sources and to distinguish variations due to biodegradation from variations due to different sources.

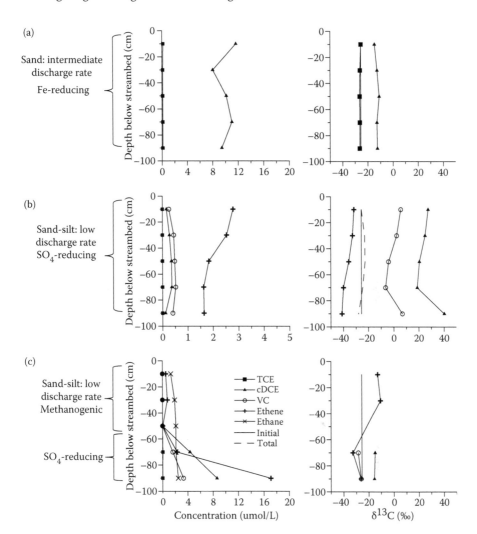

FIGURE 8.14 Concentration and carbon isotope evolution during transformation of chlorinated ethenes in streambed sediments. The groundwater flow direction is upward toward the streambed. with (a) Fe-reducing, (b) SO_4–reducing and (c) SO_4–reducing/methanogenic conditions.

8.8.1 VARIABILITY OF ISOTOPE COMPOSITION OF ORGANIC CONTAMINANTS

Chlorinated solvents from different manufacturers frequently have a different isotopic composition (Figure 8.15, Table 8.3). Variations were also observed for different batches from the same manufacturer (Shouakar-Stash, Frape, and Drimmie 2003). For PCE, a large range of $\delta^{13}C$ values has been observed so far while $\delta^{37}Cl$ are within a narrower range (Figure 8.15). For TCE and 1,1,1-trichloroethane (1,1,1-TCA), the $\delta^{13}C$ values lay within a narrower range than for PCE, while they show a considerable variation in $\delta^{37}Cl$ values. TCE is strongly enriched in ^2H probably due to isotope fractionation

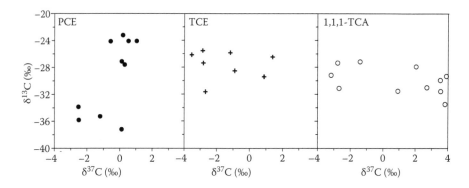

FIGURE 8.15 Carbon and chlorine isotope ratio of chlorinated solvents from different manufacturers (values from Table 8.3).

during chlorination. In contrast, 1,1,1-TCA, where the methyl group remains intact during chlorination, has a signature close to the typical signature of H in natural compounds. The chlorinated methanes are, in general, depleted in ^{13}C compared to the chlorinated ethanes and ethenes (Table 8.3), which is due to the origin of the carbon from methane that is known to be depleted in ^{13}C. During production of chlorinated methanes, chlorinated ethenes can be formed as a byproduct with a strongly ^{13}C-depleted isotope signature similarly to the chlorinated methanes. If the resulting waste is discarded, chlorinated ethene contamination with a strongly ^{13}C-depleted signature may be present.

8.8.2 DISTINGUISHING DIFFERENT SOURCES OF ORGANIC CONTAMINANTS

When carrying out source fingerprinting, potential changes in isotope ratios due to isotope fractionation has to be considered. Significant isotope fractionation usually occurs during transformation processes while other processes such as dissolution and sorption generally conserve isotope ratios. Samples from the unsaturated zone can potentially also be affected by diffusion.

For highly chlorinated ethenes, usually little biodegradation or chemical transformation occurs under aerobic conditions and isotope ratios are expected to remain constant during migration. In contrast under anaerobic conditions, reductive dechlorination frequently occurs, which is associated with significant isotope fractionation. An isotope study for source identification should include the characterization of redox conditions by measuring concentrations of dissolved oxidants (O_2, NO_3^-, SO_4^{2-}) and reduced species (Fe^{2+}, CH_4). In addition, the presence of lower chlorinated degradation products should be evaluated.

Figure 8.16 illustrates the concentration distribution and isotope ratios of PCE in a sandy aquifer where little biodegradation is expected based on redox conditions (Hunkeler, Chollet et al., 2005). The plume morphology with the different hotspots that are several 10ths of a meter apart in lateral direction suggests that the plume may originate from different sources. Groundwater samples were taken along transects of multilevel samplers 40–50 m and 220 m, respectively downgradient of the source

TABLE 8.3
Carbon, Chlorine, and Hydrogen Isotope Ratio of Chlorinated Hydrocarbons From Different Manufacturers

	δ13C (‰)				δ37Cl (‰)				δ2H (‰)			
	n	Mean	Min	Max	n	Mean	Min	Max	n	Mean	Min	Max
PCE	11	-29.1	-37.2	-23.2	10	-0.46	-2.54	1.03				
TCE	13	-29.3	-33.5	-24.5	11	0.94	-3.19	3.90	4	552	467	600
DCE	6	-25.9	-29.3	-22.2	1	0.43						
VC	2	-28.6										
1,1-TCA	8	-27.6	-31.6	-25.5	8	-1.47	-3.54	1.39	4	-3.6	-23.1	15.1
1,2-DCA	1	-30.7										
CT	3	-39.4	-47.1	-32.5	1	-0.01						
CF	4	-50.9	-63.6	-43.2	2	0.92						
DCM	6	-38.1	-53.6	-31.5	5	1.64	0.27	2.34				

Sources: Beneteau, K. M., R. Aravena, and S. K. Frape, *Organic Geochemistry*, 30, 739–53, 1999; Holt, B. D., N. C. Sturchio, T. A. Abrajano, and L. J. Heraty, *Analytical Chemistry*, 69, 2727–33, 1997; Jendrzejewski, N., H. G. M. Eggenkamp, and N. V. Coleman, *Analytical Chemistry*, 69, 4259–66, 1997; Jendrzejewski, N., H. G. M. Eggenkamp, and M. L. Coleman, *Applied Geochemistry*, 16, 1021–31, 2001; Shouakar-Stash, O., S. K. Frape, and R. J. Drimmie, *Journal of Contaminant Hydrology*, 60(3–4), 211–228, 2003; Van Wanderdam, E. M., S. K. Frape, R. Aravena, R. J. Drimmie, H. Flatt, and J. A. Cherry, *Applied Geochemistry*, 10, 547–52, 1995.

(a) Transect 1: 40–50 m downgradient of source

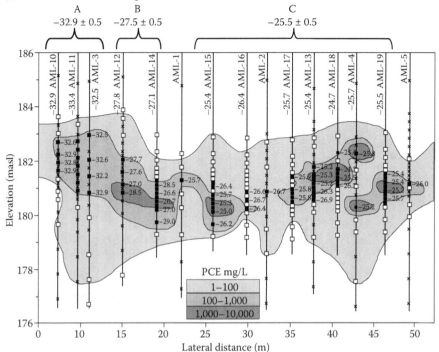

(b) Transect 2: 220 m downgradient of source

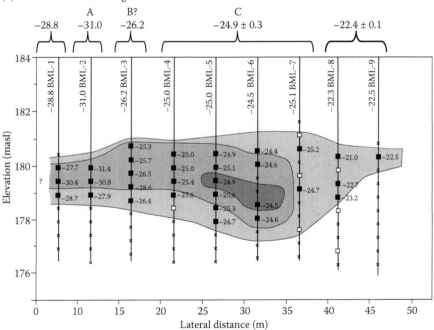

area. In transect 1, the different plume segments show a different carbon isotope composition as can also be shown using a statistical test that compares the mean isotope ratios for the different plume segments. The plume segment C is possibly related to a suspected larger PCE release event, while the second B corresponds to a second smaller release. Due to the absence of a clear hot spot typical for DNAPL accumulation, the lower concentration plume segment C may possibly originate from a more diffuse input of PCE via a septic system. In transect 2, PCE became enriched in ^{13}C compared in the left part of plume segment C likely due to biodegradation triggered by the presence of organic matter. Similarly some biodegradation also occurs in plume segment A.

If redox conditions and compound distribution indicate biodegradation, it can still be possible to distinguish between different sources if two isotopes are analyzed (e.g., carbon and chlorine) or sometimes if the overall isotope ratios of the parent and daughter products are compared (Eberts, Braun, and Jones 2008). Since reductive dechlorination involves the breaking of a carbon-chlorine bond in the initial step, a correlated shift of the carbon and chlorine isotope ratio is expected during biodegradation. Hence on a $\delta^{13}C$ versus $\delta^{37}Cl$ plot, it is possible to separate of different sources and biodegradation different sources and on isotope ratios. biodegradation.

Petroleum hydrocarbons and MTBE are usually biodegradable under oxic and anoxic conditions. Hence changes of the isotope ratio due to biodegradation have to be expected independent of the redox conditions, especially for smaller molecules. The isotope ratios of larger molecules (e.g., >C10) likely remains constant even if biodegradation occurs (see Chapter 4). Unless larger molecules are targeted, a source evaluation of petroleum hydrocarbons and MTBE should include both carbon and hydrogen isotope analysis. These isotopes are expected to show a correlated shift during biodegradation because many initial transformation processes of petroleum hydrocarbons involve changes at C-H bonds.

The dual isotope approach for fingerprinting sources is illustrated for a site with a TCE plume (Figure 8.17). The TCE contamination within the site (dashed area) is suspected to originate from two sources, a TCE source present on site and TCE originating from degradation of PCE from an off-site source. Indeed, TCE at the site in the shallow aquifer A can be separated into two groups when plotted on a $\delta^{13}C$ versus $\delta^{37}Cl$ plot, TCE with relatively constant isotopic composition (square symbols) and TCE with a wider range of values following a linear trend (round symbols). These two groups, indeed, correspond to two separate plumes that comingle at the downgradient end of the site. Sampling points in the deeper aquifer B can also associated with these two groups. The linear trend in group two can be explained by isotope fractionation during reductive dechlorination of PCE and TCE that fractionates both isotopes proportionally. The case study demonstrates that by using a

FIGURE 8.16   (Opposite) Carbon isotope ratios of PCE downgradient of a DNAPL source zone located in a sandy aquifer. Average $\delta^{13}C$ for each multilevel sampler and average $\delta^{13}C$ for different plume segments. Values in italics: Zones likely affected by biodegradation.

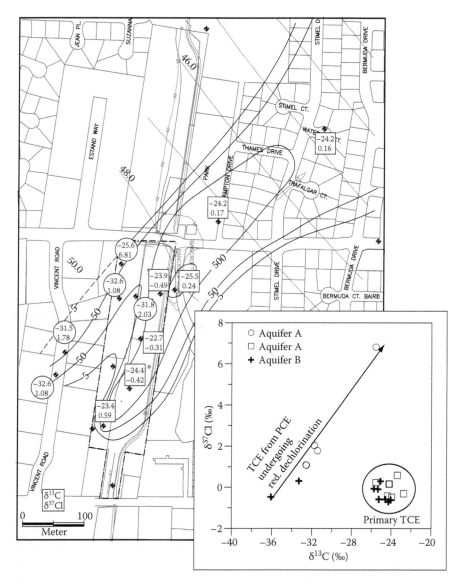

FIGURE 8.17   Carbon and chlorine isotope ratio of trichloroethene (TCE) at industrial site for sampling wells in shallow aquifer A and deeper aquifer B.

dual-isotope approach, different sources of contamination can sometimes still be clearly separated even if biodegradation occurs.

8.8.3   Sampling Strategy and Data Interpretation

Isotope analysis is usually considered when other information such as the contaminant distribution do not provide a sufficiently clear picture of the relationship between different contaminant sources and plumes or to confirm hypothesizes developed based on hydrogeological and contaminant data. The sampling strategy should be based on a conceptual site model that highlights how the different sources and plumes are potentially related. The sampling strategy should target the different sources and plume segments. The exact strategy varies depending on the actual situation (Table 8.4). If possible the NAPL source should be sampled to constrain the different source isotope signatures. However, especially for DNAPL, it is usually difficult to sample the organic phase. Therefore, often high concentration wells at or close to the source are sampled instead. There is always the possibility that different sources accidentally have the same isotopic composition or that a single source comprises several spill events with a different isotopic signature. The different pos-

TABLE 8.4
Different Scenario Where Isotopes may be Used to Evaluate the Origin of Groundwater Contamination and Corresponding Sampling Strategies

Scenario	Plume and several known source zones	Plume and one known source zone	Upgradient and downgradient pollution	Extended plume but no source zone identified yet
Question	What is contribution of the different sources to the plume(s)?	Are there additional sources contributing to the plume?	Does the site contribute to downgradient contamination?	Is plume linked to one or several sources?
Sampling	**Source characterization** NAPL samples of different sources If not available, groundwater samples in high concentration zone close to each source **Plume characterization** (only if sources have different isotopic composition) Take at least three samples in each of the plume segments presumably linked to each of the sources	**Source characterization** NAPL sample of source If not available groundwater samples in high concentration zone close to the source **Plume characterization** Take at least three samples in each plume or plume segment	Take at least three groundwater samples upgradient and downgradient of the location of the potential source	Take at least three groundwater samples in each plume or plume segments

Source: Modified after Hunkeler, D., R. U. Meckenstock, B. Sherwood Lollar, T. C. Schmidt, and J. T. Wilson, *A guide for assessing biodegradation and source identification of organic ground water contaminants using compound specific isotope analysis (CSIA).* EPA 600/R-08-148, US EPA, Office of Research and Development, Ada, Oklahoma, 2009.

TABLE 8.5
Interpretation of Isotope Data

Results	Interpretation
Isotope data partition into distinct groups each having a different isotope ratio and groups correspond to different plumes/plume segment	Different plumes/plume segment likely originate from different sources
Isotope data partition into distinct groups that are in conflict with the conceptual model	Revise conceptual model (groundwater flow regime, possibility of additional sources) Evaluate the possibility of a source consisting of different spill events
No significant difference of isotope ratios between different locations	No distinction between single source and multiple sources with the same signature possible. Possibly analyze isotope ratio of additional elements
Most of the samples are significantly different from each other without a clear pattern	Possibly occurrence of transformation processes. Evaluate spatial characteristics of trend. Consider dual isotope approach. Source consists of several spill events.

Source: Modified after Hunkeler, D., R. U. Meckenstock, B. Sherwood Lollar, T. C. Schmidt, and J. T. Wilson, *A guide for assessing biodegradation and source identification of organic ground water contaminants using compound specific isotope analysis (CSIA)*. EPA 600/R-08-148, US EPA, Office of Research and Development, Ada, Oklahoma, 2009.

sible outcomes of an isotope study for source identification and the corresponding interpretation is summarized in Table 8.5.

If one or only a few compounds are measured, as typically the case of chlorinated ethenes, usually a simple comparison of the different values combined with a statistical test to evaluate if values from different plume segments are significantly different is sufficient. If a large number of compounds is analyzed, as can be the case for petroleum hydrocarbons, a more sophisticated statistical treatment of the data should be considered (Boyd et al. 2006).

ACKNOWLEDGMENT

The authors thank Tim Buscheck for his helpful comments.

REFERENCES

Abe, Y., and D. Hunkeler. 2006. Does the rayleigh equation apply to evaluate field isotope data in contaminant hydrogeology? *Environmental Science & Technology* 40 (5):1588–96.

Abe Y., R. Aravena, B.L. Parker, and D. Hunkeler. 2009. Evaluating the fate of chlorinated ethenes in streambed sediments by combining stable isotope, geochemical and microbial methods. *Journal of Contaminant Hydrology*, 107(1–2):10–21.

Abe, Y., R. Aravena, J. Zopfi, O. Shouakar-Stash, E. Cox, J.D. Roberts, and D. Hunkeler. 2009. Carbon and chlorine isotope fractionation during aerobic oxidation and reductive dechlorination of vinyl chloride and cis-1,2-dichloroethene. *Environmental Science Technolology,* 43:101–107.

Beneteau, K. M., R. Aravena, and S. K. Frape. 1999. Isotopic characterization of chlorinated solvents: Laboratory and field results. *Organic Geochemistry* 30:739–53.

Bouchard, D., D. Hunkeler, P. Gaganis, R. Aravena, P. Höhener, and P. Kjeldsen. 2008. Carbon isotope fractionation during migration of petroleum hydrocarbon vapors in the unsaturated zone: field experiment at Værløse Airbase, Denmark, and modeling. *Environmental Science & Technology* 42:596–601.

Boyd, T. J., C. L. Osburn, K. J. Johnson, K. B. Birgl, and R. B. Coffin. 2006. Compound-specific isotope analysis coupled with multivariate statistics to source-apportion hydrocarbon mixtures. *Environmental Science & Technology* 40 (6):1916–24.

Eberts, S. M., C. Braun, and S. Jones. 2008. Compound-specific isotope analysis: Questioning the origins of a trichloroethene plume. *Environmental Forensics* 9 (1):85–95.

Gray, J. R., G. Lacrampe-Couloume, D. Gandhi, K. M. Scow, R. D. Wilson, D. M. Mackay, and B. Sherwood Lollar. 2002. Carbon and hydrogen isotopic fractionation during biodegradation of methyl tert-butyl ether. *Environmental Science & Technology* 36:1931–38.

Griebler, C., M. Safinowski, A. Vieth, H. H. Richnow, and R. U. Meckenstock. 2004. Combined application of stable carbon isotope analysis and specific metabolites determination for assessing in situ degradation of aromatic hydrocarbons in a tar oil-contaminated aquifer. *Environmental Science & Technology* 38:617–31.

Holt, B. D., N. C. Sturchio, T. A. Abrajano, and L. J. Heraty. 1997. Conversion of chlorinated volatile organic compounds to carbon dioxide and methyl chloride for isotopic analysis of carbon and chlorine. *Analytical Chemistry* 69:2727–33.

Hunkeler, D., R. Aravena, K. Berry-Spark, and E. Cox. 2005. Assessment of degradation pathways at a site with mixed chlorinated hydrocarbon contamination using stable isotope analysis. *Environmental Science & Technology* 39 (16):5975–81.

Hunkeler, D., R. Aravena, and B. J. Butler. 1999. Monitoring microbial dechlorination of tetrachloroethene (PCE) using compound-specific carbon isotope ratios: Microcosms and field experiments. *Environmental Science & Technology* 33(16):2733–38.

Hunkeler, D., R. Aravena, J. A. Cherry, and B. Parker. 2003. Carbon isotope fractionation during chemical oxidation of chlorinated ethenes by permanganate: Laboratory and field studies. *Environmental Science & Technology* 37:798–804.

Hunkeler, D., R. Aravena, and E. Cox. 2002. Carbon isotopes as a tool to evaluate the origin and fate of vinyl chloride: Laboratory experiments and modeling of isotope evolution. *Environmental Science & Technology* 36:3378–84.

Hunkeler D., N. Chollet, X. Pittet, R. Aravena, J.A. Cherry, and B.L. Parker. 2004. Effect of source variability and transport processes on carbon isotope ratios of TCE and PCE in two sandy aquifers. *Journal of Contaminant Hydrology,* 74:265–282.

Hunkeler, D., R. U. Meckenstock, B. Sherwood Lollar, T. C. Schmidt, and J. T. Wilson. 2009. *A guide for assessing biodegradation and source identification of organic ground water contaminants using compound specific isotope analysis (CSIA).* EPA 600/R-08-148, US EPA, Office of Research and Development, Ada, Oklahoma.

Imfeld, G., I. Nijenhuis, M. Nikolausz, S. Zeiger, H. Paschke, J. Drangmeister, J. Grossmann, H. H. Richnow, and S. Weber. 2008. Assessment of in situ degradation of chlorinated ethenes and bacterial community structure in a complex contaminated groundwater system. *Water Research* 42 (4–5):871–82.

Jendrzejewski, N., H. G. M. Eggenkamp, and N. V. Coleman. 1997. Sequential determination of chlorine and carbon isotopic composition in single microliter samples of chlorinated solvents. *Analytical Chemistry* 69:4259–66.

Jendrzejewski, N., H. G. M. Eggenkamp, and M. L. Coleman. 2001. Characterisation of chlorinated hydrocarbons from chlorine and carbon isotopic compositions: Scope of application to environmental problems. *Applied Geochemistry* 16:1021–31.

Kolhatkar, R., T. Kuder, P. Philp, J. Allen, and J. T. Wilson. 2002. Use of compound-specific stable carbon isotope analyses to demonstrate anaerobic biodegradation of MTBE in groundwater at a gasoline release site. *Environmental Science & Technology* 36:5139–46.

Kuder, T., J. T. Wilson, P. Kaiser, R. Kolhatkar, P. Philp, and J. Allen. 2005. Enrichment of stable carbon and hydrogen isotopes during anaerobic biodegradation of MTBE: Microcosm and field evidence. *Environmental Science & Technology* 39:213–20.

Mancini, S. A., G. Lacrampe-Couloume, H. Jonker, B. M. Van Breukelen, J. Groen, F. Volkering, and B. Sherwood Lollar. 2002. Hydrogen isotopic enrichment: An indicator of biodegradation at a petroleum hydrocarbon contaminated field site. *Environmental Science & Technology* 36:2464–70.

McKelvie, J. R., D. M. Mackay, N. R. de Sieyes, G. Lacrampe-Couloume, and B. S. Lollar. 2007. Quantifying MTBE biodegradation in the Vandenberg Air Force Base ethanol release study using stable carbon isotopes. *Journal of Contaminant Hydrology* 94 (3–4):157–65.

Meckenstock, R. U., B. Morasch, C. Griebler, and H. H. Richnow. 2004. Stable isotope fractionation analysis as a tool to monitor biodegradation in contaminated aquifers. *Journal of Contaminant Hydrology* 75:215–55.

Morrill, P. L., G. Lacrampe-Couloume, G. F. Slater, B. E. Sleep, E. A. Edwards, M. L. McMaster, D. W. Major, and B. S. Lollar. 2005. Quantifying chlorinated ethene degradation during reductive dechlorination at Kelly AFB using stable carbon isotopes. *Journal of Contaminant Hydrology* 76 (3–4):279–93.

Richnow, H. H., E. Annweiler, W. Michaelis, and R. U. Meckenstock. 2003. Microbial in situ degradation of aromatic hydrocarbons in a contaminated aquifer monitored by carbon isotope fractionation. *Journal of Contaminant Hydrology* 65 (1–2):101–20.

Richnow, H. H., R. U. Meckenstock, L. A. Reitzel, A. Baun, A. Ledin, and T. H. Christensen. 2003. In situ biodegradation determined by isotope fractionation of aromatic hydrocarbons in an anaerobic landfill leachate plume (Vejen, Denmark). *Journal of Contaminant Hydrology* 38:457–67.

Sherwood Lollar, B., G. F. Slater, B. Sleep, M. Witt, G. M. Klecka, M. Harkness, and J. Spivack. 2001. Stable carbon isotope evidence for intrinsic bioremediation of tetrachloroethene and trichloroethene at area 6, Dover Air Force Base. *Environmental Science & Technology* 35:261–69.

Shouakar-Stash, O., S. K. Frape, and R. J. Drimmie. 2003. Stable hydrogen, carbon and chlorine isotope measurements of selected chlorinated solvents. *Journal of Contaminant Hydrology* 60(3–4):211–28.

Song, D. L., M. E. Conrad, K. S. Sorenson, and L. Alvarez-Cohen. 2002. Stable carbon isotope fractionation during enhanced in situ bioremediation of trichloroethene. *Environmental Science & Technology* 36 (10):2262–68.

Spence, M. J., S. H. Bottrell, S. F. Thornton, H. H. Richnow, and K. H. Spence. 2005. Hydrochemical and isotopic effects associated with petroleum fuel biodegradation pathways in a chalk aquifer. *Journal of Contaminant Hydrology* 79 (1–2):67–88.

Steinbach, A., R. Seifert, E. Annweiler, and W. Michaelis. 2004. Hydrogen and carbon isotope fractionation during anaerobic biodegradation of aromatic hydrocarbons - A field study. *Environmental Science & Technology* 38 (2):609–16.

Van Breukelen, B. M., D. Hunkeler, and F. Volkering. 2005. Quantification of sequential chlorinated ethene degradation using a reactive transport model incorporating isotope fractionation. *Environmental Science & Technology* 39:4189–97.

Van Breukelen, B. M. 2007. Quantifying the degradation and dilution contribution to natural attenuation of contaminants by means of an open system Rayleigh equation. *Environmental Science & Technology* 41 (14):4980–85.

Van Wanderdam, E. M., S. K. Frape, R. Aravena, R. J. Drimmie, H. Flatt, and J. A. Cherry. 1995. Stable chlorine and carbon isotope measurements of selected organic solvents. *Applied Geochemistry* 10:547–52.

Zwank, L., M. Berg, M. Elsner, T. C. Schmidt, R. P. Schwarzenbach, and S. B. Haderlein. 2005. New evaluation scheme for two-dimensional isotope analysis to decipher biodegradation processes: Application to groundwater contamination by MTBE. *Environmental Science & Technology* 39 (4):1018–29.

9 Stable Isotope Fractionation of Gases and Contaminant Vapors in the Unsaturated Zone

Patrick Höhener, Daniel Bouchard,
and Daniel Hunkeler

CONTENTS

9.1 INTRODUCTION

As discussed in several chapters and in a number of recent comprehensive review articles (Meckenstock et al. 2004; Schmidt et al. 2004; Elsner et al. 2005), stable isotope data can be used to document processes at contaminated sites such as the mineralization of pollutants, chemical transformation reactions, or to identify different pollutant sources. Although the body of literature on compound-specific isotope analyses in many different pollutants has grown considerably, such analyses have almost exclusively been reported for dissolved compounds, and generally in completely water-saturated experimental systems and field sites, such as aquifers. For those dissolved compounds in the saturated zone, it has been demonstrated and generally accepted that processes other than biochemical reactions such as sorption or dissolution are not leading to a significant isotope fractionation (Hunkeler et al. 2004; Schmidt et al. 2004; Elsner et al. 2005). The widely discussed Rayleigh isotope distillation model is thus used in saturated-zone studies to assess the extent of degradation that occurred on a flow path from a contaminant source to the location of sampling (Van Breukelen, Hunkeler, and Volkering 2005; Abe and Hunkeler 2006; Van Breukelen and Prommer 2008). It has been proposed that isotopic shifts that are significantly larger than 1‰ for carbon isotopes are considered to be conclusive for documenting degradation in the saturated zone (Slater 2003), however due to small variations in the isotopic composition of the source and the complexity of systems, a more conservative baseline should be changes larger than 2‰. This includes the analytic precision (typically 0.5‰ for stable carbon isotopes) and sampling errors.

In the unsaturated zone, it has been demonstrated that gas-phase diffusion significantly contributes to changes in the isotopic composition of CO_2 (Cerling et al. 1991; Davidson 1995; Amundson et al. 1998; Barth, Olsson, and Kalin 1999) and O_2 (Hendry, Wassenaar, and Birkham 2002; Lee et al. 2003), with offsets of more than 4.4‰ for ^{13}C in CO_2 (Cerling et al. 1991). Furthermore, a recent field study documented that diffusion significantly influenced the stable carbon isotope signal of nine organic compounds diffusing from an emplaced kerosene source in the unsaturated zone (Bouchard et al. 2005; Bouchard et al. 2008b). The low molecular weight compounds were much more affected by diffusion and in case of the *n*-hexane, the diffusing *n*-hexane vapors became depleted in ^{13}C by more than −5‰ compared to the *n*-hexane in the buried kerosene source during early gas-phase migration. The correct assessment of the extent of degradation in the unsaturated zone is therefore, biased by gas-phase diffusion at least in the early phase of the vapor plume formation. Based on these observations, we conclude that isotopic shifts in the unsaturated zone cannot be related only to degradation, and that the Rayleigh isotope distillation model is not applicable in the unsaturated zone to assess the extent of degradation between a source and a sampling point.

Models describing isotope variations of CO_2 (Thorstenson et al. 1983; Cerling 1984; Cerling et al. 1991) or O_2 (Hendry, Wassenaar, and Birkham 2002; Lee et al. 2003) in the unsaturated zone have been developed. These models are very distinct from the Rayleigh model. Bouchard and coworkers (2005, 2008a) extended these models for the description of carbon isotope fractionation of volatile organic compounds (VOCs) in the unsaturated zone. The aim was to discriminate between the effects of diffusion and degradation on the isotopic shifts of VOCs of interest. This VOC isotope model, more complex than the models for CO_2 and O_2, was shown to predict reasonably well the isotope field data measured in a volatile hydrocarbons plume created by an artificially emplaced spherical petroleum hydrocarbon source in a sandy soil in Denmark (Bouchard et al. 2005, 2008b), and also evolutions measured in a laboratory column (Bouchard et al. 2008a).

This chapter reviews all these recent developments, going from the early work on the behavior of isotopes in natural soil gases like CO_2 and O_2, and adding our recent findings on stable isotopes in volatile organic pollutants (petroleum hydrocarbons) to establish a general isotope model for gases or vapors in the unsaturated zone. In order to achieve this goal, all processes governing the isotopic composition of these compounds in the unsaturated zone shall be discussed in detail. We will exclude the isotope pattern in methane and H_2O from our analysis since these compounds are broadly covered elsewhere (Clark and Fritz 1997; Cook and Herczeg 1999). A general analytical model approach for modeling stable isotopes in gaseous compounds in the unsaturated zone can then be proposed by gathering all appropriate processes. The approach is applicable for both natural soil gases and volatile organic pollutants and for all environmental isotopes (^2H, ^{13}C, ^{18}O, and ^{15}N) that are part of these compounds. The robustness of the model will be demonstrated by comparing isotopic ratio predictions to different sets of experimental data from a lab column and from field sites of different geometries and boundary conditions.

9.2 ISOTOPIC COMPOSITIONS OF GASEOUS COMPOUNDS IN THE UNSATURATED ZONE

The first measurements of isotopes in gaseous compounds in the unsaturated zone were the $\delta^{13}C$ measurements of soil gas CO_2 made by Galimov (1966) in several soils and at several depths in Russia. Galimov (1966) recognized that CO_2 in soil air can be derived from either atmospheric CO_2, from respiration of organic matter, or from plant roots at shallow depths. A general decline of $\delta^{13}C$ in soil gas CO_2 from values near atmospheric CO_2 (approx. $-7‰$) to values related to the respired CO_2 from soil organic matter (-20 to $-30‰$ in soils with C3-plants) was generally found. Thorstenson and coworkers (1983) were the first to develop a mathematical model describing variations in $\delta^{13}C$ (as well as ^{14}C) with depth in the unsaturated zone. Cerling later refined this model and presented two diffusion-reaction models for $\delta^{13}C$ in soil gas CO_2 under steady-state conditions: one (Cerling 1984) for constant and one (Quade, Cerling, and Bowman 1989; Cerling et al. 1991) for exponentially declining CO_2 production with depth. The models highlighted the importance of fractionation by gas-phase diffusion and predicted an offset of $+4.4\%_0$ between the $\delta^{13}C$ in soil gas CO_2 and the carbon from the respired organic matter. This offset is caused by the difference in diffusion velocity between $^{12}CO_2$ and $^{13}CO_2$, which have diffusion coefficients different by $4.4\%_0$ (Quade, Cerling, and Bowman 1989; Cerling et al. 1991). Although it was later shown that the offset between soil gas CO_2 and respired organic matter is not always constant under all boundary conditions (Davidson 1995), the works of Cerling and his research group, reviewed comprehensively by Amundson et al. (1998), explained a fractionation process needed for the global carbon cycling models and had an enormous impact on paleoclimatology and paleoecology. The modeling principles in these early models are also underlying the model described later in Section 9.4.

Microbial respiration is the largest global turnover process of O_2, and isotopic variations in $\delta^{18}O$ in O_2 help to understand this cycle. Respiration fractionates between ^{16}O and ^{18}O isotopes and this process is known as the Dole effect (Lane and Dole 1956). Despite the obvious importance of O_2 for subsurface microbial respiration, only a few studies presented depth profiles of O_2 and $\delta^{18}O$ in O_2 in the unsaturated zone. The research group of Hendry and Wassenaar contributed most actively to the understanding of such profiles, either measured in mesocosms (Hendry et al. 2001, 2002) or in waste rock piles (Birkham et al. 2003; Lee et al. 2003). The researchers proposed a mathematical model for profiles of $\delta^{18}O$ in O_2 in the unsaturated zone (Hendry et al. 2002). They showed that kinetic fractionation of O_2 by microbial respiration through the Dole effect may be negligible when O_2 is scarce and oxygen is consumed irrespective of its isotope composition. This may be the case in soil aggregates (Angert, Luz, and Yakir 2001). However, in case of high microbial respiration rates, a diffusive fractionation is occurring due to the faster diffusion velocity of $^{16}O-^{16}O$ compared to $^{18}O-^{16}O$ (Hendry et al. 2002; Lee et al. 2003) along the concentration gradient between atmospheric O_2 and soil O_2. With increasing soil depth, the $\delta^{18}O$ in O_2 becomes more negative in soils with microbial respiration. This will be discussed in more detail in Section 9.5.2. In contrast, in a desert sand with low microbial respiration, the $\delta^{18}O$ in O_2 was very similar to that in atmospheric O_2 (Severinghaus et al. 1996).

Vapor migration of volatile organic pollutants has become of increasing interest at many contaminated sites where the assessment of health risks or remediation actions is necessary. The products most frequently studied are petroleum hydrocarbons (alkanes, benzene, toluene, and xylenes), chlorinated solvents (chlorinated methanes, ethanes, ethenes), and chlorofluorocarbons (Wiedemeier et al. 1999). These compounds are mostly spilled in the form of nonaqueous phase liquid (NAPL), either denser or lighter than water (Pankow and Cherry 1996). The NAPL spills on ground surface or from leaking storage tanks result in gravity-driven migration and in NAPL residual held by capillary forces in the pore space of the unsaturated zone. The NAPL residual creates the source zone from which gaseous and aqueous plumes are generated. Migration of gaseous pollutants occurs in three dimensions away from such source zones. Wilson (1997) sketched idealized geometrical settings of potential source zones for volatile pollutants, which are all clearly different from the normal one-dimensional setting for the diffusion of natural soil gases. In only a few

field studies, the spatial–temporal variation of volatile pollutants has been investigated in the unsaturated zone (e.g., Ostendorf and Kampbell 1991; Conant, Gillham,
and Mendoza 1996; Franzmann et al. 1999; Lahvis, Baehr, and Baker 1999), and in
even fewer studies, the isotopic composition of the migrating compound has been
measured. The most complete study was the controlled spill experiment at Værløse,
Denmark, were both the evolution of the vapor plume concentrations around a buried kerosene source zone (Broholm et al. 2005; Christophersen et al. 2005; Höhener
et al. 2006) and the $\delta^{13}C$ in nine selected hydrocarbons measured (Bouchard et al.
2008b). Results of this field experiment will be shown and interpreted in Section 9.5.5.
In another study on gasoline vapors at leaking storage tank sites (Stehmeier et al.
1999), it was found that the carbon in volatile hydrocarbons at 4.4 m soil depth was
more depleted in ^{13}C than at 2.3 m depth, but no straightforward interpretation could
be given.

9.3 GOVERNING PARAMETERS IN THE UNSATURATED ZONE

9.3.1 ADVECTION

The unsaturated zone is a porous medium composed of three phases: air, water,
and solids. In the case of a site contaminated with non aqueous phase liquids
(NAPLs), the NAPL constitutes a fourth phase. The air, the water, and the
NAPL phase are mobile and form contaminant plumes in the unsaturated zone.
Advective transport of water and NAPL is the major transport process for these
fluids in both, the unsaturated zone and the saturated zone. Advection is defined
as the transport of a conserved scalar quantity that is transported in a vector field.
Gravity and capillary effects are the main driving forces for the advective transport of water and NAPL. For the soil air phase, advection is usually much less
important as transport process (Scanlon, Nicot, and Massmann 2000). However,
gravity-driven advection may occur at sites where pollutant vapors with greater
specific vapor density than that of soil air migrate downward (Falta et al. 1989).
Furthermore, pressure-induced advective soil gas transport can occur under specific natural pumping situations driven by barometric pressure changes in the air
(Elberling et al. 1998) or in engineered systems like soil vapor extraction schemes
(Kirtland, Aelion, and Stone 2005). However, in the context of isotope studies,
it is important to note that advective transport is not considered to fractionate
isotopes (Hunkeler et al. 2004) since molecules are simply carried along with the
moving phase independently of their mass. Advection-dominated reactive transport of isotopes in the unsaturated zone can thus be modeled using classical reactive transport models.

9.3.2 DIFFUSION

Under most natural settings, gas-phase diffusion is the dominating transport process
for gases in the unsaturated zone, whereas gas-phase advection is mostly absent
(Scanlon, Nicot, and Massmann 2000). It is thus necessary to understand in detail
the nature of gas-phase diffusion and the way it creates isotope fractionation when

one wants a detailed understanding of isotopes in gaseous compounds in the unsaturated zone. Gas-phase diffusion is a process that is about four orders of magnitude faster than liquid-phase diffusion (Schwarzenbach, Gschwend, and Imboden 2003). Diffusion is the movement of compounds from higher chemical potential (concentration, vapor pressure) to lower chemical potential and occurs due to Brownian motion of molecules. The diffusion velocity depends on molecular mass, and thus, gas-phase diffusion creates fractionation of isotopes with differing masses. For a dilute diffusing gaseous compound in a bulk phase, Fick's law is valid and widely applied (Scanlon, Nicot, and Massmann 2000). However, it must be noted that some organic vapors created at contaminated sites (such as trichloroethene or gasoline vapors) may be significant enough to violate the condition of a dilute gaseous compound as assumed by Fick's law. In that case, the application of the Stefan–Maxwell equations may be necessary for modeling gas-phase diffusion (Baehr and Bruell 1990). Other deviations from Fick's law occur in situations where two gases diffusing in the opposite directions react with each other, (e.g., when hydrocarbon vapors and oxygen are both consumed by biodegradation). Then, an advective soil gas movement may be created and Fick's law must be adjusted to correct for this effect (Van de Steene and Verplancke 2006).

Gas-phase diffusion in soil air is slower with respect to gas-phase diffusion in free air due to the increased path length (tortuosity) in soil. Effective diffusion coefficients D_e, often also called intrinsic diffusion coefficients, account for the reduction in the diffusion effective cross-sectional area and the increased path length in soils and are used to model gas-phase diffusion in the vadose zone (Werner, Grathwohl, and Höhener 2004). Their dependency on the structure and volume of the gas-filled pore space has been investigated in many soils, and a number of theoretical relationships have been proposed recently to quantify tortuosity and D_e (Wang, Reckhorn, and Grathwohl 2003). Based on a comparison with field measurements (Werner, Grathwohl, and Höhener 2004), the relationship derived by Moldrup and coworkers (2000; see also Section 9.4.3) has been found to provide estimations with the least error.

9.3.3 PARTITIONING

Gaseous compounds diffusing through the unsaturated zone are partitioning into the other phases (water, solids, and NAPLs). Partitioning compounds with a low mass fraction in the soil gas will migrate more slowly than conservative gases, and it is important to know the extent of partitioning for transient transport problems.

Partitioning of vapors between air and NAPLs is subject to a slight isotope fractionation. The phenomenon is well known as the equilibrium isotope effect. An offset between the $\delta^{13}C$ or δ^2H of a pure organic liquid and the δ values of the vapors is measurable (see Chapter 3). This pattern is created by different vapor pressures of the compounds with the heavy and light isotopes. Interestingly, the isotopically heavy nonpolar organics generally have a slightly higher vapor pressure than their light counterparts (Höhener, Bouchard, and Hunkeler et al. 2008). This results in a positive vapor–liquid enrichment factor (see Section 9.4.4 and Table 9.1). The effect is sometimes called the inverse isotope effect, since it is inverse to the one observed for very polar solvents (water and alcohols). The magnitude of the enrichment factor

TABLE 9.1

Numbers of Total Carbon Isotopes (nI) and Numbers of Carbon Isotopes at Reactive Positions in Typical Environmental VOCs and CO_2, and Resulting Percentage of Each Subspecies I, II, and III Calculated with Equations 9.1 Through 9.3 and an Initial $\delta^{13}C$ of –30‰. Also Shown is the Ratio of the Gas-Phase Diffusion Coefficient of Heavy Subspecies II or III to the Light Subspecies I, D_h/D_l, and the Equilibrium Isotope Effect (Ratio of Vapor Pressures of Heavy and Light Compounds, P_h/P_l')

		n-Hexane	MCH[a]	Benzene	Toluene[b]	Toluene[c]	TRI[a]	MTBE[a]	CO_2
nI		6	7	6	7	7	2	5	1
nI_{react}		2	5	6	1	5	2	1	1
$C_{0,I}$	%	93.4	92.4	93.5	92.4	92.4	97.8	94.5	98.9
$C_{0,II}$	%	4.4	2.2	0	6.5	2.2	0	4.4	0
$C_{0,III}$	%	2.2	5.4	6.5	1.1	5.4	2.2	1.1	1.1
D_h/D_l'[d]	‰	–1.4	–1.1	–1.7	–1.3	–1.3	–0.7	–1.4	–4.4
P_h/P_l'[e]	‰	—	—	+0.17	+0.2	+0.2	+0.31	—	—

[a] MCH = Methylcyclohexane, see Figure 9.1; TRI = Trichloroethene, MTBE = Methyl *tert*-butyl ether.

[b] Degradation pathway: toluene methyl group oxidation.

[c] Toluene ring oxidation.

[d] Calculated with Equation 9.4 and displayed in ‰, using $D_h/D_l' = 1,000*(D_h/D_l - 1)$

[e] Equilibrium isotope effect (ratio of vapor pressures of heavy and light compounds), expressed in ‰, with $P_h/P_l' = 1,000*(P_h/P_l - 1)$. (Data for 25°C from Harrington, R. R., S. R. Poulson, J. I. Drever, P. J. S. Colberg, and E. F. Kelly, *Organic Geochemistry*, 30, 765–75, 1999; Huang, L., N. C. Sturchio, T. Abrajano, L. J. Heraty, and B. D. Holt, *Organic Geochemistry*, 30, 777–85, 1999.)

for a hydrophobic compound between air and its pure phase or between any other NAPL should be quite similar.

It is widely accepted that the partitioning of gases from soil air into soil water is linear, instantaneous, and is described by Henry's law (Washington 1996). Air–water partitioning is not considered to lead to significant isotope fractionation (Slater et al. 1999). However, hydrophobic volatile compounds, especially linear alkanes, have a tendency to accumulate at the gas–water interface (Hoff et al. 1993a, 1993b). In fine grained soils with relatively low water content, the surface area of the air–water interface is very high, and the accumulation of hydrophobic compounds on this interface becomes an important process (Kim, Annable, and Rao 2001). In contrast, the gas-water interfacial area in sands and gravels is too small to be an important accumulation surface (Kim, Annable, and Rao 2001; Costanza-Robinson and Brusseau 2002). No relevant experiment has been carried out so far to demonstrate whether this deposition on the gas–water interface leads to isotope fractionation. Nonetheless, based on different physical properties between light and heavy isotopes such as a slightly higher vapor pressure for heavy compounds (Narten and Kuhn 1961), very small isotope fractionation can be expected.

Partitioning of compounds between the aqueous and solid phases is frequently considered to be linear and instantaneous and is then described by the well-known linear partition coefficient K_d (the coefficient of a linear Freundlich sorption isotherm). For gases of small molecular weight (high vapor pressure), sorption to soil solids is usually negligible (Werner and Höhener 2003). However, for gaseous hydrophobic compounds exhibiting lower partial pressures such as n-hexane, sorption may be important, especially sorption to organic matter. The K_d can either be measured in a representative soil sample or can be estimated from soil organic matter content f_{om} and the octanol–water partitioning coefficient K_{ow} through relations such as given in Schwarzenbach, Gschwend, and Imboden (2003). Although in single step sorption experiments, fractionation of isotopes was as small as the reproducibility limit of the analytical isotope techniques, a recent study on multistep sorption/desorption cycles of monoaromatic hydrocarbons on an organic rich matrix reported minor fractionation of stable carbon isotopes (Kopinke et al. 2005). It must also be stressed that quite often, sorption is nonlinear and noninstantaneous. For instance, desorption of trichloroethene was found to be kinetically controlled by slow diffusion processes at the grain scale (Grathwohl and Reinhard 1993). As diffusion is involved in desorption, it is highly probable that this leads to isotope fractionation. This has not yet been experimentally confirmed.

For the description of soil gas transport, one needs to clearly distinguish between steady-state and transient-state situations. At steady state, enough time has elapsed that all partitioning processes are in equilibrium and only equilibrium isotope effects due to partitioning are expected. During transient diffusion, however, this assumption may not be valid. Transient diffusion problems in unsaturated porous media need the knowledge of the mass fraction of the compound in soil air since only this fraction is mobile. In parts of the literature on reactive soil gas transport, capacity factors (Grathwohl 1998) are used that account for linear partitioning. Capacity factors are analogous to the retardation factors used in the literature for the saturated zone. However, it should be noted that when using conventional notation in unsaturated zone models, the capacity factor for conservative gases are not like the retardation factors in the saturated zone, but are equal to the volumetric air content of the unsaturated zone (and thus, < 1).

9.3.4 BIODEGRADATION

Biodegradation leads to kinetic isotope fractionation for a wide variety of pollutants, and fractionation factors for saturated conditions have been thoroughly reviewed and summarized (see Chapter 4). The main questions for the unsaturated zone are whether the kinetic fractionation factors from the saturated literature are transferable to the unsaturated zone, and which kinetics govern the biodegradation rates of pollutants in the unsaturated zone. In principle, it can be assumed that all microorganisms are living in the aqueous phase, including those in the unsaturated zone, and that there is no reason to assume that kinetic isotope fractionation during biodegradation is different from that in the saturated zone. Only recently, fractionation factors for the biodegradation of selected petroleum hydrocarbons in experiments with unsaturated soil were measured (Bouchard, Hunkeler, and Höhener 2008c). Although the factors compared well with the ones from the saturated zone literature, the database is still too limited to prove that this holds for all compounds. Regarding the biodegradation kinetics of volatile compounds in the unsaturated zone, there are more experimental data. Höhener et al. (2003) gave an overview of the possible relationships between the biodegradation rate in the aqueous phase of the soil and the concentration of the volatile compound in soil gas. These relationships include Henry's law combined with either Monod kinetics or zero- or first-order kinetics. At small aqueous concentrations, first-order kinetics is expected. Volatile gaseous pollutants (especially those with high Henry's law coefficients) are mostly present in small aqueous concentrations and therefore, it is justified to use first-order kinetics in reactive models for the unsaturated zone. This has been experimentally shown in laboratory batch and column experiments (Höhener et al. 2003), in a lysimeter study (Pasteris et al. 2002), and in the field (Höhener et al. 2006).

9.4 ISOTOPE MODEL FOR GASEOUS COMPOUNDS IN THE UNSATURATED ZONE

In the previous section, the governing processes in the unsaturated zone have been discussed in detail, and both, gas-phase diffusion as well as degradation, have been identified as key fractionating processes for stable isotopes in gaseous compounds. In the following, an analytical model is proposed that includes these key processes and relies on generally accepted assumptions. The model is based on analytical transport equations for steady state or transient state. The subspecies approach discussed in Chapter 8 is used to handle the gaseous compound in three isotopic subspecies (groups of isotopomers with identical behavior relative to diffusion or reaction). The model predicts the reactive transport of the gaseous compound using three analytical equations derived from the general transport equation, and summation of the results obtained for each subspecies results in the prediction of the total concentrations of the gaseous compound at any time and distance from the source. The stable isotope ratios are finally obtained from the ratios of the concentrations of the heavy and light subspecies as a function of time and space.

9.4.1 COMPOUND PROPERTIES AND DIVISION INTO SUBSPECIES (ISOTOPOMERS)

The distribution of the isotope of interest (e.g., carbon, hydrogen, oxygen, etc.) within the compound has to be considered first. The natural abundances of stable carbon, hydrogen, and oxygen isotopes are low (see Chapter 1, Table 1.1). The statistical distribution of isotopes in small organic compounds is likely to allow only one heavy isotope per single compound molecule, since the chances that two heavy isotopes are in the same molecule of the volatile compound are very small (Elsner et al. 2005). However, in molecules that incorporate one heavy isotope, it is assumed that this heavy isotope is randomly distributed at all positions in the compound. This is valid for the majority of pollutants, with very few exceptions when a specific chemical manufacturing process of a chemical leads to a weighted distribution of isotopes in different parts of the compound (Elsner et al. 2005). The isotope of interest may be placed either at a position where the initial chemical reaction is cleaving a chemical bond (this position is termed reactive position) or at another position in the molecule where no initial reaction can take place. Since kinetic isotope fractionation that is influencing the isotope signal of the remaining compound is created by the rate-limiting reaction step, which is predominantly the initial reaction at the reactive site of the molecule (Elsner et al. 2005), the compound molecules with heavy isotopes at nonreactive sites do not undergo kinetic fractionation. These molecules do, however, diffuse more slowly in the gas phase due to their increased molecular mass. As first described by Bouchard and coworkers (2005, 2008a) for *n*-hexane, we need to subdivide the compound into three isotopic subspecies (Figure 9.1):

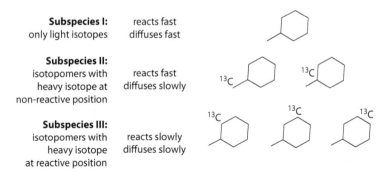

FIGURE 9.1 Grouping of methylcyclohexane into three isotopic subspecies (groups of isotopomers) according to the presence of a single heavy isotope, ^{13}C, and its position within the molecule. Isotope fractionation due to diffusion affects subspecies II and III, whereas kinetic isotope fractionation is expected for subspecies III only since the initial enzymatic reaction is a ring oxidation (Data from Koma, D., Y. Sakashita, K. Kubota, Y. Fujii, F. Hasumi, S. Y. Chung, and M. Kubo, *Applied Microbiology and Biotechnology*, 66, 92–99, 2004.)

Subspecies I: Molecule with only light isotopes of interest
Subspecies II: Molecule with one heavy isotope of interest at any nonreactive position
Subspecies III: Molecule with one heavy isotope of interest at a reactive position.

Figure 9.1 depicts the three theoretical subspecies for methylcyclohexane (MCH). This compound is supposed to undergo aerobic biodegradation following an initial oxidation of a C–H bond of one of five carbons of the ring. The two other carbons implicated in the connection of the methyl group to the ring are not involved in the initial reaction (Koma et al. 2004). Subspecies II comprises two carbon isotopomers and subspecies III comprises 3 carbon isotopomers (not five, due to symmetry, see Figure 9.1). To quantify the relative abundance of the three subspecies, Equations 9.1–9.3 are used. These equations are generalized forms of the equations for stable carbon isotopes presented previously (Bouchard et al. 2005):

$$C_{0,I} = \frac{C_l - C_h(nI - 1)}{nI} \tag{9.1a}$$

$$C_{0,II} = \frac{C_h(nI - nI_{react})}{nI} \tag{9.1b}$$

$$C_{0,III} = \frac{C_h nI_{react}}{nI}. \tag{9.1c}$$

$C_{0,I}$, $C_{0,II}$, $C_{0,III}$ are the initial concentrations of subspecies I, II, and III, respectively, C_l and C_h are the concentrations of the light and heavy isotopes, respectively (Equations 9.2a and b), nI is the total number of positions the isotope I can have in the molecule, and nI_{react} is the number of reactive positions at which kinetic isotope fractionation occurs. C_l and C_h are obtained from the total initial concentration C_0 (in moles atoms) via:

$$C_l = C_0 / (1 + R_0) \tag{9.2a}$$

$$C_h = C_0 R_0 / (1 + R_0) \tag{9.2b}$$

with $R_0 = C_h/C_l$. R_0 is obtained from the measured initial isotopic signal of the compound, which is usually given in the delta notation ($\delta^{13}C$, δ^2H, or $\delta^{18}O$) and relates to R_0 via:

$$R_0 = R_{standard}\left(\frac{\delta^{13}C}{1000} + 1\right) \tag{9.3}$$

with $R_{standard}$ is C_h/C_l in the international standard (Chapter 1, Table 1.2).
In Table 9.1, illustrative calculations were made for the three molecule subspecies of stable carbon isotopes in environmentally important volatile compounds. The reactive

positions of many pollutants are known for many compounds (see the University of Minnesota Biocatalysis/Biodegradation Database [UM-BBD]). However, for toluene, several different pathways may apply for one compound, and the subdivision in sub-species must be done with a special emphasis on the predominant biochemical path-way at the site studied. Note that for molecules where $nI = nI_{react}$ (e.g., benzene, CO_2, or O_2), the subspecies II is not occurring (Table 9.1) and transport calculations need only to include subspecies I and III.

9.4.2   FRACTIONATION FACTORS FOR DIFFUSION AND DEGRADATION

Gas-phase diffusion coefficients in air are available in literature either from measured data (e.g., Lugg 1968) or from estimations based on molar volumes (Fuller, Schettler, and Giddings 1966). They do not refer specifically to isotopically defined molecules, but rather to the compound at the mean natural abundance of light and heavy iso-topes. As can be seen in Table 9.1, the vast majority of compound molecules are sub-species I molecules, and it can be assumed that gas-phase diffusion coefficients from the literature are describing the diffusive transport of the subspecies I molecules (thus $D_m \cong D_l$). To estimate the diffusion coefficient D_h of the heavy subspecies II and III relative to that of D_l, the following relation is used (Craig 1953; Jost 1960), which is valid for diffusing gaseous compounds at dilute concentration in air:

$$\frac{D_h}{D_l} = \sqrt{\frac{M_{wl}(M_{wh} + M_a)}{M_{wh}(M_{wl} + M_a)}} \tag{9.4}$$

where M_{wh} and M_{wl} are the molecular weights of the heavy and light molecules and M_a is that of average air ($M_a = 28.8$ g mol^{-1}). Values for D_h/D_l, expressed in per mil like biochemical enrichment factors, are given in Table 9.1. The differences range from -0.6 to $-1.7\permil$ for VOCs and are as high as $-4.4\permil$ for CO_2 or $-7.3\permil$ for O_2. Note that D_h/D_l for VOCs tend to increase in the vicinity of evaporating sources since the average molecular weight of air containing VOC vapors can be higher than 28.8 g mol^{-1}.

Enrichment factors ε for the kinetic isotope effect associated with degradation of a variety of compounds under various redox conditions have been summarized in Chapter 4. The reported ε values are traditionally referring to bulk (average of whole compound) isotopic shifts. Elsner and coworkers (2005) discussed the significance of the traditionally determined enrichment factors, which they name ε_{bulk} and point out that another entity, $\varepsilon_{reactive\ position}$ is more relevant for understanding the mechanisms behind kinetic isotope fractionation in the environment. The $\varepsilon_{reactive\ position}$ is quantify-ing the apparent kinetic isotope effect of the reaction at the reactive position of the compound, and is the entity that is needed for the model approach used in this work. The $\varepsilon_{reactive\ position}$ is related to the ε_{bulk} by the approximation (Elsner et al. 2005):

$$\varepsilon_{reactive\ position} \approx \varepsilon_{bulk}\ nI/nI_{react} \tag{9.5}$$

Elsner et al. (2005) also give the exact equation for the assessment of $\varepsilon_{reactive\ position}$, which needs to take into account the originally measured δ values as a function of concentration. For a number of cases where the original data were available, the ε_{bulk} were reevaluated and reported as $\varepsilon_{reactive\ position}$ (Elsner et al. 2005).

The rate of reaction of the heavy subspecies III molecule versus the rate of the subspecies I and II molecules is finally obtained via:

$$\frac{k_h}{k_1} = \frac{K_h}{K_l} = \alpha_{reactive\ position} = \frac{\varepsilon_{reactive\ position}}{1000} + 1 \qquad (9.6)$$

9.4.3 ANALYTICAL REACTIVE TRANSPORT MODELS

The soil is described as a homogeneous porous medium with uniform and constant properties consisting of soil air, soil water, and the solid matrix. The water is immobile and all solid surfaces are wetted. It is assumed that gas-phase diffusion is the only relevant transport mechanism and that transport can be adequately modeled by Fick's law. The governing equations for diffusion-dominated transport of a subspecies i in a porous medium with uniform and constant properties and degradation can be written as:

$$\partial/\partial t\, C_i = D\nabla^2 C_i - r(C_i) + \phi \qquad (9.7)$$

where C_i is gas-phase concentration of subspecies i, ∇ stands for the Laplace operator, D is the unspecified gas-phase diffusion coefficient, $r(C_i)$ is a kinetic representation of degradation or respiration in the gas phase, and ϕ is a production term (e.g., CO_2 production from soil respiration).

In this manuscript, degradation (or respiration in case of O_2) is either represented as a first-order or zero-order kinetic rate law:

$$r(C_i) = k\, C_i \qquad (9.8a)$$

$$r(C_i) = K \qquad (9.8b)$$

where k and K are first-order or zero-order rate coefficients applying for the gas phase.

As discussed previously, the biodegradation rates are mostly known as aqueous-phase rates measured in saturated experimental systems, and when one wants to convert, for instance, an aqueous first-order rate k_w to the apparent rate for the gaseous phase k, one needs to use (Pasteris et al. 2002):

$$k = \theta_w\, k_w/H. \qquad (9.8c)$$

The partitioning of the gaseous compound follows instantaneous equilibrium and is described in a NAPL-free system using the air-water partitioning coefficient, referring to Henry's law constant H, and the solid-water partitioning coefficient K_d. The mass fraction of the compound in the soil air f_a is calculated as (Werner and Höhener 2002)

$$f_a = \cfrac{1}{1 + \cfrac{K_d \rho_s (1 - \theta_t)}{H\theta_a} + \cfrac{\theta_w}{H\theta_a}} \qquad (9.9)$$

where θ_a, θ_w, and θ_t denote the air-filled, water-filled, and total porosity, respectively, and ρ_s denotes the density of the solid. Under transient diffusion, only the fraction f_a of molecules migrates in soil, whereas the fraction $1 - f_a$ is immobile.

Steady-state diffusive transport is described by the effective diffusion coefficient D_e, which relates to D_m as:

$$D_e = \tau\, \theta_a\, D_m \qquad (9.10)$$

τ is the tortuosity coefficient, which can be measured in situ (Werner, Grathwohl, and Höhener 2004) or estimated, for example, using the Millington–Quirk relationship $\tau = \theta_a^{2.33}/\theta_t^2$ (1961) or the Moldrup relationship $\tau = \theta_a^{2.5}/\theta_t$ (Moldrup et al.2000). Transient diffusive transport is modeled using the sorption-affected diffusion coefficient D_s as:

$$D_s = \tau f_a D_m. \qquad (9.11)$$

We refer here to the terminology as defined in (Werner and Höhener 2003; Werner, Grathwohl, and Höhener 2004). Solutions for the general diffusive transport equations and a number of varying boundary conditions (Figure 9.2) have been compiled from the literature (Table 9.2).

9.4.4 BOUNDARY CONDITIONS

A short discussion of the boundary conditions applying to the modeling is needed because of their large importance on the results. The conditions at the soil–atmosphere boundary are very simple: either a constant concentration in atmosphere or a no-flow boundary condition in case of sites with a sealed soil surface. A no-flow boundary condition is also applied to the soil air–groundwater boundary, since transfer fluxes of gases across the water table are dominated by aqueous diffusion and are virtually zero. A more detailed discussion is needed, however, for the boundary condition at a volatilizing source of pollutants. If the mass of liquid pollutants in the source is very large, then the isotopic composition of the source during volatilization can be regarded as constant. However, for small source masses, depletion of compounds is expected to occur as volatilization of VOCs from the source progresses. A Fickian equation can be used to link the change in mass of a compound in the source with the concentration in a finite linear system such as a column. If one assumes simultaneous equilibrium with linear exchange between gas and liquid phase in a source, and a linear concentration profile in the soil system (due to absence of biodegradation), the equation expressing the change in mass is (Bouchard et al. 2008b).

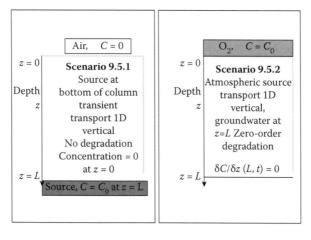

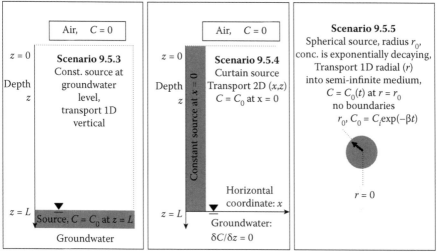

FIGURE 9.2   Schematic illustration of the five scenarios for reactive-diffusive transport in the unsaturated zone modeled in this work in Sections 9.5.1 to 9.5.5.

$$\frac{\partial M_i}{\partial t} = -\frac{AD_m \theta_a \tau C_{a,i}}{L} \tag{9.12}$$

where M_i is the mass of the compound i in the source, and A is the cross section area of the volatilizing flux. Using Raoult's law to determine the partial pressure of a compound in the headspace above a mixture, Equation 9.12 becomes:

$$\left(V_g + \frac{M_{tot}}{C_{sat}} \right) \frac{\partial C_{a,i}}{\partial t} = -\frac{AD_m \theta_a \tau C_{a,i}}{L}. \tag{9.13}$$

TABLE 9.2

Overview of Analytical Solutions Applying for the Diffusion Models Discussed in This Work

Geometry/ Reactions Included	Boundary Conditions	Steady-State Solution ($t = $ infinite)	Transient-State Solution	Reference
1D, coordinate z Length L, C_0 at $z = L$ no production or degradation Scenario 9.5.1	$C = 0$ at $z = 0$, $t > 0$ $C = 0$ at $z > 0$, $t = 0$ $C = C_0$ at $z = L, t > 0$	$C = C_0 z/L$ (Equation A)	$\dfrac{C}{C_0} = 1 - \dfrac{L-z}{L} - \dfrac{2}{\pi}\sum_{n=1}^{\infty}\dfrac{1}{n}\sin\left[\dfrac{n\pi(L-z)}{L}\right]\exp\left[\dfrac{-n^2\pi^2 Dt}{L^2}\right]$ (Equation B)	Pasteris et al. 2002
1D, coordinate z Length L, C_0 at $z = 0$ (in air) with zero-order degradation Scenario 9.5.2	$C = C_0$ at $z = 0, t > 0$ $C = C_0$ at $z > 0, t = 0$ $\delta C/\delta z$ $(z = L, t) = 0$	$C = C_0 - \dfrac{K}{D}\left(Lz - \dfrac{z^2}{2}\right)$ (Equation C)	—	Hendry, Wassenaar, and Birkham 2002
1D, model as above, but with first-order degradation Scenario 9.5.3	$C = 0$ at $z = 0$, $t > 0$ $C = 0$ at $z > 0$, $t = 0$ $C = C_0$ at $z = L, t > 0$	$C = C_0 \dfrac{\sinh\sqrt{\dfrac{k}{D}}z}{\sinh\sqrt{\dfrac{k}{D}}L}$ (Equation D)	—	Pasteris et al. 2002

Description	Boundary conditions	Equation	Reference
1D, model as above, but no degradation, constant zero-order production φ (Cerling model)	$C = C_0$ at $z = 0, t > 0$ $C = C_0$ at $z > 0, t = 0$ $\delta C/\delta z$ $(z = L, t) = 0$ $\varphi(z) = const.$	$C = C_0 + \dfrac{\varphi}{D}\left(Lz - \dfrac{z^2}{2}\right)$ (Equation E)	Cerling 1984
2D (coord. x and z), curtain source (Figure 9.2) in unsaturated zone of thickness L, first-order degradation Scenario 9.5.4	$C = C_0$ at $x = 0, z$ $C = 0$ at x, $z = 0$ $\delta C/\delta z(z = L, x > 0) = 0$	$$C = \frac{4C_0}{\pi}\sum_{n=1}^{\infty}\left(\frac{-1^{(n-1)}\cos\dfrac{\pi(2n-1)(L-z)}{2L}\,\exp\left(-x\sqrt{\dfrac{\pi^2(2n-1)^2}{4L^2}+\dfrac{k}{D}}\right)}{2n-1}\right)$$ (Equation F)	Wilson 1997
1D radial, spherical source with radius r_0 in semi-infinite porous medium, r is radial coordinate, first-order degradation Scenario 9.5.5	$C = 0$ at $r > 0$, $t = 0$ $C = 0$ at $r = \infty, t > 0$ $C = C_0$ at $r = r_0, t > 0$	$C(r, t = \infty) = \dfrac{C_0 r_0}{r}\, e^{\sqrt{\frac{k}{D'}}(r - r_0)}$ (Equation G) for nondecaying source at infinite time $$C(r,t) = \frac{C_0 e^{-\beta t} r_0}{2r}\left[e^{-\sqrt{\frac{(k-\beta)}{D}}(r-r_0)}\,\operatorname{erfc}\left(\frac{(r-r_0)-2\sqrt{D(k-\beta)}\,t}{2\sqrt{Dt}}\right) \right. $$ $$\left. + e^{\sqrt{\frac{(k-\beta)}{D}}(r-r_0)}\,\operatorname{erfc}\left(\frac{(r-r_0)+2\sqrt{D(k-\beta)}\,t}{2\sqrt{Dt}}\right)\right]$$ (Equation H) decaying source, $C_0 = C_{0,init}\,\exp^{(-\beta t)}$, at $r = r_0$, $t > 0$	Bouchard et al. 2005

a   Extension of NAPL $\gg L$

The volume of the gas mixture (V_g, in m^3) with its concentration at saturation (C_{sat}, in moles/m^3) and the number of moles in the liquid mixture (M_{tot}) are assumed to remain constant over time. The solution to Equation 9.13 simply reads

$$C_{a,i}(t) = C_{sat,i} \exp^{-\beta t} \tag{9.14a}$$

where the source decay rate β is a constant described by

$$\beta = \frac{AD_m\theta_a\tau}{(V_g + M_{tot}/C_{sat})L}. \tag{9.14b}$$

The term source decay rate refers to the rate at which the strength of the source is disappearing. Since isotopically heavier nonpolar compounds generally have a slightly higher vapor pressure, denoted as P_h (Jancso and Van Hook 1974), than the pressure P_l of their light counterparts, a proportionally larger amount of heavier compound is expected to volatilize from the liquid source compared to the lighter compound to form the saturated vapor in the chamber. Taking into account both the differences in vapor pressures and in Fickian fluxes, the ratio of the source decay rates β_h and β_l of two isotopes simply becomes (Bouchard et al, 2008b; Höhener, Bouchard, and Hunkeler 2008)

$$\frac{\beta_h}{\beta_l} = \frac{P_hD_h}{P_lD_l}. \tag{9.15}$$

Due to the small contribution of P_h/P_l in Equation 9.15 compared to D_h/D_l (Table 9.1), it can be neglected for ^{13}C in hydrocarbons. Note that this is not the case for chlorinated solvents and ^{2}H in hydrocarbons (Höhener, Bouchard, and Hunkeler 2008). The global β is attributed to ^{12}C-molecules (β_l) and β_h for ^{13}C-molecules can be derived as indicated above with Equation 9.15.

9.4.5 Implementation of the Model

The software Maple 9.01 (Waterloo Maple Inc., Waterloo, Canada) has been used in this study for the transport calculations. For each subspecies C_I through C_{III}, the basic transport equation is established, differing for each subspecies I–III in different diffusion coefficients and different reaction rate constants

$$C_I(x, y, z, t) = f(C_{0,I}, D_l, k_l), \quad C_{II}(x, y, z, t) = f(C_{0,II}, D_h, k_l), \text{ and}$$
$$C_{III}(x, y, z, t) = f(C_{0,III}, D_h, k_h),$$

where f is a function taken from the compilation of analytical solutions given in Table 9.2.

Concentration profiles are then computed for each subspecies. The transport equations for the subspecies are subsequently added to compute the profile of the total concentration C_{tot}

$$C_{tot}(x, y, z, t) = (C_I(x, y, z, t) + C_{II}(x, y, z, t) + C_{III}(x, y, z, t))*nI \qquad (9.16)$$

Finally, isotopic ratios and the delta notation are obtained from

$$R_{(x,y,z,t)} = (C_{II}(x, y, z, t) + C_{III}(x, y, z, t))/(C_{tot}(x, y, z, t)$$
$$- C_{II}(x, y, z, t) - C_{III}(x, y, z, t)) \qquad (9.17)$$

$$\delta_{(x,y,z,t)} = \left(\frac{R_{(x,y,z,t)}}{R_{ref}} - 1 \right)1000. \qquad (9.18)$$

9.5   MODELED EXAMPLE SCENARIOS

In this section the model will be applied in five scenarios that were selected and ordered with increasing complexity. The first scenario (9.5.1) is describing transient diffusion of CO_2 in a column filled with dry sand, devoid of biological reactions and kinetic isotope fractionation, and thus, only diffusion leads to fractionation of CO_2. Subsequently, steady-state scenarios including kinetic isotope fractionation in 1D (9.5.2 and 9.5.3) and 2D (9.5.4) are discussed, and in the end, a transient-state scenario is modeled in radial coordinates. Annotated MAPLE worksheets for the five scenarios are downloadable for free on www.unine.ch/chyn/isotopes

9.5.1   SCENARIO: TRANSIENT CO_2 DIFFUSION IN AN ABIOTIC SAND COLUMN

A column experiment simulating transient diffusion of CO_2 through dry soil was performed by Barth and coworkers (1999). The vertical column of 60 cm length and 7.8 cm diameter was filled with fine sand with a mean grain size of 3 mm. The column was open to laboratory air (room temperature) at the top. A 5L reservoir with pure CO_2 was connected at time zero to the column bottom end. The CO_2 concentrations and the $\delta^{13}C$ of CO_2 were measured along four sampling ports during 3.6 hours after the start of the experiment. The gas samples were transferred to a GC-MSD-IRMS by means of syringes with thick needles to avoid supplemental fractionation during sampling. Then $\delta^{13}C$ was reported relative to the Vienna Pee Dee Belemnite (VPDB) standard with a precision of 0.2‰.

The data were not modeled in the original publication. The calculations with our model are based on the transport Equation B (Table 9.2) and the following parameters: $nI = nI_{react} = 1$ (C in CO_2), $D_m = 1.36$ m²d⁻¹ at 20°C (Prichard and Currie 1982), and a tortuosity of 0.693 calculated from the Moldrup relationship $\tau = \theta_a^{2.5}/\theta_t$, with $\theta_t = \theta_a = 0.48$, as typical for loosely packed homogeneous sand. Since $nI = nI_{react}$, the model was calculating the diffusive transport for only two subspecies, $^{12}CO_2$ and $^{13}CO_2$. The measured profiles of the total CO_2 concentration as well as the $\delta^{13}C$ in CO_2 are shown in Figure 9.3. For better comparability, the total concentrations of the CO_2 profiles are shown as relative concentrations C/C_0, unlike in the original publication. The CO_2 profile at $t = 65$ min shows a slight curvature, whereas the later profiles

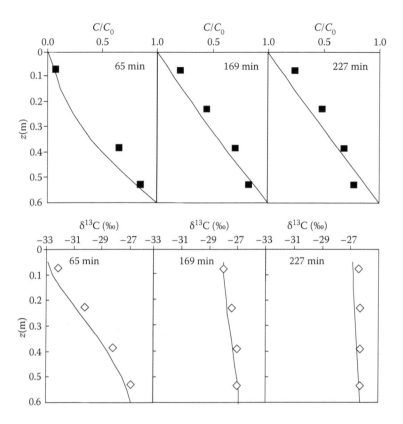

FIGURE 9.3 Illustration of isotope effects in scenario 9.5.1: Relative concentration of CO_2 (top) and $\delta^{13}C$ in CO_2 (bottom) during transient diffusive CO_2 flux in a dry sand column. Data (symbols) from (Barth, J. A. C., A. H. Olsson, and R. M. Kalin, *Geoenvironmental engineering: Ground contamination: Pollutant management and remediation*, Thomas Telford, London, 1999), model results (lines): this work.

rather are straight lines from C_0 to 0 (Figure 9.3), as expected from the steady-state solution (Table 9.2, Equation A). The $\delta^{13}C$ in CO_2 is shifting from the values of $-28.8‰$ in the CO_2 (Barth, Olsson, and Kalin 1999) to values as negative as $-32.1‰$ at the opposite end of the soil column after 65 minutes. Note that the laboratory air had a $\delta^{13}C$ of between -11.7 and $-19.2‰$ (Barth, Olsson, and Kalin 1999) and cannot be at the origin of these very negative values. In the absence of biochemical reactions, these large shifts in $\delta^{13}C$ are uniquely created by diffusion.

In the CO_2 reservoir, the isotopic signal shifted from the initial $-28.8‰$ to $-26.3‰$, due to either dilution with lab air, or due to depletion of the $^{12}CO_2$ through enhanced flux to the atmosphere. Our model approach did not attempt to model the changes occurring in the reservoir, and thus, adjusted C_0 for each time of the experiment to the values measured at the lowest sampling port, assuming that this value was close to the true C_0. The modeled curves matched the measured data satisfactorily. An offset of about $0.5‰$ is noted for the $\delta^{13}C$ at $t = 65$ min, probably due to improper knowledge of C_0 for this curve. However, the shift of $-5.2‰$ along the column of only 0.6 m length is predicted correctly by the model. It is thus, demonstrated experimentally and theoretically how diffusion alone can lead to significant isotope shifts in the unsaturated zone. The experiment shows that such an isotope shift is only created during transient diffusion and is later disappearing when steady-state diffusion is achieved. This is also intelligible from the solutions of the diffusive model (Equations A and B, Table 9.2), where only the transient state equation but not the steady state equation depends on D.

9.5.2   Scenario: Steady-State O_2 Flux into Biotic Sand Mesocosm with Zero-Order Degradation

This scenario is describing and interpreting experimental measurements of O_2 concentration and $\delta^{18}O$ performed by Hendry and coworkers (2002). A mesocosm of 2.4 m diameter was filled with moist sand from a C-horizon of a Saskatchewan soil and compacted to yield a porosity of 0.426. The sand was irrigated for 6 + years, and the depth of the water table was set to 3.2 m. The system was at steady state with respect to water and gas fluxes. The vertical distribution of soil water content followed a typical curve with an increase near the water table. The mean volumetric water content was 0.136. The O_2 concentration and the $\delta^{18}O$ in O_2 were measured on day 2,254 after mesocosm filling as reported in Hendry et al. (2002).

The model in this study used the transport Equation C (Table 9.2) and the following parameters: $nI = nI_{react} = 2$, $D_m = 1.54$ m² d⁻¹ at 20°C, $H = 32.6$, $K = 0.31$ μg O_2 (g soil day)⁻¹ (Hendry et al. 2002), and a tortuosity of 0.263 calculated from the Moldrup model with θ_t and θ_w given above. The model was calculating only subspecies I ($^{16}O_2$) and subspecies III ($^{18}O^{16}O$).

This steady-state scenario demonstrates a shift in $\delta^{18}O$ in O_2 due to diffusion and zero-order consumption. The scenario had been modeled previously by both an analytical model assuming constant moisture and a constant biodegradation rate with depth and a numerical model, which was accounting for moisture and rate variations with

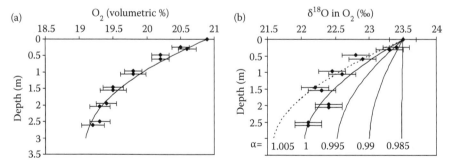

FIGURE 9.4   Illustration of isotope effects in scenario 9.5.2: (a) O_2 concentrations: data from (Hendry, M. J., L. I. Wassenaar, and T. K. Birkham, *Geochimica Cosmochimica Acta*, 66, 3367–74, 2002.) in symbols, model results (from this work) as a line; (b) Measured $\delta^{18}O$ (symbols from Hendry, M. J., L. I. Wassenaar, and T. K. Birkham, *Geochimica Cosmochimica Acta* 66, 3367–74, 2002.) and modeled $\delta^{18}O$ in O_2 for varying fractionation factors α ranging from 0.985, 0.9, 0.995, 1 to 1.005. The solid line for α = 1 (no biological fractionation) is the best model fit.

depth (Hendry et al. 2002). The analytical model proposed by Hendry and coworkers is basically the same as the model used in this study, except that an iterative approach was used to find α. The measured data suggest a decrease in O_2 with depth and a shift in $\delta^{18}O$ in O_2 from +23.5‰ in the air above ground to +22.1‰ at 2.6 m depth (Figure 9.4). This shift can correctly be modeled only when α is set to 1.000 (Figure 9.4b), which means that no kinetic isotope fractionation by degradation is active (Hendry et al. 2002). The isotopic shift in $\delta^{18}O$ in this scenario is thus caused by fractionation by gas-phase diffusion. Fractionation by biological O_2 consumption, also known as the Dole effect, was observed in a variety of other experimental systems in the absence of diffusion, with α varying from 0.971 to 0.999. Under the experimental condition of the mesocosm, the Dole effect seems not to be active. This may be explained by the fact that O_2 consumption potentially takes place at very low O_2 concentrations within soil aggregates, where microorganisms do not discriminate anymore between heavy and light O_2 (Hendry et al. 2002).

9.5.3 SCENARIO: STEADY-STATE DIFFUSION OF A VOLATILE ORGANIC COMPOUND (VOC) FROM THE SATURATED ZONE TO THE ATMOSPHERE WITH FIRST-ORDER DEGRADATION

This scenario is describing the one dimensional steady flux of a VOC from the bottom of a vadose zone to the atmosphere (Figure 9.2). This mimics, for example, volatilization from a LNAPL floating on groundwater. A theoretical scenario was modeled, assuming a LNAPL source with n-hexane at 5 m depth. The model used the transport equation D (Table 9.2) and the following parameters for ^{13}C: $nI = 6$, $nI_{react} = 2$, $\varepsilon_{reactive\ position} = -10.43$ (Bouchard et al. 2005), $D_m = 0.643$ m^2 d^{-1} at 25°C, $\theta_a = 0.35$ and a tortuosity of 0.45. The first-order degradation rate constant k was varied from 0.001 to 10 d^{-1}. Subspecies I to III were used.

The model results show total concentrations (Figure 9.5a) in distinct profiles with increasing curvature as a function of increasing k, likewise as the profiles measured experimentally in a lysimeter (Pasteris et al. 2002) and a column (Höhener et al. 2003)

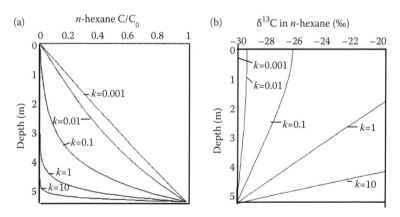

FIGURE 9.5 Illustration of isotope effects in scenario 9.5.3: (a) Relative concentrations of n-hexane and (b) $\delta^{13}C$ in n-hexane (‰) during steady-state diffusive flux. The biodegradation rate k ranges from 0.001 to 10 d^{-1}.

study. The results show an increasing fractionation of n-hexane in soil air with increasing degradation rate (Figure 9.5b). At the lowest degradation rate ($0.001\ d^{-1}$, which signifies quasi absence of degradation), the $\delta^{13}C$ is constant with depth. Increasing the degradation rate leads to an increase in $\delta^{13}C$, from the initial value of $-30\%o$ to higher than $-20\%o$ (Figure 9.5b). However, it must be noted that the $\delta^{13}C$ higher than $> 20\%o$ would be difficult to measure in practice since the total concentrations near the soil surface are very small (Figure 9.5a) when degradation is high, and the concentrations would thus be below the quantification limits of GC-IRMS systems.

9.5.4 SCENARIO: STEADY-STATE DIFFUSION OF VOLATILE ORGANIC COMPOUND (VOC) FROM A CURTAIN SOURCE WITH FIRST-ORDER DEGRADATION

This scenario has been chosen to compare a 2D steady-state model to the previously described 1D steady-state scenarios 9.5.2 and 9.5.3. A curtain source fully penetrating the vadose zone allows diffusion of a VOC in x and z direction (Figure 9.2) as

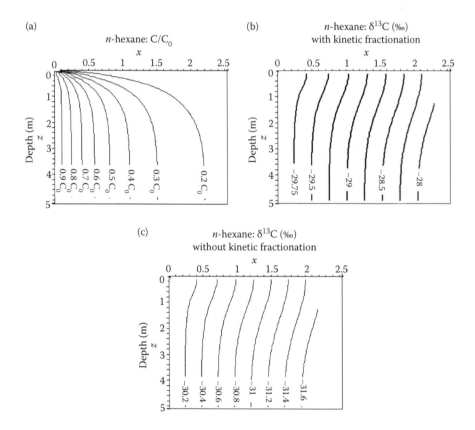

FIGURE 9.6 Illustration of isotope effects in scenario 9.5.4: Steady-state diffusion of n-hexane from a curtain source into soil, with first-order biodegradation $k = 0.1\ d^{-1}$. (a) Relative concentration, (b) $\delta^{13}C$ ($\%o$) with kinetic isotope fractionation of $\varepsilon = -10.43\%o$, and (c) $\delta^{13}C$ in $\%o$ in the absence of kinetic isotope fractionation ($\varepsilon = 0$).

in (Wilson 1997). The soil surface is permeable, first-order biodegradation is active
in the soil, and groundwater represents an impervious boundary for the gas-phase
diffusion at the bottom. The diffusion coefficients in x and z direction are chosen to
be equal. The model was run for ^{13}C in n-hexane, with $nI = 6$, $nI_{react} = 2$, Equation
F from Table 9.2, $\varepsilon_{reactive\ position}$ either set to $-10.43\%o$ (Bouchard et al. 2005) or to 0,
$D_m = 0.649$ m^2 d^{-1}, $\tau = 0.45$, depth to groundwater $L = 5$ m, and first-order biodegradation rate $k = 0.1$ d^{-1}. The modeled results for n-hexane (Figure 9.6b) show a positive isotopic shift of $\delta^{13}C$ from -30 to $-28\%o$ within 2.5 m lateral distance from the
curtain when kinetic fractionation is occurring. When $\varepsilon_{reactive\ position}$ is set to $-10.43\%o$
2.5 m is the distance within which the concentration fell to somewhat less than 20%
of C_0. In the absence of kinetic fractionation, with $\varepsilon_{reactive\ position}$ of 0, the shifts are
negative, reaching $-31.7\%o$ at 2.5 m lateral distance. The isotopic shifts are not very
dependent on depth, although they are somewhat less marked near the soil surface.
An inverse relationship with the contours of the total concentrations is observed
(Figure 9.6c). When the biodegradation rate is set to zero (data not shown), the $\delta^{13}C$
is $-30\%o$ everywhere. The isotopic shifts in this 2D steady-state scenario and the 1D
scenario 9.5.3 show that in the absence of degradation, steady-state diffusion does
not lead to isotope fractionation. The isotopic shifts in the 2D scenario are of the
same magnitude than the 1D scenario.

9.5.5   SCENARIO: TRANSIENT DIFFUSION OF VOLATILE ORGANIC COMPOUND
(VOC) FROM A SPHERICAL SOURCE WITH FIRST-ORDER DEGRADATION

For this scenario, experimental data that were measured in a homogeneous field
site are compared to the model predictions. At the Værløse field in Denmark, a
cylindrical source made of artificial kerosene was emplaced into a sandy soil
(Broholm et al. 2005; Christophersen et al. 2005; Höhener et al. 2006). The source
had 0.5 m thickness, 0.35 m radius and was placed between 0.8 and 1.3 m depth
in the vadose zone that had a total thickness of approx. 3.2 m above the water
table. The experiment was run for more than a year. Data of $\delta^{13}C$ in MCH from the
first 114 days of volatilization or biodegradation (Bouchard et al. 2008b) are modeled in this scenario. The model used to calculate this scenario for ^{13}C was using
$nI = 7$, $nI_{react} = 5$, Equation H from Table 9.2, $\varepsilon_{reactive\ position} = -1.55\%o$ (Bouchard et
al. 2005), $D_m = 0.61$ m^2 d^{-1} at 15°C, $H = 17.6$ (Gaganis, Kjeldsen, and Burganos
2004), $K_d = 0.22$ L kg^{-1}, $f_a = 0.95$, $\tau = 0.31$, decay rates of the source for light and
heavy subspecies $\beta_l = 0.014$ d^{-1} and β_h 0.013984 d^{-1} (Broholm et al. 2005), $r_0 = 0.35$
m (Christophersen et al. 2005), and a first-order biodegradation rate $k = 0.24$ d^{-1}
(Höhener et al. 2006). Partitioning and sorption were not assumed to influence the
isotope signals.

The $\delta^{13}C$ of MCH measured in the field were first depleted in ^{13}C (Figure 9.7) on
days 3 and 6 after source emplacement, with up to 4‰ decrease in the $\delta^{13}C$ value. On
day 22, the $\delta^{13}C$ were throughout at $-28.1\%o$, identical to the initial isotopic signal of
MCH in the source. On days 45 and 114, the $\delta^{13}C$ of MCH in the soil gas was enriched
in ^{13}C. The model data agreed with the field data, predicting also a flat profile for
day 22, and a general increase of $\delta^{13}C$ thereafter due to source depletion. The mass
of MCH on day 114 was only about 25% of the initial mass (Broholm et al. 2005),

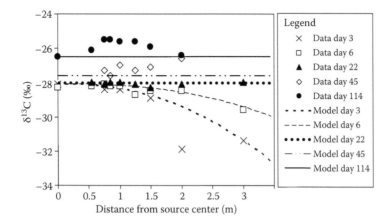

FIGURE 9.7 Illustration of isotope effects in scenario 9.5.5: Transient diffusion of methylcyclohexane from a spherical source into soil, with first-order biodegradation $k = 0.24$ d^{-1}. Symbols: measured values from the Værløse field study (Bouchard, D., D. Hunkeler, P. Gaganis, R. Aravena, P. Höhener, and P. Kjeldsen, *Environmental science & Technology*, 42, 596–601, 2008b), lines: model predictions, this work.

and our data suggest that a preferential loss of MCH depleted in ^{13}C had occurred, shifting the remaining MCH to a more enriched $\delta^{13}C$ value. The ability of the model to reproduce this shift is promising in view of using this shift as a tool for the assessment of source depletion (Höhener, Bouchard, and Hunkeler 2008). Bouchard and coworkers (2008a) produced model simulations using the same theoretical approach, but a numerical transport model obtained almost similar model predictions. The example shows that on a short scale (3 m) and within a short time (3–6 days), gas-phase diffusion creates important isotope shifts, and a correct assessment of the biodegradation effect is only achievable when the influence of source depletion on isotopes is accurately understood.

9.6 CONCLUSIONS

In the unsaturated zone, two processes can be identified that create major isotope fractionation for gaseous volatile compounds: kinetic isotope fractionation associated with degradation and gas-phase diffusion. Both processes need to be addressed carefully in models, and up to three subspecies need to be distinguished for each gaseous compound in order to account for fast and slow diffusion and fast and slow biodegradation. A general model for stable isotopes in gaseous compounds is proposed, and three steady-state and two transient-state scenarios describing the reactive-diffusive transport of volatile compounds in the unsaturated zone have been analyzed in this chapter. A summary of all effects observed in these scenarios on isotope signals is given in Table 9.3. A distinct difference is found between effects under steady-state and transient-state conditions. No fractionation is created by gas-phase diffusion under steady-state conditions in the absence of degradation (or production). This is demonstrated here for one- and two-dimensional settings of finite length, (e.g., for

TABLE 9.3

Summary of the Effects of Gas-phase Diffusion and Degradation on the Isotopic Signals in Gaseous Compounds in the Unsaturated Zone

	Predicted Isotope Effects for Diffusing Gaseous Compounds Under:	
	Steady-State Conditions	**Transient-State Conditions**
Absence of degradation or production	Isotope ratios in vapors are equal to ratios in source. Source gets enriched in the heavy isotope with time, since diffusive fluxes of light molecules are larger Scenarios: 9.5.1(final times) and 9.5.4	Diffusion creates more negative isotope signals in vapors, especially far from source, which becomes isotopically enriched with time. Scenario 9.5.1
Degradation is active but does not fractionate	A linear relationship between depletion of isotope ratios in vapors with distance from source is expected[a] Scenarios: 9.5.2 and 9.5.4 (c)	Depleted isotope ratios expected in vapors always and everywhere, with most depleted ratios in early times. Nonlinear profiles with respect to distance from source. No scenario modeled
Degradation is active and fractionates (preferential degradation of light molecules, $k_h/k_l < 1$)	The process that fractionates more strongly will govern the isotope signals: if $k_h/k_l < D_h/D_l$, then positive isotope shifts in vapors are observed, depending on the magnitude of k Scenarios: 9.5.3 and 9.5.4. (b)	Diffusion first produces more depleted isotope ratios in vapors, which later increase and become more enriched than the source, especially far from the source. The overall effect depends on the magnitudes of k_h/k_l and D_h/D_l See Bouchard, Höhener, and Hunkeler 2008a; Bouchard et al. 2008b Scenario: 9.5.5

[a] When a compound is produced, then more enriched isotope signals are observed compared with the source, as shown for CO_2 by (Cerling, T. E., D. K. Solomon, J. Quade, and J. R. Bowman, *Geochimica Cosmochimica Acta*, 55, 3403–5, 1991.)

diffusion from the water table to the atmosphere). For one-dimensional diffusion, the result is intelligible from the steady-state solution of the diffusion equation, which does not depend on D (Equation A, Table 9.2). Only in the presence of degradation, will the isotope ratios shift in such a setting due to fractionation by diffusion. In the absence of kinetic fractionation ($k_l/k_h = 1$ or $\varepsilon = 0$), for instance for respiration of O_2 (scenario 9.5.2), the isotope ratios shift to more negative values due to diffusion. It is noteworthy that when a compound is produced in such a setting (e.g., CO_2 by soil respiration), the isotope shift created by diffusion is positive since the light CO_2 molecules diffuse out of the soil (Cerling et al. 1991), and the soil CO_2 is enriched in ^{13}C with respect to the soil organic matter. When kinetic fractionation is present, the isotopic shift of the diffusing species in the unsaturated zone depends on the ratio of biological versus diffusive fractionation factors.

Table 9.2 shows that the steady-state diffusion Equations C and E are a function of the ratio of k/D, and Equation D of K/D. It follows thus that a comparison of k_h/k_l and D_h/D_l determines the direction of the isotope shifts. In steady-state scenarios with diffusion and reaction and compounds with only two diffusing subspecies ($C_{II} = 0$, e.g., O_2, CO_2, benzene), it is easily seen that isotope ratios stay constant in time and space when k_h/k_l equals D_h/D_l. However, this is not the case for compounds where three subspecies are diffusing, and no simple solution exists at which isotope shifts remain constant at any time and at any point in space. It is concluded that from a theoretical point of view, it is important to model VOCs with $nI \neq nI_{react}$ as three distinct subspecies.

The two transient scenarios 9.5.1 and 9.5.5, both well documented with experimental data, suggest a significant negative shift in $\delta^{13}C$ during initial gas-phase diffusion, due to the faster diffusion of light molecules compared to heavy molecules. Later, when the systems approach steady state, this effect gradually disappears. The time to reach steady state depends on the fraction of the diffusing compound in the soil air f_a and on the geometry and temperature of the site. Volatile organic pollutants can have mass fractions in soil air f_a in a wide range, from close to 1 for hydrophobic compounds such as n-hexane to < 0.001 for hydrophilic compounds such as MTBE or ethanol (Dakhel et al. 2003). The vapors of the latter polar compounds migrate slower than n-alkane vapors (Pasteris et al. 2002; Dakhel et al. 2003), and thus, diffusive fractionation of hydrophilic VOCs during transient vapor transport can occur during extended periods at a site. Both transient scenarios also showed that isotopic depletion of the source may occur. The sources in scenario 9.5.1 and 9.5.5 both became enriched in ^{13}C. In case of a gaseous source such as CO_2 in scenario 9.5.1, isotopic source enrichment can be easily incorporated into any model approach, since a perfect mixing of a gaseous source can be assumed. However, for petroleum sources in soils or on groundwater, it is not well understood to what extent they are fully mixed and to what extent they only erode at the edges (Conant, Gillham, and Mendoza 1996; Broholm et al. 2005). More work is needed to investigate effects at eroding sources.

This work has been illustrated using analytical models for vapor diffusion in the unsaturated zone. The concept behind the model approach is, however, also applicable for numerical models. A more rigorous investigation of fractionation due to diffusion and degradation in heterogeneous settings is possible using numerical models (Bouchard et al. 2008b).

Further experimental studies should be performed in other unsaturated permeable media to complement the studies available for sandy soils. It should be investigated whether processes like deposition on the gas-water interface or intra-particle diffusion contribute to isotope fractionation. More work is needed to make sure that source processes are incorporated correctly (i.e., when mass transfer limitation at the source boundary is occurring; Zaidel and Russo 1994). A case of special interest is the situation where a contaminated groundwater plume acts as a source for volatile compounds degassing to the unsaturated zone. More work is needed to understand the fractionation processes during the transfer of volatile compounds through the capillary zone at such sites.

ACKNOWLEDGMENT

We thank Johannes Barth and Ramon Aravena for critical reviews and helpful suggestions.

REFERENCES

Abe, Y., and D. Hunkeler. 2006. Does the Rayleigh equation apply to evaluate field isotope data in contaminant hydrogeology? *Environmental Science & Technology* 40:1588–96.

Amundson, R., L. Stern, T. Baisden, and Y. Wang. 1998. The isotopic composition of soil and soil-respired CO_2. *Geoderma* 82:83–114.

Angert, A., B. Luz, and D. Yakir. 2001. Fractionation of oxygen isotopes by respiration and diffusion in soils and its implications for the isotopic composition of atmospheric O_2. *Global Biogeochemical Cycles* 15:871–80.

Baehr, A. L., and C. J. Bruell. 1990. Application of the Stefan-Maxwell equations to determine limitations of Fick's law when modeling organic vapor transport in sand columns. *Water Resources Research* 26:1155–63.

Barth, J. A. C., A. H. Olsson, and R. M. Kalin. 1999. Carbon stable isotope fractionation of CO2 flux through soil columns. In *Geoenvironmental engineering: Ground contamination: Pollutant management and remediation*, eds. R. N. Young and H. R. Thomas, 209–13. London: Thomas Telford.

Birkham, T. K., M. J. Hendry, L. I. Wassenaar, C. A. Mendoza, and E. S. Lee. 2003. Characterizing geochemical reactions in unsaturated mine waste-rock piles using gaseous O_2, CO_2, $^{12}CO_2$, and $^{13}CO_2$. *Environmental Science & Technology* 37:496–501.

Bouchard, D., P. Höhener, and D. Hunkeler. 2008a. Carbon isotope fractionation during volatilization of petroleum hydrocarbons and diffusion across a porous medium: A column experiment. *Environmental Science & Technology* 42:7801–6.

Bouchard, D., D. Hunkeler, P. Gaganis, R. Aravena, P. Höhener, and P. Kjeldsen. 2008b. Carbon isotope fractionation during migration of petroleum hydrocarbon vapors in the unsaturated zone: Field experiment at Værløse Airbase, Denmark, and modeling. *Environmental Science & Technology* 42:596–601.

Bouchard, D., D. Hunkeler, and P. Höhener. 2008c. Carbon isotope fractionation during aerobic biodegradation of n-alkanes and aromatic compounds in unsaturated sand. *Organic Geochemistry* 39:23–33.

Bouchard, D., D. Hunkeler, P. Höhener, R. Aravena, M. Broholm, and P. Kjeldsen. 2005. Use of stable isotope analysis to assess biodegradation of petroleum hydrocarbons in the unsaturated zone. Laboratory studies, field studies, and mathematical simulations. In *Reactive transport in soil and groundwater: Processes and models*, eds. G. Nützmann, P. Viotti, and P. Aagaard, 17–38. Berlin: Springer.

Broholm, M., M. Christophersen, U. Maier, E. Stenby, P. Höhener, and P. Kjeldsen. 2005. Compositional evolution of the emplaced fuel source experiment in the vadose zone field experiment at Airbase Værløse, Denmark. *Environmental Science & Technology* 39:8251–63.

Cerling, T. E. 1984. The stable isotopic composition of modern soil carbonate and its relationship to climate. *Earth and Planetary Science Letters* 71:229–40.

Cerling, T. E., D. K. Solomon, J. Quade, and J. R. Bowman. 1991. On the isotopic composition of carbon in soil carbon dioxide. *Geochimica Cosmochimica Acta* 55:3403–5.

Christophersen, M., M. Broholm, H. Mosbaek, H. Karapanagioti, V. N. Burganos, and P. Kjeldsen. 2005. Transport of hydrocarbons from an emplaced fuel source experiment in the vadose zone at Airbase Værløse, Denmark. *Journal of Contaminant Hydrology* 81:1–33.

Clark, I., and P. Fritz. 1997. *Environmental isotopes in hydrogeology*. Boca Raton, FL: CRC Press.

Conant, B. H., R. W. Gillham, and C. A. Mendoza. 1996. Vapor transport of trichloroethylene in the unsaturated zone: Field and numerical modeling investigations. *Water Resources Research* 32:9–22.

Cook, P. G., and A. L. Herczeg. 1999. *Environmental tracers in subsurface hydrology*. Boston: Kluwer.

Costanza-Robinson, M. S., and M. L. Brusseau. 2002. Air-water interfacial areas in unsaturated soils: Evaluation of interfacial domains. *Water Resources Research* 38:1195, doi:10.1029/2001WR000738.

Craig, H. 1953. The geochemistry of stable carbon isotopes. *Geochimica Cosmochimica Acta* 3:53–92.

Dakhel, N., G. Pasteris, D. Werner, and P. Höhener. 2003. Small-volume releases of gasoline in the vadose zone: Impact of the additives MTBE and ethanol on groundwater quality. *Environmental Science & Technology* 37:2127–33.

Davidson, G. R. 1995. The stable isotopic composition and measurement of carbon in soil CO_2. *Geochimica Cosmochimica Acta* 59:2485–89.

Elberling, B., F. Larsen, S. Christensen, and D. Postma. 1998. Gas transport in a confined unsaturated zone during atmospheric pressure cycles. *Water Resources Research* 34:2855–62.

Elsner, M., L. Zwank, D. Hunkeler, and R. Schwarzenbach. 2005. A new concept linking observable stable isotope fractionation to transformation pathways of organic pollutants. *Environmental Science & Technology* 39:6896–916.

Falta, R. W., I. Javandel, K. Pruess, and P. A. Witherspoon. 1989. Density-driven flow of gas in the unsaturated zone due to the evaporation of volatile organic compounds. *Water Resources Research* 25:2159–69.

Franzmann, P. D., L. R. Zappia, T. R. Power, G. B. Davis, and B. M. Patterson. 1999. Microbial mineralisation of benzene and characterisation of microbial biomass in soil above hydrocarbon-contaminated groundwater. *FEMS Microbiology Ecology* 30:67–76.

Fuller, E. N., P. D. Schettler, and J. C. Giddings. 1966. A new method for prediction of binary gas-phase diffusion coefficients. *Industrial & Engineering Chemistry* 58:19–27.

Gaganis, P., P. Kjeldsen, and V. N. Burganos. 2004. Modeling natural attenuation of multicomponent fuel mixtures in the vadose zone: Use of field data and evaluation of biodegradation effects. *Vadose Zone Journal* 3:1262–275.

Galimov, E. M. 1966. Carbon isotopes of soil CO_2. *International Geochemistry* 3:889–98.

Grathwohl, P. 1998. *Diffusion in natural porous media: Contaminant transport, sorption/desorption and dissolution kinetics*. Boston: Kluwer.

Grathwohl, P., and M. Reinhard. 1993. Desorption of trichloroethylene in aquifer material - Rate limitation at the grain scale. *Environmental Science & Technology* 27:2360–66.

Harrington, R. R., S. R. Poulson, J. I. Drever, P. J. S. Colberg, and E. F. Kelly. 1999. Carbon isotope systematics of monoaromatic hydrocarbons: Vaporization and adsorption experiments. *Organic Geochemistry* 30:765–75.

Hendry, M. J., C. A. Mendoza, R. Kirkland, and J. R. Lawrence. 2001. An assessment of a mesocosm approach to the study of microbial respiration in a sandy unsaturated zone. *Ground Water* 39:391–400.

Hendry, M. J., L. I. Wassenaar, and T. K. Birkham. 2002. Microbial respiration and diffusive transport of O_2, $^{16}O_2$, and $^{18}O^{16}O$ in unsaturated soils: A mesocosm experiment. *Geochimica Cosmochimica Acta* 66:3367–74.

Hoff, J. T., R. Gillham, D. Mackay, and W. Y. Shiu. 1993a. Sorption of organic vapors at the air-water interface in a sandy aquifer material. *Environmental Science & Technology* 27:2789–94.

Hoff, J. T., D. Mackay, R. Gillham, and W. Y. Shiu. 1993b. Partitioning of organic chemicals at the air-water interface in environmental systems. *Environmental Science & Technology* 27:2174–80.

Höhener, P., D. Bouchard, and D. Hunkeler. 2008. Stable isotopes as a tool for monitoring the volatilization of non-aqueous phase liquids from the unsaturated zone. In *Advances in subsurface pollution of porous media. Indicators, processes and modelling,* eds. L. Candela, I. Vadillo, and F. J. Elorza, 123–35. Boca Raton, FL: CRC Press, Taylor and Francis Group.

Höhener, P., N. Dakhel, M. Christophersen, M. Broholm, and P. Kjeldsen. 2006. Biodegradation of hydrocarbons vapors: Comparison of laboratory studies and field investigations in the vadose zone at the emplaced fuel source experiment, Airbase Værløse, Denmark. *Journal of Contaminant Hydrology* 88:337–58.

Höhener, P., C. Duwig, G. Pasteris, K. Kaufmann, N. Dakhel, and H. Harms. 2003. Biodegradation of petroleum hydrocarbon vapors: Laboratory studies on rates and kinetics in unsaturated alluvial sand. *Journal of Contaminant Hydrology* 66:93–115.

Huang, L., N. C. Sturchio, T. Abrajano, L. J. Heraty, and B. D. Holt. 1999. Carbon and chlorine isotope fractionation of chlorinated aliphatic hydrocarbons by evaporation. *Organic Geochemistry* 30:777–85.

Hunkeler, D., N. Chollet, X. Pittet, R. Aravena, J. A. Cherry, and B. L. Parker. 2004. Effect on source variability and transport processes on carbon isotope ratios of TCE and PCE in two sandy aquifers. *Journal of Contaminant Hydrology* 74:265–82.

Jancso, G., and W. A. Van Hook. 1974. Condensed phase isotope effects (especially vapor pressure isotope effects). *Chemical Reviews* 74:689–750.

Jost, W. 1960. *Diffusion in solids, liquids, and gases,* 3rd ed. New York: Academic Press.

Kim, H., M. D. Annable, and P. S. C. Rao. 2001. Gaseous transport of volatile organic chemicals in unsaturated porous media: Effect of water-partitioning and air-water interfacial adsorption. *Environmental Science & Technology* 35:4457–62.

Kirtland, B. C., C. M. Aelion, and P. A. Stone. 2005. Assessing in situ mineralization of recalcitrant organic compounds in vadose zone sediments using delta C-13 and C-14 measurements. *Journal of Contaminant Hydrology* 76:1–18.

Koma, D., Y. Sakashita, K. Kubota, Y. Fujii, F. Hasumi, S. Y. Chung, and M. Kubo. 2004. Degradation pathways of cyclic alkanes in *Rhodococcus sp.* NDKK48. *Applied Microbiology and Biotechnology* 66:92–99.

Kopinke, F. D., A. Georgi, M. Voskamp, and H. H. Richnow. 2005. Carbon isotope fractionation of organic contaminants due to retardation on humic substances: Implications for natural attenuation studies in aquifers. *Environmental Science & Technology* 39:6052–62.

Lahvis, M. A., A. L. Baehr, and R. J. Baker. 1999. Quantification of aerobic biodegradation and volatilization rates of gasoline hydrocarbons near the water table under natural attenuation conditions. *Water Resources Research* 35:753–65.

Lane, G. A., and M. Dole. 1956. Fractionation of oxygen isotopes during respiration. *Science* 123:574–76.

Lee, E. S., T. K. Birkham, L. I. Wassenaar, and M. J. Hendry. 2003. Microbial respiration and diffusive transport of O_2, $^{16}O_2$, and $^{18}O^{16}O$ in unsaturated soils and geologic sediments. *Environmental Science & Technology* 37:2913–17.

Lugg, G. A. 1968. Diffusion coefficients of some organic and other vapors in air. *Analytical Chemistry* 40:1072–77.

Meckenstock, R. U., B. Morasch, C. Griebler, and H. H. Richnow. 2004. Stable isotope factionation analysis as a tool to monitor biodegradation in contaminated aquifers. *Journal of Contaminant Hydrology* 75:215–55.

Millington, R., and J. P. Quirk. 1961. Permeability of porous solids. *Transactions of the Faraday Society* 57:1200–7.

Moldrup, P., T. Olesen, J. Gamst, P. Schjonning, T. Yamaguchi, and D. E. Rolston. 2000. Predicting the gas diffusion coefficient in repacked soil: Water-induced linear reduction model. *Soil Science Society of America Journal* 64:1588–94.

Narten, A., and W. Kuhn. 1961. Der 13C/12C-Isotopieeffekt in tetrachlorkohlenstoff und in Benzol. *Helvetica Chimica Acta* 44:1474–1479.

Ostendorf, D. W., and D. H. Kampbell. 1991. Biodegradation of hydrocarbon vapors in the unsaturated zone. *Water Resources Research* 27:453–62.

Pankow, J. F., and J. A. Cherry. 1996. *Dense chlorinated solvents and other DNAPLs in groundwater.* Waterloo, Ontario: Waterloo Press.

Pasteris, G., D. Werner, K. Kaufmann, and P. Höhener. 2002. Vapor phase transport and bio-degradation of volatile fuel compounds in the unsaturated zone: A large scale lysimeter experiment. *Environmental Science & Technology* 36:30–39.

Prichard, D. T., and J. A. Currie. 1982. Diffusion coefficients of carbon dioxide, nitrous oxide, ethylene and ethane in air and their measurement. *Journal of Soil Science* 33:175–84.

Quade, J., T. E. Cerling, and J. R. Bowman. 1989. Systematic variations in the carbon and oxy-gen isotopic composition of pedogenic carbonate along elevation transects in the southern Great-Basin, United States. *Geological Society of America Bulletin* 101:464–75.

Scanlon, B. R., J. P. Nicot, and J. M. Massmann. 2000. Soil gas movement in unsaturated systems. In *Handbook of soil sciences*, ed. M. E. Sumner, A277–319. Boca Raton, FL: CRC Press.

Schmidt, T. C., L. Zwank, M. Elsner, M. Berg, R. U. Meckenstock, and S. B. Haderlein. 2004. Compound-specific stable isotope analysis of organic contaminants in natural environments: A critical review of the state of the art, prospects, and future challenges. *Analytical and Bioanalytical Chemistry* 378:283–300.

Schwarzenbach, R. P., P. M. Gschwend, and D. M. Imboden. 2003. *Environmental organic chemistry*, 2nd ed. Hoboken, NJ: Wiley.

Severinghaus, J. P., M. L. Bender, R. F. Keeling, and W. S. Broecker. 1996. Fractionation of soil gases by diffusion of water vapor, gravitational settling, and thermal diffusion. *Geochimica Cosmochimica Acta* 60:1005–18.

Slater, G. F. 2003. Stable isotope forensics: When isotopes work. *Environmental Forensics* 4:13–23.

Slater, G. F., H. S. Dempster, B. S. Lollar, and J. Ahad. 1999. Headspace analysis: A new appli-cation for isotopic characterization of dissolved organic contaminants. *Environmental Science & Technology* 33:190–94.

Stehmeier, L. G., M. M. Francis, T. R. Jack, E. Diegor, L. Winsor, and T. A. J. Abrajano. 1999. Field and in vitro evidence for in-situ bioremediation using compound-specific 13C/12C ratio monitoring. *Organic Geochemistry* 30:821–33.

Thorstenson, D. C., E. P. Weeks, H. Haas, and D. W. Fisher. 1983. Distribution of gaseous $^{12}CO_2$, $^{13}CO_2$, and $^{14}CO_2$ in the subsoil unsaturated zone of the western United States Great Plains. *Radiocarbon* 25:315–46.

UM-BBD. The University of Minnesota biocatalysis/biodegradation database, http://umbbd.msi.umn.edu/ accessed May 2009.

Van Breukelen, B. M., D. Hunkeler, and F. Volkering. 2005. Quantification of sequential chlo-rinated ethene degradation by use of a reactive transport model incorporating isotope fractionation. *Environmental Science & Technology* 39:4189–97.

Van Breukelen, B. M., and H. Prommer. 2008. Beyond the Rayleigh equation: Reactive trans-port modeling of isotope fractionation effects to improve quantification of biodegrada-tion. *Environmental Science & Technology* 42:2457–63.

Van de Steene, J., and H. Verplancke. 2006. Adjusted Fick's law for gas diffusion in soils contaminated with petroleum hydrocarbons. *European Journal of Soil Science* 57:106–21.

Wang, G., S. B. F. Reckhorn, and P. Grathwohl. 2003. Volatile organic compounds volatilization from multicomponent organic liquids and diffusion in unsaturated porous media. *Vadose Zone Journal* 2:692–701.

Washington, J. W. 1996. Gas partitioning of dissolved volatile organic compounds in the vadose zone: Principles, temperature effects and literature review. *Ground Water* 34:709–18.

Werner, D., P. Grathwohl, and P. Höhener. 2004. Review of field methods for the determination of the tortuosity and effective gas-phase diffusivity in the vadose zone. *Vadose Zone Journal* 3:1240–49.

Werner, D., and P. Höhener. 2002. Diffusive partitioning tracer test for nonaqeous phase liquid (NAPL) detection in the vadose zone. *Environmental Science & Technology* 36:1592–99.

Werner, D., and P. Höhener. 2003. In situ method to measure effective and sorption-affected gas-phase diffusion coefficients in soils. *Environmental Science & Technology* 37:2502–10.

Wiedemeier, T. H., H. S. Rifai, C. J. Newell, and J. T. Wilson. 1999. *Natural attenuation of fuels and chlorinated solvents in the subsurface*. New York: Wiley.

Wilson, D. J. 1997. Soil gas volatile organic compound concentration contours for locating vadose zone nonaqueous phase liquid contamination. *Environmental Monitoring and Assessment* 48:73–100.

Zaidel, J., and D. Russo. 1994. Diffusive transport of organic vapors in the unsaturated zone with kinetically-controlled volatilization and dissolution: Analytical model and analysis. *Journal of Contaminant Hydrology* 17:145–65.

Section IV

Isotope Emerging Areas

10 Isotopic Labeling in Environmental and Biodegradation Studies

C. Marjorie Aelion and R. Sean Norman

CONTENTS

10.1   INTRODUCTION

The natural abundance of isotopes has been the thrust of previous chapters. In this chapter we focus on the use of isotopically labeled compounds to investigate various biological processes in natural systems and contaminant biodegradation processes. Isotopically labeled compounds can be used to identify microorganisms, examine microbial growth, microbial metabolic pathways, and biodegradation of natural organic compounds and organic contaminants. The use of isotopes in environmental studies relies on the incorporation of a substrate that is highly enriched in a rare isotope, either radioactive or stable; the enriched radioactive or stable isotopic signal improves detection and quantification from the natural signal. Examples of specific radioactive and stable isotopes commonly used in tracer investigations include ^{14}C,

^{3}H, ^{35}S, ^{13}C, ^{15}N, ^{2}H, ^{37}Cl, and ^{18}O. This chapter will focus primarily on enrichments of $^{13/15}$N, ^{13}C, and their combination and more briefly on ^{3}H.

Isotopically labeled compounds can be incorporated into many types of molecules that are associated with distinct biological functions. Isotopic signatures can then be used to examine these specific functions with which the molecule is associated. For example, if either ^{13}C or ^{14}C is used as a tracer, labeling of carbon can occur in: biological molecules including carbohydrates that include monosaccharides, oligosaccharides, and polysaccharides that are used for energy during metabolism; major components of lipids including waxes, triacylglycerols, phospholipids, sphingolipids, and steroids that are important components of cellular membranes and energy storage; CO_2 and CH_4 that are gaseous end products of microbial metabolism; and DNA and proteins. If either ^{13}N or ^{15}N is used as a tracer, labeling of N can occur in amino acids and proteins that are responsible for cellular function; and in transformation products associated with the nutrient cycling in nitrification, denitrification, and dissimilatory nitrate reduction to ammonium. Tracer studies using ^{35}S-thiopurine can be used to examine protein synthesis and those using ^{3}H-thymidine can be used to examine microbial growth. Tracers are equally effective whether examining natural compounds (^{15}NH$_4$, ^{15}N$_2$, and ^{15}NO$_3$) or contaminants (^{14}C-naphthalene, a lower molecular weight PAH, or ^{14}C atrazine, a triazine pesticide).

Stable isotope probing (SIP) using primarily ^{13}C-labeled compounds is a technique that is used to identify the microorganisms in environmental samples that use a particular growth substrate. By coupling molecular biological techniques of genomics and proteomics to stable isotopic analysis of biological materials, the activities of microorganisms can be identified in an environmental sample on a molecular level, and microbial identity can be linked to function (Dumont and Murrell 2005; Singleton et al. 2005). With the on-going development and availability of new molecular biological analytical methods, applications of SIP to investigate microbial molecular processes will continue to expand.

10.2   ANALYTICAL AND AVAILABILITY CONSIDERATIONS

Once a molecule containing the element of interest is identified, it must be determined whether the molecule can be synthesized with an enriched stable or radioactive isotope. The location of the enriched element on the molecule is important. For example, carbons associated with an aromatic ring can be uniformly enriched in ^{13}C or ^{14}C, versus the enrichment of one carbon that is associated with a methyl side chain as might occur with the organic chemical toluene (Figure 10.1). Using a uniformly labeled tracer, or one labeled only at the side chain will determine what biological and chemical reactions can be investigated. For instance, ß-cleavage can be investigated using toluene labeling of the side chain as shown in Figure 10.1a, and ring cleavage may be studied with toluene labeled as shown in Figure 10.1b.

Many enriched compounds are available for purchase to use in the laboratory or at field sites in situ. The planned release of radiolabeled chemicals into the environment for ecological and biodegradation studies is not currently accepted. However stable isotopes have been used extensively in environmental studies,

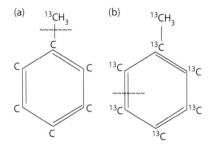

FIGURE 10.1 ^{13}C-Enrichment of toluene can occur on the (a) side chain or on the (b) aromatic ring. Different degradation processes break different bonds (~~) and can be investigated by selecting the label location.

although the natural abundance of the element will be changed in the location in which the stable isotope is released. The ease of synthesizing the enriched compound of interest impacts the cost of the compound. One limitation to the use of stable isotopes in field studies is the prohibitive cost of applying sufficient labeled, enriched tracer to large areas in the environment or large laboratory experiments (Bedard-Haughn et al. 2003).

The specific biological matrix to be studied will determine what tracers are appropriate, and what separation and detection methods are required after the tracer is incorporated into the matrix. The method to detect stable isotopes is distinct from the techniques used to quantify radioisotopes. For stable isotope analysis in environmental samples, isotope ratio mass spectrometry as described in Chapter 2 might be used to measure ^{13}C signature of the chemical of interest, while in biological matrices stable isotopes associated with SIP, physical separation of isotopes occurs along cesium gradients produced in ultra centrifuges. Experiments using radioactive labeling use liquid scintillation counting or photo imaging similar to radiography to measure isotopic content of compounds and molecular components.

Both synthesis and use of the tracers and disposal of the laboratory waste are less regulated for nonradioactive than for radioactive compounds. There is extensive federal regulation for the production, transportation, and handling of radioactive chemicals. Radiolabeled chemicals are normally used in laboratory experiments under appropriate licensure, and the radioactive waste must be disposed of properly by a licensed radioactive waste disposal facility. Radioisotopes purchase and disposal have become increasingly expensive. Most liquid radioactive waste products generated in environmental laboratories must be incinerated at a licensed radioactive waste incinerator, and low-level radioactive solid wastes are land disposed (Eisenbud and Gesell 1997). Stable isotopically labeled compounds are significantly less expensive to purchase than radioisotopes, and because no radioactive waste is generated in the laboratory, associated disposal costs are avoided.

An additional consideration is the basic assumptions that underlie isotope labeling studies, most importantly the assumption that compounds carrying the isotopic label undergo the same chemical, physical, and biological processes as the natural, unlabeled compounds. There are some studies that address this assumption. In laboratory experiments, ^{13}C-labeled and unlabeled benzene and toluene could not be separated in high performance liquid chromatography (HPLC) columns by Harrington et al. (1999) who concluded that the sorption behavior of ^{13}C-labeled and unlabeled benzene and toluene did not differ. They measured a small but consistent enrichment in the vapor phase over a broad temperature range.

Several studies have used deuterated compounds to examine the assumption that labeled and unlabeled compounds behave similarly. Poulson, Drever, and Colberg (1997) investigated whether perdeuterotoluene and perdeuterobenzene would behave similarly as their hydrocarbon counterparts in aquifers, with respect to sorption and retardation. The authors found that deuterocarbons were slightly less retarded compared to nondeuterocarbons, but concluded that the differences were so small that deuterocarbons were useful substitutes for tracing the fate of the petroleum hydrocarbons, benzene, toluene, ethylbenzene, and xylene (BTEX) in groundwater. Similar studies were performed in Australia (Thierrin et al. 1993; Thierrin, Davis, and Barber 1995) and in Germany (Fischer et al. 2006). The latter study involved the injection of approximately 200 gms each of ring-deuterated and perdeuterated toluene, and nonlabeled toluene into an aquifer that was already polluted with BTEX. The breakthrough curves of the three toluenes were not significantly different at wells 24 or 35 m downgradient from the tracer injection well. In this study, the deuterium concentration in groundwater increased, demonstrating that deuterotoluenes were mineralized in the aquifer.

It is generally accepted that the behavior of labeled compounds is virtually the same as unlabeled compounds for most physical and chemical processes, or of a sufficiently small scale to be ignored relative to biological processes of interest. However biokinetic reaction rates of labeled and unlabeled compounds may be different, a phenomenon that is discussed in detail in Chapters 4 and 8. For example the degradation rates of perdeuterotoluene and toluene were different in studies of anaerobic toluene degradation because the compound carried eight deuterium–carbon bonds that were more difficult to break than hydrogen–carbon bonds (Morasch et al. 2001).

For the ^{15}N pool dilution technique (dilution of the ^{14}N indigenous pool by the added isotope) in which isotopically labeled N is added to a natural N pool, there is the assumption that uniform mixing of the isotope and the natural pool has occurred. Uniform mixing in soils may be impacted by mass transfer limitations. Cliff et al. (2002) suggested that in the case of first-order consumption of NH_4^+, adsorption affected the consumption rate estimates but mass transfer limitations did not affect rate estimates. Watson et al. (2000) reported that microorganisms used applied ^{15}N and indigenous ^{14}N pools at different rates due to nonuniform mixing of the two pools in short-term soil incubations estimating microbial N transformation rates. They concluded that the newly applied $^{15}NH_4^+$ was more accessible to nitrifiers than the indigenous soil NH_4^+. Herrmann, Witter, and Kätterer (2005) reanalyzed these data using first-order rate constants for NH_4^+ consumption and NO_3^- immobilization to estimate gross mineralization rates and first-order rate constants for nitrification of the unlabeled NH_4^+. They also concluded that the added isotopically labeled N pool was used more rapidly than the unlabeled NH_4^+. However, Rütting and Müller (2007) suggested that the conclusions of Watson et al. (2000)

and Herrmann, Witter, and Kätterer (2005) were an artifact of the use of inappropriate kinetics. By using Michaelis-Menten kinetics instead of first-order kinetics, Rütting and Müller (2007) found no preferential use of the added $^{15}NH_4^+$ over the unlabeled NH_4^+.

The assumption that compounds carrying the isotopic label undergo the same chemical, physical processes as the natural, unlabeled compounds should be considered in light of the complexities of the soil matrix, but is generally considered appropriate. The assumption that compounds carrying the isotopic label undergo the same biological processes as the natural, unlabeled compounds should take into account the extent of the labeling of the molecule, and the use of appropriate kinetic models as was shown in the case of estimating gross N transformations.

10.3 NITROGEN

Nitrogen in marine and terrestrial systems is a growth substrate for aquatic organisms and a pollutant. Research examining and quantifying nitrogen cycling in the natural environment has been on-going for multiple decades (Jansson 1958; Hardy et al. 1968). Starting in the 1960s, stable isotopes were used to refine N-budget and cycling estimates in ecological studies and to measure phytoplankton productivity in aquatic systems and as tracers in marine aquatic food webs (Lajtha and Michener 1994). Although natural isotope abundance measurements have been used to quantify microbial processes of denitrification and nitrification, it can be difficult to discern the natural cycling of nitrogen if multiple processes are occurring simultaneously, each of which impacts the isotopic signature of the N-containing compounds (Högberg 1997; Robinson 2001). The transformation of inorganic nitrogen from nitrate to ammonium, ammonium to nitrate, and eventually to nitrogen gas is complex, and involves rapid turnover rates and multiple biological and chemical reactions (Figure 10.2). Therefore, N-tracer studies have been used to quantify specific N-transformation processes including denitrification, dissimilatory reduction of nitrate to ammonium (DNRA), nitrification, and nitrogen fixation.

10.3.1 ^{13}NITROGEN

Although rarely used, a few studies have been carried out using the radioisotope of N, ^{13}N. $^{13}NH_4^+$ has a half-life of only 10 minutes and thus, the use of this short-lived radioisotope has inherent difficulties. One complexity is that the synthesis of the ^{13}N compound of interest must be carried out at the place of experimentation that limits its use. Regardless, the technique is extremely sensitive, and the short duration of the experiments allows a focus on mechanisms and processes that occur immediately in the cells. The technique was first developed by Gersberg et al. (1976) and has been used by several investigators in both aquatic and terrestrial studies. Capone et al. (1990) were able to synthesize $^{13}NH_4^+$, which was used in experiments to measure the production of $^{13}NO_2^-$ over short time periods of 60 minutes. The results demonstrated that the $^{13}NH_4^+$ was oxidized to $^{13}NO_2^-$ within as short a time as 30 minutes by the ammonium oxidizing enrichment culture as measured by

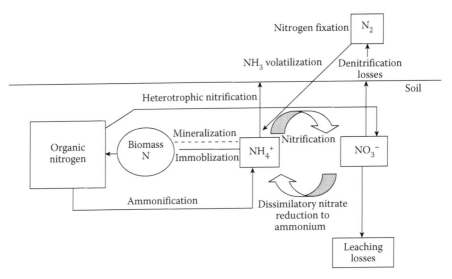

FIGURE 10.2 Nitrogen cycling in the environment is complex. Bacteria play an important role in N cycling and N turnover rates may be rapid. (Modified from Hodge, A., D. Robinson, and A. Fitter, *Trends in Plant Science*, 5, 304–8, 2000.)

radiochromatographic methods. They compared results to those using $^{15}NH_4^+$ and found similar nitrification rates.

Speir, Kettles, and More (1995a) used ^{13}N to examine the effects of amendments on production of N_2O and N_2 in soil cores after addition of $^{13}NO_3^-$ or $^{13}NH_4^+$ to investigate denitrification and nitrification, respectively. Producing the ^{13}N-labeled species requires specialized facilities and equipment, and involves multiple steps. Speir, Kettles, and More (1995a) produced the ^{13}N-labeled species by bombarding H_2O with a proton beam generated by a tandem accelerator. After irradiation of the pure water, the $^{15}NO_3$ was removed. They used NaI radiation detectors to identify the purified gases $^{13}N_2O$ and $^{13}N_2$ (Speir, Kettles, and More 1995b). The results were difficult to interpret and were contradictory in different soils. There was large variability in results based on field variability within a sample set, temporal variability between sets of the same location, and different responses to amendments in the cores. In subsequent experiments, Speir et al. (1999) used the ^{13}N technique to compare N_2O emissions in sediment cores measured using conventional gas chromatography (GC) to those measured using the ^{13}N method. The N_2O produced in the soil core slices was not sufficiently above the ambient background N_2O concentrations to generate a signal measurable by the GC method, but $^{13}N_2O$ and $^{13}N_2$ were measurable in the ^{13}N-labeled core slices, and were up to an order-of-magnitude greater than ambient background radiation levels.

The ^{13}N technique allows assessment of the immediate effects of amendments on denitrification and nitrification and is extremely sensitive. The use of ^{13}N has the potential to add to our understanding of processes that occur on these short time scales in ways that cannot be investigated with conventional techniques if the signals are low. However, these procedures are extremely specialized and methodologically complex. Perhaps due to these methodological constraints, the technique is not commonly used.

10.3.2 ^{15}NITROGEN

The use of ^{15}N-labeled compounds in the study of nitrogen cycling including N-cycle processes, N sinks, and N turnover rates, is a methodological extension of classical nitrogen cycling experiments. In classical studies, the uptake of new and regenerated N was quantified (Dugdale and Goering 1967) in phytoplankton and bacteria primarily in fresh water and marine environments. As early as the 1960s, Neess et al. (1962) described the fixation of N in lakes using ^{15}N. At this time, ^{15}N was analyzed using emission spectrometry, a technique that was less costly than mass spectroscopy (Fiedler and Proksch 1975). Analytical expertise improved for ^{15}N detection in the 1970s, and the application of mass spectrometry allowed ^{15}N to become routinely used to quantify N uptake and assimilation throughout the 1980s and 1990s (Dugdale and Wilkerson 1986; Collos 1987; Boyer, Stanley, and Christian 1994).

Some of the first uses of labeled ^{15}N in marine studies examined the turnover of nitrogen. Laws (1985) examined the nitrogen assimilation rates of bacteria, which, because most do not photosynthesize, could be measured during dark conditions. They also examined phytoplankton nitrogen demand from primary producers based on ^{15}N uptake and autotrophic protein synthesis. By adding ^{15}NH$_4$$^+$ and ^{15}NO$_3$$^-$, Laws (1985) was able to demonstrate that dissolved inorganic nitrogen (DIN) uptake was more than twice the calculated demand and that the difference was due to bacterial uptake. Boyer, Stanley, and Christian (1994) measured DIN uptake in water by adding enriched ^{15}NH$_4$Cl or K^{15}NO$_3$ during light and dark cycles. Significant dark NH$_4$$^+$ uptake occurred, while dark uptake of NO$_3$$^-$ was less than 20% that of light uptake. Boyer, Stanley, and Christian (1994) estimated that total annual DIN uptake was more than twice that of phytoplankton DIN uptake estimated from studies based on phytoplankton alone. These early studies, which represent only a few examples, indicated the importance of bacteria to N cycling in marine systems, which had previously focused on phytoplankton as the main players in N cycling.

Hadwen and Bunn (2005) assessed the impact of low concentrations of NH$_4$$^+$ addition on primary productivity in oligotrophic lakes. By adding ^{15}NH$_4$NO$_3$ to littoral sites at low concentrations, significant increases in periphyton ^{15}N were measured within the first five hours of application. No significant changes in chlorophyll *a* were measured until after 10 days suggesting that the initial consumption of the N was due to the periphyton community followed by the subsequent primary producers. By the end of the experiment after 13 days, all primary producers and primary consumers sampled were enriched in ^{15}N so it was difficult to make specific conclusions about trophic transfer of N to higher consumers. Consumers δ^{15}N did not reach isotopic equilibrium and consumers showed highly variable enrichment in ^{15}N. Several secondary consumers had limited enrichment in ^{15}N, of < 5‰. These experiments demonstrated that the duration of the experiment, and the complexity of the assimilation processes being investigated may make data interpretation difficult.

The N isotopic enrichment studies also have successfully quantified specific bacterial processes in aquatic and sediment systems. Nitrate can be reduced by bacteria to NH_4^+ by dissimilatory nitrate reduction to ammonium DNRA; Equations 10.1 and 10.2; Atlas and Bartha 1997). which retains N in the system. Nitrate also can be reduced by bacteria to N_2 by denitrification (Equation 10.3; note Equation 10.3 is not balanced) which generally occurs in anaerobic sediments (Figure 10.2). In denitrification, microbes convert NO_3^- ultimately to nitrogen gas (N_2) with the intermediate production of nitric oxide (NO_2) and the gas nitrous oxide (N_2O). Because gases are the end products of denitrification, this process can have a significant impact on removing N from aquatic systems as the gases diffuse to the atmosphere. Nitrification is an aerobic process in which NH_4^+ is converted to NO_3^- (Equations 10.4 and 10.5). The nitrification process is catalyzed two different nitrifying bacteria such as *Nitrosomonas* sp., which is capable of oxidizing ammonium to nitrite (Equation 10.4), and then by *Nitrobacter* sp., which is capable of oxidizing nitrite to nitrate (Equation 10.5). Nitrifying bacteria are obligate aerobic chemolithotrophs.

$$0.5\ NO_3^- + H^+ \rightarrow 0.5\ NO_2^- + 0.5\ H_2O \tag{10.1}$$

$$0.5\ NO_2^- + 4H^+ \rightarrow 0.5\ NH_4^+ + H_2O \tag{10.2}$$

$$NO_3^- \rightarrow NO_2^- \rightarrow NO \rightarrow N_2O \rightarrow N_2 \tag{10.3}$$

$$NH_4^+ + 1.5\ O_2 \rightarrow NO_2^- + 2H^+ + H_2O \tag{10.4}$$

$$NO_2^- + 0.5\ O_2 \rightarrow NO_3^- \tag{10.5}$$

Nitrate used in denitrification in sediments diffuses from the overlying water. Nitrification provides a source of nitrate for denitrification, a process that has been termed coupled nitrification–denitrification (CND). As environments become more contaminated with NH_4^+, CND becomes potentially more important in terms of overall N removal, as the NH_4^+ is oxidized to nitrate and eventually to N_2.

$$0.5\ NO_3^- + H^+ \rightarrow 0.5\ NO_2^- \rightarrow NO \rightarrow N_2O \nrightarrow N_2 \tag{10.6}$$

[15]N ratios of several species including N_2, N_2O, NO_3^-, and NH_4^+ may be useful to quantify specific NO_3^- reduction processes. Denitrification often is quantified using N_2O production. A chemical inhibition technique termed the acetylene block method in which the addition of acetylene inhibits the last step of conversion from N_2O to N_2 (Equation 10.6). The accumulation of N_2O can be measured using gas chromatography and is an indirect measure of complete denitrification. This nontracer technique works well but acetylene inhibits nitrification and thus the CND process is underestimated with the acetylene block technique. In addition, CND is difficult to discern using natural abundance measurements of NO_3^- and NH_4^+ because small differences in ^{15}N content are often measured in the natural abundance of NO_3^- and NH_4^+ over time, and large differences in $\delta^{15}N$ among potential sources are needed to use $\delta^{15}N$ as a natural tracer (Robinson 2001). Coupled nitrification–denitrification

also cannot be discerned simply by measuring changes in concentration of these compounds over time. Isotope enrichment can overcome some of the problems by measuring the $\delta^{15}N$ of specific N-intermediates and end products of denitrification, CND, and DNRA based on the enrichment on the initial NO_3^-.

In the isotope pairing technique used to examine denitrification, $^{15}NO_3^-$ is added to the overlying water of a sediment column and the generation of $^{15}N_2$ molecules with mass 28, 29, and 30 is followed over time as is the $\delta^{15}NO_3^-$ of remaining NO_3^- after reduction of the NO_3^- to N_2 by denitrifying bacteria. Results from several isotope pairing studies improved estimates of turnover rate for NO_3^- and generation of N_2 in denitrification. Eyre et al. (2002) compared the N_2:Ar method with the isotope pairing technique. The N_2:Ar method measures net N_2 fluxes from denitrification minus nitrification. The isotope pairing technique measures gross N_2 production by measuring all generated $^{29}N_2$ and $^{30}N_2$ using a gas chromatograph coupled to a triple-collector isotopic ratio mass spectrometer, and the $\delta^{15}NO_3^-$ of remaining NO_3^-. Eyre et al. (2002) concluded that the N_2 production rate measured using the N_2:Ar method was slightly lower than that measured using the isotope pairing technique.

Ignoring the formation of N_2O was identified as a shortcoming of the isotope pairing technique (Master, Shavit, and Shaviv 2005) that could lead to underestimates of denitrification. Although the NO_3^- is known to be converted completely to N_2, it has been shown more recently that N_2O also can accumulate, particularly in N-contaminated environments. Master, Shavit, and Shaviv (2005) examined assumptions made using the isotope pairing technique and suggested improvements to estimate denitrification with a modified isotope pairing technique by measuring the production and isotopic composition of N_2O in addition to the N_2 and NO_3^-. In a separate experimental approach that examined the fate of N_2O in sediments, Clough et al. (2006) examined the production and loss of N_2O directly by adding $^{15}N_2O$ to soil columns and simultaneously measuring the production and consumption of the N_2O.

N species may be cycled quickly. Morier et al. (2008) labeled organic mineral soil horizons with $^{15}NO_3^-$ or $^{15}NH_4^+$ and measured immobilization of the ^{15}N as quickly as one hour after ^{15}N addition in all samples. Some of the ^{15}N was incorporated into an amide structure indicating rapid biological incorporation. Because of the potential rapid reaction rates of certain transformations, therefore, the timing of the experiments becomes a critical element in the experimental design and the interpretation of the isotopic signature of the associated N-species that are measured.

10.4   CARBON

Similar to studies carried out using individual chemicals labeled with ^{15}N, individual chemicals are available for purchase that are enriched in ^{13}C and can be used to investigate biological processes. For example ^{13}C-amino acids added to a soil sample will be incorporated by bacteria and can be detected when the bacterial proteins are separated from the soil matrix. From labeled contaminants, ^{13}C can be incorporated into cellular biomass or respired and measured as ^{13}C-biomass or ^{13}C-CO_2. By knowing the initial ^{13}C added and the distribution of the ^{13}C into different biological samples, a mass balance of the added label can be carried out to determine the relative proportions of the label associated with different physical (i.e., sorption) and biological (i.e., uptake) processes. These processes including incorporation into the biological compartments of cells and specific cellular components, trophic transfer of biological materials such as fatty acids (Ruess et al. 2005), and incorporation into respired end products such as CO_2 and methane, and can be investigated in water, soil, and sediment.

10.4.1 [13]CARBON CONTAMINANT STUDIES

The studies and techniques described in the previous sections have been associated with the use of isotopes in naturally occurring compounds to examine microbial and plant processes. In addition, isotope-labeled contaminants can be used to elucidate the ability of microorganisms to biodegrade anthropogenic compounds. As with other techniques, several specific isotopes can be used. In this section only examples of ^{13}C-based compounds will be given.

Several researchers have used ^{13}C-isotopically enriched organic contaminants to examine their mineralization by soil microbes. The simplest approach, similar to that used with ^{14}C-radiolabeled tracer is to add ^{13}C-labeled compounds and monitor the production of ^{13}C-CO_2 over time. This approach was used by Morasch, Höhener, and Hunkeler (2007) who added [13]C_6-benzene, [13]C_6-napththalene, and [13]C_6-acenaphthene to sediment microcosms and monitored ^{13}C-CO_2 over time as a semiquantitative screening method to investigate intrinsic biodegradation under in situ-like conditions.

Since any C-containing compound can be followed using the ^{13}C tracer, in addition to ^{13}C-CO_2 generated over time, subsequent ^{13}C incorporation into soil organic matter can identify the relation between microbial degradation and organic matter. Richnow et al. (1998) used ^{13}C-labeled anthracene to measure anthracene's mineralization rate through the formation of ^{13}C-CO_2, and the amount of ^{13}C-anthracene fragments that were incorporated into macromolecular organic matter, particularly into bound sediment organic fractions during biodegradation. In order to measure both of these, following incubation with the ^{13}C-labeled anthracene, the soils were extracted with organic solvents. An alkaline hydrolysis was carried out on the extracts to cleave the ester-bound nonextractable residues. The extracted and bound residues were analyzed for concentration and isotopic C composition. In the extractable fraction ^{13}C-carboxylic acid were present, indicative of mineralization. The ^{13}C-labeled 2-hydroxy-3-naphthoic acids and ^{13}C-anthraquinone typical metabolites of anthracene also were present, as were other ^{13}C metabolites. Using this stepwise extraction process, Richnow et al. (1998) were able to measure the release of ^{13}C into humic material associated with the mineralization of ^{13}C-anthracene and ^{13}C metabolites. Some of these were not identifiable by isotope ratio GC/MS in which the GC is interfaced with a combustion device with a water removal assembly and coupled to an MS.

Guthrie et al. (1999) examined pyrene sequestration by the humin fraction of sediment using ^{13}C-pyrene. After incubation of surface sediment for two months with ^{13}C-pyrene, humic acids and humin were isolated and analyzed by flash pyrolysis GC/MS, which destroys the three-dimensional structure of the macromolecule.

Solution and solid-state ^{13}C MNR using cross-polarization magic angle spinning was used to evaluate the covalent and noncovalent interactions between the pyrene and the sedimentary organic matter. These combined techniques allowed direct identification of the ^{13}C-labeled pyrene and structural changes to the pyrene retained in insoluble sediment organic matrixes. Nuclear magnetic resonance (NMR) spectra of humin fractions indicated a loss of alcohols and carbohydrates and conservation of paraffinic and aromatic groups over the two-month incubation. Guthrie et al. (1999) concluded that the bound pyrene residues or those that were nonextractable were not necessarily covalently bound intermediate products of pyrene degradation. Instead, they suggested that these nonextractable fractions represented pyrene that was adsorbed or encapsulated in sediment humin. This study and a subsequent ^{13}C-pyrene study by Greenwood, Guthrie, and Hatcher (2000) using laser micropyrolysis GC/MS provided examples of isotopic labeling used in conjunction with a sophisticated analytical tool to garner additional information than that available using the nonlabeled contaminant. The ^{13}C-label and the pyrolysis GC/MS technique allowed characterization of the compositional structure of the sorbent matrix, which could not be done using traditional chemical extractions and GC/MS compound identification in cases of nonextractable contaminants.

10.5 THE USE OF STABLE ISOTOPES TO EXAMINE CELLULAR COMPONENTS

More recent innovations have expanded the utility of tracer experiments to include stable carbon tracers to examine the relation between the function of a microbial community and the phylogeny of the organisms accountable for that function. Isotopically enriched (primarily ^{13}C and ^{15}N labeled) compounds have been used for microbial identification as an alternative to traditional plating techniques. Current culturing techniques are lengthy, nonspecific, and must rely on the use of the appropriate growth medium that may select for only certain types and a small percentage of organisms in the environment. Continued development of alternatives to traditional culturing techniques, such as stable isotope probing (SIP; see Section 11.6) that uses molecular biomarkers, will enhance our knowledge base about microorganisms.

The use of ^{15}N has not been as developed as that for ^{13}C in SIP that uses molecular techniques, and most of the labeling has been used for protein solution state (Goto and Kay 2000) and solid state MNR spectroscopy. Isotopic labeling of amino acids can be used to solve the structures of small insoluble and membrane-bound proteins that cannot be crystallized (Lian and Middleton 2001). Isotopic labels are uniformly incorporated into the protein within a specific amino acid type or incorporated at specific atomic sites. Several methods exist to isotopically label proteins, which include full labeling, backbone-only labeling, methyl group labeling, amino-acid type selective isotope labeling, segmental labeling, and reverse labeling (Lian and Middleton 2001). Strauss et al. (2003) described a method for amino acid-type selective labeling of proteins for essential amino acids phenylalanine, tyrosine, and valine with ^{15}N-label incorporation rates of greater than 90%. Improvements to the synthesis of isotopically labeled molecules have increased availability of these materials to researchers and expanded the scope of research that can be carried out in the field.

10.5.1   Dual Isotopes and Dual Isotope Labeling

The use of dual isotope use and dual isotope labeling offers two approaches to examine microbial processes. Individual isotopes can be used in parallel experiments in which each isotope provides information of the process associated with that isotope. ^{15}N and ^{3}H have been used in parallel experiments to measure the products of nitrogen cycling and the biological production of N-containing molecules (^{15}N) and examine microbial activity (^{3}H). For example, Bengtsson and Annadotter (1989) used aerobic and anaerobic continuous flow microcosms to which $^{15}NO_3^-$ was added. They subsequently measured $^{15}N_2$, $^{15}N_2O$, $^{15}NH_4^+$ and ^{15}N-labeled protein amino acids. They also used [methyl-^{3}H]thymidine incorporation into bacterial cells as a measure of microbial activity in the sediment columns. Based on the transformation of the NO_3^- and the fate of the ^{15}N label, the authors concluded that 100% of the added NO_3^- was reduced. In the aerobic columns, they estimated that 80–90% of the NO_3^- was reduced by aerobic denitrification to N_2. In the anaerobic microcosm, they estimated that 35% of the NO_3^- was reduced to N_2 by denitrification, and 50% of the NO_3^- was reduced to NH_4^+. Only a small fraction of the ^{15}N was assimilated into bacterial cells and appeared only in the aerobic microcosm cells and only in the amino acid alanine. Based on results of parallel experiments using [methyl-^{3}H] thymidine, it was estimated that the [^{3}H]thymidine incorporation rate of the anaerobic microcosms was four times higher than that of the aerobic microcosms. In this example, the information garnered from the ^{15}N and ^{3}H isotopes was collected and used independently.

Another example of dual isotope use (^{15}N and ^{13}C) in which the data are more integrated than in the previous example is the investigation of the microbial contribution to soil organic matter turnover and sequestration, and subsequent interactions with plants. Inorganic nitrogen in the form of NH_4^+ and NO_3^- is taken up by microbes and immobilized into microbial tissues and more specifically into amino sugars. The ratio of organism-specific amino sugars can be used to differentiate the relative contributions of microorganisms, fungi, or bacteria for example, to the soil organic matter turnover and accumulation. He, Xie, and Zhang (2006) combined ^{15}N and ^{13}C in this unique way to study the transformation process from NH_4^+-N to amino sugar N, as well as from glucose C to amino sugar C skeleton. Either ^{15}N or ^{13}C was added to surface soils and the individual amino sugars were measured to identify the incorporation of the ^{15}N or the ^{13}C. The labels initially added to soil samples included ($^{15}NH_4)_2SO_4$ plus unlabeled glucose, and unlabeled $(NH_4)_2SO_4$ plus uniformly labeled-^{13}C-glucose. Amino sugars analyzed for ^{15}N included D-glucosamine (GluN), D-galatosamine (GalN), D-muramic acid (MurN), and D-mannosamine (ManN). He, Xie, and Zhang (2006) found that ^{15}N derived from ^{15}N-NH_4^+ and ^{13}C derived from ^{13}C-glucose were incorporated into three amino sugars, glucosamine, galatosamine, and muramic acid, during microbial synthesis of amino sugars in soils. In addition, the atom percentage excess (APE) was measured in MurN within one day of incubation and APE was measured within three days of incubation for GalN and GluN, demonstrating that microbial synthesis of the amino acids in soils can be rapid and varies for specific amino sugars synthesized. No enrichment was measured in mannosamine during microbial synthesis in either ^{15}N or ^{13}C demonstrating selectivity of microbial amino acid synthesis. The method used by He, Xie, and Zhang (2006) was highly sensitive and reproducible with coefficients of variation less than 7%. This more sensitive approach, which combined stable isotopes and GC/MS to examine microbial synthesis of amino sugars in soils, is a significant improvement over measuring concentrations of amino sugars in soils using GC/MS without using ^{15}N or ^{13}C tracers.

The studies by Bengtsson and Annadotter (1989) and He, Xie, and Zhang (2006) were carried out using individual incubations of either ^{15}N- or ^{3}H-, or ^{15}N- or ^{13}C-enriched substrates, respectively. Dual isotope labeling involves one compound that has been labeled with two isotopes, for example an amino acid that has been labeled with both ^{15}N and ^{13}C. Jones et al. (2005) provides a review of the literature using dual labeling of ^{15}N and ^{13}C amino acids to examine dissolved organic nitrogen in soils and uptake by plants in terrestrial systems. The dual label technique allows distinction between N uptake directly by plants, and uptake by plants subsequent to microbial mineralization. If the plants take up the dual-labeled amino acid directly, then the ^{15}N and ^{13}C will both be present in the plant (Figure 10.3). If microbes mineralize the amino acid prior to plant assimilation, only the ^{15}N will be present in the plants (Figure 10.4; Jones et al. 2005).

In these studies, the dual-labeled amino acid of choice is often ^{15}N-^{13}C-glycine due to commercial availability although Hodge, Robinson, and Fitter (2000) suggest it is a relatively poor substrate for microbial growth. Certain difficulties arise in these experiments that must be addressed, such as the ability to determine initial concentrations of each N-species so that isotope pool dilution (see Section 11.2) can be quantified. Hodge et al (2000) found no ^{13}C enrichment in plants when grown in

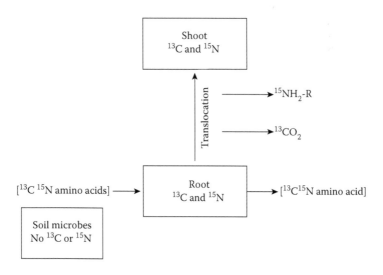

FIGURE 10.3   Direct ^{15}N-DON and ^{13}C uptake with no microbial utilization is verified by the presence of both ^{15}N-DON and ^{13}C in the plant roots and shoots. (Modified from Jones, D. L., J. R. Healey, V. B. Willett, J. F. Farrar, and A. Hodge, *Soil Biology & Biochemistry*, 37, 413–23, 2005.)

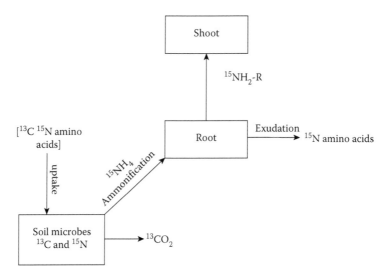

FIGURE 10.4 Indirect ^{15}N-DON uptake after microbial utilization of dual ^{15}N-and ^{13}C-labeled amino acids is verified by the presence of only ^{15}N in the plant roots and shoots. (Modified from Jones, D. L., J. R. Healey, V. B. Willett, J. F. Farrar, and A. Hodge, *Soil Biology & Biochemistry*, 37, 413–23, 2005.)

microcosms with five types of dual [15]N-^{13}C-labeled patches with varying C:N ratios. The plant material was not enriched in ^{13}C, but capture of the ^{15}N was measured in the roots suggesting that the plants took up the N after mineralization by the microbial community. Streeter, Bol, and Bardgett (2000) used in situ dual-labeled glycine-2-[15]N-^{13}C to determine whether temperate grassland grasses were able to take up the amino acid N directly. They found that both ^{15}N and ^{13}C were detected in shoot material and thus, concluded that plants were able to take up organic nitrogen directly from soil and that uptake was greatest after 24 hours. However they were not able to distinguish the proportion of added glycine-N that was taken up as intact amino acid or as mineralized N, nor the relative amount of the ^{15}N and ^{13}C that was transformed by microorganisms in the soil.

Despite differences in the conclusions about the ability of plants to take up intact amino acids in these two studies, the use of dual-labeled substrate technique effectively differentiated between intact plant uptake of organic nitrogen and uptake after microbial mineralization. The commercial synthesis of additional dual-labeled compounds, which then becomes available for purchase at reasonable cost, will allow more experiments of this type to be conducted, in which elements can be followed through chemical and biological processes with high analytical sensitivity.

10.6   STABLE ISOTOPE PROBING (SIP)

SIP is a general term used to describe a method wherein environmental samples are exposed to stable isotope enriched substrates. Over time, microorganisms that are capable of utilizing the labeled compounds as growth substrates will incorporate the stable isotope into their biomass and specific cellular components. Thus, the incorporation of stable isotopes into biomass can be used to determine the specific components of a microbial community involved in a particular function, such as the biodegradation of xenobiotic compounds or biogeochemical cycling. To date, three main cellular components have been used as molecular markers to examine the incorporation of stable isotopes into biomass. These markers are polar lipid derived fatty acids, or phospholipids fatty acids (PLFAs), deoxyribonucleic acids (DNA), and ribonucleic acids (RNA). While the general method of SIP remains similar for all three biomarkers, the types of molecular analyses and interpretation of results differ. A set of review articles published during 2006 in *Current Opinions in Biotechnology* has examined several major applications of SIP in microbial ecology. A significant portion of this chapter is derived from this group of articles.

10.6.1   PHOSPHOLIPID FATTY ACIDS (PLFAs)–STABLE ISOTOPE PROBING (SIP)

SIP was initially developed using PLFAs as a biological marker. In a seminal study, researchers exposed sediment microbial communities to $^{13}CH_4$ and subsequently extracted and analyzed the labeled PLFA using isotope ratio mass spectrometry (Boschker et al. 1998). Due to the unique pattern of PLFA among bacterial phylogenetic groups, these researchers were able to identify bacterial groups responsible for using labeled methane as a carbon source. This technique was subsequently adapted to examine microbial communities involved in the degradation of organic contaminants. For instance, Hanson et al. (1999) pulsed yolo silt loam with ^{13}C-toluene and then subsequently examined the PLFA composition of the microbial community. This approach was able to link the degradation of toluene in this system to specific genera within the *Actinomycetales*. Other studies have used PLFA-SIP to identify bacteria involved in the degradation of toluene within a mixed waste scenario and to understand styrene degradation by industrial biofilters (Alexandrino, Knief, and Lipski 2001; Pelz et al. 2001). While PLFA-SIP has been successfully applied to many systems to examine how microbial communities function, the use of PLFA as a molecular marker does not provide detailed resolution at the genus or species level. More extensive analysis of stable isotope-labeled microbial communities requires more sensitive biological markers, such as nucleic acids.

10.6.2   DEOXYRIBONUCLEIC ACIDS (DNA)–STABLE ISOTOPE PROBING (SIP)

The basic premise behind DNA-SIP is that nucleic acids that have been labeled with stable isotopes have a higher buoyancy density so that they can be separated from nonlabeled DNA, thus, allowing for the identification of microorganisms capable of utilizing labeled substrates. The ability to alter the buoyant density of DNA by the incorporation of stable isotopes was first demonstrated by examining the uptake of ^{15}N or ^{14}N labeled nitrogen sources by *Escherichia coli* (Meselson and Stahl 1958). It was later discovered that ^{13}C, ^2H, ^{15}N, and ^{18}O could also be incorporated into nucleic acids and heavy DNA subsequently separated from light DNA through cesium chloride density-gradient centrifugation (Radajewski et al. 2000; Radajewski, McDonald, and Murrell 2003; Buckley et al. 2007; Schwartz

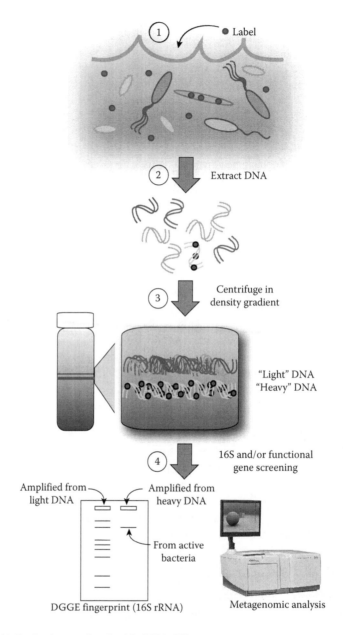

FIGURE 10.5 Basic steps involved in DNA-SIP.

2007). Since nucleotides contain far fewer nitrogen atoms than carbon or hydrogen, this finding significantly increased the overall sensitivity of the DNA-SIP process and most SIP experiments have used ^{13}C-enriched compounds. An overview of the basic steps involved in DNA-SIP is shown in Figure 10.5. First, a labeled substrate (e.g., $^{13}CH_3OH$, $^{13}CH_4$) is added to environmental samples, such as water, soil, or sediment. During an incubation period, active bacteria that are capable of utilizing the labeled compound as a growth substrate will incorporate the stable isotope into biomass, including nucleic acids.

The second step involves the extraction of nucleic acids from the microbial community using DNA extraction protocols that have been optimized for a given type of environmental sample. The DNA extract will contain labeled DNA from microorganisms actively utilizing the labeled substrate as well as nonlabeled DNA from other microbial community members. Following DNA extraction, the pooled nucleic acids are subjected to density-gradient centrifugation. Due to differences in the buoyant density of labeled and nonlabeled DNA, the pool of DNA will be separated into a nonlabeled light fraction and a heavy labeled fraction. The heavy DNA fraction theoretically contains the collective genomes of microorganisms capable of utilizing the initial labeled substrate as a growth substrate. This DNA fraction can then be collected and analyzed using a host of molecular-based methods to determine the phylogenetic distribution of substrate-utilizing microorganism or the functional genes involved in the breakdown to the substrate.

The bacterial 16S rRNA gene has been the most commonly used molecular marker combined with DNA-SIP for the profiling of complex microbial communities involved in the breakdown of labeled substrates. This approach involves the amplification of 16S rRNA genes from pooled DNA samples with subsequent amplicon analysis using microbial fingerprint methods such as denaturing gradient gel electrophoresis (DGGE), terminal restriction fragment length polymorphism (T-RFLP), and clone library construction. Similar approaches can also be used to identify the functional genes involved in the breakdown of the labeled substrates. For instance, DNA-SIP has been used to examine the genes that code for key enzymes of methanotrophs (pmoA, mmoX), methylotrophs (mxaF), methyl halide degraders (cmuA), and denitrifiers (nosZ), particulate monooxgenases (pmoA), soluble monooxygenases (mmoX), methanol dehydrogenases (mxaF), methyl halide transferases (cmuA), and nitrous oxide reductases (nosZ) (Friedrich 2006). Furthermore, SIP is now being coupled with metagenomic approaches to understand the complete genetic capacity of complex microbial communities. Because the metagenomic approach does not depend on prior knowledge of gene sequences, it has the potential to allow for the discovery of novel phylotypes and gene sequences that are often overlooked using traditional molecular approaches.

10.6.3 RIBONUCLEIC ACIDS (RNA)–STABLE ISOTOPE PROBING (SIP)

While DNA-SIP has often been used to discover novel bacteria and genetic pathways involved in the degradation of many labeled compounds, this process requires long incubation times to allow for incorporation of the labeled substrate into the DNA. An alternative approach is to use RNA as the molecular marker (Manefield et al. 2002a; Manefield et al. 2002b). RNA is synthesized at a much faster rate than DNA, providing rapid incorporation of labeled substrates into the RNA. The labeling of RNA not only allows shorter incubation times, but also provides a marker for determining qualitative and quantitative changes in gene expression occurring almost immediately following the addition of the labeled substrate.

Except for a few caveats, the steps involved in the assimilation and incorporation of stable isotope labeled substrates into RNA are similar to those described for DNA-SIP. One of the first caveats is that RNA is significantly less stable than DNA and its extraction and purification requires rapid and rigorous procedures to prevent RNA degradation. This point is not trivial in that even partial degradation of RNA can significantly bias the results of subsequent molecular analysis. A second caveat is that, unlike labeled DNA, differences in the buoyant density of labeled RNA prevent its separation using cesium chloride gradients. Instead, labeled RNA is separated using cesium trifluoracetate gradient centrifugation. However, due to varying transcript sizes, labeled RNA cannot simply be separated into heavy and light fractions as can labeled DNA. The gradient distribution of labeled RNA requires that molecular analysis be performed on more fractions throughout the gradient as compared to DNA-SIP. Once fractions are recovered, labeled RNA is generally converted to its stable complementary DNA (cDNA) using a reverse transcriptase protocol and subsequently analyzed using molecular methods similar to those described for DNA-SIP.

While RNA-SIP is a challenging procedure, it has the potential to provide great insight into the genetic pathways involved in bacterial-mediated contaminant degradation and has been applied in various ways. For instance, RNA-SIP has been used in activated sludge pools and in soil ecosystems to determine microbial community members responsible for phenol and pentachlorophenol degradation, respectively (Manefield et al. 2002a; Mahmood, Paton, and Prosser 2005; Manefield et al. 2005). Bacteria involved in the degradation of the methyl halides MeBr and MeCl in soil microcosms, including *Burkholderia*, *Rhodobacter*, *Lysobacter*, and *Nocardiodes*, were identified by analysis of 16S rRNA gene sequences amplified from the ^{13}C-DNA fractions (Miller et al. 2004). Also, by ^{13}C-CO_2 labeling photosynthetically active plants, researchers have used RNA-SIP to examine the utilization of labeled photosynthate carbon by rhizosphere bacteria (Griffiths et al. 2004; Lu et al. 2005).

As SIP research becomes more commonplace, limitations of the DNA and RNA technique and experiments should be evaluated. First, there is the possibility of preferential uptake and utilization of labeled versus unlabeled substrates. Furthermore, many SIP experiments are carried out in microcosms using high concentrations of labeled substrates. While these studies may provide suggestive evidence of bacterial activity, microcosms do not represent natural conditions and the high level of labeled substrate required to visualize labeled nucleic acids may cause changes in microbial community structure that would otherwise not occur under in situ conditions. Finally, it is important to understand that microbial cells must be actively replicating for DNA to be labeled with stable isotopes. Thus, results generated from microbial community fingerprinting methods using labeled DNA must be carefully interpreted. Since transcription does not require actively replicating microbial cells, RNA could be used as an alternative to DNA as an informative biomarker for microbial communities.

10.7   CONCLUSION

Cellular components are made up primarily of C, N, O, H, and chemicals upon which microorganisms act also are made up primarily of these elements. These C-, N-, O-, and H-containing compounds are used by bacteria to derive energy and are biodegraded or incorporated into cellular biomass. By careful selection of the isotope and the chemical of interest, enrichment with stable and radioactive isotopes of C, N, O, and H is a useful tool for examining these processes. Specific radioactive and stable isotope-enriched chemicals can be used to investigate the cycling of N in the natural environment, the biodegradation of N and C-based anthropogenic chemicals, and the incorporation of N and C into cellular components. The basic assumption is that the labeled compounds behave similarly to the natural nonlabeled compounds, which is generally accepted except in special cases.

The use of SIP to determine the functional diversity of microbial communities using culture-independent methods is a powerful approach that has advanced the field of microbial ecology. However, when planning SIP experiments, a number of caveats must be considered to minimize bias in data interpretation. Nonetheless, SIP techniques provide insight into the functional diversity of the greater than 99% of viable but nonculturable microorganisms. Lastly, and most excitingly, combining SIP with the rapidly advancing field of metagenomics and next-generation DNA sequencing will provide researchers with the tools required to discover novel genes involved in many crucial biogeochemical cycles as well as novel biotransformation pathways.

ACKNOWLEDGMENTS

We thank Melissa Engle and Erin Biers for assistance with the preparation of the figures. This work was supported by an ERIG Award from the University of South Carolina.

REFERENCES

Alexandrino, M., C. Knief, and A. Lipski. 2001. Stable-isotope-based labelling of styrene-degrading microorganisms in biofilters. *Applied Environmental Microbiology* 67: 4796–804.

Atlas, R. M., and R. Bartha. 1997. *Microbial ecology: Fundamentals and applications,* 4th ed. Redwood City, CA: Benjamin/Cummings Publishing, Inc.

Bedard-Haughn, A., J. W. van Groenigen, and C. van Kessel. 2003. Tracing ^{15}N through landscapes: Potential uses and precautions. *Journal of Hydrology* 272:175–90.

Bengtsson, G., and H. Annadotter. 1989. Nitrate reduction in a groundwater microcosm determined by ^{15}N gas chromatography-mass spectrometry. *Applied Environmental Microbiology* 55:2861–70.

Boschker, H. T. S., S. C. Nold, P. Wellsbury, D. Bos, W. de Graaf, R. Pel, R. J. Parkes, and T. E. Cappenberg. 1998. Direct linking of microbial populations to specific biogeochemical processes by ^{13}C-labelling of biomarkers. *Nature* 392:801–4.

Boyer, J. N., D. W. Stanley, and R. R. Christian. 1994. Dynamics of NH_4^+ and NO_3^- uptake in the water column of the Neuse River Estuary, North Carolina. *Estuaries* 17:361–71.

Buckley, D. H., V. Huangyutitham, S-F. Hsu, and T. A. Nelson. 2007. Stable isotope probing with $^{15}N_2$ reveals novel noncultivated diazotrophs in soil. *Applied Environmental Microbiology* 73:3196–204.

Capone, D. G., S. G. Horrigan, S. E. Dunham, and J. Fowler. 1990. Direct determination of nitrification in marine waters by using the short-lived radioisotope of nitrogen, ^{13}N. *Applied Environmental Microbiology* 56:1182–84.

Cliff, J. B., P. J. Bottomley, R. Haggerty, and D. D. Myrold. 2002. Modeling the effects of diffusion limitations on nitrogen-15 isotope dilution experiments with soil aggregates. *Soil Science Society of America Journal* 66:1868–77.

Clough, T. J., F. M. Kelliher, Y. P. Wang, and R. R. Sherlock. 2006. Diffusion of ^{15}N-labelled N_2O into soil columns: A promising method to examine the fate of N_2O in subsoils. *Soil Biology & Biochemistry* 38:1462–68.

Collos, Y. 1987. Calculations of ^{15}N uptake rates by phytoplankton assimilating one or several forms of nitrogen. *Applied Radiation and Isotopes* 38:275–82.

Dugdale, R. C., and J. J. Goering. 1967. Uptake of new and regenerated forms of nitrogen in primary productivity. *Limnology and Oceanography* 12:196–206.

Dugdale, R. C., and F. P. Wilkerson. 1986. The use of ^{15}N to measure nitrogen uptake in eutrophic oceans: Experimental conditions. *Limnology and Oceanography* 31:673–89.

Dumont, M. G., and J. C. Murrell. 2005. Stable isotope probing—linking microbial identity to function. *Nature Reviews Microbiology* 3:499–504.

Eisenbud, M., and T. Gesell. 1997. *Environmental radioactivity from natural, industrial, and military sources,* 4th ed. San Diego, CA: Academic Press.

Eyre, B. D., S. Rysgaard, R. Dalsgaard, and P. B. Christensen. 2002. Comparison of isotope pairing and N_2:Ar methods for measuring sediment denitrification: Assumptions, modifications and implications. *Estuaries* 25:1077–87.

Fiedler, R., and G. Proksch. 1975. The determination of ^{15}N by emission and mass spectroscopy in biochemical analyses: A review. *Analytical Chemistry Acta* 78:1–62.

Fischer, A., J. Bauer, R. U. Meckenstock, W. Stichler, C. Griebler, P. Maloszewski, M. Kastner, and H. H. Richnow. 2006. A multitracer test proving the reliability of Rayleigh equation-based approach for assessing biodegradation in a BTEX contaminated aquifer. *Environmental Science & Technology* 40:4245–52.

Friedrich, M. W. 2006. Stable-isotope probing of DNA: Insights into the function of uncultivated microorganisms from isotopically labeled metagenomes. *Current Opinion in Biotechnology* 17:59–66.

Gersberg, R., K. Krohn, N. Peek, and C. R. Goldman. 1976. Denitrification studies with ^{13}N-labelled nitrate. *Science* 192:1229–31.

Goto, N. K., and L. E. Kay. 2000. New developments in isotope labeling strategies for protein solution NMR spectroscopy. *Current Opinion in Structural Biology* 10:585–92.

Greenwood, P. F., E. A. Guthrie, and P. G. Hatcher. 2000. The in situ analytical pyrolysis of two different organic components of a synthetic environmental matrix doped with [4,9-^{13}C]pyrene. *Organic Geochemistry* 31:635–43.

Griffiths, R. I., M. Manefield, N. Ostle, N. McNamara, A. G. O'Donnell, M. J. Bailey, and A. S. Whiteley. 2004. $^{13}CO_2$ pulse labeling of plants in tandem with stable isotope probing: Methodological considerations for examining microbial function in the rhizosphere. *Journal of Microbiological Methods* 58:119–29.

Guthrie, E. A., J. M. Bortiatynski, J. D. H. Van Heemst, J. E. Richman, K. S. Hardy, E. M. Kovach, and P. G. Hatcher. 1999. Determination of [^{13}C]pyrene sequestration in sediment microcosms using flash pyrolysis-GC-MS and ^{13}C NMR. *Environmental Science & Technology* 33:119–25.

Hadwen, W. L., and S. E. Bunn. 2005. Food web responses to low-level nutrient and ^{15}N-tracer additions in the littoral zone of an oligotrophic dune lake. *Limnology and Oceanography* 50:1096–105.

Hanson, J. R., J. L. Macalady, D. Harris, and K. M. Scow. 1999. Linking toluene degradation with specific microbial populations in soil. *Applied Environmental Microbiology* 65:5403–8.

Hardy, R. W. F., R. P. Holsten, E. K. Jackson, and R. C. Burns. 1968. The acetylene ethylene assay for N_2-fixation: Laboratory and field evaluation. *Plant Physiology* 43: 1185–207.

Harrington, R. R., S. R. Poulson, J. I. Drever, P. J. S. Colberg, and E. F. Kelly. 1999. Carbon isotope systematics of monoaromatic hydrocarbons: Vaporization and adsorption experiments. *Organic Geochemistry* 30:765–75.

He, H., H. Xie, and X. Zhang. 2006. A novel GC/MS technique to assess ^{15}N and ^{13}C incorporation into soil amino sugars. *Soil Biology & Biochemistry* 38:1083–91.

Herrmann, A., E. Witter, and T. Kätterer. 2005. A method to assess whether "preferential use" occurs after ^{15}N ammonium addition; implication for the ^{15}N isotope dilution technique. *Soil Biology & Biochemistry* 37:183–86.

Hodge, A., D. Robinson, and A. Fitter. 2000. Are microorganisms more effective than plants at competing for nitrogen? *Trends in Plant Science* 5:304–8.

Hodge, A., J. Stewart, D. Robinson, B. S. Griffiths, and A. H. Fitter. 2000. Competition between roots and soil microorganisms for nutrients from nitrogen-rich patches of varying complexity. *Journal of Ecology* 88:150–64.

Högberg, P. 1997. Tansley review No. 95, ^{15}N natural abundance in soil-plant systems. *New Phytologist* 137:179–203.

Jansson, S-L. 1958. Tracer studies on nitrogen transformations in soil with special attention to mineralization-immobilization relationships. *Annals of the Royal Agricultural College of Sweden* 24:101–36.

Jones, D. L., J. R. Healey, V. B. Willett, J. F. Farrar, and A. Hodge. 2005. Dissolved organic nitrogen uptake by plants: An important N uptake pathway. *Soil Biology & Biochemistry* 37:413–23.

Lajtha, K., and R. Michener. (eds.). 1994. *Stable isotopes in ecology and environmental science (ecological methods and concepts).* Cambridge, MA: Blackwell Scientific Publications.

Laws, E. A. 1985. Analytic models of NH_4^+ uptake and regeneration experiments. *Limnology and Oceanography* 30:1340–50.

Lian, L. Y., and D. A. Middleton. 2001. Labelling approaches for protein structural studies by solution-state and solid-state NMR. *Progress in Nuclear Magnetic Resonance Spectroscopy* 39:171–90.

Lu, Y., T. Lueders, M. W. Friedrich, and R. Conrad. 2005. Detecting active methanogenic populations on rice roots using stable isotope probing. *Environmental Microbiology* 7:326–36.

Mahmood, S., G. I. Paton, and J. I. Prosser. 2005. Cultivation-independent *in situ* molecular analysis of bacteria involved in degradation of pentachlorophenol in soil. *Environmental Microbiology* 7:1349–60.

Manefield, M., R. I. Griffiths, M. B. Leigh, R. Fisher, and A. S. Whiteley. 2005. Functional and compositional comparison of two activated sludge communities remediating coking effluent. *Applied Microbiology* 7:715–22.

Manefield M., A. S. Whiteley, R. I. Griffiths, and M. J. Bailey. 2002a. RNA stable isotope probing, a novel means of linking microbial community function to phylogeny. *Applied Environmental Microbiology* 68:5367–73.

Manefield, M., A. S. Whiteley, N. Ostle, P. Ineson, and M. J. Bailey. 2002b. Technical considerations for RNA-based stable isotope probing: An approach to associating microbial diversity with microbial community function. *Rapid Communications in Mass Spectrometry* 16:2179–83.

Master, V., U. Shavit, and A. Shaviv. 2005. Modified isotope pairing technique to study N transformations in polluted aquatic systems: Theory. *Environmental Science & Technology* 39:1749–56.

Meselson, M., and F. W. Stahl. 1958. The replication of DNA in *Escherichia coli*. *Proceedings of the National Academy of Sciences USA* 44:671–82.

Miller, L. G., K. L. Warner, S. M. Baesman, R. S. Oremland, I. R. McDonald, S. Radajewski, and J. C. Murrell. 2004. Degradation of methyl bromide and methyl chloride in soil microcosms: Use of stable C isotope fractionation and stable isotope probing to identify reactions and the responsible microorganisms. *Geochimica Cosmochemica Acta* 68:3271–83.

Morasch B., P. Höhener, and D. Hunkeler. 2007. Evidence for in situ degradation of mono- and polyaromatic hydrocarbons in alluvial sediments based on microcosm experiments with ^{13}C-labeled contaminants. *Environmental Pollution* 148:739–48.

Morasch, B., B. Schink, H. H. Richnow, and R. U. Meckenstock. 2001. Stable hydrogen and carbon isotope fractionation during microbial toluene degradation. Mechanistic and environmental aspects. *Applied Environmental Microbiology* 67:4842–49.

Morier, I., P. Schleppi, R. Siegwolf, H. Knicker, and C. Guenat. 2008. N-15 immobilization in forest soil: A sterilization experiment coupled with (15)CPMAS NMR spectroscopy. *European Journal of Soil Science* 59:467–75.

Neess, J. C., R. C. Dugdale, V. A. Dugdale, and J. Goering. 1962. Nitrogen metabolism in lakes. 1. Measurement of nitrogen fixation with [15]N. *Limnology and Oceanography* 7:163–69.

Pelz, O., A. Chatzinotas, A. Zarda-Hess, W.-R. Abraham, and J. Zeyer. 2001. Tracing toluene-assimilating sulfate-reducing bacteria using ^{13}C-incorporation in fatty acids and whole-cell hybridisation. *FEMS Microbiology Ecology* 38:123–31.

Poulson, S. R., J. I. Drever, and P. J. S. Colberg. 1997. Estimation of K_{oc} values for deuterated benzene, toluene, and ethylbenzene, and application to ground water contamination studies. *Chemosphere* 35, 2215–24.

Radajewski S., P. Ineson, N. R. Parekh, and J. C. Murrell. 2000. Stable-isotope probing as a tool in microbial ecology. *Nature* 403:646–49.

Radajewski, S., I. R. McDonald, and J. C. Murrell. 2003. Stable-isotope probing of nucleic acids: A window to the function of uncultured microorganisms. *Current Opinion in Biotechnology* 14:296–302.

Richnow, H. H., A. Eschenbach, B. Mahro, R. Seifert, P. Wehnmg, P. Albrecht, and W. Michaelis. 1998. The use of ^{13}C-labelled polycyclic aromatic hydrocarbons for the analysis of their transformation in soil. *Chemosphere* 36:2211–24.

Robinson, D. 2001. $\delta^{15}N$ as an integrator of the nitrogen cycle. *Trends in Ecology Evolution* 16:153–62.

Ruess, L., A. Tiunov, D. Haubert, H. H. Richnow, M. M. Haggblom, and S. Scheu. 2005. Carbon stable isotope fractionation and trophic transfer of fatty acids in fungal based soil food chains. *Soil Biology & Biochemistry* 37:945–53.

Rütting, T., and C. Müller. 2007. [15]N tracing models with a Monte Carlo optimization procedure provide new insights on gross N transformations in soils. *Soil Biology & Biochemistry* 39:2351–61.

Schwartz, E. 2007. Characterization of growing microorganisms in soil by stable isotope probing with $H_2^{18}O$. *Applied Environmental Microbiology* 73:2541–46.

Singleton, D. R., S. N. Powell, R. Sangaiah, A. Gold, L. M. Ball, and M. D. Aitken. 2005. Stable-isotope probing of bacteria capable of degrading salicylate, naphthalene, or phenanthrene in a bioreactor treating contaminated soil. *Applied Environmental Microbiology* 71:1202–09.

Speir, T. W., H. A. Kettles, and R. D. More. 1995a. Aerobic emissions of N_2O and N_2 from soil cores: Measurement procedures using ^{13}N-labelled NO_3^- and NH_4^+. *Soil Biology & Biochemistry* 27:1289–98.

Speir, T. W., H. A. Kettles, and R. D. More. 1995b. Aerobic emissions of N_2O and N_2 from soil cores: Factors influencing production from ^{13}N-labelled NO_3^- and NH_4^+. *Soil Biology & Biochemistry* 27:1299–1306.

Speir, T. W., J. A. Townsend, R. D. More, and L. F. Hill. 1999. Short-lived isotopic method to measure nitrous oxide emissions from a soil under four low-fertility management systems. *Soil Biology & Biochemistry* 31:1413–21.

Strauss, A., F. Bitsch, B. Cutting, G. Fendrich, P. Graff, J. Liebetanz, M. Zurini, and W. Jahnke. 2003. Amino–acid-type selective isotope labeling of proteins expressed in Baculovirus-infected insect cells useful for NMR studies. *Journal of Biomolecular NMR* 26:367–72.

Streeter, T. C., R. Bol, and R. D. Bardgett. 2000. Amino acids as a nitrogen source in temperate upland grasslands: The use of dual labeled (^{13}C, ^{15}N) glycine to test for direct uptake by dominant grasses. 2000. *Rapid Communications in Mass Spectrometry* 14:1351–55.

Thierrin, J., G. B. Davis, and C. Barber. 1995. A ground-water tracer test with deuterated compounds for monitoring in situ biodegradation and retardation of aromatic hydrocarbons. *Ground Water* 33:469–75.

Thierrin, J., G. B. Davis, C. Barber, B. M. Patterson, F. Pribac, T. R. Power, and M. Lambert. 1993. Natural degradation rates of BTEX compounds and naphthalene in a sulphate reducing groundwater environment. *Hydrological Sciences Journal* 38:309–22.

Watson, C. J., G. Travers, D. J. Kilpatrick, A. S. Laidlaw, and E. Riordan. 2000. Overestimation of gross N transformation rates in grassland soils due to non-uniform exploitation of applied and native pools. *Soil Biology & Biochemistry* 32:2019–30.

11 Combined Use of Radiocarbon and Stable Carbon Isotopes in Environmental and Degradation Studies

C. Marjorie Aelion

CONTENTS

11.1 INTRODUCTION

Carbon is the backbone of life and is ubiquitous in the land, sea, and atmosphere. Carbon is cycled between organic matter and CO_2 in the atmosphere. So carbon that is incorporated into plants and animals may have originated as CO_2 in the atmosphere and may return to CO_2 in the atmosphere. Stable isotope analysis of carbon, and radiocarbon analysis can be useful tools for understanding biotic and abiotic processes; where the carbon has been, what processes have acted upon it, and how it has changed chemically. In addition to process information, radiocarbon analysis can be used to garner temporal information, such as how long the carbon has been separated from active cycling in biological and chemical systems. This is based on the premise that the radiocarbon, unlike the stable carbon isotope, bears a component of time in itself. The ^{14}C data allows one to classify carbon sources based on knowledge that while living, entities actively exchange carbon with the atmosphere. This exchange ends when biological activity ends. The time scale over which information can be garnered is governed by the radioactive half-life, the time required for the radioactive isotope to decay to half of its initial activity due to radioactive decay. Radioactive half-lives vary tremendously among elements (Table 11.1).

Naturally occurring radiocarbon and stable carbon isotopic analysis can be complementary when applied to different molecules that are connected by biological or chemical processes. For example, in an uncontaminated or contaminated environment, microbial end products such as CO_2 can be assessed using naturally occurring radiocarbon, and individual contaminants or carbon sources from which CO_2 may have been generated can be assessed using stable carbon isotopes. Alternatively the

---

TABLE 11.1

Naturally Occurring and Artificial Radioisotopes, Their Half-Lives and Major Sources

Radioisotope	Half-Life	Major Emissions	Major Sources
^{212}Polonium	3.04×10^{-7} s	α	Natural, Th-series
^{222}Radon	3.825 d	$\alpha, < 1\% \gamma$	Natural, U-series
^{131}Iodine	8.04 d	β, γ	Fission product, Nuclear weapons
^{32}Phosphorus	14.3 d	β	Natural, U-series
^{210}Polonium	138.4 d	$\alpha, < 1\% \gamma$	Natural, U-series
^{60}Cobalt	5.27 y	β, γ	Nuclear weapons, Reactors
^{3}Hydrogen	12.33 y	β	Natural, Cosmic, Fission product
^{90}Strontium	28.8 y	β	Nuclear weapons
^{137}Cesium	30.17 y	β, γ	Fission product
^{241}Americium	432.2 y	α, γ	Decay of ^{241}Pu
^{226}Radium	1,622 y	α, γ	Natural, U-series
^{14}Carbon	5,730 y	β	Natural, Cosmic, Nuclear weapons
^{239}Plutonium	2.4×10^4 y	α	Fission product, Nuclear weapons
^{234}Uranium	2.5×10^5 y	$\alpha, < 1\% \gamma$	Natural, U-series
^{40}Potassium	1.28×10^9 y	β, γ	Natural, Primordial, Nonseries
^{238}Uranium	1.25×10^{10} y	$\alpha, < 1\% \gamma$	Natural, U-series

Source: Adapted from Aelion, C. M, *Natural remediation of environmental contaminants: Its role in ecological risk assessment and risk management,* SETAC Press, Society of Environmental Toxicology and Chemistry, Pensacola, FL, 2000.

naturally occurring radiocarbon and stable carbon isotopic signature can be measured on the same molecule such as on the CO_2 itself, or a specific contaminant. Combining the CO_2 stable carbon, radiocarbon, and contaminant concentration measurements allow pieces of the biodegradation puzzle to be more precisely deciphered because these measurements provide complementary information.

11.2   INTRODUCTION TO RADIOCARBON

Since the discovery of radioactivity just over 100 years ago by Marie Sklodowska Curie, Pierre Curie, and Henri Becquerel who jointly were awarded the Nobel Prize for physics in 1903, radioisotopes have been used in several scientific fields as disparate as medicine and warfare. The initial use of *cis*-radioisotopes for dating the Earth was theorized soon after its discovery. As early as 1905, scientists including Rutherford, carried out age determinations on uranium-bearing minerals based on the formation of helium. Boltwood used the idea that lead was the end product of the decay of uranium to make age determinations on uranite using U/Pb ratios (Faure 1986). Today the identity of naturally occurring radioactive elements and their decay series are well known, including environmentally important radioisotopes, such as radiocarbon (^{14}C) and tritium (^{3}H). The radiocarbon dating technique is commonly used and best known in geological and archeological dating, and has provided an accurate and precise method of estimating groundwater age and residence time (Hanshaw et al. 1967; Vogel 1970; Lloyd 1981) within systems of certain age limits. For analyses of great ages, corrections may be required and models may be used to estimate the levels of uncertainties.

Naturally occurring radiocarbon is produced in the transitional region of the atmosphere between the stratosphere and the troposphere. A nuclear reaction between nitrogen nuclei (^{14}N) and cosmic ray-produced neutrons generates ^{14}C and protons (Figure 11.1). The ^{14}C is then oxidized to $^{14}CO_2$ in the atmosphere and enters the carbon cycle through its availability to photosynthetic biota. The extra neutron in the

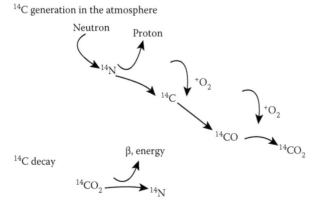

FIGURE 11.1   Generation of $^{14}CO_2$ in the atmosphere and its radioactive decay.

^{14}C nucleus makes the atom unstable, leading to β decay producing ^{14}N, a β^- particle and energy.

Radiocarbon techniques have been used in several different fields including geology, hydrology, atmospheric sciences, geochemistry, ecology, and microbiology. Because carbon is ubiquitous in the flora and fauna of the natural environment, radiocarbon was identified as a useful tool for geological and biological studies as early as the 1950s and 1960s. These early studies using naturally occurring radiocarbon focused on understanding unpolluted natural environments. Living plants and animals are in ^{14}C equilibrium with the atmosphere due to constant CO_2 exchange during photosynthesis and metabolism. But after death, exchange stops and ^{14}C incorporated into biological tissues decays with a half-life of 5730 ($\pm$40) years (Figure 11.2). Elapsing of 10 half-lives generally is considered to generate radioactive-depleted conditions for the element. Thus, fossil carbon petroleum or coal that has undergone multiple half-lives is distinguished from modern carbon in materials such as living biomass, or products of living biomass such as root exudates, by its depleted ^{14}C content (Figure 11.3). Materials may weather over geologic time and contribute abiotically to the naturally occurring radiocarbon signature of carbon-based molecules in certain environmental compartments, for instance carbonates in soils and aquatic sediments.

Naturally occurring radiocarbon may be a highly sensitive contaminant tracer in the environment if the contaminant of interest is carbon-based and derived from fossil sources. While carbon is ubiquitous in nature, the radioactive isotope, ^{14}C, exists in small amounts in nature and is only about $< 2 \times 10^{-10}\%$ of the total C in the environment (Table 11.2). The radiocarbon technique has the advantage that small amounts of carbon can provide information about the origin of the carbon used to generate biota, the CO_2 in air and vadose zone soils, and dissolved organic carbon (DOC) and dissolved inorganic carbon (DIC) in surface water and groundwater. The

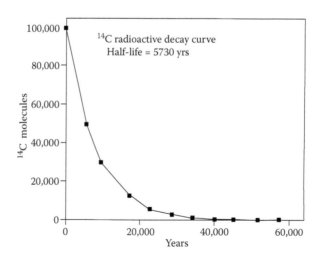

FIGURE 11.2   Classic radioactive decay curve for ^{14}C that has a half-life of 5730 ± 40 years as determined by (Godwin, H, *Nature*, 195, 984, 1962.)

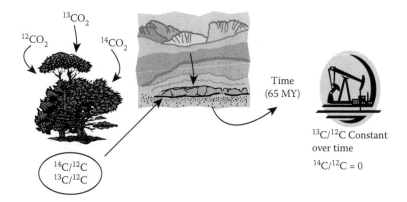

FIGURE 11.3 Plant materials decay and eventually produce petroleum. Petroleum and contaminants generated from petroleum are depleted in radiocarbon because the ^{14}C has undergone decay of many half-lives.

TABLE 11.2
Relative Abundances of Isotopes of Environmental Interest

Isotope	Protons (#)	Neutrons (#)	Abundance (%)	Form
^{12}C	6	6	98.9	Stable
^{13}C	6	7	1.07	Stable
^{14}C	6	8	Trace (2×10^{-10})	Radioactive
^{14}N	7	7	99.634	Stable

^{14}C data can resolve whether groundwater is in contact with modern biota or isolated in deep geologic repositories. Recent groundwater that is in contact with aquifer solids devoid of carbonate minerals will be enriched in ^{14}C DIC and DOC relative to groundwater DIC derived from dissolution of carbonate minerals that are depleted in ^{14}C.

Radiocarbon has been used to identify sources of contaminants in the environment, and most recently to identify biodegradation of contaminants in the environment. ^{14}C can be used to determine if plant organic matter or microbial end products, CO_2 or methane, have been generated from recent carbon or from ancient carbon in soil and groundwater. Petroleum hydrocarbons contain no radiocarbon because of their ancient origin, thus, the resulting CO_2 and methane generated from their complete degradation will also be devoid of radiocarbon. Ancient CO_2 and methane from natural carbon inputs are readily discernible from plant root respiration. Modern plants and recent humus found in most soils that are equilibrated metabolically with ^{14}C in the atmosphere are modern in ^{14}C activity. Any synthesized anthropogenic compound derived from petroleum or other ancient source of carbon (e.g., perchloroethylene [PCE], and trichloroethylene [TCE]) will also be depleted in ^{14}C. The CO_2 and methane produced by bacteria degrading these contaminants thus will be depleted in radiocarbon (Figure 11.4).

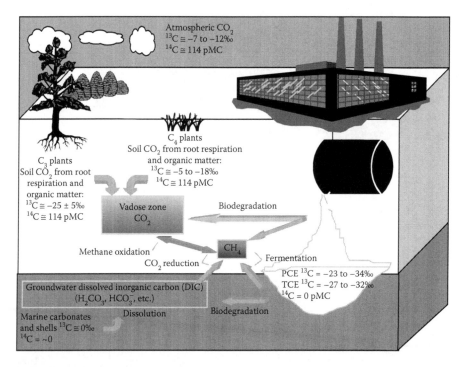

FIGURE 11.4   Values of ^{14}C and ^{13}C of various contaminants and end products as might be found at an uncontaminated and a contaminated site.

Although fractionation processes affect ^{14}C as they do ^{13}C, there is no need to incorporate them specifically into the ^{14}C analysis, except in studies for which a very accurate age dating of carbon material is intended. If radiocarbon-depleted material is oxidized to CO_2, the resulting $^{14}CO_2$ also will be depleted in radiocarbon. If the same source material is degraded to methane, the methane will be depleted in ^{14}C. For environmental studies such as those discussed later in this chapter, ^{14}C data depict the age of the carbon source, whereas process-specific fractionation can be elucidated by ^{13}C data. The consistency of the radiocarbon signal simplifies tracking of the carbon of interest through several biotransformations in the environment. However, information about the microbial processes acting on the carbon, such as specific enzyme systems or electron acceptors that are responsible for the biodegradation process, is obtained by ^{13}C data.

Combined naturally occurring radiocarbon and stable carbon isotopes have been used in several ways as will be illustrated in this chapter. These disparate uses include determining the conversion of grass shrub in the arid southwest United States (Biedenbender et al. 2004), elucidating the origin and age of groundwater (Drimmie, Aravena, and Wassenaar 1991), identifying sources of volatile organic compounds (VOCs) in urban air to distinguish between biomass and fossil mass inputs (Klouda et al. 1996), and examining biodegradation of recalcitrant chlorinated solvents in contaminated groundwater (Kirtland, Aelion, and Stone 2005).

11.3   RADIOCARBON NOTATION

The term radioactivity was coined by Marie Sklodowska Curie based on her research with radium. The Curie (1 Ci = 3.7×10^{10} disintegrations per second) was initially used as the unit of radioactivity and later replaced by the SI unit Becquerel (1 Bq = one disintegration per second). Both units of radioactivity are based on disintegrations per time during radioactive decay.

Because radiocarbon has been used in many fields to examine ancient and modern carbon, a standardization of the nomenclature and reporting of data is needed. In early scientific studies most laboratories used count rates to determine carbon radioactive decay. The count rates are given in disintegrations per minute per g of carbon. The use of correction factors, normalization for isotope fractionation, and other considerations were applied to convert counting data to determine age and geochemical parameters of environmental samples (Stuiver and Polach 1977).

The beta decay of ^{14}C atoms has traditionally been used to date samples up to 35,000 years old (Libby, Anderson, and Arnold 1949; Broecker 2004). Conventional radiocarbon ages are calculated from a measurement of an isotopic ratio and are not true chronological ages of the sample. The initial $^{14}C/^{12}C$ ratio of the sample must be known to calculate a true calendar age. For natural systems ultimately deriving C from atmospheric CO_2, the initial $^{14}C/^{12}C$ is unknown due to changes in atmospheric radiocarbon that has varied throughout recorded history. The atmospheric production rate of ^{14}C and variations of the carbon reservoirs and rate of transfer between different reservoirs have affected the atmospheric ^{14}C concentration. Conventional radiocarbon ages assume constancy of ^{14}C atmospheric levels throughout time (Stuiver and Polach 1977).

Conventional ages are normally cited in years before present (BP) based on a measured ^{14}C content relative to an absolute modern standard. Present is considered AD 1950 therefore AD 1950 equals 0 yrs BP. The international modern carbon standard is the radiocarbon activity measured in 1950 of a sample of wood grown in the northern hemisphere in 1890. The wood grown in 1890 represents the pre-Industrial Era and therefore significant amounts of fossil fuel burning that added radiocarbon-depleted CO_2 to the atmosphere had not occurred. The year 1950 represents the pre-Atomic Era when no ^{14}C had been added to the atmosphere from the testing and detonation of nuclear bombs.

For actual measurements in the laboratory, oxalic acid prepared by the National Bureau of Standards (NBS; renamed the National Institute of Standards and Technology [NIST]) is used as the modern standard. The internationally accepted radiocarbon dating reference value is 95% of the activity in AD 1950 of the NBS oxalic acid (Goh 1991a) normalized to $\delta^{13}C = -19\%o$ with respect to the Pee Dee Belemnite (PDB; Olsson 1970). The correction to $\delta^{13}C = -19\%o$ is carried out to account for fractionation that occurs during sample preparation for ^{14}C counting. Currently, the original oxalic material used has been exhausted and a second oxalic standard material has been prepared by NBS.

For pre-1950s geological and archeological age dating, the sample's calculated age based on radiocarbon measurement is independent of the year of sampling because both the sample and the standard oxalic acid lose their ^{14}C at the same rate, and data are reported as conventional radiocarbon age in years BP (Stuiver and Polach 1977;

Stuiver, Pearson, and Brazunias 1986). For geochemical samples, many of which are modern post-1950s, the sample's calculated age based on radiocarbon measurement is dependent of the year of sampling and measurement of oxalic acid. The international modern carbon standard corresponds to a ^{14}C decay rate of 0.227 Bq per gram of carbon (Rafter Radiocarbon Laboratory 2008). Because the modern standard is defined for 1950, measurements taken at a later time must be corrected for the measured oxalic activity decay since 1950. For example in 2000 this would correspond to 0.0225 Bq per gram carbon due to radioactive decay (Rafter Radiocarbon Laboratory 2008).

As carbon is cycled from one environmental compartment to another a certain degree of fractionation will occur. Fractionation may be due to chemical reactions, evaporation, condensation, or diffusion. This abiotic ^{14}C fractionation can be accounted for by measuring and correcting for stable carbon isotopic fractionation. By convention, the ^{13}C isotopic fractionation in all samples is taken into account by normalizing to $\delta^{13}C = -25\%_o$ with respect to the PDB standard (Stuiver and Polach 1977; Faure 1986; Clark and Fritz 1997). Thus the correction to $\delta^{13}C = -25\%_o$ is used to correct the ^{14}C activity of the sample in question (Fritz and Fontes 1980; Mook 1980).

^{14}C data are reported in the literature using several notations and units based in part on conventions of different disciplines. Units of percent modern carbon (pMC) are used when the measured activity of the sample is to be given as a percentage of the modern standard activity; the standard again being 0.95 times the specific activity of NBS oxalic acid or 95% of the activity in AD 1950 (Stuiver and Polach 1977). δ^{14}C is the deviation of a measured sample ^{14}C activity from the modern standard activity expressed in a ratio ($^{14}C/(^{13}C + {}^{12}C)$), without any correction for fractionation (Rafter Radiocarbon Laboratory 2008) and expressed in units of per mil ($\%_o$). δ^{14}C is expressed as $\%_o$ greater than or less than the standard. Δ^{14}C is the deviation of a measured sample ^{14}C activity from the modern standard activity expressed as the ^{14}C ratio of a material relative to the modern standard after correction for fractionation to $\delta^{13}C = -25\%_o$ and is expressed in units of per mil ($\%_o$).

Notations such as δ^{14}C and Δ^{14}C should be used if samples have been age corrected. Notations such as D^{14}C and d^{14}C also are seen in the literature, and should be used when age correction is not possible as in some oceanographic studies; however this convention is not always followed (Stuiver and Polach 1977).

True chronological ages are obtained by calibrating the conventional age using samples of trees whose chronological ages are known exactly by tree ring counting, or dendochronology. Bard et al. (1990) examined the possible errors associated with the ^{14}C dating method by comparing ancient sample ages obtained using analysis of ^{14}C by accelerator mass spectrometry (AMS) to those obtained using analysis of $^{234}U/^{230}Th$ by thermal ionization mass spectrometry of cores obtained from corals. The accuracy of the U/Th method was verified back to 10,000 years by comparing it to dendrochronological estimates. Beyond 9000 years BP the ^{14}C age estimates were systematically younger than the U/Th age estimates (Figure 11.5; Bard et al. 1990), and these discrepancies were attributed to changes in the cosmogenic nuclide production due to a gradual decrease in the Earth's magnetic field, and changes in the carbon cycle during these time periods (Bard et al. 1990). However errors in the U/Th age estimates exist due to the Th ionization efficiency, and the mass separation of the magnet sector for the $^{234}U/^{238}U$ ratio estimate (Bard et al. 1990).

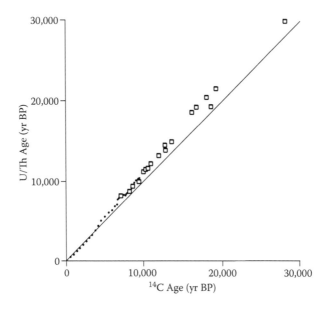

FIGURE 11.5   Th/U ages versus ^{14}C ages calculated by AMS and thermal ionization MS (squares) and dendochronological calibration (circles). (Modified from Bard, E., B. Hamelin, R. G. Fairbanks, A. Zindler, M. Arnold, and G. Mathieu, *Nuclear Instruments & Methods in Physics Research Section B-Beam Interactions with Materials and Atoms*, 52, 461–68, 1990, NIMPR. With permission.)

Reporting of conventional ages in years BP is based on measured ^{14}C compared with a contemporary standard and this value is reported with an estimate of the total analytical precision on the measurement (Taylor, Stuiver, and Reimer 1996). Taylor and colleagues give an excellent discussion of the components of the precision of the reported ^{14}C ages, and the radiocarbon calibration process used to convert ^{14}C time expressed in terms of a conventional radiocarbon age described by Stuiver and Polach (1977) to calibrated time, as calibrated years BP. The analytical precision is comprised of the counting error derived from the variances associated with the sample, background, and contemporary standard count rates (expressed as ±1 standard deviation (σ)), and additional variance is associated with experimental conditions such as sample dilution errors and instrument fractionation effects (termed noncounting effects; Taylor, Stuiver, and Reimer 1996). For most decay counting systems, counting error may dominate, whereas for AMS system, the noncounting statistics error may be dominant (Taylor, Stuiver, and Reimer 1996). Taylor, Stuiver, and Reimer suggest that Bard et al. (1993) have successfully extended the period for which calibration might be possible back to 21,950 calendar years BP by using cores drilled from coral formations, while dendochronology based on tree ring segments documents time back to about 9,800 years.

11.4   RADIOCARBON IN THE ENVIRONMENT

Concentrations of radionuclides in the atmosphere are constantly changing due to human activities (Aelion 2000). For example, recent atmospheric ^{14}CO$_2$ is greater than 100 pMC due to anthropogenic additions of [14]C. Atmospheric CO_2 has been highly contaminated with artificial ^{14}C since the mid-1950s from above-ground nuclear bomb tests (Figure 11.6; Broecker 2004). The countries contributing most to ^{14}C atmospheric contamination were the United States, the former Soviet Union, the United Kingdom, France, and China. Maximum increases in atmospheric ^{14}C content occurred in 1961 and 1962 and reached radiocarbon values as high as ~170 to 200 pMC in tropospheric CO_2 (Goh 1991b). Bomb ^{14}C was estimated to account for approximately 70% of the total atmospheric ^{14}C at that time and has been depleting at a rate of 6.1% per year (Goh 1991b). Bomb ^{14}C entering the stratosphere is cycled into plants (Figure 11.3), and from plants to the remaining compartments of the environment including soil, water, animals, and microorganisms. [14]C concentrations in plant tissue reached a peak (~200 pMC) in 1964 and have declined thereafter (Mazor 1991).

Human activities also may cause a decline in atmospheric ^{14}C concentrations. Although not as dramatic and immediate as atmospheric additions of radiocarbon caused by nuclear bomb testing between 1961 and 1962, the burning of fossil fuels has decreased the radiocarbon signal in the troposphere by adding large quantities of ancient, radiocarbon-free CO_2 (Figure 11.7; Key 2001). CO_2 exchange with the oceans also decreases [14]CO_2 content in the atmosphere. Currently the concentration of radiocarbon in the atmosphere is decreasing due to the moratorium of nuclear bomb testing and continued fossil fuel use, which in turn causes a decrease in the radiocarbon content of living flora and fauna in equilibrium with the atmosphere.

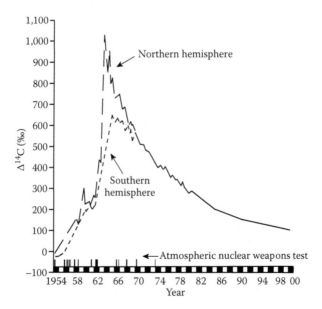

FIGURE 11.6   The location-dependent increase in atmospheric [14]CO_2 due to the addition of ^{14}C-bomb carbon in the early 1960s, and the subsequent decline due to oceanic uptake, plant respiration, and ^{14}C-depleted fossil fuel CO_2 addition to the atmosphere. (Modified from Broecker, W. S., *Treatise on geochemistry*, Elsevier-Pergamon, Oxford, 2004. With permission.)

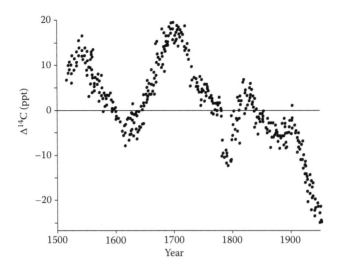

FIGURE 11.7 Variations in atmospheric $\Delta^{14}C$ from 1511 to 1954 measured in tree rings by M. Stuiver 1970. (Modified from Key, R. M., *Encyclopedia of ocean sciences,* Academic Press, London, 2001. With permission.)

The Suess effect is used to describe this phenomenon in which radiocarbon-free CO_2 dilutes the $^{14}CO_2$ in the atmosphere, has been expanded to include similar dilution of ^{13}C by deleted $^{13}CO_2$ in fossil fuels, and expanded to include other large carbon reservoirs such as oceans and soils. The need to correct for the Suess effect for both $\delta^{13}C$ (Verburg 2007) and ^{14}C (Tans, De Jong, and Mook 1979) has been recognized.

11.5 RADIOCARBON AND STABLE CARBON STUDIES IN UNCONTAMINATED ENVIRONMENTS

Radiocarbon and stable carbon isotopes have been measured in many environmental compartments including river water, groundwater, ocean water, terrestrial and aquatic plants, bacteria, soil gas, soil, and sediment. Within these environmental compartments a plethora of carbon-based molecules has been measured for combined radiocarbon and stable carbon isotopes. Chemical samples include DOC (Karltun et al. 2005), DIC (Druffel et al. 2008), POC (Hwang, Druffel, and Bauer 2006) soil and sediment total organic carbon (TOC), sediment organic matter (Glaser and Zech 2005; Longworth et al. 2007), individual organic contaminants (Slater et al. 2006), and biodegradation end products (Aelion, Kirtland, and Stone 1997). Biological samples include total lipids (Roland, McCarthy, and Guilderson 2008), archaeal lipids (Shah et al. 2008), several classes of lipids (Pearson et al. 2001; Petsch, Eglinton, and Edwards 2001), total hydrolysable amino acids and carbohydrates (Wang and Druffel 2001), phospholipid fatty acids (Slater et al. 2006; Wakeham et al. 2006), acid-insoluble organic fractions from zooplankton and phytoplankton (Wang and Druffel 2001), and organs of marine organisms such as worms (Teuten, King, and Reddy 2006).

The early studies that combined the use of naturally occurring radiocarbon and stable carbon isotopes examined the natural environment, in particular the transport, sources, and sinks of carbon in unpolluted environments including sedimentation and carbon deposition rates, the origin of sedimentary particles, carbon in marine systems, and groundwater mixing and recharge–discharge history. Several examples will be discussed in this chapter to illustrate the variety of applications of combined radiocarbon and stable carbon isotopes to the natural environment. However not all can be included so the reader is directed to the literature for additional information.

11.5.1   Surface Water Studies

In unpolluted waters, DIC and DOC have been used to examine carbon turnover and to date fresh and marine waters. Wigley, Plummer, and Pearson (1978) provided a theoretical discussion of DIC in natural systems and stressed the importance of considering all sources and sinks of carbon. Although they focused their example on groundwater systems, they recognized that the concept could be applied equally well to any aqueous environment including marine and freshwater systems. They identified major sources of carbonates to groundwater and ocean water that include the dissolution of the carbonate minerals calcite ($CaCO_3$) commonly referred to as limestone, and dolomite ($CaMg(CO_3)_2$), which dilute the ^{14}C signal due to their radiocarbon-depleted signatures. In addition to Wigley, Plummer, and Pearson's geochemical processes (1978), sources and sinks for DIC may originate from reactions involving biological processes. Oxidation of organic matter including some fermentation reactions and sulfate reduction, may contribute to DIC, while reduction of organic matter and CO_2 to methane via methanogenesis may reduce DIC concentrations.

Wigley, Plummer, and Pearson (1978) used mathematical equations based on conservation of mass to interpret ^{13}C and ^{14}C data in natural water systems. They considered both CO_2 and methane under two scenarios in groundwater systems from biological and geological perspectives. The first scenario, which they discounted, was that produced methane was in ^{14}C isotopic equilibrium with produced DOC at all times. The second scenario, which they felt was supported by popular opinion in the literature, was that produced methane was biologically fractionated and immediately isolated from DOC with no isotopic exchange reactions. Therefore as early as 1978, based on theoretical considerations and an understanding of groundwater systems, it was determined that biological processes had a significant impact on DIC production and consumption, and therefore should be accounted for in groundwater ^{14}C dating.

Gulin, Polikarpov, and Egorov (2003) combined the use of naturally occurring radiocarbon and ^{13}C of the methane-derived carbonates and the substrate material, methane and seawater bicarbonates, to evaluate the origin and estimate the age of microbial carbonate structures growing in methane seeps in anoxic water of the Black Sea. These carbonate areas were discovered in 1989 and have subsequently been found to occur at many depths, including approximately 1500 to 2000 m. Previous ^{14}C-dating estimated the age of the carbonates in the base of the microbial buildups from ~5100 years old at water depths of 230 m, to 17,500 years at water depths of 1738 m (Gulin, Polikarpov, and Egorov 2003). Using a two end-member mixing model that did not account for anaerobic bacterial oxidation of methane, Gulin, Polikarpov, and Egorov (2003) estimated that the methane seeping from the shallowest carbonate structure at 230 m was 8,500 ± 250 to 10, 600 ± 300 years BP, and that the age of the methane seeping from the deeper carbonate structure at 1,738 m seep was 29,200 ± 900 to 36,500 ± 1100 years BP. The ^{14}C dating showed a gradual increase in the age of the carbonate structures with water depth (Figure 11.8; Gulin, Polikarpov, and Egorov 2003); the estimated time of origin of the deepest carbonate structure was 5316 years and that of the shallowest carbonate structure was 2888 years.

The ^{14}C-dating of the microbial carbonate structures alone could not distinguish their origin because the radiocarbon signal combines all carbon sources including methane-derived carbon and carbon from bicarbonates dissolved in seawater. The stable carbon isotopic analyses of the methane-derived carbonates and their original components, methane and seawater bicarbonates, were used to evaluate the contribution of methane and seawater bicarbonates to the carbonate material of the microbial structures. Gulin, Polikarpov, and Egorov (2003) assumed that the Black Sea microbial buildups were formed from anaerobic bacterial oxidation of methane seeping up from the bottom sediments. The carbonates of the buildups were depleted in ^{13}C relative to the seawater bicarbonate. The $\delta^{13}C$ of the carbonate was $-36‰$, $\delta^{13}C$ of the seawater bicarbonate was $-2‰$ at 200 m and $-6.3‰$ at 2,000 m, and $\delta^{13}C$ of methane was $-58.2‰$ to $-68.3‰$. Gulin, Polikarpov, and Egorov(2003) calculated the percentage of the seeping methane that contributed to the microbial structures based on the stochiometric reaction of the anaerobic oxidation of methane that produces methane-derived hydrogen carbonate that precipitates as $CaCO_3$ with an equimolar amount of seawater bicarbonate. Based on these calculations, Gulin, Polikarpov, and

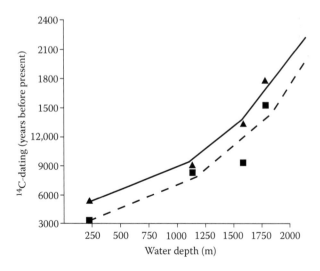

FIGURE 11.8 Radiocarbon age of the base (triangles) and the middle (squares) of the carbonate chimneys of the NW Black Sea slope. Fitted lines were used to extrapolate to the deeper depths. (Modified from Gulin, S. B., G. G. Polikarpov, and V. N. Egorov, *Marine Chemistry*, 84, 67–72, 2003. With permission.)

Egorov (2003) calculated that 48–60% of the carbon in the carbonate structures was derived from methane seeping from the sediments.

Like DIC, DOC is important in understanding carbon cycling in natural environments and the contribution of biological activities to this cycling. DOC in freshwater consists primarily of fulvic and humic acids. DOC arises from plant decay and associated microbial activity, for example the production of organic acids during biodegradation. Raymond and Bauer (2001) summarized available data of naturally occurring $\Delta^{14}C$ and $\delta^{13}C$ of particulate organic carbon (POC) and DOC from rivers, estuaries, and coastal ocean waters. They concluded that in general, riverine DOC was enriched in ^{14}C relative to POC in rivers and estuaries. POC exported from rivers was ^{14}C-depleted, which was attributed to the addition of old soil horizons, sedimentary fossil carbon, sorbed petroleum, and petroleum byproducts. Although DOC in river water was enriched in ^{14}C relative to POC, the opposite occurred in coastal marine waters. River and estuarine DOC become enriched in ^{13}C and depleted in ^{14}C during transport because heterotrophic bacteria preferentially used the younger or ^{14}C-rich DOC fraction as the materials make their way from upland to coastal areas.

11.5.2 SOIL GAS AND GROUNDWATER STUDIES

Surface waters equilibrate rapidly with the atmosphere unlike soil gas and groundwater, which may have some exchange with the atmosphere (soil gas) or be significantly isolated from the atmosphere (deep groundwater). Depending on the environmental compartment being examined, physical processes such as diffusion, transport, and biological degradation processes can impact CO_2 and methane concentrations and their stable carbon and radiocarbon values. Kunkler (1969) is attributed as the first to perform ^{14}C analyses of CO_2 gases from depths of 24 m and 86 m in the unsaturated zone of an aquifer. The focus of the study was physical CO_2 diffusion.

Thorstenson et al. (1983) described some mathematical modeling considerations for the carbon partitioning for the seasonal depth distribution of CO_2 in unsaturated soil gas and for the time and depth distribution of $^{14}CO_2$. They also considered the importance of biologically generated $^{14}CO_2$. They concluded that $^{14}CO_2$ diffusion was a major mechanism of transport through the unsaturated zone, and each isotope of C diffused in response to its particular sources and sinks. Thus, depth profiles of $^{12}CO_2$, $^{13}CO_2$, and $^{14}CO_2$ differed due to their individual sources and sinks including root respiration and oxidation of organic matter. Concentrations of $^{14}CO_2$ measured in the unsaturated zone were 10 to 100 times greater than those measured in the atmosphere. Based on ^{14}C data, this distribution was attributed to a combination of biologically generated $^{14}CO_2$ in the shallow soil and rapid diffusion that allowed penetration of $^{14}CO_2$ into the subsurface.

Subsequently, Dörr and Münnich (1986) used soil $^{14}CO_2$ measurements to look at soil CO_2 flux from the decomposition of organic matter to the atmosphere in shallow soil depths of 2–20 cm. Plant root respiration $^{14}CO_2$ is essentially equal to that of the atmosphere. The $^{14}CO_2$ from the decomposition of organic matter can be either depleted if the organic material is old or enriched if the organic material has been impacted by nuclear bomb contamination. The resulting $^{14}CO_2$ signature of the soil gas will be a combination of these two sources in relative proportion. In summer when plant respiration is high, the ^{14}C of the CO_2 soil gas was close to atmospheric levels. In winter, soil gas [14]CO_2 was depleted in ^{14}C because there was limited root respiration, and increased bacterial decomposition of organic matter even though there was slower turnover of organic matter. The difference in Δ^{14}C between summer and winter was as high as 50‰ at a site with comparatively rapid turnover of carbon (resulting in values closer to atmospheric values) to 100‰ in soils at a site with comparatively slow turnover of carbon.

Because groundwater is an important source of drinking water, research has been carried out to identify quantity and quality of this valuable resource. The source of water for recharge to the upper Floridian coastal aquifer was examined by Landmeyer and Stone (1995) using ^{14}C and δ^{13}C of DIC and tritium to identify modern recharge by groundwater, which was in equilibration with the atmosphere that has been contaminated by ^{14}C from nuclear-weapons testing. They measured highly depleted ^{14}C DIC values of 0.5–1.4 pMC, stable δ^{13}C DIC values of –2.25 to –2.78‰ and tritium concentrations at or below the detection limit. Using data from these three separate isotopes, they estimated the groundwater age to be 16,000 years BP and that the δ^{13}C values were close to those of aquifer carbonates (assumed to be 0‰). Thus limited recent modern recharge had occurred suggesting that overuse of the groundwater could lead to recharge from the surficial aquifer and saltwater sources (Landmeyer and Stone 1995) due to proximity of the aquifer to the coast.

In a subsequent study, combined radiocarbon, tritium, and contaminant concentrations were used by Stone, Devlin, and Moore (1997) to estimate aquifer vulnerability to contamination. They used nitrate, a common pollutant in agricultural areas as an indicator of contamination from surface sources, and compared nitrate to [14]CO_2 of soil gas, and groundwater age estimates. The authors assumed that finding tracers of recent origin (elevated concentrations of bomb ^{14}C, ^{3}H, and nitrate) in groundwater implied aquifer vulnerability compared to aquifer systems devoid of recent tracers. They found that higher vulnerability occurred in the middle and lower coastal plain aquifers. But many shallow aquifers appeared to have low vulnerability provided by geologic protection, evidenced by low concentrations of nitrate and background tritium concentrations, and very old groundwater determined by depleted ^{14}C, 3.9 ± 0.1 pMC with a corrected age of 19,000 years. These studies did not take into account possible biological consumption of nitrate via microbial processes such as denitrification as nitrate travels through the soil profile.

Early research was carried out by Haas et al. (1983), who used the combined stable carbon and radiocarbon technique to investigate subsurface unsaturated soil gas at shallow depths and groundwater DIC. They measured soil CO_2 in the vadose zone using a nest of gas probes spaced 3 m apart vertically in the unsaturated zone and DIC extracted from groundwater from unpolluted systems from 2 m to 13 m depths. Essentially no methane was measured in the unsaturated gas samples. Seasonal variations in CO_2 concentrations were measured in the shallow probes and also were observed in the ^{13}C data. These were attributed to depleted values in spring and summer due to influx of biogenic CO_2 during the growth season, compatible with the Calvin cycle vegetation of the area. The ^{14}C concentrations were constant and showed limited seasonal variation, despite varying CO_2 concentrations. [14]C activities were close to postbomb levels at the surface and generally decreased with depth, from 5 to 8 PMC per meter depth. ^{14}C activities were relatively constant within each depth sampled. The chemical oxidation of lignite decreased the ^{14}C concentrations because of its radiocarbon-free contribution to CO_2. In areas with lignite present, $^{14}CO_2$ from 5.8 m depths was highly depleted in ^{14}C, at approximately 10 PMC. In areas without lignite the $^{14}CO_2$ of samples collected at 5.8 m and 8.5 m depths was approximately 85 and 55 PMC, respectively, and surface $^{14}CO_2$ was 110–120 PMC. While fluctuations in ^{14}C activity and CO_2 concentrations occurred seasonally in the vadose zone, groundwater ^{14}C activity was stable and varied by only < 2 PMC. Thus geochemical processes in the aquifer varied on time scales that were longer than those impacting the near-surface unsaturated zone.

11.6   RADIOCARBON AND STABLE CARBON STUDIES IN CONTAMINATED ENVIRONMENTS

The studies described above, many of which were published over 20 years ago, set the stage for the continued use of combined radiocarbon and stable carbon to examine the physical, geochemical, and biological transformations of contaminants in the environment. Most of the previously described studies focused on natural carbon turnover, in other words groundwater DIC and DOC as traditionally done in the fields of geology and geochemistry. It is not surprising that the use of radiocarbon in pollution studies originated from these fields. The next logical step was to identify sources of pollutant in environmental media, and production of biodegradation and end products that has expanded interest in the use of combined ^{14}C and ^{13}C to the fields of biogeochemistry, environmental chemistry, biology, and microbiology.

Although tremendous potential exists because of its innate sensitivity and comprehensive mineralization assessment, naturally occurring radiocarbon has not been as well developed as stable carbon isotopes for application to contaminant biodegradation at a field or laboratory scale. With the development of AMS, small amounts of CO_2 and methane can be analyzed and provide information about the origin of the carbon used to generate gaseous end products such as CO_2 and methane. However part of the limitation of the radiocarbon analysis is precisely due to the sophistication of the instrumentation and its associated expense. Radiocarbon analyses by AMS currently cost approximately 10 times those of stable carbon. Regardless, radiocarbon data overcome several of the limitations of ^{13}C applications and interpretation of the data as are illustrated in the following section.

The combined use of ^{14}C and ^{13}C can be applied to the study of gaseous end products of degradation in contaminated environments, the investigation of the specific contaminants and their transformation products, and finally to the biological components of organisms in uncontaminated and contaminated environments (see Section 11.5). Addressing each of these areas is not feasible in one chapter. In the following sections, the application of combined ^{14}C and ^{13}C will focus primarily on the measurement of end products of contaminant degradation, CO_2 and methane, and to a smaller degree on the compound specific application to contaminants.

11.6.1 USE OF COMBINED ¹⁴C AND ¹³C ANALYSIS FOR GASEOUS END PRODUCTS

Contaminant degradation can cause stable isotopic shifts in metabolic end products including CO_2 to lower $\delta^{13}C$ values than the parent compound. However, that shift alone has not been adequate to definitively estimate biodegradation because shifts are generally small for aerobic processes (< 6‰), and there can be significant overlap in stable isotope ratios of plant root respired CO_2 (see Chapter 5). Anaerobic processes that generate methane cause larger shifts in $\delta^{13}C$ (see Chapter 6). Thus the DIC produced by the acetate reduction pathway will be enriched in ¹³C compared to the substrate, and the $\delta^{13}C$ of the methane generated will be depleted compared to the substrate (Figure 11.9; Clark and Fritz 1997) resulting in limited overlap with root-respired CO_2. In field studies in which multiple contaminants and microbial degradation processes are possible and in which plants may contribute to CO_2 and DIC through root-respiration, the combined use of naturally occurring radiocarbon and stable carbon isotopes may elucidate microbial degradation processes better than either measure individually.

11.6.1.1 Surface Water Studies

Over 40 years ago, Rosen and Rubin (1965) described the use of radiocarbon to distinguish between industrial pollution and natural organic matter, primarily readily oxidized organics in animal and vegetable matter in garbage and wastewater. Their objective was to develop a method to differentiate between organic wastes such as municipal wastewater that carries a high biological oxygen demand (BOD), and

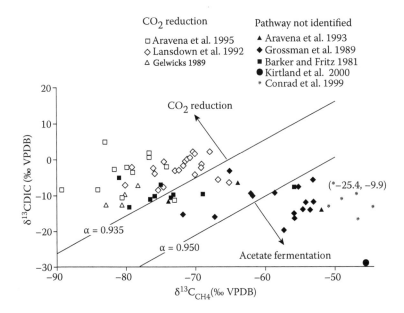

FIGURE 11.9 Enrichment of ¹³CO_2 and depletion of ¹³CH_4 due to CO_2 reduction and acetate fermentation processes. (Modified from Clark, I. D., and P. Fritz, *Environmental isotopes in hydrogeology*, 1st ed., CRC Press, Boca Raton, FL, 1997. With permission.)

industrial pollution that has limited BOD but may be highly toxic to aquatic organisms. The samples in their study were collected from surface water supplies that were impacted by different land uses and their respective surface water discharges. These surface water discharges ranged from entirely industrial, to mixed municipal and petroleum refinery, to predominantly municipal. Thus some of the study sites contained natural organic materials that were contemporary or modern in radiocarbon, and others contained fossil carbon materials such as petroleum, coal, and natural gas that were depleted in radiocarbon. From the ^{14}C data, Rosen and Rubin (1965) qualitatively differentiated between recent and ancient carbon sources to surface water, and identified the distribution of carbon of several different organic pollutants and natural organic matter in surface water (Figure 11.10). In order to measure the carbon in these different surface water systems, an exhaustive extraction of organic adsorbate from an activated carbon filter was used. The extracted samples were converted to acetylene and analyzed using the acetylene ^{14}C dating method of the U.S. Geological Survey (Suess 1954). The authors noted that possible errors were associated with the adsorption and desorption of the extraction technique, the ^{14}C sample treatment and counting technique, the assumed contemporary activity baseline, as well as the unknown contribution of photosynthesis to the total organic content of the stream. Despite these unknowns and limited analytical capabilities, Rosen and Rubin (1965) surmised that as sampling proficiency increased and size of sample needed decreased, radiocarbon analysis could be used to determine not only the origin of bulk contaminants, but also

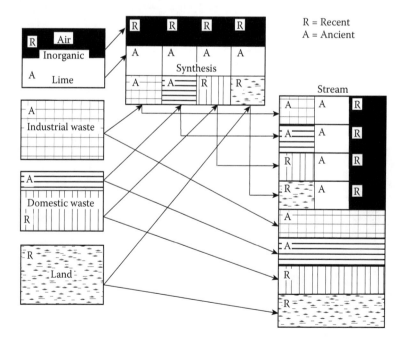

FIGURE 11.10 Generic distribution of recent (R) and ancient (A) ^{14}C in surface waters due to variation of sources. (Adapted from Rosen, A. A., and M. Rubin, *Journal of the Water Pollution Control Federation*, 37, 1302–7, 1965.)

whether a specific chemical originated from industrial waste or originated from the natural biological material such as vegetation.

This research approach was expanded by Spiker and Rubin (1975) who combined radiocarbon and stable carbon isotope analysis of DOC of several river waters in the United States, primarily in Washington, DC and Maryland to investigate petrochemical, domestic waste, and natural decaying matter inputs. They suggested that several biological inputs to DOC, including municipal wastes, soil runoff, agricultural and feedlot drainage, and bioproduction within the aquatic system were modern in their carbon signature. They assumed that petrochemical industrial wastes contained no ^{14}C and that the sources of intermediate ^{14}C signature were negligible. These simplifying assumptions allowed a two component model to be used to calculate the percentage of modern DOC, and percentage of fossil DOC in the river water. Based on ^{14}C data, the authors estimated that the freshwater rivers were approximately 0–37% fossil DOC but there were significant changes at the same site at different sampling times. For example, computed percentage of fossil DOC was ~7% in samples collected in April 1972, and ~30% in samples collected in December 1972 from the Potomac River in Washington, DC. The increase in modern carbon DOC was attributed to increased runoff during flood conditions in December that contributed primarily modern DOC to the water. The $\delta^{13}C$ values of the river water DOC samples ranged from –24.6 to –34.1‰ and were more depleted in ^{13}C than those of terrestrial plants that was attributed to the input of more highly fractionated petroleum products. However Spiker and Rubin (1975) recognized that the $\delta^{13}C$ range for terrestrial and fluvial plants overlapped with that of petroleum products, so that the differences in $\delta^{13}C$ values were not as distinct as those of the two-component ^{14}C model. In their study there was no correlation between $\delta^{13}C$ and ^{14}C of the DOC. Regardless, the ^{14}C DOC method was concluded to be a useful environmental indicator in surface waters.

11.6.1.2   Soil Gas and Groundwater Studies

In surface waters, chemicals and gases generally are in equilibrium with the atmosphere. Thus volatile chemicals and their degradation products do not accumulate relative to atmospheric concentrations and the resulting ^{14}C signatures do not deviate significantly from those of atmospheric signatures. As geochemists and geologists have long known, in unpolluted and polluted groundwater and subsurface soil environments isolated geologically from the atmosphere, chemicals and gases accumulate and are not in equilibrium with the atmosphere. In addition, processes in groundwater and subsurface soil environments occur on longer time scales than those in surface water systems, which results in a more stable environment and facilitates the use of combined ^{14}C and ^{13}C.

Spiker and Rubin (1975), as part of the study previously described, were one of the early research groups to investigate groundwater DOC that was impacted by industrial and municipal pollution. They measured depleted radiocarbon signatures in groundwater DOC from a California site and calculated that the groundwater DOC originated from 45 to 65% fossil carbon due primarily to railroad industrial waste sources. The $\delta^{13}C$ values of the groundwater DOC samples ranged from –18.8 to –38.3‰. The authors surmised that anaerobic production of methane was a possible source of variation in the $\delta^{13}C$ values. Methane concentrations were not measured in the study and conclusions about biological activity were inferred rather than investigated directly. Although their stable isotope data showed distinct differences between surface water and groundwater, and between groundwater samples from different wells at the same site, a limited data set was available and conclusions about the biological contribution to isotope fractionation could not be made. More recent developments in the field since the Spiker and Rubin (1975) study was published, include the incorporation of additional information on biological processes occurring in groundwater.

Suchomel, Kreamer, and Long (1990) carried out one of the first studies to investigate biodegradation of volatile contaminants in groundwater and subsurface soils using combined radiocarbon and stable carbon isotopes at a contaminated field site. They investigated the aerobic biodegradation using soil gas samples from an organic solvent disposal site. Primary contaminants included TCE and its daughter products 1,1-dichloroethylene (DCE), and 1,1,1-trichloroethane (TCA). They measured nitrogen, CO_2, O_2, and argon using gas chromatography, and because O_2 and argon have the same response factor, calculated the percentage of O_2 in the sample by assuming the ratio of nitrogen and argon was constant. Using discrete sampling depths in the vadose zone, they verified that in the contaminated area there were elevated CO_2 levels (> 15%) and depleted O_2 (as low as 1%). Generally O_2 concentrations were in the 4% range suggesting that the system was primarily aerobic. One objective of the study was to verify the origin of the CO_2 that had several potential chemical and biological origins including plant root respiration, chemical and biological decomposition of organic matter, atmospheric venting of the upper soil surface, infiltration of water percolating through the vadose zone, methane oxidation and microbial fermentation, and biodegradation of organic contaminants.

The $\delta^{13}CO_2$ values from aerobic oxidation of the organic material in the contaminated area were approximately –23.27 to –24.70‰, which were close to the $\delta^{13}C$ of petroleum-derived CO_2 (–25‰). The uncontaminated site soil gas $\delta^{13}CO_2$ values were –17.97 to –19.93‰. The number of fractionating steps for the degradation of solvent synthesis was unknown. However Suchomel, Kreamer, and Long (1990) assumed that because the chlorinated solvents were derived from petroleum, the $\delta^{13}C$ value close to that of petroleum indicated that biological aerobic degradation of the contaminant had occurred. The study was carried out in the arid southwest United States with estimated background root respired $\delta^{13}CO_2$ values of approximately –16.7‰ and soil organic matter ranging from –22 to –27‰. These $\delta^{13}C$ values spanned the $\delta^{13}C$ values measured in soil gas in both the uncontaminated and the contaminated areas. Based on these values, CO_2 in the uncontaminated area was attributed to plant root respiration and a minor component to oxidation of organic matter. From the $\delta^{13}C$ data from the contaminated site, Suchomel, Kreamer, and Long (1990) could conclude only that biodegradation in the field could not be precluded. In contrast, ^{14}C values of CO_2 soil gas ranged from 7.7 to 8.6 pMC in the contaminated wells, and from 106.6 to 112.3 pMC in the uncontaminated wells, which is slightly lower in pMC than atmospheric $^{14}CO_2$. The ^{14}C data gave the most dramatic indication that the solvents had been biodegraded and these data combined with CO_2 and O_2 concentrations, aerobic degradation of the contaminants was demonstrated.

Many subsequent studies using stable and radiocarbon analyses to examine microbial degradation of organic pollutants followed the lead of Suchomel, Kreamer, and Long (1990). Aelion, Kirtland, and Stone (1997) used radiocarbon and stable carbon analyses at a former gas station contaminated by leaking underground storage tanks to provide direct quantified evidence of in situ aerobic petroleum hydrocarbon biodegradation. The site was undergoing remediation by air sparging/soil vapor extraction (AS/SVE) in which atmospheric air was injected below the water table creating aerobic conditions at the site, which otherwise presumably would have been anaerobic due to microbial consumption of the available oxygen. The AS/SVE system was turned off for the study. Vadose zone soil gas $^{14}CO_2$ ranged from 15.9 pMC to 35.0 pMC, and groundwater ^{14}C DIC values ranged from 30.7 to 47.7 pMC. In general, the vadose zone CO_2 was more depleted than the DIC in groundwater, which also was reflected in samples of the AS/SVE exhaust gas (21.8 pMC). Based on the $^{14}CO_2$ values and the CO_2 concentrations, a two-member mixing equation was used to calculate the percent of the CO_2 that was attributable to hydrocarbon degradation. It was calculated that 59 to 87% of the CO_2 was from aerobic petroleum biodegradation and the remainder was from natural carbon sources.

The $\delta^{13}C$ values were not as indicative of microbial degradation as the ^{14}C values, primarily because there was not sufficient distinction between the control site vadose zone $\delta^{13}CO_2$ values and those from the contaminated site (Figure 11.11; Aelion, Kirtland, and Stone 1997). $\delta^{13}C$ values ranged from –22.0 to –27.9‰ in the contaminated vadose zone soil gas CO_2, averaged –23.9‰ in the uncontaminated vadose zone CO_2, and –22.9‰ in the uncontaminated groundwater DIC. In contrast, the differences between the control site and contaminated site vadose zone $^{14}CO_2$ were clearly discernible. Also, ^{14}C vadose zone soil gas CO_2 was correlated with $\delta^{13}C$-CO_2 values indicating a relation of depleted ^{13}C-CO_2 with increased aerobic petroleum biodegradation (Figure 11.12; Aelion, Kirtland, and Stone 1997). Given the small number of samples and the relatively small range in $\delta^{13}C$ values, however, the correlation was not sufficient for predictive value. Regardless, ^{14}C analysis in this study clearly illustrated biodegradation even in samples of a range of CO_2 concentrations (3% to 16% CO_2 (v/v)). This study also illustrated some of the difficulties of relying solely on stable carbon isotopes of CO_2 at contaminated field sites located in vegetated areas.

Conrad et al. (1997) carried out a study using combined radiocarbon and stable carbon isotopes to examine anaerobic degradation of petroleum hydrocarbons at a coastal naval air station in an area contaminated by leaking underground storage tanks. They measured ^{14}C DIC of groundwater and CO_2 of the soil gas. $\delta^{13}C$ values were –22.8‰ in uncontaminated soil gas, –11.4‰ in uncontaminated groundwater DIC, and –17.5 to –25.7‰ in soil organic matter. Respective ^{14}C values were 107 pMC, 97 pMC, and 92 pMC. CO_2 of contaminated soil gas samples averaged –19.5‰ and ^{14}C averaged 14 pMC at high CO_2 concentrations of 10 to 15%. Measured methane concentrations were 4–5%. Based on calculations of the relative proportion of methane production, including CO_2 reduction to methane versus acetic acid fermentation, Conrad et al. (1997) calculated that ~20% conversion of the CO_2 to methane via CO_2 reduction would produce the observed $\delta^{13}CO_2$ values, or ~37% of CO_2 would have to be produced by acetic acid fermentation to result in the observed $\delta^{13}CO_2$ values. The $\delta^{13}C$ and ^{14}C of methane were not measured in the study.

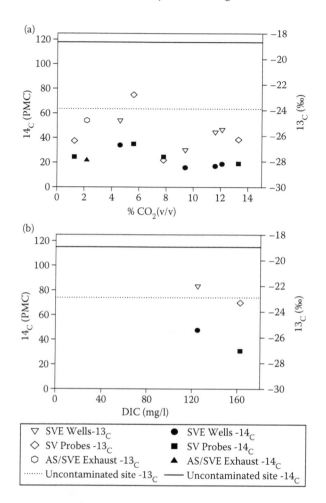

FIGURE 11.11 ^{13}C and ^{14}C CO_2 and DIC from a petroleum-contaminated aquifer and an uncontaminated control site. (Modified from Aelion, C. M., B. C. Kirtland, and P. A. Stone, *Environmental Science & Technology*, 31, 3363–70, 1997. With permission.)

Conrad et al. (1999) used combined stable carbon and radiocarbon in contaminated methanogenic environments to examine degradation of VOCs at a gas station by measuring the stable carbon isotope ratio, ^{14}C and δD of methane. The metabolic pathways for methane production in freshwater and marine systems have been investigated extensively using $\delta^{13}C$ and δD ratios of methane to define CO_2 reduction, acetate fermentation, and thermogenic methane (Whiticar, Faber, and Schoell 1986; Coleman et al. 1996). By compiling the data (Figure 11.13), measured $\delta^{13}C$ and δD ratios of methane can be plotted to identify the pathway that generated measured methane (Kirtland, Aelion, and Stone 2000). During acetate fermentation, three-quarters of the hydrogen is derived from the methyl group of acetate and the remaining hydrogen is derived from the associated water (Pine and Barker 1956). CO_2 reduction derives all the hydrogen from the surrounding water and it is less depleted

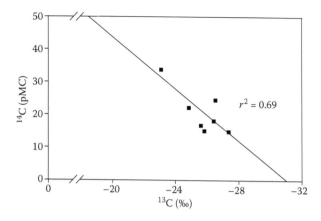

FIGURE 11.12 Inverse correlation between ^{14}C and ^{13}C CO_2 from the vadose zone of a petroleum-contaminated aquifer. (Modified from Aelion, C. M., B. C. Kirtland, and P. A. Stone, *Environmental Science & Technology,* 31, 3363–70, 1997. With permission.)

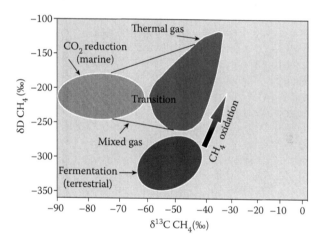

FIGURE 11.13 Typical isotopic values of deuterium and carbon of methane from methane oxidation, CO_2 reduction and fermentation, and thermogenic sources. Oxidation processes enrich the carbon and the deuterium. (Modified from Whiticar, M. J., E. Faber, and M. Schoell, *Geochimica Cosmochimica Acta,* 50, 693–709, 1986; Coleman, D. D., C. Liu, K. C. Hackley, and S. R. Pelphrey, *Environmental Geoscience,* 2, 95–103, 1996, and Kirtland, B. C., C. M. Aelion, and P. A. Stone, *Bioremediation Journal,* 4, 187–201, 2000. With permission.)

in deuterium δD than that of the acetate. Thus the resulting methane derived from CO_2 reduction will be less depleted in δD than the methane derived from fermentation. Conrad et al. (1999) used these underlying relations in combination with the radiocarbon assessment to identify biodegradation processes. Methane in the center of the gasoline plume was at high concentrations, approximately 20% of the total soil gas and the ^{14}C of the methane, soil gas CO_2 and DIC were less than 10 PMC. These depleted radiocarbon signals confirmed that the degradation of the gasoline was the source of the methane and soil gas CO_2 as opposed to degradation of natural organic matter or dissolution of carbonate shells.

The $\delta^{13}C$ of the soil gas methane measured by Conrad et al. (1999) ranged from –52.2 to –25.5‰ but the majority of samples fell in the –50 to –40‰ range. The δD of methane ranged from –365 to –274‰, which based on compiled $\delta^{13}C$ and δD data of Whiticar, Faber, and Schoell (1986) was primarily attributed to acetate fermentation (Figure 11.13). The ^{13}C of the CO_2 and DIC in the center of the contaminant plume was enriched (–4.9 to –8.6‰) suggesting that acetate fermentation was the source of the CO_2. At the plume edges where contamination was limited, the ^{13}C of CO_2 was depleted relative to the contaminated area ($\delta^{13}C$ –26‰) reflective of aerobic degradation of the petroleum hydrocarbons that had a $\delta^{13}C$ value of –25‰. Other factors that could impact the isotopic signatures of the CO_2 soil gas and DIC, such as oxidation of methane by methane utilizing bacteria, fractionation of CO_2 and methane associated with diffusion, and CO_2 exchange with groundwater DIC were considered to be negligible.

Kirtland, Aelion, and Stone (2000) measured methane and CO_2 in the vadose zone of a gasoline-contaminated site and their respective ^{14}C, $\delta^{13}C$, and δD values. The site was originally aerobic and Ar injection into the subsurface was used to generate anaerobic conditions. Methane concentrations were significantly lower than those reported by Conrad et al. (1999) and ranged from 0.2 to 2.5% of the total soil gas. Methane was highly depleted in ^{13}C ($\delta^{13}C$ –45.6‰) and D (δD –316‰), typical of signals produced by fermentation. Methane oxidation was not evident based on $\delta^{13}C$ and δD results. The ^{14}C of CO_2 (24.6 pMC) clearly indicated that the CO_2 was derived from petroleum hydrocarbon mineralization and ^{14}C of methane was 2.6 pMC indicating that essentially all methane was generated from microbial processes.

These previous studies (Aelion, Kirtland, and Stone 1997; Conrad et al. 1997; Kirtland, Aelion, and Stone 2000) examined CO_2 and methane production at petroleum-contaminated sites. Petroleum hydrocarbons are generally considered to be biodegradable. In the field study of Kirtland, Aelion, and Stone (2000), concentrations of methane were lower than might be anticipated at contaminated sites undergoing natural attenuation because the system was made anaerobic artificially and degradation had not been occurring over a long time period under these conditions. Regardless, the use of naturally occurring radiocarbon was a particularly sensitive indicator of anaerobic biodegradation. Similarly, combined radiocarbon and stable carbon isotopes can be useful for contaminants that are relatively recalcitrant or less readily degradable, such as organic solvents. Although Suchomel, Kreamer, and Long (1990) found extensive degradation based on CO_2 production at the chlorinated solvent contaminated site, these contaminants have been shown to degrade to a lesser degree at other contaminated sites. In situations in which extensive degradation occurs, traditional monitoring techniques such as gas chromatography can be used to quantify microbial biodegradation and end products. Traditional monitoring techniques are ineffective if concentrations of biodegradation and end products are near the detection limits. The small signal, however, is discernible using naturally occurring radiocarbon analyzed by AMS even with potentially low biodegradation rates.

Kirtland et al. (2003) measured chlorinated solvents, $\delta^{13}CO_2$, $^{14}CO_2$, and DIC in the vadose zone and groundwater of a contaminated Department of Energy nuclear weapons facility field site undergoing remediation by soil vapor extraction (SVE). Biodegradation of the TCE and carbon tetrachloride was measured after the SVE system was shut down. The groundwater was aerobic with dissolved oxygen concentrations that ranged from 2.4 to 9.2 mg L^{-1}. The DIC in groundwater was depleted in ^{14}C (76 pMC), compared to DIC in the control well (121 pMC), but the DIC in groundwater was only enriched in ^{13}C by 1‰ compared to the DIC in the control well. The elevated ^{14}C value of the control well reflected the Department of Energy's nuclear weapons facility location, where radioactive releases to the environment had occurred historically increasing the concentrations of several radioisotopes including ^{14}C in the soil gas, soil, and groundwater. There was a discernible decrease (up to 12 pMC) in $^{14}CO_2$ over time suggesting that the radiocarbon measurement was sufficiently sensitive to indicate biodegradation of the chlorinated solvents over the 60 day duration of the experiment. Based on $\delta^{13}C$ values of the CO_2 it was difficult to discern biological activity because root respired CO_2 was a significant contributor to the total CO_2 present in the vadose zone. In this case the $\delta^{13}C$ of the contaminants themselves was a better indicator than that of the CO_2 for microbial degradation. The $\delta^{13}C$ value of the daughter product DCE was isotopically enriched compared to the parent compound by approximately 4–12‰. This result suggested that a substantial amount of the DCE had been degraded either via oxidation in the aerobic region of the vadose zone or via reductive dechlorination in the anaerobic microniches followed by complete oxidation to CO_2 in the aerobic zones. However, actual concentrations of the daughter products were low indicating that biodegradation was not extensive.

Continuing this research on chlorinated solvent biodegradation, Kirtland, Aelion, and Stone (2005) examined the biodegradation of PCE at a contaminated commercial industrial site also undergoing AS/SVE remediation. After the system was shut down, they measured $\delta^{13}C$ and ^{14}C of CO_2 in vadose zone sediment, $\delta^{13}C$, ^{14}C, and δD of methane, $\delta^{13}C$ of the parent PCE, and metabolites TCE and cis-1,2-dichloroethylene (1,2-DCE). Vadose zone soil gas CO_2 (approximately 1% v/v of soil gas), and methane (~10–70 ppm) concentrations were low, and O_2 concentrations ranged from 18 to 20% v/v of soil gas, thus conditions were primarily aerobic. The $\delta^{13}CO_2$ values in the source area wells (–23.5‰) were significantly depleted in ^{13}C compared to the marginally contaminated perimeter wells (–18.4‰) and background wells (–16.9‰), which was a good indicator of biodegradation.

One benefit of the $^{14}CO_2$ data is that degradation rates can be calculated which cannot be garnered from ^{13}C data alone. Because samples were taken over time for approximately 60 days, the degradation rates could be calculated in a two-step process. The CO_2 production rate was calculated using ^{14}C values to correct CO_2 concentrations to the amount attributed to only chlorinated solvent mineralization as follows

$$K_{CO_2\,CHC} = \frac{CO_{2\,(f)}\left(1 - \dfrac{pMC_{sample\,(f)}}{pMC_{unc}}\right) - CO_{2\,(0)}\left(1 - \dfrac{pMC_{sample\,(0)}}{pMC_{unc}}\right)}{\left(t_{(f)} - t_{(0)}\right)}, \quad (11.1)$$

where K_{CO_2CHC} is the CO_2 production rate (% d^{-1}) from CHC mineralization, CO_2 is the measured vadose zone concentration (%) at the initial (0) and final (f) time points, pMC$_{sample}$ is the measured ^{14}C PMC value of vadose zone CO_2 at the initial (0) and final (f) time points, pMC$_{unc}$ is the average measured^{14}C pMC value from the uncontaminated site (control well, 116.8 pMC), and t is time (day) at initial (0) and final (f) sampling points.

Using the mass–mole relation and the corrected rate of CO_2 production in the vadose zone (Equation 11.1), and modifying the equation for aerobic petroleum hydrocarbons biodegradation in soil (Hinchee and Ong 1992) for chlorinated solvents, the estimated aerobic chlorinated solvent biodegradation rate was calculated as

$$K = \frac{K_{CO2\,CHC}\ A\ D_{CO_2}\ C_{DCE}}{100} \times 1.33, \qquad (11.2)$$

where K is the biodegradation rate (mg-DCE kg soil^{-1} d^{-1}), K_{CO2CHC} is the corrected CO_2 production rate (% d^{-1}) from Equation 11.1, A is the air volume per mass of soil (1 kg^{-1}), D_{CO_2} is the density of CO_2 at 20° (1829.6 mg l^{-1}), and C_{DCE} is the DCE/CO_2 mass ratio (1.101). Site-specific sediment parameters used in these calculations were bulk density (1.29 l kg^{-1}), porosity (0.47), and the air-filled pore space (0.36 l kg^{-1}). A 25% conversion efficiency of substrate carbon to biomass was assumed, which is a conservative estimate based on reported ranges of 33 to approximately 50% (Sharabi and Bartha 1993; Shen and Bartha 1996; Atlas and Bartha 1997). Other assumptions included that diffusion of O_2 and CO_2 in the vadose or saturated zone was limited (Hinchee and Ong 1992), the activity of $^{14}CO_2$ was the same for the control and contaminated site with respect to plant root respiration, CHCs contained no ^{14}C, and no other ^{14}C-depleted source contributed to vadose zone CO_2. Using Equations 11.1 and 11.2 as described, 1,2-DCE degradation rates were calculated to be 0.003–0.01 mg DCE kg soil^{-1} day^{-1} based on ^{14}C of CO_2 in the contaminated wells and an uncontaminated control site, assuming no other ^{14}C-depleted source contributed to vadose zone CO_2 (Hinchee and Ong 1992).

Methane concentrations were measurable indicating anaerobic microniches at this field site. Indications show that 74% of methane measured in vadose zone sediments was derived from in situ biodegradation of chlorinated hydrocarbons based on ^{14}C values of methane of 29 pMC. Methane were significantly enriched in ^{13}C (–1.3 to –1.6‰) relative to what was expected for fermentation. The δD value of methane was –256‰, which suggested that methane was produced via fermentation, and that microbial oxidation of methane may have fractionated both the carbon and hydrogen of methane resulting in their enrichment in the remaining methane.

Landfill systems present a particularly appropriate study site to use combined isotopic techniques, because of the significant fractionation of methane and CO_2 during biological degradation. Hackley, Liu, and Coleman (1996) examined the isotopic composition of municipal landfill leachate (δD and δO in water and δ^{13}C and ^{14}C in DIC), and CO_2 (δ^{13}C) and methane (δD and δ^{13}C) produced in the landfill. They used radiocarbon and stable carbon isotopes to determine whether contamination in the subsurface was from the municipal landfill or from other local sources. CO_2 was enriched in ^{13}C to values as high as + 20‰ and δ$^{13}CH_4$ values from this and several other landfill studies compiled by Hackley, Liu, and Coleman (1996) ranged from −48.5 to −60.0‰, indicating that methane was produced by acetate fermentation. The ^{14}C of the landfill leachate DIC also was enriched to 120 to 170 pMC, which they attributed to increased radiocarbon content in the atmosphere caused by nuclear devices, and the subsequent presence in organic materials decomposing in the landfills. Regardless, given the expected ^{14}C values in equilibrium with the atmosphere at the time of the sampling, these ^{14}C values would appear to be sufficiently high that other sources of ^{14}C contamination to the landfill may have occurred. Hackley, Liu, and Coleman (1996) suggested that this was the case for tritium values in municipal landfills that were too high to be explained by input from contemporaneous precipitation. Although a benefit of the radiocarbon and tritium analysis is its sensitivity, this study also illustrates that this can be a drawback because samples may become contaminated easily. Landfills are particularly susceptible to unknown additions of materials that may affect both the leachate and subsequent naturally occurring radiocarbon and stable carbon isotopic signatures of microbial end products.

11.6.2 USE OF COMBINED ^{14}C AND ^{13}C ANALYSIS FOR SPECIFIC CONTAMINANTS IN CONTAMINATED ENVIRONMENTS

Most of the studies described in the previous section have focused on the use of naturally occurring radiocarbon and stable carbon isotopes for end products of microbial degradation. As discussed in other chapters of this book (Chapters 4 and 8), stable carbon isotopes are highly fractionated during the biodegradation of certain individual contaminants (e.g., TCE and other chlorinated solvents) and are not significantly fractionated during the degradation of other contaminants (e.g., Polychlorinated biphenyls (PCBs)). In theory, ^{14}C analysis of any carbon-containing entity, combined with ^{13}C analysis, can be useful for examining degradation of contaminants in terrestrial and aquatic environments and overcoming some of these limitations in interpretation of ^{13}C data. The same principles discussed in previous sections on the combined use of ^{14}C and ^{13}C apply in these cases and will not be repeated here.

The use of naturally occurring radiocarbon and stable carbon isotopes for individual chemical contaminants, microbial intermediate compounds, and biological components is growing. Because of analytical improvements in the separation of chemicals by high-performance liquid chromatography and two-dimensional preparative capillary gas chromatography coupled to analysis using isotope ratio mass spectrometry (^{13}C) or AMS (^{14}C; Reddy et al. 2004), many chemicals can now be analyzed for ^{14}C and ^{13}C with limited interferences and at smaller concentrations. However many of the studies are associated with natural, uncontaminated marine systems. Relatively few studies have been carried out in contaminated environments to identify the role of biodegradation by examining radiocarbon and stable carbon in individual chemicals, or in biological materials or organisms inhabiting those contaminated environments. The scope of this chapter does not allow a complete and detailed examination of this body of literature, but a few recent applications of combined ^{14}C and ^{13}C studies measuring individual contaminants or biological materials associated with contaminated environments will be highlighted to direct the reader to that literature.

Mandalakis et al. (2004) carried out one of the few studies that examined the apportionment of natural versus anthropogenic sources of polycyclic aromatic hydrocarbons (PAHs) to urban sediment. They measured 20 individual PAHs in surface sediments to calculate the percentage of PAHs in the Stockholm area that originated from fossil fuel combustion versus biofuel combustion such as wood products. They concluded that the predominance of PAHs in sediment originated from fossil sources. The contribution of biofuels to PAH contamination was estimated at $17 \pm 9\%$ (Mandalakis et al. 2004).

A few studies have extracted microbial phospholipids, which are part of the cell membrane lipid bilayer from contaminated areas, to assess whether there has been incorporation of the depleted radiocarbon. Slater et al. (2005) measured the $\Delta^{14}C$ and $\delta^{13}C$ of bacterial phospholipid fatty acids (PLFA) (See Chapter 10, section 10.6.1) to characterize degradation in salt marsh sediments contaminated with petroleum hydrocarbons. The $\Delta^{14}C$ of the bacterial PLFA was similar to that of the sediment that was solvent-extracted representing sedimentary organic material only, and matched less closely with unextracted sediment residues representing both the petroleum hydrocarbon and natural sedimentary organic material, particularly at depths of approximately 14–17 cm. The $\delta^{13}C$ values of the PLFAs were distinguishable from both the extracted sediment and unextracted sediment at 4‰ depleted compared to sediment residues. The source of carbon for the cells was not identified, but was assumed to originate from a common source because the $\delta^{13}C$ values did not vary significantly by depth from 0 to 20 cm. The authors concluded from the PLFA ^{14}C results that biodegradation of the petroleum had not occurred, and that bacteria preferentially used the natural organic matter as a carbon source and not the radiocarbon-depleted petroleum hydrocarbons.

Slater et al. (2006) isolated PLFA from bacteria in the rocky intertidal zone impacted by an oil spill. Of the four groups of microbial PLFA that were analyzed for $\Delta^{14}C$, results suggested that only the PLFA of the heterotrophic bacteria were depleted in $\Delta^{14}C$. This depletion corresponded to biodegradation of the n-alkane fraction and the removal of that fraction was attributed to biodegradation and not physical removal due to evaporation of water washing of the area. Wakeham et al. (2006) also measured PLFA extracted from sediment from a petroleum-contaminated salt marsh, and compared results to those from an uncontaminated salt marsh. They measured depleted $\Delta^{14}C$ PLFA ($+ 25 \pm 19‰$) extracted from the contaminated sediment compared to the $\Delta^{14}C$ PLFA ($+ 101 \pm 12‰$) extracted from the uncontaminated sediment. Based on the $\Delta^{14}C$ data they calculated that 6 to 10% of the C in the bacterial PLFAs at the contaminated site was derived from petroleum.

11.7 CONCLUSION

Natural and anthropogenic carbon cycling is complex and affected by differing transport processes, bacterial activities and sources, and sinks of organic carbon. Spatial scales in environmental research can range from large ecological units to units of individual enzymes. Combined naturally occurring radiocarbon and stable carbon isotope analysis can be used at all scales, and to estimate relative contributions of different microbial processes to contaminant degradation. They have been used to determine sources, sinks, and residence times of organic matter in rivers, estuaries, and oceans, sources and biodegradation of environmental contaminants, and the impact of bacteria on groundwater geochemistry.

Regardless of the microbial process (aerobic, anaerobic, methanogenic based on acetate fermentation or CO_2 reduction, or subsequent oxidation of methane produced), the radiocarbon measured is reflective of the radiocarbon content of the starting material. Depleted radiocarbon indicates that the product was generated from degradation of radiocarbon-depleted sources. However limitations exist in the use of radiocarbon alone. Naturally occurring radiocarbon cannot assess which microbial process contributed to the DIC generated from contaminant degradation. Also, all sources that contribute to the $^{14}CO_2$ signature must be assessed because any radiocarbon-free sources of CO_2 will dilute the signal from the biodegradation of the contaminant of interest. Additional limitations of the radiocarbon measure are associated with costs and sample processing. However it is likely that analytical costs will decrease as the instruments' costs decrease and they are acquired by more facilities. The use of $\delta^{13}C$ of DIC and CO_2 has limitations when examining degradation of contaminants if plant root-respired $\delta^{13}CO_2$ overlaps with the $\delta^{13}CO_2$ from contaminant degradation. The $\delta^{13}C$ of individual contaminants will not have this interference.

Radiocarbon analysis is expensive and requires sophisticated analytical equipment. However, radiocarbon and stable carbon are useful for identifying sources of carbon in natural systems, identifying significant aerobic and anaerobic microbial degradation pathways, and providing an additional degree of specificity to studies of organic matter cycling. In the future as more instrumentation is developed with additional sample inlet systems and compound separation capabilities, more and more compounds will be amenable to combined ^{14}C and ^{13}C analysis. The combined stable carbon and radiocarbon signals provide complementary information and will continue to be developed in future biodegradation studies.

ACKNOWLEDGMENTS

We thank Drs. Thomas Boyd and Greg Slater for helpful suggestions during review of this chapter. Ms. Melissa Engle for help with the figures and Dr. Thomas Boyd for use of his Figures 11.1 and 11.3. This work was supported by an ERIG Award from the University of South Carolina.

REFERENCES

Aelion, C. M. 2000. Natural remediation of radionuclides. In *Natural remediation of environmental contaminants: Its role in ecological risk assessment and risk management*, eds. C. M. Swindoll, R. G. Stahl Jr., and S. J. Ells, 273–95. Pensacola, FL: SETAC Press, Society of Environmental Toxicology and Chemistry.

Aelion, C. M., B. C. Kirtland, and P. A. Stone. 1997. Radiocarbon assessment of aerobic petroleum bioremediation in the vadose zone and groundwater at an AS/SVE site. *Environmental Science & Technology* 31:3363–70.

Aravena, R., B. G. Warner, D. J. Charman, L. R. Belyea, S. P. Mathur, and H. Dinel. 1993. Carbon isotopic composition of deep carbon gases in an ombrogenous peatland, northwestern Ontario, Canada. *Radiocarbon* 35:271–76.

Aravena, R., L. I. Wassenaar, and L. N. Plummer. 1995. Estimating ^{14}C groundwater ages in a methanogenic aquifer. *Water Resources Research* 31:2307–17.

Atlas, R. M., and R. Bartha. 1997. *Microbial ecology: Fundamentals and applications,* 4th ed. Redwood City, CA: The Benjamin/Cumming Publishing Company Inc.

Bard, E., M. Arnold, R. G. Fairbanks, and B. Hamelin. 1993. [230]Th-[234]U and ^{14}C ages obtained by mass spectrometry on corals. *Radiocarbon* 35:191–99.

Bard, E., B. Hamelin, R. G. Fairbanks, A. Zindler, M. Arnold, and G. Mathieu. 1990. U/Th and ^{14}C ages of corals from Barbados and their use for calibrating the ^{14}C time scale beyond 9000 years B.P. *Nuclear Instruments & Methods in Physics Research Section B-Beam Interactions with Materials and Atoms* 52:461–68.

Barker, J. F., and P. Fritz. 1981. The occurrence and origin of methane in some groundwater flow systems. *Canadian Journal of Earth Sciences* 18:1802–16.

Biedenbender, S. H., M. P. McClaran, J. Quade, and M. A. Weltz. 2004. Landscape patterns of vegetation change indicated by soil carbon isotope composition. *Geoderma* 119:69–83.

Broecker, W. S. 2004. Radiocarbon. In *Treatise on geochemistry*, eds. H. D. Holland and K. K. Turekian, 245–60. Oxford: Elsevier-Pergamon.

Clark, I. D., and P. Fritz. 1997. *Environmental isotopes in hydrogeology*, 1st ed. Boca Raton, FL: CRC Press.

Coleman, D. D., C. Liu, K. C. Hackley, and S. R. Pelphrey. 1996. Isotopic identification of landfill methane. *Environmental Geoscience* 2:95–103.

Conrad, M. E., P. F. Daley, M. L. Fischer, B. B. Buchanan, T. Leighton, and M. Kashgarian. 1997. Combined ^{14}C and δ^{13}C monitoring of *in situ* biodegradation of petroleum hydrocarbons. *Environmental Science & Technology* 31:1463–69.

Conrad, M. E., A. S. Templeton, P. F. Daley, and L. Alvarez-Cohen. 1999. Isotopic evidence for biological controls on migration of petroleum hydrocarbons. *Organic Geochemistry* 30:843–59.

Dörr, H., and K. O. Münnich. 1986. Annual variations of the ^{14}C content of soil CO_2. *Radiocarbon* 28:338–45.

Drimmie, R. J., R. Aravena, and L. I. Wassenaar. 1991. Radiocarbon and stable isotopes in water and dissolved constituents, Milk River aquifer, Alberta, Canada. *Applied Geochemistry* 6:381–92.

Druffel, E. R. M., J. E. Bauer, S. Griffin, S. R. Beaupré, and J. Hwang. 2008. Dissolved inorganic radiocarbon in the North Pacific Ocean and Sargasso Sea. *Deep-Sea Research Part I-Oceanographic Research Papers* 55:451–59.

Faure, G. 1986. *Principles of isotope geology.* New York: John Wiley & Sons.

Fritz, P., and J.-C. Fontes. 1980. *Handbook of environmental isotope geochemistry.* Amsterdam: Elsevier.

Gelwicks, J. T. 1989. The stable carbon isotope biogeochemistry of acetate and methane in freshwater environments. PhD diss., Indiana University.

Glaser, B., and W. Zech. 2005. Reconstruction of climate and landscape changes in a high mountain lake catchment in the Gorkha Himal, Nepal during the Late Glacial and Holocene as deduced from radiocarbon and compound-specific stable isotope analysis of terrestrial, aquatic and microbial biomarkers. *Organic Geochemistry* 36:1086–98.

Godwin, H. 1962. Half-life of radiocarbon. *Nature* 195:984.

Goh, K. M. 1991a. Carbon dating. In *Carbon isotope techniques,* eds. D. C. Coleman and F. Fry, 125–45. San Diego, CA: Academic Press, Inc.

Goh, K. M. 1991b. Bomb carbon. In *Carbon isotope techniques,* eds. D. C. Coleman and F. Fry, 147–51. San Diego, CA: Academic Press, Inc.

Grossman, E. L., B. K. Coffman, S. J. Fritz, and H. Wada. 1989. Bacterial production of methane and its influence on ground-water chemistry in east-central Texas aquifers. *Geology*, 17:495–99.

Gulin, S. B., G. G. Polikarpov, and V. N. Egorov. 2003. The age of microbial carbonate structures grown at methane seeps in the Black Sea with an implication of dating of the seeping methane. *Marine Chemistry* 84:67–72.

Haas, H., D. W. Fisher, D. C. Thorstenson, and E. P. Weeks. 1983. $^{13}CO_2$ and $^{14}CO_2$ measurements on soil atmosphere sampled in the subsurface unsaturated zone in the western Great Plains of the US. *Radiocarbon* 25:301–14.

Hackley, K. C., C. L. Liu, and D. D. Coleman. 1996. Environmental isotope characteristics of landfill leachates and gases. *Ground Water* 34:827–36.

Hanshaw, B. B., R. Meyer, W. Back, and I. Friedman. 1967. Radiocarbon determinations applied to ground water hydrology. In *Isotope techniques in the hydrologic cycle,* ed. G. E. Stout, 117–18. Geophysical Monograph 11. Washington, DC: American Geophysical Union.

Hinchee, R. E., and S. K. Ong. 1992. A rapid in situ respiration test measuring aerobic biodegradation rates of hydrocarbons in soil. *Air & Waste Management Association* 42:1305–12.

Hwang, J., E. R. M. Druffel, and J. E. Bauer. 2006. Incorporation of aged dissolved organic carbon (DOC) by oceanic particulate organic carbon (POC): An experimental approach using natural carbon isotopes. *Marine Chemistry* 98:315–22.

Karltun, E., A. F. Harrison, A. Alriksson, C. Bryant, M. H. Garnett, and M. T. Olsson. 2005. Old organic carbon in soil solution DOC after afforestation—evidence from ^{14}C analysis. *Geoderma* 127:188–95.

Key, R. M. 2001. Radiocarbon. In *Encyclopedia of ocean sciences,* eds. J. H. Steele, K. K. Turekian, and S. A. Thorpe, 2338–53. London: Academic Press.

Kirtland, B. C., C. M. Aelion, and P. A. Stone. 2000. Monitoring natural attenuation of petroleum using soil gas and isotopic assessment. *Bioremediation Journal* 4:187–201.

Kirtland, B. C, C. M. Aelion, and P. A. Stone. 2005. Assessing in situ mineralization of recalcitrant organic compounds in vadose zone sediments using $\delta^{13}C$ and ^{14}C measurements. *Journal of Contaminant Hydrology* 76:1–18.

Kirtland, B. C., C. M. Aelion, P. A. Stone, and D. Hunkeler. 2003. Isotopic and geochemical assessment of in situ biodegradation of chlorinated hydrocarbons. *Environmental Science & Technology* 37:4205–12.

Klouda, G. A., C. W. Lewis, R. A. Rasmussen, G. C. Rhoderick, R. L. Sams, R. K. Stevens, L. A. Currie, D. J. Donahue, A. J. T. Jull, and R. L. Seila. 1996. Radiocarbon measurements of atmospheric volatile organic compounds: Quantifying the biogenic contribution. *Environmental Science & Technology* 30:1098–1105.

Kunkler, J. L. 1969. The sources of carbon dioxide in the zone of aeration of the Bandelier Tuff, near Los Alamos, New Mexico, Professional Paper 650-13, U.S. Geological Survey, Washington, DC.

Landmeyer, J. E., and P. A. Stone. 1995. Radiocarbon and $\delta^{13}C$ values related to ground-water recharge and mixing. *Ground Water* 33:227–34.

Lansdown, J. M., P. D. Quay, and S. L. King. 1992. CH_4 production via CO_2 reduction in a temperate bog: A source of ^{13}C-depleted CH_4. *Geochimica Cosmochimica Acta* 56:3493–3503.

Libby, W. F., E. C. Anderson, and J. R. Arnold. 1949. Age determination by radiocarbon content: Worldwide assay of natural radiocarbon. *Science* 109:227–28.

Lloyd, J. W. 1981. Environmental isotopes in groundwater. In *Case studies in groundwater resources evaluation,* ed. J. W. Lloyd, 113. Oxford: Clarendon Press.

Longworth, B. E., S. T. Petsch, P. A. Raymond, and J. E. Bauer. 2007. Linking lithology and land use to sources of dissolved and particulate organic matter in headwaters of a temperate, passive-margin river system. *Geochimica Cosmochimica Acta* 71:4233–50.

Mandalakis, M., Ö. Gustafsson, C. M. Reddy, and L. Xu. 2004. Radiocarbon apportionment of fossil versus biofuel combustion sources of polycyclic aromatic hydrocarbons in the Stockholm metropolitan area. *Environmental Science & Technology* 38:5344–49.

Mazor, E. 1991. *Applied chemical and isotopic groundwater hydrology.* New York: Halsted Press.

Mook, W. G. 1980. Carbon-14 in hydrological studies. In *Handbook of environmental isotope geochemistry*, eds. C. P. Fritz and J.-Ch. Fontes, 49–74. Amsterdam: Elsevier.

Olsson, I. U. 1970. The use of oxalic acid as a standard. In *Radiocarbon variations and absolute chronology*, ed. I. U. Olsson. Nobel symposium 12th Proc., 17. New York: John Wiley and Sons.

Pearson, A., A. P. McNichol, B. C. Benitez-Nelson, J. M. Hayes, and T. I. Eglinton. 2001. Origins of lipid biomarkers in Santa Monica Basin surface sediment: A case study using compound specific $\Delta^{14}C$ analysis. *Geochimica Cosmochimica Acta* 65:3123–37.

Petsch, S. T., T. I. Eglinton, and K. J. Edwards. 2001. ^{14}C-dead living biomass: Evidence for microbial assimilation of ancient organic carbon during shale weathering. *Science* 292:1127–31.

Pine, M. J., and H. A. Barker. 1956. Studies on methane fermentation XII. The pathway of hydrogen in the acetate fermentation. *Journal of Bacteriology* 71:644–48.

Rafter Radiocarbon Laboratory: GNS Science Limited. 2008. Radiocarbon glossary, Institute of Geological and Nuclear Sciences, Ltd. http://www.gns.cri.nz/nic/rafterradiocarbon/glossary.htm (accessed September 25, 2008).

Raymond, P. A., and J. E. Bauer. 2001. Use of ^{14}C and ^{13}C natural abundances for evaluating riverine, estuarine, and coastal DOC and POC sources and cycling: A review and synthesis. *Organic Geochemistry* 32:469–85.

Reddy, C. M., L. Xu, G. W. O'Neil, R. K. Nelson, T. I. Eglinton, D. J. Faulkner, R. Norstrom, P. S. Ross, and S. A. Tittlemier. 2004. Radiocarbon evidence for a naturally produced bioaccumulating halogenated organic compound. *Environmental Science & Technology* 38:1992–97.

Roland, L. A., M. D. McCarthy, and T. Guilderson. 2008. Sources of molecularly uncharacterized organic carbon in sinking particles of three ocean basins: A coupled $\Delta^{14}C$ and $\delta^{14}C$ approach. *Marine Chemistry* 111:199–213.

Rosen, A. A., and M. Rubin. 1965. Discriminating between natural and industrial pollution through carbon dating. *Journal of the Water Pollution Control Federation* 37:1302–7.

Shah, S. R., G. Mollenhauer, N. Ohkouchi, T. I. Eglinton, and A. Pearson. 2008. Origins of archaeal tetraether lipids in sediments: Insights from radiocarbon analysis. *Geochimica Cosmochimica Acta* 72:4577–94.

Sharabi, N., and R. Bartha. 1993. Testing of some assumptions about biodegradability in soil as measured by carbon dioxide evolution. *Applied Environmental Microbiology* 59:1201–5.

Shen, J., and R. Bartha. 1996. Metabolic efficiency and turnover of soil microbial communities in biodegradation tests. *Applied Environmental Microbiology* 62:2411–15.

Slater, G. F., R. K. Nelson, B. M. Kile, and C. M. Reddy. 2006. Intrinsic bacterial biodegradation of petroleum contamination demonstrated *in situ* using natural abundance, molecular-level ^{14}C analysis. *Organic Geochemistry* 37:981–89.

Slater, G. F., H. K. White, T. I. Eglinton, and C. M. Reddy. 2005. Determination of microbial carbon sources in petroleum contaminated sediments using molecular ^{14}C analysis. *Environmental Science & Technology* 39:2552–58.

Spiker, E. C., and M. Rubin. 1975. Petroleum pollutants in surface and groundwater as indicated by the carbon-14 activity of dissolved organic carbon. *Science* 187:61–4.

Stone, P. A., R. J. Devlin, and J. P. Moore. 1997. Aquifer vulnerability in the Piedmont and Coastal Plain: Some environmental tracer evidence of recharging. In *Proceedings of the 1997 Georgia water resources conference,* ed. K. J. Hatcher. Athens, GA, March 20–22, 1997, 497–500. Athens, GA: Institute of Ecology, University of Georgia.

Stuiver, M. 1970. Evidence for the variation of atmospheric ^{14}C content in the late quaternary. In *Late cenozoic glacial ages,* ed. K. K. Turekian, 55–70. New Haven, CT: Yale University Press.

Stuiver, M., G. W. Pearson, and T. Brazunias. 1986. Radiocarbon age calibration of marine samples back to 9000 calendar years before present. *Radiocarbon* 28:980–1021.

Stuiver, M., and H. A. Polach. 1977. Discussion: Reporting of ^{14}C data. *Radiocarbon* 19:355–63.

Suchomel, K. H., D. K. Kreamer, and A. Long. 1990. Production and transport of carbon dioxide in a contaminated vadose zone: A stable and radioactive carbon isotope study. *Environmental Science & Technology* 24:1824–31.

Suess, H. E. 1954. Natural radiocarbon measurements by acetylene counting. *Science* 120:5–7.

Tans, P. P., A. F. M. De Jong, and W. G. Mook. 1979. Natural atmospheric ^{14}C variation and the Suess effect. *Nature* 280:826–28.

Taylor, R. E., M. Stuiver, and P. J. Reimer. 1996. Development and extension of the calibration of the radiocarbon time scale: Archaeological applications. *Quaternary Science Reviews* 15:655–68.

Teuten, E. L., G. M. King, and C. M. Reddy. 2006. Natural ^{14}C in *Saccoglossus bromophenolosus* compared to ^{14}C in surrounding sediments. *Marine Ecology Progress Series* 324:167–72.

Thorstenson, D. C., E. P. Weeks, H. Haas, and D. W. Fisher. 1983. Distribution of gaseous $^{12}CO_2$, $^{13}CO_2$, and $^{14}CO_2$ in the sub-soil unsaturated zone of the western US Great Plains. *Radiocarbon* 25:315–46.

Verburg, P. 2007. The need to correct for the Suess effect in the application of $\delta^{13}C$ in sediment of autotrophic Lake Tanganyika, as a productivity proxy in the anthropocene. *Journal of Paleolimnology* 37:591–602.

Vogel, J. C. 1970. Carbon-14 dating of groundwater. In *Isotope hydrology.* International Atomic Energy Agency Symposium 129, March 9–13, 225–39. Vienna: International Atomic Energy Agency.

Wakeham, S. G., A. P. McNichol, J. E. Kostka, and T. K. Pease. 2006. Natural-abundance radiocarbon as a tracer of assimilation of petroleum carbon by bacteria in salt marsh sediments. *Geochimica Cosmochimica Acta* 70:1761–71.

Wang, X.-C., and E. R. M. Druffel. 2001. Radiocarbon and stable carbon isotope compositions of organic compound classes in sediments from the NE Pacific and Southern Oceans. *Marine Chemistry* 73:65–81.

Whiticar, M. J., E. Faber, and M. Schoell. 1986. Biogenic methane formation in marine and freshwater environments: CO_2 reduction vs. acetate fermentation-isotopic evidence. *Geochimica Cosmochimica Acta* 50:693–709.

Wigley, T. M. L., L. N. Plummer, and F. J. Pearson Jr. 1978. Mass transfer and carbon isotope evolution in natural water systems. *Geochimica Cosmochimica Acta* 42:1117–39.

12 Nontraditional Stable Isotopes in Environmental Sciences

Christopher S. Romanek, Brian Beard,
Ariel D. Anbar, and C. Fred T. Andrus

CONTENTS

12.1   INTRODUCTION

Stable isotopes are used in two ways in environmental investigations: (1) they are introduced as tracers to map the flow of mass through a system, or (2) the stable isotope compositions of materials are measured to infer the processes responsible for isotope distribution patterns at the natural abundance level. The latter usage requires some fundamental knowledge about the factors that influence stable isotope partitioning in nature. Much of this information has been gathered over the last half century for the biologically active elements, hydrogen (H), carbon (C), nitrogen (N), oxygen (O), and sulfur (S). The mass difference between the rare and most common isotope is relatively large for these elements, which results in isotope fractionation during physicochemical and biological reactions and permits distinctions to be made concerning the origin and modification of materials over time.

With recent technological advances in instrumentation (e.g., multicollector inductively coupled plasma mass spectrometers) and techniques (e.g., double spike), the stable isotope composition of a new array of nontraditional stable isotopes is being measured with unprecedented precision. In a review of the literature, Coplen et al. (2002) report detectable isotope variation (not due to radioactive decay) for 20 of these elements, including boron (B), calcium (Ca), chlorine (Cl), chromium (Cr), copper (Cu), iron (Fe), lithium (Li), magnesium (Mg), molybdenum (Mo), palladium (Pd), selenium (Se), silicon (Si), tellurium (Te), thallium (Tl), and zinc (Zn). Among these, Li, Ca, Cl, Cr, Fe, Mo, and Se are reported in this chapter because they display significant isotope variation in nature, their isotope systematics are reasonably well known and they are of direct relevance to environmental studies. Much of the information presented here is drawn from the book *The Geochemistry of Non-Traditional Stable Isotopes* edited by Johnson et al. (2004). The reader is referred to this book for additional details on the subject matter discussed as well as other emerging stable isotope studies (e.g., Cu and Zn) that may be useful in environmental investigations.

Elements are families of atoms that have a common number of protons in their nucleus; each family may be further subdivided into smaller groupings of isotopes depending on the number of neutrons contained in the nucleus of each atom. Two basic types of isotopes exist in nature, those that have a stable nuclear configuration and those that do not. Elementary particles that contain too few or too many neutrons are inherently unstable and spontaneously decay into new elements over time. This attribute defines the radioisotopes, while stable isotopes comprise particles that do not decay over geological time scales. As a general rule, the number of stable isotopes of an element increases with atomic number. For example, the relatively volatile elements H, C, and N have two stable isotopes while O has three and S has four (tin has the most, 10 stable isotopes in all).

12.1.1   MASS FRACTIONATION PROCESSES

Stable isotopes are fractionated in nature during physicochemical and biological reactions by mass-dependent and mass-independent processes. Mass-dependent isotope fractionation occurs because of small changes in the bond energies of molecules that contain isotopes of an element of differing mass. Vibrational frequencies for molecules containing the isotope of greater mass are relatively low and this makes them slightly more stable. The effect of mass-dependent fractionation can be readily observed by comparing isotope ratios for an element that has at least three isotopes. For example, the $^{57}Fe/^{54}Fe$ and $^{56}Fe/^{54}Fe$ isotope ratios (reported as the delta values $\delta^{57}Fe$ and $\delta^{54}Fe$, respectively; see below) for terrestrial materials produce a linear array on a cross plot with a slope of ~3/2 because the ratio of the relative mass difference between ^{57}Fe and ^{54}Fe (~3 amu), and ^{56}Fe and ^{54}Fe (~2 amu) is ~3/2. Mass-independent reactions, which occur rarely in nature, do not follow this rule because the mechanism of isotope fractionation responds to symmetry operations in unusual chemical reactions (e.g., see Heidenreich and Thiemens 1983) and other effects (e.g. nuclear volume; Bigeleisen 1996). The magnitude of mass-dependent fractionation decreases as atomic mass increases because relative mass differences between isotopes do not increase linearly with the average atomic mass of the elements. This is one reason why elements of lower mass, such as H, C, and N display wider ranges in isotope composition than elements of higher mass (e.g., Fe).

The isotope composition of natural materials also depends on the nature and extent of the reactions that produce them. During some physicochemical reactions, atoms and isotopes continually exchange between reactants and products, even after molecular assemblages attain chemical equilibrium. Exchange reactions that proceed at similar rates in both directions, i.e., at isotope equilibrium, display an *equilibrium isotope effect* (see Chapter 3), where reactants and products have predictable isotope compositions. Knowledge of the expected equilibrium isotope effect for physicochemical reactions is a powerful tool that is widely used in the earth and environmental sciences. Often times, equilibrium exchange reactions are temperature dependent and this forms the basis for the discipline called isotope paleothermometry.

In a second class of reactions, mass flows preferentially from reactant to product and isotopes are partitioned according to their relative rate of reaction. These reactions display a kinetic isotope effect where the reactant of lower relative mass is converted to product at a slightly faster rate (see Chapter 3). Kinetic isotope effect invariably enriches the product in the isotope of lower mass and leaves the reactant enriched in the isotope of higher mass. Such partitioning behavior is a hallmark of most biological reactions.

Early studies that measured stable isotopes in natural materials focused primarily on the light elements, H, C, N, O, and S because mass differences between the isotopes of these elements are relatively large (up to 100% in the case of H). Isotope fractionation can be significant under these conditions and relatively large differences in isotope composition have been measured in samples using a gas source isotope ratio mass spectrometer (IRMS).

12.1.2 NOTATIONS AND CONVENTIONS

It is difficult to measure *absolute* isotope abundances in terrestrial materials with the precision required to discriminate between them because of the very small differences that exist in nature. This obstacle is overcome analytically by bracketing the isotope measurement of a sample with the identical measurement of a standard reference material of known or assumed isotope content. The convention most often used for reporting stable isotope data is delta notation (see Chapter 1, Equation 1.2), where the isotope ratio of the sample is normalized to that of a standard.

In this way, instrumental mass discrimination associated with the analysis of the sample and standard cancel each other out and extremely precise isotope compositions may be achieved. For most applications, isotope ratios are determined by placing the rare isotope in the numerator position and the major isotope in the denominator position. This is done to facilitate mathematical manipulations in mixing equations and other relations. For elements having multiple isotopes, the choice for reporting isotope ratios is dependent primarily on the level of precision that is required. Measuring the isotope ratio of the highest and lowest atomic mass for an element will yield the largest difference in isotopic composition among materials because mass-dependent fractionation drives variability. There are instances, however, where this comparison is not practical, such as when one isotope is at very low abundance or where isobaric interferences degrade analytical precision.

Two very useful terms that are widely used for defining isotope relationships are the fractionation factor (α) and the Δ value (see Chapter 3). The Δ value is used to describe the difference in delta value (δ) between any two substances. This value approximates the function $1000 \ln\alpha$ closely when α values are relatively close to unity and δ values are close to zero. This relationship is useful because theoretical relationships follow the form of the expression $1000 \ln\alpha$.

12.2 INSTRUMENTATION

Three types of mass spectrometers are commonly used to analyze the stable isotope composition of environmental compounds: (1) the gas source IRMS, (2) the thermal ionization mass spectrometer (TIMS), and (3) the multicollector inductively coupled plasma mass spectrometer (MC-ICP-MS). All of these mass spectrometers have similar analyzers that consist of a magnetic sector that is used to produce a mass spectrum, and multiple faraday buckets to collect different mass ion beams simultaneously. Multicollection helps to avoid imprecision caused by ion beam instability. The main differences between these different mass spectrometers are how the sample is introduced into the source and how the sample is converted to ions. For IRMS analysis, samples are converted to (or equilibrated with) a gas that is admitted to the source of the mass spectrometer for the simultaneous analysis of isotope abundances. The technique of thermal ionization mass spectrometry was developed for the analysis of elements that are not easily volatilized. This technique involves depositing the element of interest as a solid on a filament, which is then loaded into the source of an IRMS for analysis. A current is applied to the filament that volatilizes the element of interest for isotope measurement.

With the advent of the MC-ICP-MS, a valuable new tool has become available for the measurement of isotope ratios in natural materials. With this technique, an aerosol sample (liquid or solid) is introduced to the source of the mass spectrometer where an argon plasma atomizes and ionizes particles for simultaneous or sequential isotope analysis. This technique has revolutionized the way geochemists view the natural world as relatively small differences (~0.05‰) in the isotope composition of elements can be routinely measured with a high throughput (20 samples per day) of small sized samples (~100 nanograms of analyte).

12.2.1 ISOTOPE RATIO MASS SPECTROMETERS (IRMSS)

Traditionally, stable isotope compositions have been measured using a gas source IRMS (Chapter 2). For this instrument, a gas (e.g., CH_3Cl for Cl isotope measurement) is admitted to the source region of the mass spectrometer where molecular ions are created by electron impact. The ions are focused into a beam and accelerated to a near uniform kinetic energy through a voltage potential. From there, the ion beam travels through an evacuated flight tube to a magnetic sector for mass discrimination. Ion beams of distinct mass to charge ratio emerge from the mass analyzer and are captured in precisely positioned faraday cups by manipulating the voltage potential of the source region and the field strength of the magnet.

For IRMS, instrumental mass fractionation (mass bias) occurs primarily when gas is introduced to the source of the mass spectrometer. Reference and sample gases are admitted alternatively in dynamic switching mode or they are injected as a pulse in a flowing stream of carrier gas in continuous flow mode. In either mode, the sample and reference gas are mass-fractionated similarly so these effects are canceled and highly accurate and precise results may be achieved.

12.2.2 THERMAL IONIZATION MASS SPECTROMETERS

The stable isotope composition of nonvolatile elements is determined using a thermal ionization mass spectrometer. With this type of analysis, the element of interest is quantitatively extracted from the sample matrix and purified in the laboratory using wet chemical techniques. The analyte is loaded on a filament and placed inside the source of the mass spectrometer where it is heated by ohmic heating, the sample is vaporized, and then ionized by electrons produced by the filament. Ions may be generated by a single filament or by an additional filament that is heated to a high temperature (2000°C) to produce a high flux of electrons with the sample being vaporized at a lower temperature. Once ionized, the elementary particles of interest are focused into an ion beam, accelerated, and mass analyzed as described above.

In TIMS analysis, instrumental mass bias of up to 5‰ per mass unit may occur because isotopes of lower relative mass evaporate from the filament preferentially. This effect may vary from one filament to the next or over the course of a single analysis so corrections must be applied to the data to achieve meaningful results. There are three methods to correct for instrumental mass bias. One method is internal normalization of isotope ratios to a constant isotopic composition. Internal normalization is commonly used for detecting isotopic variations that are produced by radiogenic decay where, for example, the decay of ^{87}Rb to ^{87}Sr results in Sr having variable $^{87}Sr/^{86}Sr$ ratios as a function of the age of the sample and its Rb/Sr ratio. Internal normalization of Sr isotope ratios to a constant $^{88}Sr/^{86}Sr$ ratio allows one to measure $^{87}Sr/^{86}Sr$ ratios to a precision of ~0.01‰. However, internal normalization removes any naturally occurring mass-dependent variations and hence cannot be used to study mass-dependent isotopic variations. A second method involves empirically determining an instrumental mass bias value by analysis of standards of known isotopic composition. This method is commonly used to determine the isotopic composition of Pb, as well as other elements, such as Fe for which isotopic standards are available (Taylor, Maeck, and De Bièvre 1992; Taylor et al. 1993), but this method is not able to obtain precision beyond ~0.5‰ per amu. The most robust method of instrumental mass bias correction is by the addition of two isotope spikes of known composition (double spike method) to the sample. This method can result in a precision of 0.05‰ but can only be used with elements that have four or more isotopes.

12.2.3 Inductively Coupled Plasma Mass Spectrometers

Our understanding of stable isotope systematics has increased greatly with the introduction of MC-ICP-MS. The very high ionization efficiency, stable mass bias, and high sample throughput have opened new analytical opportunities in the earth and environmental sciences. The main drawback of MC-ICP-MS systems is the magnitude of the mass bias, which is significantly greater (one to two orders of magnitude) than for TIMS instruments. Because these instruments are so new, various aspects of MC-ICP-MS are discussed in greater detail below.

The MC-ICP-MS has five main parts: (1) a sample introduction system, (2) an ion source, (3) an ion guide, (4) an energy and mass analyzer, and (5) a collector system. The most common type of sample delivery system is a nebulizer that produces an aqueous aerosol that is aspirated into a spray chamber where the aerosol particles are sorted by size prior to introduction to the source (Monaster et al. 1998a,b). The analyte is introduced through a Fassel-type torch that sits upstream of an induced argon plasma (8000 K) that is generated by radio frequency power. Plasma dynamics promote the formation of solid particles that are subsequently vaporized, atomized, and ionized. The ionization efficiency for most metallic elements is 100%. Since the plasma is composed primarily of Ar atoms (mostly neutral atoms but also charged species and electrons) that are in constant collision, analyte ions tend to travel at similar velocities so the kinetic energies of the particles (and instrumental mass bias) are proportional to the mass differences between analytes (Maréchal, Télouk, and Albarède 1999). The dynamics of ion transfer also give rise to space charge effects that permit analytes of higher relative mass to be transferred more efficiently through the instrument, resulting in the measurement of isotope values that are higher than true values. Space charge effects are particularly important when sample and standard matrices vary in concentration and composition.

Special analytical considerations must be met with ICP-MS instruments because positive gas pressures are required to sustain the plasma while a vacuum is required for mass analysis. These two environments are regulated by positioning apertures (sampling and skimmer cones) within the source and using differential pumping to maintain the appropriate pressure in different regions of the instrument (Douglas and Tanner 1998). On the high vacuum side of the skimmer cone there is a series of lenses that function to collimate and focus the ion beam. Ions of variable kinetic energy enter the mass analyzer where they may be filtered using an electrostatic field and then refocused for mass separation using a magnetic sector and directed to an array of collectors. In some instruments, the electrostatic filter is replaced by a collision cell that thermalizes and reduces the energy spread of the ions.

12.2.3.1 Mass Bias Correction in Multicollector Inductively Coupled Plasma Mass Spectrometer (MC-ICP-MS)

Instrumental mass bias in plasma ionization is large and is a function of atomic mass. For example, heavy elements (e.g., U) fractionate by ~3‰ per mass and lighter elements (e.g., Mg) fractionate more (~50‰ per mass). The magnitude of instrumental mass bias, however, is roughly constant during an analysis. Several methods have been developed to correct for it. The simplest method is standard bracketing, similar to the methods developed for gas source isotope ratio mass spectrometry. An alternate method is to spike the sample with an element of similar atomic mass to determine the instrumental mass fractionation. However, it is critical to recognize that the instrumental mass bias for differing elements is not identical and one must apply a correction factor when inferring the mass bias factor for one element from another (Maréchal, Télouk, and Albarède 1999). An additional method, which is also widely used in TIMS, is the double spike technique. This method has afforded the opportunity for new avenues of research in nontraditional stable isotope analysis (e.g., Johnson and Beard 1999). Two isotopes that occur at relative low abundance are added in known proportion to a sample prior to the chemical separation of the element of interest for isotope analysis. In this way, natural mass dependent isotope fractionation can be discriminated from instrumental mass bias caused during the analysis. The fundamental advantage of this technique is that the added isotopes follow the exact same fractionation behavior as the isotopes of interest during the analysis. This technique depends critically on accurate and precise knowledge of the concentration ratio of the spike for it to be a useful analytical tool. The technique has practical advantages over other analytical procedures such as sample standard bracketing and element spike methods, because instrumental and natural mass discrimination effects are evaluated simultaneously on a single sample and the method does not require a quantitative yield in chemical separation prior to analysis.

12.3 STABLE ISOTOPE SYSTEMATICS OF SELECTED ELEMENTS

The stable isotope systematics of lithium, chlorine, calcium, selenium, chromium, iron, and molybdenum are briefly discussed below as they pertain to environmental studies (Table 12.1). Lithium isotope distributions show the greatest range in isotope composition known to exist in terrestrial samples. Lithium isotope analysis may gain its widest acceptance in environmental studies as an anthropogenic indicator based on the fact that industrial processing imparts an unusual isotope composition on Li that is not observed in natural samples. Chlorine isotopes have been used in environmental studies for similar reasons. In addition, the industrial manufacturers of chlorinated organic solvents can potentially be identified based on their stable isotope composition. Calcium isotopes display a more limited range of values in natural samples but systematic variations in biological tissues provide a novel way to trace the flow of contaminants through food webs in impacted ecosystems. Additionally, Ca may be useful as either a paleothermometer or indicator of pCO_2 in studies of past climatic change.

Next, stable isotope systematics are presented for the redox sensitive elements Se, Cr, Fe, and Mo. These are elements of relatively high atomic mass that exist in various oxidation states and bonding environments. Since the bioavailability of these elements depends heavily on these factors, stable isotope compositions may shed light on exposure routes and uptake pathways for living organisms. Moreover, microorganisms use these elements to exploit electron transfer processes to gain energy from the environment for metabolism and growth so the stable isotope composition of these elements may provide insights into microbiological processes that may control the mobility of metal contaminants in surface and subsurface environments.

12.3.1 Lithium

The relatively large mass difference for the stable isotopes of lithium (~16% between ^6Li and ^7Li) makes Li an attractive element to explore for isotope variation in natural and anthropogenic materials. Comprising ~7.5% of the terrestrial inventory, ^6Li is the rarer of the two stable isotopes. The isotope ratios of lithium

---

TABLE 12.1

Summary of Nontraditional Stable Isotopes Reported in This Chapter

Isotopes (Minor/Major)	Other Minor Stable Isotopes[a]	Technique	Standards
$^6Li/^7Li$	None	TIMS, MC-ICP-MS	SRM-8545
$^{37}Cl/^{35}Cl$	None	IRMS, TIMS	SMOC
$^{44}Ca/^{40}Ca$	^{42}Ca, ^{43}Ca, ^{46}Ca, ^{48}Ca	TIMS; MC-ICP-MS	None
$^{80}Se/^{76}Se$	^{74}Se, ^{77}Se, ^{78}Se, ^{82}Se	IRMS, TIMS, MC-ICP-MS	None
$^{53}Cr/^{52}Cr$	^{50}Cr, ^{54}Cr	TIMS	SRM-979
$^{56}Fe/^{54}Fe$	^{57}Fe, ^{58}Fe	TIMS, MC-ICP-MS	IRMM-014
$^{97}Mo/^{95}Mo$	^{92}Mo, ^{94}Mo, ^{96}Mo, ^{98}Mo, ^{100}Mo	TIMS, MC-ICP-MS	None

[a] Double spike technique can be used when the total number of isotopes ≥ 4.

are measured using various instrumentation (TIMS and ICP-MS) to a precision of 0.5‰ or better. Care must be taken when evaluating Li isotope compositions because both $^6Li/^7Li$ and $^7Li/^6Li$ ratios are commonly reported in the literature. The International Union for Pure and Applied Chemistry recommends the latter convention (e.g., Coplen 1996), which is opposite the convention of placing the rare isotope in the numerator of an isotope ratio. Nevertheless, delta values (δ^6Li and δ^7Li) can be related to each other by switching the sign for most natural abundance values that fall between −40 and +40‰ (Tomascak 2004). Lithium derived from spodumene deposits in North Carolina serves as the international standard L-SVEC (now NIST SRM-8545) for the calibration of all Li isotope analyses (Flesch, Anderson, and Svec 1973).

Lithium is a trace constituent of natural groundwaters (up to 100 ppb; Wedepohl 1970) while many industrial wastes (e.g., landfill refuse, fly ash, agriculture byproducts, petrochemical derivatives) contain elevated levels of Li (Davidson and Bassett 1993; Vengosh et al. 1994; Bassett et al. 1995; Komor 1997). Anthropogenic sources of Li are expected to have distinct isotope compositions because industrial processing by ion exchange and electrochemical separations tend to preferentially remove ^6Li from feedstocks. Historically, ^6Li was used for the production of tritium in nuclear processing facilities while the ^7Li-enriched remainder made its way into the commercial sector where its environmental impacts have been felt.

The isotope fractionation of Li is most highly expressed in surface environments where temperatures are relatively low and Li bonding environments can differ. In aqueous solution, Li is highly hydrated and displays tetrahedral coordination, while in solids it commonly displays octahedral coordination. Both theoretical calculations and experimental data suggest ^6Li is preferentially favored in crystalline solids, with the magnitude of the fractionation being temperature dependent. However, nonreversible sorption processes tend to obscure these relationships, resulting in solids (e.g., clay minerals and evaporates) having δ^7Li values that are similar to the fluids with which they equilibrate.

The δ^7Li value of the bulk earth (1.5–5.6‰) is determined primarily from the measurement of mid-ocean ridge basalts (Figure 12.1a). Clastic rocks and sediments have a wide range of values (−10–20‰), while seawater has a range of values from 29–33‰. Seawater is depleted in ^6Li because surface weathering results in the preferential release of ^7Li from sediments, contributing to the average δ^7Li value of 21‰ for river water. Huh et al. (1998) suggested that relatively high δ^7Li values for river water may be explained by the degree of weathering in differing riverine catchments, with basins that experience the greatest weathering having the lowest dissolved δ^7Li values. Although this explanation does not adequately explain δ^7Li values observed for all basins of the world, it is consistent with isotope fractionation expressed during ion exchange reactions.

Application of Li isotope systematics in the environmental sciences may benefit from the fact that industrial processing of Li has resulted in anthropogenic compounds having Li isotope compositions that lie far outside the range observed for natural materials (Figure 12.1b), therefore, accidental releases of Li-bearing pollutants should be easily detected in environmental samples. Gellenbeck and

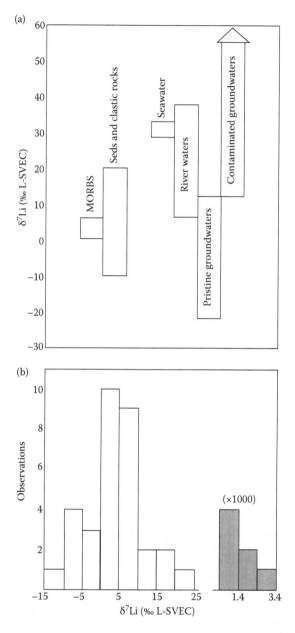

FIGURE 12.1 (a) Range of isotope compositions observed for lithium for mid-ocean ridge basalts (MORBS: representing the bulk earth), the weathering products of igneous rocks (sediments and clastic rocks), and natural waters (Modified from Tomascak, P. B., *Geochemistry of non-traditional stable isotopes*, The Mineralogical Society of America, Washington DC, 2004). (b) Distribution pattern of δ⁷Li values measured from 39 reagents acquired from various manufacturers (Modified from Qi, H. P., T. B. Coplen, Q. Z. Wang, and Y. H. Wang, *Analytical Chemistry,* 69, 4076–78, 1997). Note break in x-axis where seven samples have δ⁷Li values ranging from 434 to 3013‰ (shaded bars).

Bullen (1991) showed that relatively high δ^7Li values (+19‰) in groundwater could be attributed to point sources derived from cow manure leachate. Also, Tomascak et al. (1995a,b) showed that the ^7Li content of a stream draining agricultural lands increased in the early stages of a rainfall event, suggesting that loosely bound Li, possibly of anthropogenic origin, was preferentially mobilized from soils and aquifer materials. Hogan and Blum (2003) measured the chemistry and isotope composition of groundwaters from a landfill situated on a estuarine tidal flat in New York. They compared the Li (and boron) isotope composition and [B]/[Sr] ratio of the waters to determine the origin of mixed waters derived from landfill leachate, meteoric, and marine sources. They concluded that relatively high δ^7Li values (22–54‰), which could not be explained using mass balance mixing models, were attributed to isotope exchange during the interaction of aqueous Li with clay minerals that resulted in the depletion of ^6Li in groundwater over time.

12.3.2 CHLORINE

Approximately 76% of the chlorine that exists in nature is ^{35}Cl while the remainder is ^{37}Cl. The relative low mass and high mass difference (5.7%) between these two isotopes results in isotope fractionation that can be exploited to address environmental problems, including the release of chlorinated solvents to natural environments. The stable isotopes of chlorine are measured using both TIMS and gas source IRMS instrumentation. Early studies achieved accurate and precise results by converting environmental samples to methyl chloride followed by measurement on a gas source IRMS instrument. Kaufmann (1984) and Kaufmann et al. (1984) used the technique in dynamic switching mode to measure the Cl-isotope composition of a variety of natural materials including seawater, groundwater, and evaporates Figure 12.2a). In an effort to expand the capability of the analysis, Holt et al. (1997) and Jendrzejewski, Eggenkamp, and Coleman (1997) developed techniques to measure the chlorine and carbon isotope composition of aliquots of sample material. Most recently, Shouakar-Stash, Drimmie, and Frape (2005) introduced a continuous flow IRMS technique to reduce sample analysis time while maintaining acceptable analytical precision. In a different approach, Xiao and Zhang (1992), Magenheim et al. (1994), Stewart (2000), Numata, Nakamura, and Gamo (2001), and Numata et al. (2002) developed sensitive positive ion TIMS techniques to analyze microgram quantities of Cl enabling the analysis of a broader array of samples. These cumulative efforts have resulted in a significant body of work that documents Cl-isotope fractionation in nature, particularly associated with the origin, movement, and evolution of crustal fluids (for a more lengthy review, see Stewart and Spivak 2004).

Theoretical isotope fractionation relations for chlorine were first explored by Urey and Greiff (1935), Urey (1947), and more recently by Richet, Bottinga, and Javoy (1977). Schauble, Rossman, and Taylor (2003) demonstrated that the magnitude of chlorine isotope fractionation should vary in proportion to the oxidation state of chlorine and the atoms to which it bonds. This latter effect is important for environmental studies of chlorinated hydrocarbons, and for the inorganic and biological processes by which chlorinated hydrocarbons are transformed in nature.

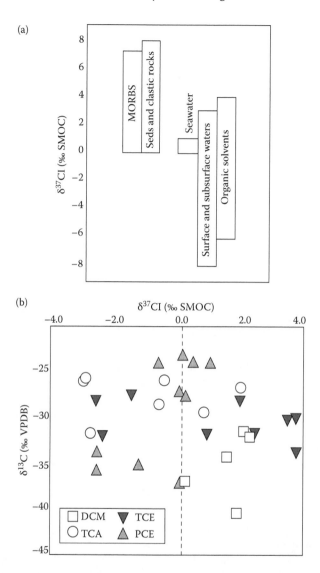

FIGURE 12.2 (a) Range of isotope compositions observed for chlorine of MORBS, weathering products of igneous rocks (sediments and clastic rocks), natural waters and organic solvents (Modified from Stewart, M. A., and A. J. Spivak, *Geochemistry of non-traditional stable isotopes*, The Mineralogical Society of America, Washington DC, 2004). (b) Chlorine and carbon stable isotope compositions of various organic solvents (DCM: dichloromethane; TCA: trichloroethane; TCE: trichloroethene; PCE: tetrachloroethene). (Modified from Jendrzejewski, N., H. G. M. Eggenkamp, and M. L. Coleman, *Applied Geochemistry,* 16, 1021–31, 2001.)

The Cl isotope composition of materials is determined in comparison to seawater (standard mean ocean Cl or SMOC), which is analytically homogeneous. Most crustal materials have $\delta^{37}Cl$ values that range from −1 to +7.2‰ (Long et al. 1993). This variability can be attributed to the interaction of fluids with Cl-bearing rocks and minerals, which tend to leave the solid enriched and the fluid depleted in ^{37}Cl (Magenheim et al. 1995). Eggenkamp (1994) showed that continental fluids are generally within 1‰ of seawater, with variations being attributed to diffusion and the exchange of Cl between circulating fluids and Cl-bearing minerals including evaporites.

The use of chlorine isotopes in the environmental sciences has centered on identifying sources, delineating degradation pathways, and determining the extent of chemical and biological transformations of chlorinated organic compounds (e.g., trichloroethene [TCE]; polychlorinated biphenyls [PCBs]; chlorofluorocarbons [CFCs]). Chlorinated hydrocarbons (e.g., TCE) have been used widely as dry cleaning fluids and degreasing agents while PCBs were widely used as insulators for electrical components and CFCs were commonly used as a refrigerant. Environmental releases of these compounds into the air, soil, and groundwater have resulted in the contamination of valued natural resources (Fisher, Rowan, and Spalding 1987).

Van Warmersdam et al. (1995) were the first to show that Cl (and C) isotopes could be used as a fingerprint for various manufacturers of chlorinated solvents. They observed $\delta^{37}Cl$ values that ranged from −2.9 to 4.0‰ and $\delta^{13}C$ values that ranged from −40.0 to −23.2‰ for perchloroethene (PCE), TCE, and 1,1,1 trichloroethane (TCA) from four manufacturers. When compared in a Cl-C isotope cross plot, solvents from each manufacturer were isotopically distinct. Differences in Cl isotope composition were attributed to manufacturing processes as natural sources of Cl have a limited range of $\delta^{37}Cl$ values. Jendrzejewski, Eggenkamp, and Coleman (2001) extended the list of solvents to include dichloromethane (DCM) and showed that with only one exception all $\delta^{37}Cl$ values of the DCM solvents were positive, suggesting they are highly fractionated during organic synthesis (Figure 12.2b). Shouakar-Stash, Frape, and Drimmie (2003) demonstrated that δD values for chlorinated hydrocarbons are wide ranging and distinct as well. Industrial chlorination reactions result in the progressive increase in the δD value of solvents due to kinetic effect as C–H bonds are broken and replaced by C–Cl bonds. They showed that inorganic reductive dechlorination in natural environments imprints a distinctively low H isotope signature on TCE derived from PCE as H from pore water is incorporated in the solvent.

Sturchio et al. (1998) demonstrated how the chlorine isotope composition of contaminated groundwater can be used to evaluate the natural attenuation (i.e., degradation) of TCE under aerobic conditions. They observed an inverse relationship between the concentration and $\delta^{37}Cl$ value of dissolved TCE in a groundwater plume, and showed that a kinetic effect characterizes the natural degradation of solvent in the subsurface. Measured $\delta^{37}Cl$ values for aqueous Cl were inconsistent with a simple model for TCE degradation, suggesting that additional processes such as volatilization, solubilization, and sorption must be considered to better understand the degradation of TCE over time.

In culture experiments, Heraty et al. (1999) determined kinetic isotope fractionation factors for Cl and C for the aerobic degradation of dichloromethane (DCM) by methylotrophic bacteria. Their results showed distinct increases in the $\delta^{37}Cl$ and

$\delta^{13}C$ values of residual DCM with consumption of the solvent. Numata et al. (2002) determined the biological pathways of PCE and TCE dechlorination to cis-1,2-dichloroethene (cDCE) by anaerobic microorganisms, and Paneth (2003) reviewed kinetic isotope fractionations for Cl associated with various enzymatic dehalogenation reactions. While the factors controlling isotope fractionation by microorganisms may be complex, the application holds much promise for characterizing the effects of bioremediation in contaminated aquifers.

Reddy et al. (2000) showed that PCBs (Araclor, Clophen, Phenoclor) from various suppliers display a relatively limited range in $\delta^{37}Cl$ values (–3.4 to –2.1‰). Differences in $\delta^{37}Cl$ values were not correlated with the percentage of Cl in each mixture indicating that isotope compositions were not related to congener composition but the way the compounds were manufactured. Bulk PCB extracted from contaminated sediments displayed a similar range of values except for a few samples, which had significantly lower $\delta^{37}Cl$ values (to –4.5‰). The anomalous sediments were relatively low in PCB concentration but enriched in highly Cl-substituted congeners. They suggested the natural processes that degraded the lower chlorinated congeners were influenced by a kinetic isotope effect. Alternatively, the congeners were isotopically distinct and the differences in isotopic composition were due to the removal of congeners of low $\delta^{37}Cl$ value. Compound specific isotope ratio analysis of two congeners from unaltered PCB revealed no difference in $\delta^{37}Cl$ value. In a subsequent study, Drenzek et al. (2004) showed that little Cl isotope fractionation accompanies the reductive dechlorination of 2, 3, 4, 5-tetrachlorobiphenyl by microorganisms. These results suggest other physicochemical factors may be responsible for the alteration of PCBs in contaminated environments.

Isotope studies have also been undertaken on inorganic Cl-bearing pollutants such as perchlorates, which are widely used as explosives and rocket fuels (Gulick, Lechevallier, and Barhorst 2001; Urbansky 2002). Ader et al. (2001) developed a technique for the stable isotope analysis of Cl from perchlorate that was subsequently used by Coleman et al. (2003) and Sturchio et al. (2003) to characterize Cl-isotope fractionation for the microbiological degradation of perchlorate. They showed that a very strong kinetic isotope effect influences the reduction of ClO_4^- to aqueous Cl^- during microbial respiration (Δ value of 14–17‰). This distinctive isotope fractionation has great potential in remedial investigations of perchlorate-contaminated groundwaters.

12.3.3 CALCIUM

Calcium is not of great concern as an environmental contaminant; nevertheless, it is included here because Ca isotopes potentially provide insight into the biomagnification of contaminants in much the same way as nitrogen isotopes are used (Cabana and Rasmussen 1994). Calcium has six stable isotopes: ^{40}Ca (96.98%), ^{42}Ca (0.642%), ^{43}Ca (0.133%), ^{44}Ca (2.056%), ^{46}Ca (0.003%), and ^{48}Ca (0.182%). The relatively low atomic mass and large mass spread among these isotopes permit the effects of mass-dependent fractionation to be expressed in nature.

Russell, Papanastassiou, and Tombrello (1978) reported the first precise Ca isotope ratio measurements using a double-spike TIMS technique. Calcium isotope compositions are typically reported as either $^{44}Ca/^{40}Ca$ (as $\delta^{44/40}Ca$) or $^{44}Ca/^{42}Ca$ ($\delta^{44/42}Ca$) (Eisenhauer et al. 2004) and delta values have been reported relative to the isotope composition of a variety of Ca salts including NIST SRM-915a, a highly purified calcium carbonate, as well as Ca from natural seawater, and an estimate of bulk-earth composition. Hippler et al. (2003) and Sime, De La Rocha, and Galy (2005) discussed methods for the conversion between the different measured Ca isotope ratios and normalizing standards. There is no consensus on the use of $^{44}Ca/^{40}Ca$ or $^{44}Ca/^{42}Ca$ however, ^{40}Ca is the stable product of ^{40}K decay and hence the abundance of ^{40}Ca is also a function of the age of the sample and its K/Ca ratio (see DePaolo [2004] for a discussion). Additionally, analytical methods that employ MC-ICP-MS tend to favor use of the $^{44}Ca/^{42}Ca$ ratio because of the significant isobaric interference of $^{40}Ar^+$ produced by the Ar plasma on ^{40}Ca.

Low temperature calcium isotopic studies have been conducted to evaluate paleoenvironmental conditions such as paleotemperature or pCO_2 as well as a trophic level indicator. Skulan, DePaolo, and Owens (1997) defined a $\delta^{44/40}Ca$ value for the bulk earth of zero that corresponds closely to the value of igneous rocks Figure 12.3a). This designation is somewhat problematic because no international standard reference material has been designated, yet significant differences in isotope composition exist between terrestrial volcanic rocks. Skulan, DePaolo, and Owens (1997) were the first to observe a systematic decrease in the $\delta^{44/40}Ca$ value of biological tissues through food chains, with apex predators in both terrestrial and marine food webs having the lowest skeletal $\delta^{44/40}Ca$ values within an ecosystem (Figure 12.3b). Collectively, these data suggest that Ca sequestered internally in mineralized tissue is lower in ^{44}Ca than assimilated food resources.

To better understand the factors that control the biological fractionation of Ca isotopes, Skulan and DePaolo (1999) analyzed the Ca isotope composition of soft and mineralized tissues of organisms. They determined that skeletal tissue was 1.3‰ lower in $\delta^{44/40}Ca$ value than dietary intake while soft tissue was more variable but nearly identical in isotopic composition to the diet. Skulan and DePaolo (1999) developed a box model that assumes a kinetic isotope fractionation of 1.5‰ for Ca fixation into bone and showed that the $\delta^{44/40}Ca$ values of both soft and mineralized tissues are dependent on: (1) the amount of available Ca in the diet, (2) the rate of bone growth, and (3) the rate of bone loss. If the dietary supply of Ca is significantly greater than the rate at which Ca is sequestered in bone then isotope fractionation is expressed to the fullest in bone, and the isotope composition of the soft tissue approximates that of the diet. On the other hand, if dietary Ca supply nearly equals that of bone sequestration, as in the case of a rapidly growing individual, the $\delta^{44/40}Ca$ value of bone will approach that of the diet and the soft tissue will be 1.5‰ higher in $\delta^{44/40}Ca$ value. Finally, if there is a significant amount of bone loss compared to dietary supply of Ca, then the soft tissue will approach the isotope composition of the bone.

This model has significant metabolic and ecotoxicological implications for environmental studies. For a healthy growing organism, the $\delta^{44/40}Ca$ value of bone should be approximately 1.5‰ lower than the soft tissue and the diet. If Ca uptake is limited, then bone will more closely approximate the diet and the $\delta^{44/40}Ca$ value of soft tissue will increase. In healthy adults, where bone growth and remodeling (recrystallization) both occur, bone and soft tissues should not be affected, but if significant bone loss occurs, the $\delta^{44/40}Ca$ value of soft tissue and bone should be relatively low. These observations may be used to better understand the uptake and trace the flow of contaminants through food webs to determine the effects of contaminants on Ca metabolism (e.g., Rodriguez-Navarro et al. 2002, 2006) for organisms living in impacted environments.

Workers have also begun to investigate Ca isotope fractionation associated with uptake of Ca by plants and how this Ca utilization coupled with Ca atmospheric deposition affects the pH buffering capacity of soils (Wiegand et al. 2005; Schmitt and Stille 2005). In general plants fractionate Ca during uptake, which results in the soil exchangeable Ca pool having higher $\delta^{44/40}Ca$ as compared to leaf matter. Moreover, these plant Ca isotope pools have different Ca isotope compositions as compared to atmospheric Ca inputs (e.g., precipitation), which allow workers to quantify Ca fluxes in soil environments.

The use of Ca as an isotope thermometer is being evaluated by: (1) conducting experiments in which foraminifera are grown in seawater at different temperatures in a laboratory and then analyzed for their Ca isotope composition, (2) empirical studies that analyze biogenic calcian carbonate (aragonite and calcite) from different organisms for comparisons to other paleotemperature proxies (e.g., $\delta^{18}O$ values, Mg/Ca thermometer), and (3) controlled abiotic precipitation experiments of calcian carbonate to determine isotope fractionation. The results of these investigations yield inconsistent results. Initial experiments conducted using the planktonic foraminifera *Globigerinoides sacculifer* indicate that the $\delta^{44/40}Ca$ value of the foraminiferal test was highly dependent on the temperature of seawater, where a 2.5‰ change was produced over a 10°C range in seawater (Nägler et al. 2000). Experiments conducted by this and other groups using cultured specimens of the planktonic foraminifera *Orbulina universa*, abiotic carbonate precipitation, or culturing of corals indicate the magnitude of the temperature dependence is small (~0.2‰ in $\delta^{44/40}Ca$ value over a 10°C range: Gussone et al. 2003; Marriott et al. 2004; Fietzke et al. 2004; Böhm et al. 2006). In an attempt to integrate these observations, Gussone et al. (2003) suggested some species of foraminifera dehydrate the Ca aquocomplex before carbonate mineral precipitation and this produces a significant temperature dependence on Ca isotope fractionation. In contrast, other foraminiferal species may precipitate carbonate by a mechanism that is more similar to abiotic precipitates where a minimal temperature dependence is demonstrated (Gussone et al. 2003). However, Ca isotope fractionation associated with the precipitation of carbonate from a solution is likely to be influenced by the rate of carbonate precipitation. Lemarchand, Wasserburg, and Papanastassiou (2004) showed that a 1.4‰ range in $^{44}Ca/^{40}Ca$ ratios in calcite can be produced by varying carbonate growth rate at a constant temperature. Extrapolating these data to infinitesimally slow growth rates, Lemarchand, Wasserburg, and Papanastassiou (2004) inferred that the equilibrium fractionation factor between calcite and aqueous Ca is −1.5‰. However, this equilibrium fractionation factor only includes data from experiments that were not stirred during carbonate precipitation. If one extrapolates data for stirred experiments, a separation factor of −0.4‰ (as $\Delta^{44}Ca_{calcite-Ca(aq)}$) would be inferred.

The uncertainty in Ca isotope fractionation in carbonates is also manifest in studies that have used naturally occurring biogenic carbonates. For example, studies by Nägler et al. (2000), Gussone et al. (2004) and Hippler, Eisenhauer, and Nägler

(2006) showed that the calcium isotope composition of *G. sacculifer* from different age samples display a 2‰ range in $\delta^{44/40}Ca$ values. Moreover, the Ca isotope ratio of *G. sacculifer* tracks other paleotemperature proxies including $\delta^{18}O$ values and Mg/Ca ratios (Nägler et al. 2000; Gussone et al. 2004). These relationships have lead Hippler, Eisenhauer, and Nägler (2006) to propose that Ca isotope composition of *G. sacculifer* can be used to infer sea surface temperatures (SST) in degrees C using the relationship: $\delta^{44}Ca = 0.22 \times SST - 4.88$, which is similar to the relationship identified by Nägler et al. (2000) in culture experiments. In marked contrast, Sime, De La Rocha, and Galy (2005) found there was no correlation between the Ca isotope composition of *G. sacculifer* and $\delta^{18}O$ values over a 10°C temperature range. Calcium isotope analysis of biogenic carbonate from other organisms yields inconsistent results with respect to temperature. For example, the Ca isotope composition of carbonate from *Vaccinites ultimus* (a mollusk) tracks $\delta^{18}O$ and Mg/Ca paleotemperature proxies (Immenhauser et al. 2005), whereas other workers found no measurable control of Ca isotope composition by temperature for a variety of planktonic and benthic foraminiferal species (De La Rocha and DePaolo 2000; Chang et al. 2004; Sime, De La Rocha, and Galy 2005). Finally, other workers have highlighted that naturally occurring foraminifera define a small temperature dependency of 0.1–0.3‰ in $\delta^{44/40}Ca$ values for a 10°C change in temperature (Kasemann et al. 2008 and Griffith et al. 2008), which is similar to the temperature dependency of the Ca isotope fractionation found in culture experiments (exclusive of *G. sacculifer*) and abiotic precipitation experiments (see above). These workers highlight that the Ca isotope compositions measured in an organism will be a reflection of the seawater Ca composition and temperature, the amount of isotope exchange between seawater Ca and aqueous Ca that an organism transports from seawater and is used for calcification, and the isotopic fractionation associated with biogenic calcian carbonate formation (e.g., Böhm et al. 2006; Griffith et al. 2008; Kasemann et al. 2008). Indeed, the recognition of up to 4‰ difference in $\delta^{44/40}Ca$ values of different carbonate layers within a single foraminifera test as revealed by ion microprobe analysis highlights the extreme control that Ca biomineralization processes are likely to play in controlling the Ca isotope composition of biogenic carbonates (Kasemann et al. 2008).

A second significant Ca isotope research endeavor has focused on evaluating steady state behavior of the Ca isotope composition of the world's oceans and its implication for weathering and carbonate precipitation in the geologic past (e.g., Zhu and MacDougall 1998; De La Rocha and DePaolo 2000; Schmitt, Chabaux, and Stille 2003a,b; DePaolo 2004; Soudry et al. 2004; Fantle and DePaolo 2005; Kasemann et al. 2005; Tipper, Galy, and Bickle 2006; Fantle and DePaolo 2007; Farkaš et al. 2007; Sime et al. 2007). An emerging picture from this research is that if one assumes a constant Ca isotope fractionation between carbonate and seawater then the Ca isotope composition of seawater has changed through time (e.g., DePaolo 2004; Farkaš et al. 2007). It is possible that changes in the Ca isotope composition of seawater have been driven by changes associated with the Ca isotope composition of material delivered to the oceans, such as riverine or atmospheric inputs, that the inferred changes are an artifact of assuming a constant carbonate seawater fractionation factor, or that these changes are related to the Ca concentration in seawater and hence a reflection of weathering rates and the partial pressure of CO_2. The inversion of the Ca isotope record of carbonates to the inferred Ca concentration of seawater is strongly dependent on the inferred fractionation factor between seawater Ca and carbonate, and the Ca isotope composition of the Ca fluxes delivered to the oceans (e.g., Sime et al. 2007), and thus, these inversions will require careful characterization of the Ca isotope fractionation associated with weathering and transport of Ca (e.g., Tipper, Galy, and Bickle 2006; Ewing et al. 2008; Jacobson and Holmden 2008), the possible influence of diagenesis on the measured Ca isotope composition of carbonate (e.g., Fantle and DePaolo 2007), and on the controls of Ca isotope fractionation between carbonate and seawater (Fantle and DePaolo 2005; 2007).

12.3.4 SELENIUM

Selenium has six stable isotopes: ^{74}Se (0.889%), ^{76}Se (9.366%), ^{77}Se (7.635%), ^{78}Se (23.772%), ^{80}Se (49.607%) and ^{82}Se (8.731%). While Se is an essential element for all living things, it can also be toxic to organisms (Cooper and Glover 1974; Ganther 1974; Lemly 1998; Skorupa 1998). Selenium is a constituent of most shales Figure 12.4a) and the soils that form from them, and it is concentrated in evaporative waters so it is a common contaminant of surface and groundwaters in arid environments. Anthropogenic sources of Se include fly ash so coal-fired power plants are a common point of entry for industrial releases to the environment (Lemly 1985). As a redox sensitive element, the bioavailability of selenium depends highly on its oxidation state, which varies from VI to −II. Abiotic transformation reactions involving Se generally proceed slowly so multiple forms of Se may exist in nature simultaneously (e.g., SeO_4^{2-}, Se^0). Microorganisms exploit Se as an electron acceptor to extract energy from the environment for metabolism and growth and these redox reactions are important pathways for the uptake of Se in food webs.

Early work on the characterization of Se isotopes involved the conversion of Se to SeF_6 for analysis by gas source IRMS (Krouse and Thode 1962; Rees and Thode 1966). Wachsmann and Heumann (1992) developed a TIMS technique that requires less material (~500 ng Se) but instrumental mass bias limits the acquisition of relatively precise results. This problem was overcome by Johnson et al. (1999) who used a double spike composed of ^{82}Se and ^{74}Se to make precise Se isotope measurements (e.g., $^{80}Se/^{76}Se$ ratios can be measured to a precision of ±0.2‰). Most recently, Rouxel et al. (2002) developed a MC-ICP-MS technique that permits the analysis of ~20 ng Se at a comparable level of precision. Sample storage and preparation requires care due to the volatility and reactivity of Se (Johnson and Bullen 2003). Presently there are no certified reference materials for Se although a NIST Se concentration standard solution (SRM3149) has been used by several laboratories. Using this standard, Hagiwara (2000) showed that shale and sediment have a range of $\delta^{80}Se$ values from −5 to 3‰ (Figure 12.4a).

The fractionation of Se isotopes in nature is complex because the relevant redox reactions involve multiple charge transfers and oxidation states that proceed inorganically and biologically at distinct reaction rates (Figure 12.4b). The kinetics and isotope fractionation of each step must be known to fully understand the significance of measured isotope compositions for Se-bearing compounds.

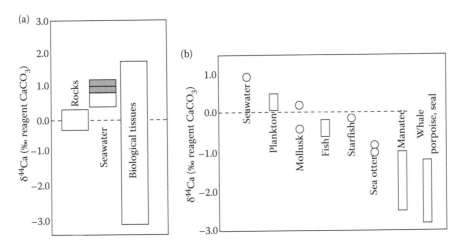

FIGURE 12.3 (a) Range of stable isotope compositions observed for calcium (as $\delta^{44/40}Ca$). Filled box for seawater represents present day average value of $0.95 \pm 0.1‰$ while larger box is last 20 Ma record (Fantle, M. W., and D. J. DePaolo, *Earth and Planetary Science Letters,* 237, 102–17, 2005). All other data are taken from DePaolo, D. J., *Geochemistry of non-traditional stable isotopes,* The Mineralogical Society of America, Washington DC, 2004 and (b) Ca isotope composition of skeletal tissue or shells of marine organisms. Circles are individual analyses while rectangles are data ranges. Modified from Skulan, J., and D. J. DePaolo, *Proceedings of the National Academy of Sciences,* 96, 13709–713, 1999 and Clementz, M. T., P. Holden, and P. L. Koch, *International Journal of Osteoarchaeology,* 13, 29–36, 2003.

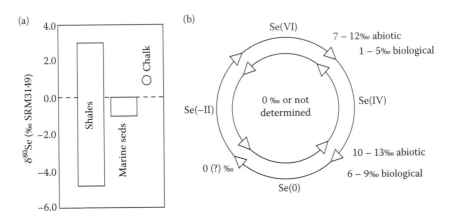

FIGURE 12.4 (a) Range of stable isotope compositions observed for selenium. Circle is an individual analysis while rectangles are data ranges. (Modified from Johnson, T. M., and T. D. Bullen, *Geochemistry of non-traditional stable isotopes,* The Mineralogical Society of America, Washington DC, 2004.) (b) Inorganic and biological fractionation of Se isotopes in the Se cycle. Isotope fractionation for Se oxidation pathways are placed inside the circles while fractionation for reduction pathways are placed outside the circles.

12.3.4.1 Se Reduction

Very few natural examples of the inorganic reduction of Se are documented in the literature. Johnson and Bullen (2003) demonstrated that green rust, a ferrous–ferric oxide that contains sulfate, may participate as an electron donor for the reduction of Se(VI) to Se(IV) with an attendant kinetic isotope fractionation of 7.4‰ (for $\delta^{80}Se$). The reduction of Se(IV) is not kinetically inhibited compared to Se(VI), nevertheless various inorganic experiments have demonstrated isotope fractionations that range from 10.0 to 12.8‰ (Krouse and Thode 1962; Rees and Thode 1966; Webster 1972). Rashid and Krouse (1985) conducted inorganic reduction experiments and developed a two-step reaction model that provides consistent results for all of the existing data. The model demonstrates that even in relatively simple inorganic systems, kinetic effect can significantly influence Se isotope systematics.

Kinetic isotope fractionation associated with the microbiological reduction of Se has received considerable attention. Herbel et al. (2000) grew anaerobic heterotrophic bacteria in batch culture and showed that several strains are capable of fractionating Se up to 4‰ during the reduction of Se(VI) to Se(IV), while fractionations up to 9‰ were observed for the microbial reduction of Se(IV) to elemental Se. To more closely approximate natural conditions, Ellis et al. (2003) conducted anaerobic batch experiments with Se-contaminated sediments and recorded similar results, with fractionations of up to 3.1‰ for Se(VI) reduction and 5.6‰ for Se(IV) reduction. These results suggest that inorganic and biologically induced isotope fractionations may be expressed in Se contaminated environments.

Johnson, Bullen, and Zawislanski (2000) measured the isotope composition of waters and sediments of the San Francisco, California estuary and found little difference in $\delta^{80}Se$ values among oxidized and reduced forms of Se. Since the concentrations of Se were relatively low (2 nmolL^{-1}), the possibility exists that kinetic effect was not expressed in this environment so Herbel et al. (2002) performed a similar investigation in highly contaminated (320 nmolL^{-1}) wetlands used to manage Se-rich agricultural drainage. Their results were similar to those of Johnson, Bullen, and Zawislanski (2000), suggesting that microbial Se reduction was not the primary mechanism for Se accumulation in these environments. The observed isotope data are more consistent with the incorporation of Se in algae and plants, followed by decay and burial as the primary mode of Se accumulation in sediments.

12.3.4.2 Other Biological Processes

Small but measurable isotope fractionations have been observed during the assimilation of Se(VI) and Se(IV) by primary producers (1.0 to 3.0‰; Hagiwara 2000) when Se concentrations are relatively high. There is no evidence for isotope fractionation associated with the volatilization of methylated forms of Se by cyanobacteria (Johnson et al. 1999).

12.3.4.3 Inorganic Oxidation and Sorption

Experiments have failed to demonstrate measurable isotope fractionation during the inorganic oxidation of Se(IV) to Se(VI). Although experiments have yet to be performed, isotope fractionation during the oxidation of elemental Se is unlikely given the fact that it is thermodynamically stable under most surface conditions. Likewise, little isotope fractionation probably occurs during the oxidation of Se(-II), especially if this species behaves similarly to dissolved sulfide (Canfield 2001). Finally, Johnson et al. (1999) found little evidence for isotope fractionation associated with the sorption of Se onto hydrous ferric oxides.

12.3.5 CHROMIUM

Chromium has four stable isotopes: ^{50}Cr (4.345%), ^{52}Cr (83.785%), ^{53}Cr (9.506%), and ^{54}Cr (2.364%). Chromium is commonly found in mafic rocks and minerals and it is used in electroplating, leather tanning, and wood preservation as well as a biofilm inhibitor (Nriagu and Niebor 1988). The release of Cr in surface environments primarily occurs through industrial or mining activities. Chromium is a redox sensitive element that occurs in the VI and III oxidation state in most surface environments, with Cr(VI) being the most soluble form as the oxyanion CrO_4^{2-}. The more reduced form, Cr(III), is more common in reducing environments; it is highly insoluble, sorbs strongly onto solid surfaces, and is much less toxic. And like Se, the oxidized form of Cr is exploited as an electron acceptor by some microorganisms (Ehrlich 1996).

Early interest in chromium isotopes was largely driven by cosmochemists interested in using the short-lived ^{53}Mn-^{53}Cr geochronometer to better understand the early history of our solar system (Lee and Tera 1986). Cosmochemical studies are largely conducted by TIMS and internal normalization to a Cr isotope ratio to document ^{53}Cr anomalies caused by decay of the now extinct ^{53}Mn radionuclide. Ellis, Johnson, and Bullen (2002) developed a double spike using ^{50}Cr and ^{54}Cr to characterize natural occurring mass dependent variations of Cr isotopes. This double spike study reported Cr isotope compositions as $^{53}Cr/^{52}Cr$ ratios using $\delta^{53}Cr$ delta notation relative to the isotope composition of an isotope standard from NIST (SRM-979).

Ellis, Johnson, and Bullen (2002) measured the Cr isotope composition of basalts, Cr ores, chemical reagents, and Cr plating baths and found a range of $\delta^{53}Cr$ values from −0.07 to 0.37‰, suggesting that little isotope fractionation occurs over a wide range of formation conditions. Alternatively, they observed relatively high and wide ranging $\delta^{53}Cr$ values for two Cr groundwater plumes (1.1 to 5.8‰). To explain these results, they performed batch experiments using reagent grade magnetite and natural sediments to characterize kinetic isotope fractionation during Cr reduction. They determined a kinetic fractionation of 3.4‰ ($\delta^{53}Cr$) for the reduction of Cr(VI) to Cr(III) using both substrates. Replicate experiments performed before and after autoclaving provided similar results suggesting that microorganisms do not influence isotope fractionation, or that biological and abiotic reduction pathways are isotopically indistinguishable. Most recently, Johnson, Bullen et al. (2005) documented a kinetic isotope fractionation of 4.1‰ for the microbial reduction of Cr(VI) to Cr(III) in laboratory experiments so both biological and abiotic isotope fractionation pathways appear feasible.

Ellis, Johnson, and Bullen (2002) attributed relatively high $\delta^{53}Cr$ values in groundwater plumes to progressive Cr reduction primarily because groundwater samples having the highest $\delta^{53}Cr$ values also displayed the lowest Cr concentrations. Johnson and Bullen (2004) made additional bulk concentration and isotope measurements of

Cr in one of the two plumes Ellis, Johnson, and Bullen (2002) studied and combined the data in a mass balance model to determine that 20 to 80% of the Cr(VI) originally released to the environment was reduced in the plumes. In addition, they studied a third groundwater plume where relatively high δ^{53}Cr values were observed but no relation with Cr concentration was observed. They suggested the Cr source at this site was enriched in ^{53}Cr and that little to no reduction occurred within the plume.

Additional factors that may influence Cr isotope distribution patterns are Cr(III) oxidation and sorption reactions. Isotope fractionation during Cr(III) oxidation has yet to be determined experimentally while theoretical calculations suggest an equilibrium isotope fractionation (1000 lnα) of 7.6‰ for the Cr(VI) – Cr(III) system at 25°C (Schauble, Rossman, and Taylor 2004; Ottonello and Zuccolini 2005). Ellis, Johnson, and Bullen (2004) showed that sorption does not significantly fractionate Cr isotopes.

These results show promise for the use of Cr isotopes as a natural abundance isotope tracer but additional work is needed to better understand kinetic and equilibrium isotope fractionation relations in abiotic and biological systems.

12.3.6 IRON

Iron isotopes have received considerable attention recently because of their potential use as a diagnostic biomarker as well as general interest in Fe as a bioessential element and a paleoredox indicator. Significant effort has been expended to better understand Fe isotope fractionation under a variety of conditions and as a result our knowledge of the environmental chemistry of Fe has expanded greatly.

Iron has four stable isotopes: ^{54}Fe (5.84%), ^{56}Fe (91.76%), ^{57}Fe (2.12%), and ^{58}Fe (0.28%). As a redox-sensitive element, Fe typically occurs in the (III) and (II) oxidation state in oxidized and reduced environments, respectively. The concentration of dissolved Fe(III) is relatively low compared to Fe(II) in their respective environments due to the low solubility of solid Fe(III) oxides, except under low pH conditions. The relative mass difference among the stable isotopes and contrasting redox behavior results in significant isotope variability among materials in surface environments.

Iron isotope data are reported as ^{57}Fe/^{54}Fe, ^{57}Fe/^{56}Fe, or ^{56}Fe/^{54}Fe ratios, where δ^{56}Fe and δ^{57}Fe refers to isotope variations in ^{56}Fe/^{54}Fe and ^{57}Fe/^{54}Fe, respectively and ^{57}Fe/^{56}Fe variations are reported as $\delta^{57/56}$Fe (see Beard and Johnson 2004). Analysis of ^{58}Fe is generally not performed because of the relatively low natural abundance of this isotope and interference from ^{58}Ni (^{58}Ni is the most abundant Ni isotope and Ni is commonly used in sampler and skimmer cones in MC-ICP-MS). The relationships between isotope ratios (or delta values) is relatively straightforward because all known isotope fractionations are mass-dependent (Figure 12.5a). The choice of reference standards used to determine isotope compositions varies between laboratories, and this has led to some confusion regarding interlaboratory comparisons (see Table 1 from Beard and Johnson 2004) but interlaboratory comparisons are easily made through the internationally recognized iron isotope standard IRMM-014.

Isotope measurements can be made using the double spike technique with TIMS and MC-ICP-MS instruments. More commonly, MC-ICP-MS measurements are made using sample-standard bracketing with, or without, an element spike. While

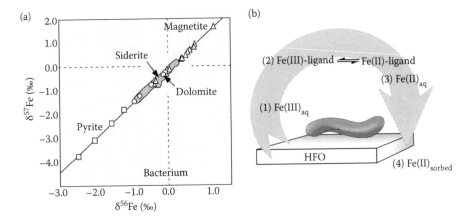

FIGURE 12.5 (a) Stable isotope compositions for Fe from mineral separates taken from the Transvaal banded iron formation. Average values are plotted from Table 2 of (Johnson, C. M., B. L. Beard, N. J. Beukes, C. Klein, and J. M. O'Leary, *Contributions to Mineralogy and Petrology*, 144, 523–47, 2003.) The data form an array of slope 3/2 as predicted for mass-dependent fractionation. Also note the isotope fractionation expected for the relative distribution of minerals (squares: pyrite; circles: dolomite; triangles: magnetite; shaded field: siderite). (b) Schematic illustrating the origin of Fe isotope fractionation during bacterial dissimilatory iron reduction. The dissimilatory iron reducing bacteria (DIRB) produces aqueous Fe(II) ($Fe_{II}aq$) during the metabolic transfer of electrons to the ferric substrate (dark gray box). The $Fe_{II}aq$ interacts with the ferric substrate and produces a Fe(III) reactive surface layer (light gray rectangle) on the ferric substrate by an electron transfer and atom exchange mechanism (shown in the dashed rectangle where superscripts i and j denote different Fe isotopes). The low $\delta^{56}Fe$ values of the DIRB produced $Fe(II)_{aq}$ is counterbalanced by the production of an Fe(III) reactive layer that has a high $\delta^{56}Fe$ value (Modified from Crosby, H. A., E. E. Roden, C. M. Johnson, and B. L. Beard, *Geobiology*, 5, 169–89, 2007.)

TIMS analysis offers the advantages of few isobaric interferences, analysis times are relatively long and sample requirements are large compared to MC-ICP-MS. Iron isotope analysis by MC-ICP-MS requires a firm understanding of the isobaric interferences originating from Ar isobars that include $^{40}Ar^{14}N$ at ^{54}Fe, $^{40}Ar^{16}O$ at ^{56}Fe, and $^{40}Ar^{16}OH$ at ^{57}Fe. A variety of techniques have been used to minimize or eliminate Ar isobars, and in general these techniques allow one to determine Fe isotope compositions to an external precision of ±0.05 to 0.1‰ (2σ standard deviation) for $\delta^{56}Fe$ on 0.1 to 10 micrograms of Fe (Belshaw et al. 2000; Weyer and Schwieters 2003; Beard et al. 2003a,b; Arnold et al. 2004; Dideriksen, Baker, and Stipp 2006). The accuracy of MC-ICP-MS results depends on the sample and standard being chemically similar so significant sample preparation is required to eliminate matrix effects (e.g., Albarède and Beard 2004). Quantitative yields are required when preparing samples to avoid analytical artifacts such as significant isotope fractionations that have been observed using wet chemical techniques including ion exchange columns (Anbar et al. 2000). The analytical issues are well understood and overall agreement between laboratories is excellent (Beard et al. 2003a,b; Poitrasson et al. 2004; Weyer et al. 2005; Dideriksen, Baker, and Stipp 2006).

12.3.6.1   Origin of Fe Isotope Variations

Theoretical estimates of isotope fractionation for a variety of iron-bearing minerals and aqueous Fe complexes have been made that predict fractionations of up to 10‰ (as $\delta^{56}Fe$: Polyakov and Mineev 2000; Schauble, Rossman, and Taylor 2001; Schauble 2004; Anbar, Jarzecki, and Spiro 2005; Polyakov et al. 2007; Hill and Schauble 2008). In general, reduced iron species and minerals are predicted to have lower $\delta^{56}Fe$ values compared to oxidized species.

Experimentation has confirmed the direction of these predicted isotopic fractionations, however, the overall magnitude of experimentally measured fractionations are less (e.g., Beard and Johnson 2004). There are some exceptions to this general prediction where, for example, pyrite or Fe-Ni sulfides are predicted to have higher $^{56}Fe/^{54}Fe$ ratios compared to oxidized minerals such as hematite or goethite (Polyakov and Mineev 2000), however, natural data on pyrite from sedimentary rocks and ore samples indicate that pyrite has among the lowest $\delta^{56}Fe$ values measured for natural samples (Johnson et al. 2003; Rouxel, Fouquet, and Ludden 2004; Rouxel, Bekker, and Edwards 2005; Graham et al. 2004; Rouxel et al. 2008).

Experimental investigations of Fe isotope fractionation have focused on isotopic fractionation between aqueous Fe(III) and Fe(II) species (Johnson et al. 2002; Welch et al. 2003) and between aqueous Fe and mineral phases (Skulan, Beard, and Johnson 2002; Wiesli, Beard, and Johnson 2004; Butler et al. 2005; Bergquist and Boyle 2006). Iron isotope fractionation and isotope exchange rates between aqueous Fe(III) and Fe(II) have been investigated at two pH values (2 and 5.5) and over a range of chloride concentrations (0 to 111 mM Cl). Isotopic exchange between Fe(III) and Fe(II) occurs rapidly (10's of seconds); at 22°C, Fe isotope fractionation ($\Delta^{56}Fe_{Fe(III)-Fe(II)}$) is +2.9‰, while at 0°C $\Delta^{56}Fe_{Fe(III)-Fe(II)}$ is +3.6‰ (Welch et al. 2003). Because there is no measurable difference in the Fe(III)–Fe(II) fractionation factor at different Cl concentrations and pH conditions, it does not appear that different aqueous Fe(II) and Fe(III) species (e.g., OH or Cl complexes) exert a significant control over Fe isotope fractionation under most natural conditions, although at higher Cl concentrations, such as in some brines, there may be an important Cl speciation control on Fe isotope fractionation (Hill and Schauble 2008). The experimentally determined $Fe(III)(H_2O)_6$-$Fe(II)(H_2O)_6$ fractionation factor of 2.9‰ at 22°C has been confirmed by theoretical calculations (Schauble, Rossman, and Taylor 2001; Anbar, Jarzecki, and Spiro 2005; Hill and Schauble 2008). Since the isotope exchange rate between Fe(II) and Fe(III) appear to be rapid, experimental measurements are thought to represent equilibrium isotope fractionation. In contrast, isotope exchange rates between fluids and minerals are significantly slower (days to years; Skulan, Beard, and Johnson 2002; Poulson, Johnson, and Beard 2005) so experimental results may approach equilibrium isotope exchange conditions or they may represent kinetic isotopic fractionations. Kinetic isotope fractionation accompanies the rapid precipitation of hematite from aqueous Fe(III) in acid hydrolysis experiments where $\Delta^{56}Fe_{Fe(III)-hematite}$ is +1.3‰ (Skulan, Beard, and Johnson 2002) because in experiments at slower precipitation rates, the fractionation between aqueous Fe(III) and hematite is smaller. Skulan, Beard, and Johnson (2002) extrapolated the Fe(III)-hematite experiments to a zero precipitation rate and inferred that the equilibrium $\Delta^{56}Fe_{Fe(III)-hematite}$ is $-0.1‰$. In experiments that simulated the flocculation of Fe from river water upon entry into the sea, Bergquist and Boyle (2006) found that the Fe isotope composition of a ferric oxide/hydroxide flocculate is $\sim0.2‰$ greater in $^{56}Fe/^{54}Fe$ compared to the Fe isotope composition of river water. This flocculation experiment is consistent with the equilibrium fractionation factor for aqueous Fe(III) and iron oxide inferred by Skulan, Beard, and Johnson (2002). Wiesli, Beard, and Johnson (2004) precipitated siderite from aqueous Fe(II) solutions at different rates and determined an equilibrium fractionation factor ($\Delta^{56}Fe_{Fe(II)-siderite}$) of $+0.5‰$. Experimental results for FeS precipitated from Fe(II) solutions indicate that the initial $\Delta^{56}Fe_{Fe(II)-FeS}$ value is $+0.9‰$, but that upon aging for several days $\Delta^{56}Fe_{Fe(II)-FeS}$ becomes $+0.3‰$, and this value is thought to represent the equilibrium fractionation between aqueous Fe(II) and FeS. A $\Delta^{56}Fe_{Fe(II)-FeS}$ value of $0.3‰$ closely matches the fractionation factor between pore fluids and FeS measured in continental shelf sediments (Severmann et al. 2006). These experimental results show that the largest isotopic variations are produced by redox changes and that although fluid mineral fractionations are small they can have an important control on Fe isotope compositions. For example, in banded iron formations (BIFs) the $\delta^{56}Fe$ value of minerals increase from pyrite to Fe carbonate to magnetite and hematite, the latter two having Fe isotope compositions with overlapping ranges (Johnson et al. 2003; Johnson, Beard, and Roden 2008).

A number of partial dissolution experiments have been conducted to evaluate the aqueous Fe isotope composition that may be produced during weathering processes. Partial dissolution of iron bearing minerals by mineral acids (e.g., HCl) does not produce any Fe isotope fractionation between mineral and dissolved Fe (e.g., Skulan, Beard, and Johnson 2002; Beard et al. 2003a; Wiederhold et al. 2006). In contrast, partial dissolution experiments using a variety of organic ligands indicate there can be significant isotope fractionations between minerals and dissolved iron. Brantley, Liermann, and Bullen (2001) and Brantley et al. (2004) showed that iron dissolved from amphibole using a variety of organic ligands has a lower $^{56}Fe/^{54}Fe$ ratio (up to 1‰) compared to the original amphibole, and that the magnitude of the fractionation is positively correlated to the binding strength of the organic ligand. In these experiments the degree of partial dissolution was low (<1%), and it was believed to occur through an incongruent dissolution process. In other partial dissolution experiments that used oxalic acid to dissolve goethite through a ligand promoted reaction mechanism, Wiederhold et al. (2006) showed that at <1% partial dissolution, dissolved Fe had a lower $\delta^{56}Fe$ value (up to 1.7‰) than goethite, while in longer term experiments that involved higher degrees of partial dissolution there was no isotopic fractionation or dissolved Fe had a higher $^{56}Fe/^{54}Fe$ ratio (up to 0.5‰) than goethite. Ligand promoted incongruent dissolution was documented in partial dissolution experiments using the reacted goethite from long-term, high-degree partial dissolution experiments. Iron leached from goethite using HCl had a lower $^{56}Fe/^{54}Fe$ ratio than aqueous Fe generated using oxalic acid as an extractant, while bulk goethite had an intermediate Fe isotope composition (Wiederhold et al. 2006). Additionally, Dideriksen, Baker, and Stipp (2008) showed by experimentation that there is an equilibrium isotopic fractionation between aqueous Fe(III) and aqueous Fe(III) bound to a siderophore, where at 25°C the aqueous Fe(III) has a $^{56}Fe/^{54}Fe$ ratio that is 0.6‰ greater than the Fe(III) bound to the siderophore. In nature these partial dissolution processes, coupled with bacterial dissimilatory iron reduction result in important iron isotope fractionation during soil formation processes (Fantle and DePaolo 2004; Emmanuel, Matthews, and Teutsch 2005; Thompson et al. 2007; Wiederhold 2007a,b).

Rouxel et al. (2003) showed that the Fe isotope composition of altered oceanic crust of mid-ocean ridge hydrothermal systems is either higher or lower than fresh igneous rocks, which have a very narrow range in Fe isotope compositions and this can be used as a baseline for comparison with other natural materials from surface environments (Beard et al. 2003a). In other studies, clastic sedimentary rocks containing little carbon (i.e., organic carbon or carbonates) have the same Fe isotope composition as igneous rocks despite the fact that these sedimentary rocks have higher ratios of ferric to total Fe. These data demonstrate that in most sedimentary environments, Fe oxidizes in place due to its low solubility and, thus, there is no net isotope fractionation of Fe in the bulk rock (Beard and Johnson 2004; Yamaguchi et al. 2005). Importantly, these data show that Fe isotope variability in sedimentary rocks and soils is not a reflection of clastic material with variable Fe isotope compositions. Rather, one can conclude that Fe isotope variability must be a result of secondary processes that are associated with the redistribution of Fe by the Fe geochemical cycle.

Numerous experiments have highlighted some novel isotopic fractionations associated with diffusion of aqueous Fe and fractionation between different aqueous Fe complexes. In one set of experiments, Fe was allowed to diffuse from a pure solution into nitric acid and it was found that samples having low Fe concentration (distal from the pure Fe solution) had $^{56}Fe/^{54}Fe$ ratios that were 0.3‰ lower than high Fe concentration samples (Rodushkin et al. 2004). This diffusion experiment indicates that aqueous transport may be a viable method to produce minor Fe isotope variations. In another set of experiments, small fractionations (0.1–0.2‰) were documented between different ferric chloro-complexes based on differences in the Fe isotope composition of Fe sorbed onto anion exchange resin and free aqueous Fe in 2–6 M HCl solutions (Anbar et al. 2000; Roe, Anbar, and Barling 2003). In other experiments that involved the separation of Fe complexes using anion exchange resin, large fractionations (~10‰) were observed between Fe(II) complexed with tris(2,2'-bipyridine) and Fe(III) in 6M HCl (Mathews, Zhu, and O'Nions 2001). Interestingly, the $[Fe^{(II)}(bipy)_3]^{2+}$ had a higher $^{56}Fe/^{54}Fe$ ratio compared to the Fe(III) chloro complex. Additionally, workers have shown that there are measurable equilibrium Fe isotope fractionations (0.2–0.4‰ in $\delta^{56}Fe$ values) between magnetite and olivine at high temperatures (600–800°C; Shahar, Young, and Manning 2008) and between pyrrohtite and ferric rich rhyolitic melts (Schuessler et al. 2007).

The combination of these experimental programs and analytical methods that calculate equilibrium fractionation factors from spectroscopic data have made the Fe isotope system the most mature of the nontraditional isotope systems. It is now possible to interpret Fe isotope measurements within an expanding database of experimentally calibrated or theoretically calculated fractionation factors, similar to that which has been developed for O isotopes. These experiments and calculation methods clearly establish the importance of redox processes in producing Fe isotope variability and also highlight that fluid mineral fractionation, and in some important cases, aqueous Fe ligands such as Cl or organic complexes are able to produce important Fe isotope fractionations (see reviews by Anbar 2004; Beard and Johnson 2004; Johnson et al. 2004; Dauphas and Rouxel 2006; Anbar and Rouxel 2007; Johnson, Beard, and Roden 2008).

12.3.6.2 Biological Systems

It is well known that the redox sensitive nature of Fe is exploited by microorganisms to extract energy from the environment for metabolism and growth (e.g., Nealson 1983; Lovley et al. 1987; Myers and Nealson 1988; Ghiorse 1989; Lovley 1991; Nealson and Saffarini 1994). The work done to make this determination provides a firm foundation upon which microbiologically mediated Fe transformation reactions may be evaluated for characteristic isotope fractionations.

Johnson et al. (2004) identified three major microbiological pathways by which Fe isotopes may be fractionated in nature: (1) dissimilatory Fe(III) reduction (Lovley 1987; Nealson and Myers 1990), (2) lithotrophic or phototrophic Fe(II) oxidation (Widdel et al. 1993; Ehrlich 1996; Heising and Schink 1998; Heising et al. 1999; Emerson 2000; Straub, Benz, and Schink 2001), and (3) assimilatory Fe metabolism (Bazylinski and Moskowitz 1997). For microorganisms to use any of these three strategies, they must catalyze biochemical reactions that otherwise proceed more slowly (or not at all) in their absence. Little work has been done on Fe isotope fractionation by assimilatory Fe metabolism and the one study that has been conducted suggests that there is no measurable Fe isotope fractionation during the production of intracellular magnetsomes by magnetotactic bacteria (Mandernack et al. 1999). Two studies have been conducted on Fe isotope fractionation during oxidation of Fe by bacteria (Croal et al. 2004; Balci et al. 2006) and these studies are discussed below in a general section on Fe oxidation. In this section we focus on dissimilatory Fe reducing bacteria because this is the best studied example of Fe isotope fractionation associated with biological processes. Additionally, we discuss new Fe isotope work that is focusing on evaluating Fe isotope fractionation by higher order organisms.

12.3.6.2.1 Fe Isotope Fractionation in Higher Order Organisms

Iron isotopes are fractionated in higher order organisms during the utilization of Fe nutrients. For example, in humans, blood has a lower δ^{56}Fe than dietary Fe sources by up to 2.6‰ (Walczyk and von Blanckenburg 2002). Moreover, the Fe isotope composition of blood from females is not as fractionated as compared to males, relative to dietary Fe isotope compositions, which has been suggested to imply that females either uptake Fe from dietary sources more efficiently or fractionate Fe during uptake differently as compared to males (Walczyk and von Blanckenburg 2002). The utilization of Fe by plants proceeds by one of two strategies (strategies 1 and 2; Guelke and von Blanckenburg 2007) that differ in their degree of Fe isotope fractionation. In plants that utilize a strategy 1 Fe uptake process, Fe(III) is chelated by acidification followed by reduction of Fe, which results in these plants having δ^{56}Fe values that are up to 1.6‰ less than the Fe isotope composition of the soil in which they are growing. In contrast, strategy 2 plants utilize Fe without a reduction step and tend to have the same Fe isotope composition of Fe in soil (Guelke and von Blanckenburg 2007).

12.3.6.2.2 Dissimilatory Fe Reduction

Experimental studies of dissimilatory Fe reduction (DIR) have shown that a variety of environmental materials can act as electron acceptors including dissolved Fe(III) and hydrous ferric oxide (HFO) as well as other more crystalline materials (e.g., goethite, hematite, magnetite, Fe(III)-clay minerals; Lovley 1991; Gates, Wilkinson, and Stucki 1993; Kostka and Nealson 1995; Kostka et al. 1996; Neal et al. 2003; Kim et al. 2004). The end product of DIR is Fe(II) that commonly forms magnetite and siderite under various environmental conditions (Lovley and Phillips 1988; Roden and Lovley 1993; Fredrickson et al. 1998; Zhang et al. 1998, 2001; Roden, Leonardo, and Ferris 2002; Zachara et al. 2002). Based on batch culture experiments, the aqueous Fe(II) produced by DIR has a $^{56}Fe/^{54}Fe$ ratio that is 0.5–2‰ less than the original ferric substrate (Beard et al. 1999, 2003a; Icopini et al. 2004; Johnson, Roden et al. 2005; Crosby et al. 2005, 2007). As theoretical and empirical studies of inorganic systems suggest, the key step to understanding Fe isotope fractionation during DIR is the transformation reaction that occurs during electron transfer. Based on experiments that used goethite or hematite as the ferric substrate and the microorganisms *Geobacter sulfurreducens* and *Shewanella putrefaciens*, Crosby et al. (2005, 2007) demonstrated that there was a 3‰ isotope fractionation ($^{56}Fe/^{54}Fe$) between the aqueous Fe(II) and the outermost layer of ferric iron on the oxide surface, which was explained by equilibrium isotope fractionation between aqueous Fe(II) and the solid reactive ferric iron substrate (Figure 12.5b). Sodium acetate leaches of the hematite or goethite substrates revealed that sorbed Fe(II) had a $^{56}Fe/^{54}Fe$ ratio that was 0.4‰ (hematite) or 0.8‰ (goethite) greater than aqueous Fe(II). Although sorbed Fe(II) had a slightly different Fe isotope composition compared to the aqueous Fe(II), sorption alone was insufficient to produce Fe(II) with the low $\delta^{56}Fe$ values measured in these DIR experiments. Instead, it appears the isotope fractionation required electron and atom exchange between ferrous and ferric iron within a reactive surface layer of the iron oxide (Figure 12.5b).

As mentioned earlier, the end product of DIR is aqueous Fe(II), which may react with other dissolved constituents to produce siderite and/or magnetite. These solids may preserve an Fe isotope legacy of the processes responsible for their formation. Johnson, Roden et al. (2005) reported little difference between the $\delta^{56}Fe$ value for pure siderite and aqueous Fe(II) in DIR experiments when these phases were assumed to reach isotope equilibrium, while siderite was ~1.0‰ higher than aqueous Fe(II) for experiments where kinetic fractionation was expressed. The substitution of foreign ions into the siderite structure appears to strongly influence the isotope systematics with $\delta^{56}Fe$ values for systems containing Ca-substituted siderite being ~1.0‰ greater than their pure counterparts.

Calculated and measured Fe isotope fractionations for the microbial magnetite–aqueous Fe(II) system range from 4.2 to 0.0‰ (Mandernack et al. 1999; Polyakov and Mineev 2000; Schauble, Rossman, and Taylor 2001; Johnson et al. 2003, 2004), which is nearly as broad as the entire range of terrestrial $\delta^{56}Fe$ values observed in nature. Based on long-term (~1 year) DIR experiments that produced magnetite by reaction between ferrihydrite and aqueous Fe(II), Johnson, Roden et al. (2005) inferred that the equilibrium fractionation between Fe(II) and magnetite is –1.3‰.

12.3.6.3   Fe(II) Oxidation

The effects of iron oxidation have been evaluated in natural and laboratory settings where Fe-rich anoxic fluids were oxidized by exposure to atmospheric oxygen (Bullen et al. 2001) and in laboratory settings using circumneutral anaerobic photo-synthetic iron oxidizing bacteria (Croal et al. 2004) and acidophilic bacteria (Balci et al. 2006). In these experiments, the isotopic contrast between aqueous Fe and the precipitated ferric oxide is 1–3‰ whether precipitation was mediated by biotic or abiotic oxidation. However, this fact should not be construed to suggest that the utility of Fe isotope studies is diminished for oxidative processes. For example, Fe isotope fractionation can occur during Fe oxidation depending on the degree of completeness of the oxidation. Complete oxidation of a solution will result in no net isotope fractionation, and this appears to be common during weathering and the sedimentary transport of clastic sediments in an oxic atmosphere (Beard and Johnson 2004). In contrast, partial oxidation will produce iron oxides that have high $^{56}Fe/^{54}Fe$ ratios that are typified by BIFs the oldest sedimentary rock sequences (3.8 Ga) from Greenland (Dauphas et al. 2004). The presence of Fe oxides with high $^{56}Fe/^{54}Fe$ ratios implies that oxidation took place close to the oxic/anoxic interface where oxidation would not go to completion.

   In general, the oxidation of Fe(II) solutions is followed by precipitation of Fe(III)-bearing iron oxide/hydroxide minerals. Due to the rapid isotope exchange between Fe(II) and Fe(III), an equilibrium isotope fractionation is established between aqueous Fe species. In contrast, the precipitation of iron oxide is a rapid process and the isotope exchange kinetics are slow, which results in a kinetic isotopic fractionation. Beard and Johnson (2004) proposed a conceptual model of iron oxidation that predicts the resulting net Fe isotope fractionation between aqueous Fe and iron oxide is a combination of the equilibrium fractionation between aqueous Fe(II) and Fe(III) (2.9‰ at 25°C; Welch et al. 2003) and a kinetic fractionation between aqueous Fe(III) and iron oxide (as large as 1.3‰; Skulan, Beard, and Johnson 2002). Thus, the model ultimately predicts an isotope fractionation between aqueous Fe(II) and ferric oxide that is as large as ~3‰ and decreases to lower values if there is a strong kinetic fractionation between Fe(III) and iron oxide. A numerical integration approach for predicting these isotopic fractionations is presented in Beard and Johnson (2004) and a solution in closed form is presented in Dauphas and Rouxel (2006). This conceptual framework for predicting Fe isotope fractionation during oxidation has been validated experimentally by Balci et al. (2006) using iron-oxidizing acidophilic bacteria to conduct iron oxidation experiments at low pH conditions suitable to allow Fe isotope analysis of the $Fe(II)_{aq}$, $Fe(III)_{aq}$, and ferric oxide pools. Balci et al. (2006) found that the difference in isotope composition between aqueous Fe(III) and Fe(II) was approximately 3‰. Moreover, ferric oxide had a $\delta^{56}Fe$ value that was either equal to or less than aqueous Fe(III).

   Teutsch et al. (2005) documented Fe isotope compositions associated with the remediation of a polluted groundwater aquifer as a function of time after oxygenated water had been pumped into the aquifer. Groundwater removed from the aquifer following injection of oxygenated water had a low Fe concentration (0.1 ppm Fe) and a low $\delta^{56}Fe$ value of –2.5‰. Over time, the Fe concentration and $\delta^{56}Fe$ values of groundwater increased, and after three days the groundwater had levels similar to those prior to injection of oxic water (1.8 ppm Fe and a δ^{56}Fe value of $-0.5\permil$). Based on the results of conservative tracer experiments, Teutsch et al. (2005) suggested that the homogenous oxidation of Fe (II) cannot explain the isotopic and concentration changes observed in aqueous Fe, and that sorption of Fe onto iron oxides was responsible for the observed changes. Considering the work of Crosby et al. (2005) who documented that sorbed Fe(II) does not have an Fe isotope composition significantly different from aqueous Fe(II), the reaction mechanism probably included atom and electron exchange and the production of a reactive ferric oxide pool at the iron oxide surface with a high ^{56}Fe/^{54}Fe ratio.

12.3.6.4   Fe Isotope Composition of Iron Oxides and Fluids

Recent research regarding the isotopic composition of iron oxides highlights that iron oxides tend to have low ^{56}Fe/^{54}Fe ratios (Emmanuel et al. 2005; Teutsch et al. 2005; Severmann et al. 2006; Staubwasser, von Blanckenburg, and Schoenberg 2006). Because iron oxides have high ^{56}Fe/^{54}Fe ratios relative to aqueous Fe(II), iron oxides with low δ^{56}Fe values must form from fluids that experience complete oxidation or those with unusually low δ^{56}Fe values. Based on partial leaching studies of semiarid and forest soils, Emmanuel et al. (2005) determined that iron oxides have δ^{56}Fe values between -1 and $-2\permil$. Additional partial dissolution studies of marine sediments and aquifers have concluded that the low δ^{56}Fe values measured for iron oxides (-0.3 to $-0.7\permil$) are a product of DIR (Severmann et al. 2006; Staubwasser, von Blanckenburg, and Schoenberg 2006). These low δ^{56}Fe values for iron oxides are thought to be produced by the generation of aqueous Fe(II) with a low ^{56}Fe/^{54}Fe ratio followed by oxidation to iron oxide, with repeated cycles of DIR and oxidation.

The Fe isotope analysis of fluids from most oxic environments is an analytical challenge because of the low Fe concentration of the fluids. Hydrothermal fluids from mid-ocean ridge spreading center vents have been analyzed and their δ^{56}Fe values vary between -0.8 and $-0.3\permil$ (Sharma, Polizzotto, and Anbar 2001; Beard et al. 2003b). Based on Fe isotope analysis of altered oceanic crust, it appears that isotope fractionation associated with water rock interaction and precipitation of high temperature sulfides is responsible for the low δ^{56}Fe values of these vent fluids (e.g., Rouxel et al. 2003, 2008). In two studies on the anoxic waters of terrestrial springs or aquifers, the Fe isotope composition of the vent fluids were $\sim-0.5\permil$, which matches the Fe isotope composition of mid-ocean ridge vent fluids (Bullen et al. 2001; Teutsch et al. 2005). These fluids with moderately negative δ^{56}Fe values ($-0.5\permil$) tend to have relatively high Fe concentrations (>1 ppm Fe). As discussed above, additional oxidation of these anoxic fluids will decrease the δ^{56}Fe value and Fe concentration of the fluid. The low Fe concentration of partially oxidized fluids makes it difficult to determine if they are an important source of Fe with low δ^{56}Fe values. Analysis of unfiltered water from rivers from the Pacific Northwest (Alaska to California) indicates that riverine Fe has δ^{56}Fe values of -0.9 to $0\permil$; samples with relatively low ^{56}Fe/^{54}Fe ratios are associated with lower Fe concentrations (Fantle and DePaolo 2004). Analysis of filtered river water from the Amazon Basin shows that riverine Fe varies between $+0.3$ to $-0.4\permil$. Sediment pore water has also been analyzed for its Fe isotope composition at two sites along the continental margin of California (Severmann et al. 2006). At one site the system is dominated by Fe sulfide derived from bacterial sulfate reduction and pore fluids have a low Fe concentration (~1.5 ppm Fe) and $\delta^{56}Fe$ values of -0.5 to $+0.3‰$. At a different site, where there are moderate amounts of Fe sulfide and active DIR, the pore fluids have high Fe concentrations (15 ppm) and $\delta^{56}Fe$ values between -3 and $-1‰$. Similar findings for low $\delta^{56}Fe$ value pore fluids from sites that involve DIR have been found along the Amazon shelf region in South America (Bergquist and Boyle 2006). It appears that fluids with the lowest $\delta^{56}Fe$ values and highest Fe concentrations are associated with DIR. In contrast, fluids with high Fe concentration (>1 ppm) and $\delta^{56}Fe$ values that range between -0.8 and $-0.3‰$ are associated with anoxic fluids from groundwater and hydrothermal vents. Such fluids probably obtain their Fe isotope compositions from water rock interactions or as products from bacterial sulfate reduction. Anoxic fluids that have interacted with an oxidant tend to have lower $^{56}Fe/^{54}Fe$ ratios compared to reduced fluids, but these fluids tend to show a general positive correlation between Fe concentration and $\delta^{56}Fe$ values, which reflects loss of oxidized Fe with high $^{56}Fe/^{54}Fe$ ratios.

Increasingly, workers are recognizing the importance of dissimilatory iron reduction as a source of large quantities of iron with low $\delta^{56}Fe$ values. For example, the low $\delta^{56}Fe$ values measured in magnetite from late Archean to early Proterozoic BIFs are best explained by a dominance of dissimilatory iron reducing bacteria at a time in Earth's history when the basins in which these BIFs formed were anoxic and sulfide concentrations were lower than Fe concentrations (Johnson, Beard, and Roden 2008). Moreover, in many ocean basins Fe isotope studies reveal that the shelf to basin Fe shuttle models (e.g., Raiswell and Anderson 2005) appear to be driven by dissimilatory iron reducing bacteria that produce large quantities of reactive Fe with low $\delta^{56}Fe$ values (e.g., Severmann et al. 2008; Fehr et al. 2008). Indeed, the supply of reactive Fe with low $\delta^{56}Fe$ values predicted by these shelf-to-basin shuttle models may explain the origin of Archean sedimentary pyrite with low $\delta^{56}Fe$ values formed in analogous ancient basins (Severmann et al. 2008) in contrast to an alternative model in which Fe with low $^{56}Fe/^{54}Fe$ is produced by the partial oxidation of $Fe(II)_{aq}$ (Rouxel, Bekker, and Edwards 2005).

The cumulative body of knowledge on Fe isotope systematics demonstrates that significant isotope fractionation accompanies key reactions involved in Fe cycling in natural environments. The factors that control Fe isotope compositions are complex and varied, but the level of detail Fe isotope analysis provides cannot be achieved readily by other methods. Much of the work on Fe isotopes has been oriented toward a better understanding of biological and abiotic mechanisms in pursuit of the identification of a diagnostic fingerprint for both. Much additional work is required before a clear consensus is formed regarding key fractionation pathways and isotope signatures, but the application of this knowledge to other redox sensitive contaminants will benefit future environmental investigations.

12.3.7 MOLYBDENUM

The stable isotopes of molybdenum have drawn attention in environmental research for two primary reasons. First, this element's distinctive redox chemistry makes it a potential probe for changes in seafloor oxygenation through time, an important variable in paleoceanographic and paleoclimatological research. Mo is relatively unreactive in oxygenated, aqueous solutions, where it is predominantly present as the molybdate anion (MoO_4^{2-}). Hence, although it is a trace constituent of the upper crust (average abundance of 1–2 ppm; Taylor and McLennan 1985), Mo is the most abundant transition metal in the modern, oxygenated oceans (~105 nmol kg^{-1}; Morris 1975; Bruland 1983; Collier 1985). In contrast, Mo is removed from solution in anoxic-sulfidic ("euxinic") settings, due to formation of particle-reactive oxythiomolybdates ($MoO_{4-x}S_x^{2-}$; Helz et al. 1996; Erickson and Helz 2000), as well as in "suboxic" settings ($0 < O_2 < 5$ µmol kg^{-1}; Poulson et al. 2006). This bimodal character has led to the use of Mo abundances in sediments to trace paleoredox conditions (e.g., Emerson and Huested 1991; Crusius et al. 1996; Dean, Gardner, and Piper 1997; Dean, Piper, and Peterson 1999; Yarincik et al. 2000; Zheng et al. 2000; Calvert, Pedersen, and Karlin 2001; Werne et al. 2002; Lyons et al. 2003; Sageman et al. 2003; Algeo and Maynard 2004). However, the relationship between elemental abundances in sediments and overlying water is often complicated, and elemental abundances may be modified after deposition. Hence, many Mo isotope investigations have focused on the possible use of this isotope system as a complement to elemental analyses to trace changes in ocean redox through time (e.g., Barling, Arnold, and Anbar 2001; Siebert et al. 2003; Arnold, Weyer, and Anbar 2004).

The second reason for environmental interest in Mo is the importance of this element in biology, particularly as a cofactor in enzymes critical for the global biogeochemical cycling of nitrogen (Stiefel 1997; Frausto da Silva and Williams 2001). These include the most common form of nitrogenase, the enzyme that converts N_2 to biologically useful NH_3 and some forms of nitrate reductase, the class of enzymes responsible for reduction of NO_3^- to NO_2^- for either nutrient assimilation or dissimilatory energy production. Molybdenum is also important for the sulfur cycle as a constituent of sulfite reductase and DMSO reductase. The bioavailability of this element has likely varied through time (Anbar and Knoll 2002). The stable isotopes of Mo may therefore prove useful in tracing or reconstructing the biological use of this element in modern or ancient times, or may shed light on the mechanisms of Mo uptake and intracellular trafficking.

12.3.7.1 Isotope Systematics

Mo has seven naturally occurring stable isotopes: ^{92}Mo (14.84%), ^{94}Mo (9.25%), ^{95}Mo (15.92%), ^{96}Mo (16.68%), ^{97}Mo (9.55%), ^{98}Mo (24.13%) and ^{100}Mo (9.63%). Most data have been obtained using MC-ICP-MS and either a Zr or Pd element spike (Anbar, Knab, and Barling 2001; Malinovsky et al. 2005) or an isotope double spike (Siebert, Nägler, and Kramers 2001). A small amount of data has been obtained using sample standard bracketing (Pietruszka, Walker, and Candela 2006) and double spike TIMS (Wieser and DeLaeter 2003). These various methods produce data with precision ranging from ±0.05 – 0.25‰.

Natural mass dependent Mo isotope variations span a range of ~1‰/amu mass difference, usually reported as $\delta^{97/95}$Mo or $\delta^{98/95}$Mo. There is presently no international Mo reference standard so data are typically reported relative to in-house laboratory standards calibrated against seawater (Barling, Arnold, and Anbar 2001;

Siebert et al. 2003). The laboratory standards used to date are close to the apparent average $\delta^{97/95}$Mo of the continental crust.

12.3.7.2 Natural Variations

Most Mo isotope studies have focused on variations in oceanographic settings. The Mo ocean budget is dominated by riverine sources with removal occurring in euxinic settings exemplified by the Black Sea and ferromanganese oxides (Bertine and Turekian 1973), which are both enriched in authigenic Mo. Analytical efforts are therefore, focused on materials potentially representative of the continental crust, seawater, and the various types of sedimentary sinks.

Representative crustal materials include granites and basalts and molybdenites from continental hydrothermal ore deposits. No significant variations have been observed in igneous rocks, including granites subjected to partial leaching (Siebert et al. 2003). However, variations are noted in continental molybdenites (Anbar, Knab, and Barling 2001; Wieser and DeLaeter 2003; Malinovsky et al. 2005; Pietruszka, Walker, and Candela 2006), although the mean value is close to that of igneous rocks Figure 12.6). As such, they are presently used as laboratory standards. Rivers carry Mo with an apparent average isotopic composition of ~0.5‰ (Archer and Vance, 2008), suggesting some Mo isotope fractionation during weathering.

Warm fluids emerging from hydrothermal systems on the flanks of mid-ocean ridges are a net source of Mo to the oceans (Wheat and Mottl 2000). The Mo

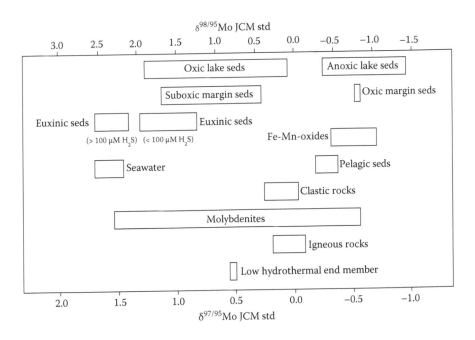

FIGURE 12.6 Stable isotope compositions for Mo from various terrestrial materials (Redrawn from Anbar, A. D., and O. Rouxel, *Annual Review of Earth and Planetary Sciences,* 35, 717–46, 2007.)

isotope composition of these fluids is slightly higher than igneous rocks, and mixing relationships point to an end member hydrothermal fluid with a $\delta^{97/95}$Mo value of ~0.5‰ (McManus et al. 2002).

In comparison to these sources, Mo in seawater (Figure 12.6) has a relatively high $\delta^{97/95}$Mo value that is shifted to ~+1.6‰ relative to crustal rocks (Barling, Arnold, and Anbar 2001; Siebert et al. 2003). This isotope composition is invariant with depth to ~3000 m in the world's oceans to within ±0.1‰. Such uniformity is consistent with the residence time of Mo in the modern ocean (~800,000 years). These data suggest that Mo undergoes significant isotope fractionation upon deposition in sedimentary sinks. Large and systematic isotope variations are observed in a diverse suite of ferromanganese oxide sediments, including Fe nodules from the Pacific and Atlantic Oceans, and crusts from the Pacific, Atlantic, and Indian Oceans (Figure 12.6). The Mo isotope composition of all these samples is remarkably uniform (range of ±<0.25‰), with the mean isotope composition being offset from the $\delta^{97/95}$Mo value of crustal rocks by ~−0.5‰. In contrast, Mo in sediments from the Black Sea and the Cariaco Basin have $\delta^{97/95}$Mo values that are >1‰ higher than crustal rocks, with values as high as 1.6‰ in Unit I and 1.2‰ in Unit II of the Black Sea. The $\delta^{97/95}$Mo values in ~14,000 year old sediments of the Cariaco Basin are comparable to Unit II of the Black Sea.

More variability is observed in suboxic settings. Sedimentary $\delta^{97/95}$Mo values cover a large range, essentially filling the gap between crustal rocks and euxinic sediments (Siebert et al. 2003; Siebert et al. 2006; Poulson et al. 2006). The concentration of Mo in pore waters of these sediments, and $\delta^{97/95}$Mo values become progressively higher, ultimately reaching values nearly 1‰ higher than seawater—among the highest $\delta^{97/95}$Mo value measured to date.

12.3.7.3   Fractionating Processes

The largest Mo isotope fractionation yet known results from adsorption of dissolved Mo to Mn oxide surfaces. In laboratory experiments in which dissolved Mo and Mn oxides are equilibrated, dissolved Mo is ~1‰/amu higher than adsorbed Mo (Barling and Anbar 2004; Wasylenki, Anbar, and Gordon 2006a). This effect probably explains the offset between Mo isotopes in seawater and ferromanganese oxides given that the magnitudes observed in nature and laboratory are indistinguishable. The fractionation is insensitive to pH or temperature, although a slight sensitivity to oxide composition is inferred from natural variations between Atlantic and Pacific crusts (Siebert et al. 2003; Barling and Anbar 2004). The cause of this fractionation is not known, but it may result from isotope fractionation between MoO_4^{2-}, which dominates Mo speciation in seawater, and one or more scarce, dissolved Mo species, such as $Mo(OH)_6^0$ or MoO_3^0, that sorb more readily to environmental particles (Siebert et al. 2003; Tossell 2005). Alternatively, fractionation may result from isotope exchange between MoO_4^{2-} and surface-bound Mo species (Barling and Anbar 2004).

Smaller but significant Mo isotope effects are inferred during the removal of Mo to sediments underlying suboxic waters based on observations in reducing sedimentary basins (McManus et al. 2002; Siebert et al. 2003; Nägler et al. 2005; Poulson et al. 2006; Siebert et al. 2006). These effects are not yet understood mechanistically and have yet to be studied in the laboratory. Mo isotope fractionation in these settings may be closely coupled to Mn cycling. In these systems, Mn oxide particles settling from overlying oxic waters may be reduced and dissolved. Depleted Mo may then be introduced if these particles carry with them adsorbed, fractionated Mo. This Mo may then enter sediment pore waters where it is immobilized in sulfide-rich sediments below the suboxic zone.

The small magnitude of Mo isotope fractionation between seawater and euxinic sediments is thought to reflect quantitative or near-quantitative removal of Mo from the water column (Barling, Arnold, and Anbar 2001; Siebert et al. 2003). Small fractionations may reflect extensive but nonquantitative removal, as might be expected in euxinic basins that have short deep water renewal times but are in contact with Mo-rich oceans (Algeo and Lyons 2006). Such effects could be generated during ligand exchange when MoO_4^{2-} reacts with H_2S to form $MoO_{4-x}S_x^{2-}$ or during Mo reduction.

Biological isotope effects appear to have only a minor effect on the global Mo isotope budget, but there are reports of preferential uptake of isotopes of lower mass (~0.25‰/amu) by N-fixing bacteria (Nägler, Mills, and Siebert 2004; Liermann et al. 2005; Wasylenki et al. 2006b).

12.3.7.4 Synthesis and Applications

With respect to the ocean system, these data suggest a first-order model in which Mo enters the oceans primarily from continental sources with an average $\delta^{97/95}Mo$ value ~0‰; Mo is then fractionated during removal to Mn oxides and suboxic sediments, resulting in an enriched reservoir of Mo dissolved in seawater. Removal of Mo to euxinic sediments appears to impart little isotope effect. From a paleoceanographic perspective, this model implies that $\delta^{97/95}Mo_{seawater}$ values should vary with the extent of seafloor oxygenation; if there were no Mn-oxide deposition or suboxic sedimentation, then the $\delta^{97/95}Mo_{seawater}$ value would be ~0‰ in contrast to 1.6‰ today (Barling, Arnold, and Anbar 2001; Siebert et al. 2003). This concept underlies the use of the Mo isotope system as a global paleoredox indicator (e.g., Arnold et al. 2004).

There are at least three major caveats to this model. First, continental rocks being weathered may not be isotopically uniform, and so changes in weathering regimes could lead to changes in $\delta^{97/95}Mo_{seawater}$ values independent of ocean redox. Second, the extent of Mo isotope fractionation during weathering and transport to the oceans is not known but is apparently not negligible (Archer and Vance 2008); and most importantly, third, that suboxic sedimentary environments, once thought to be of negligible importance to the Mo element and isotope budgets (Morford and Emerson 1999; Barling, Arnold, and Anbar 2001) may account for as much as 60% of Mo removal (McManus et al. 2005). If so, an effect on the ocean isotope budget is likely because Mo isotopes are significantly fractionated in pore waters and sediments from such settings (McManus et al. 2002; Poulson et al. 2006). Relative to seawater Mo, the direction of this effect is the same as fractionation by removal to Mn-oxide sediments, and so changes in $\delta^{97/95}Mo_{seawater}$ values due to contraction of oxic deposition could be partially countered by expansion of suboxic deposition.

Despite these caveats, Mo isotope compositions of ancient sediments are being cautiously used to study ancient ocean redox variations, using both ferromanganese sediments and black shales to reconstruct ancient seawater Mo isotope variations. For example, Siebert et al. (2003) proposed that the uniformity of Mo isotope compositions over the last 60 million years is consistent with invariance of $\delta^{97/95}Mo_{seawater}$ values at a temporal resolution of 1 to 3 million years through the Cenozoic, potentially indicating that ocean oxygenation has varied less than 10% from modern values over this time. Although larger ocean redox perturbations are possible as a response to glacial pCO_2 drawdown, their duration does not approach the ~10^6 year timescale of crust sampling and the mean ocean residence time of Mo.

Mo isotope paleorecords also can be constructed from Mo in euxinic sediments, subject to the assumption that there is little fractionation between Mo in sediments and in contemporaneous seawater. For such studies, it is important to demonstrate that sampled shales represent a persistently euxinic water column to minimize the possibility of Mo isotope artifacts arising from nonquantitative scavenging. Recent studies have documented that nominal "black shales" are not necessarily deposited under euxinic water conditions analogous to the modern Black Sea (Werne et al. 2002) and that $\delta^{97/95}Mo$ values are systematically offset in settings that are not persistently euxinic (Gordon et al. 2009). With these concerns in mind, Arnold, Weyer, and Anbar (2004) analyzed $\delta^{97/95}Mo$ values in black shales from the McArthur Basin, Australia (1.4–1.6 Ga) to assess the possibility of expanded ocean euxinia during this time (Canfield 1998; Anbar and Knoll 2002). Isotopic compositions were distinctly lower than modern seawater and sediments, consistent with the euxinic hypothesis. Siebert et al. (2005) measured $\delta^{97/95}Mo$ values in black shales from several locations in South Africa and Zimbabwe that ranged in age from 2.2 to 3.25 Ga. They observed generally increasing $\delta^{97/95}Mo$ values through time, hovering around 0‰ before 3 Ga. These data are consistent with increasing ocean oxygenation during this time, as expected from other proxies of atmospheric O_2. Despite these exciting results, the use of Mo isotopes to quantify ocean oxygenation remains a challenge because of uncertainties in the Mo ocean element budget described above. Further work is required on modern systems and on younger (Paleozoic and Cenozoic) sediments where depositional context may be better understood. The recent work of Pearce et al. (2008) is a step in this direction.

Mo isotopes may also be useful to constrain local redox conditions. Data from reduced sediment cores and ancient black shales indicate distinct Mo isotope compositions associated with local depositional (and early diagenetic) environments (e.g., suboxic to anoxic to euxinic). In effect, $\delta^{97/95}Mo$ values may monitor the extent of reducing conditions at the time of deposition (Gordon et al. 2009; Poulson et al. 2006).

Studies of the use of Mo isotopes to examine biological utilization are in their infancy. Although the effects are small, they may prove useful in examining the biological pathways of Mo uptake (Liermann et al. 2005) and in identifying the biological use of Mo in ancient sediments (Wasylenki, Anbar, and Gordon 2006a). For example, in experiments in which N-fixing soil bacteria were grown under Mo-limited conditions, Liermann et al. (2005) hypothesized that the magnitude of Mo isotope fractionation may be related to the production of extracellular Mo-chelating ligands. While rare today, Mo limitation would have been common in ancient oceans prior to the development of a fully oxygenated environment (Anbar and Knoll 2002). Hence, the evaluation and extension Mo isotope hypotheses will point the way to future biological applications of this system.

12.4 CONCLUSIONS

With the development and improvement of analytical instrumentation and techniques, stable isotope ratios can now be measured with unparalleled precision for elements that were once thought to have fixed compositions. These breakthroughs provide scientists with a new tool for determining natural processes and the origin and fate of environmental contaminants. Some metals such as Li or Cr may have characteristic isotope patterns that can be linked to anthropogenic origins. Subsequent isotope fractionation patterns may provide a key to the natural attenuation or modification of environmental contaminants over time. Other isotope fractionation patterns (e.g., Fe, Mo) may prove more valuable indirectly by providing information on the physicochemical potential or conditions (e.g., redox) that influence the mobility (e.g., Cr, Se) or bioavailability (e.g., Ca) of elements of interest. The systematics of nontraditional stable isotopes are in their infancy to the extent that international standards are not yet available for some elements (e.g., Ca, Se). Much work still needs to be done to fully characterize the isotope and kinetic effects associated with physicochemical and biological processes (e.g., phase change, sorption, reaction) that control the isotope composition of natural materials, both from empirical and theoretical perspectives. And in the near term, whole new isotope systems will emerge (e.g., Hg, Ni, Sb) that require basic characterization prior to their application in environmental studies.

ACKNOWLEDGMENTS

This chapter would not be possible without the dedicated efforts of numerous individuals who have meticulously and rigorously measured the composition of nontraditional stable isotopes in natural materials. Thanks are given to Brian Jackson, Laura Janecek, Carl Strojan, Tom Bullen, Diane Newman, Patrick Höhener, C. Marjorie Aelion, Edwin Schauble, Reinhard Kozdon, and Andrew Czaja for help with this review. This work was supported in part by the U.S. Department of Energy through Financial Assistance Award No. DE-FC09-96-SR18546 to the University of Georgia Research Foundation.

REFERENCES

Ader, M., M. L. Coleman, S. P. Doyle, M. Stroud, and D. Wakelin. 2001. Methods for the stable isotopic analysis of chlorine in chlorate and perchlorate compounds. *Analytical Chemistry* 73:4946–50.

Albarède, F., and B. Beard. 2004. Analytical methods for non-traditional isotopes. In *Geochemistry of non-traditional stable isotopes*, eds. C. M. Johnson, B. L. Beard, and F. Albarède, 113–52. Washington DC: The Mineralogical Society of America.

Algeo, T. J., and T. W. Lyons. 2006. Mo-total organic carbon covariation in modern anoxic marine environments: Implications for analysis of paleoredox and paleohydrographic conditions. *Paleoceanography* 21:PA1016, doi:10.1029/2004PA001112.

Algeo, T. J., and J. B. Maynard. 2004. Trace element behavior and redox facies analysis of core shales of upper Pennsylvanian Kansas-type cyclothems. *Chemical Geology* 206:289–318.

Anbar, A. D. 2004. Iron stable isotopes: Beyond biosignatures. *Earth and Planetary Science Letters* 217:223–36.

Anbar, A. D., A. A. Jarzecki, and T. G. Spiro. 2005. Theoretical investigation of iron isotope fractionation between $Fe(H_2O)_6^{3+}$ and $Fe(H_2O)_6^{2+}$: Implications for iron stable isotope geochemistry. *Geochimica Cosmochimica Acta* 69:825–37.

Anbar, A. D., K. A. Knab, and J. Barling. 2001. Precise determination of mass-dependent variations in the isotopic composition of molybdenum using MC-ICPMS. *Analytical Chemistry* 73:1425–31.

Anbar, A. D., and A. H. Knoll. 2002. Proterozoic ocean chemistry and evolution: A bioinorganic bridge? *Science* 297:1137–42.

Anbar, A. D., J. E. Roe, J. Barling, and K. H. Nealson. 2000. Nonbiological fractionation of iron isotopes. *Science* 288:126–28.

Anbar, A. D., and O. Rouxel. 2007. Metal stable isotopes in paleoceanography. *Annual Review of Earth and Planetary Sciences* 35:717–46.

Archer, C., and D. Vance. 2008. The isotopic signature of the global riverine molydenum flux and anoxia in the ancient oceans. *Nature Geoscience* 1:597–600.

Arnold, G. L., A. D. Anbar, J. Barling, and T. W. Lyons. 2004. Molybdenum isotope evidence for widespread anoxia in mid-proterozoic oceans. *Science* 304:87–90.

Arnold, G. L., S. Weyer, and A. D. Anbar. 2004. Fe isotope variations in natural materials measured using high mass resolution multiple collector ICPMS. *Analytical Chemistry* 76:322–27.

Balci, N., T. D. Bullen, K. Witte-Lien, W. C. Shanks, M. Motelica, and K. W. Mandernack. 2006. Iron isotope fractionation during microbially stimulated Fe(II) oxidation and Fe(III) precipitation. *Geochimica Cosmochimica Acta* 70:622–39.

Barling, J., and A. D. Anbar. 2004. Molybdenum isotope fractionation during adsorption by manganese oxides. *Earth and Planetary Science Letters* 217:315–29.

Barling, J., G. L. Arnold, and A. D. Anbar. 2001. Natural mass-dependent variations in the isotopic composition of molybdenum. *Earth and Planetary Science Letters* 193:447–57.

Bassett, R. L., P. M. Buszka, G. R. Davidson, and D. Chongdiaz. 1995. Identification of groundwater solute sources using boron isotopic composition. *Environmental Science & Technology* 29:2915–22.

Bazylinski, D. A., and B. M. Moskowitz. 1997. Microbial biomineralization of magnetic iron minerals: Microbiology, magnetism and environmental significance. *Geomicrobiology* 35:181–223.

Beard, B. L., and C. M. Johnson. 2004. Fe isotope variations in the modern and ancient earth and other planetary bodies. In *Geochemistry of non-traditional stable isotopes*, eds. C. M. Johnson, B. L. Beard, and F. Albarède, 319–57. Washington DC: The Mineralogical Society of America.

Beard, B. L., C. M. Johnson, L. Cox, H. Sun, K. H. Nealson, and C. Aguilar. 1999. Iron isotope biosignatures. *Science* 285:1889–92.

Beard, B. L., C. M. Johnson, J. L. Skulan, K. H. Nealson, L. Cox, and H. Sun. 2003a. Application of Fe isotopes to tracing the geochemical and biological cycling of Fe. *Chemical Geology* 195:87–117.

Beard, B. L., C. M. Johnson, K. L. Von Damm, and R. L. Poulson. 2003b. Iron isotope constraints on Fe cycling and mass balance in oxygenated Earth oceans. *Geology* 31:629–32.

Belshaw, N. S., X. K. Zhu, Y. Guo, and R. K. O'Nions. 2000. High precision measurement of iron isotopes by plasma source mass spectrometry. *International Journal of Mass Spectrometry* 197:191–95.

Bergquist, B. A., and E. A. Boyle. 2006. Iron isotopes in the Amazon River system: Weathering and transport signatures. *Earth and Planetary Science Letters* 248:54–68.

Bertine, K. K., and K. K. Turekian. 1973. Molybdenum in marine deposits. *Geochimica Cosmochimica Acta* 37:1415–34.

Bigeleisen, J. 1996. Nuclear size and shape effects in chemical reactions. Isotope chemistry of the heavy elements. *J. American Chemical Society* 118:3676–3680.

Böhm, F., N. Gussone, A. Eisenhauer, W-C. Dullo, S. Reynaud, and A. Paytan. 2006. Calcium isotope fractionation in modern scleractinian corals. *Geochimica Cosmochimica Acta* 70:4452–62.

Brantley, S. L., L. Liermann, and T. D. Bullen. 2001. Fractionation of Fe isotopes by soil microbes and organic acids. *Geology* 29:535–38.

Brantley, S. L., L. J. Liermann, R. L. Guynn, A. Anbar, G. A. Icopini, and J. Barling. 2004. Fe isotopic fractionation during mineral dissolution with and without bacteria. *Geochimica Cosmochimica Acta* 68:3189–204.

Bruland, K. W. 1983. Trace elements in seawater. In *Chemical oceanography*, eds. J. P. Riley and R. Chester, 157–220. London: Academic Press.

Bullen, T. D., A. F. White, C. W. Childs, D. V. Vivit, and M. S. Schulz. 2001. Demonstration of significant abiotic iron isotope fractionation in nature. *Geology* 29:699–702.

Bulter, I. B., C. Archer, D. Vance, A. Oldroyd, and D. Rickard. 2005. Fe isotope fractionation on FeS formation in ambient aqueous solution. *Earth and Planetary Science Letters* 236:430–42.

Cabana, G., and J. B. Rasmussen. 1994. Modeling food-chain structure and contaminant bio-accumulation using stable nitrogen isotopes. *Nature* 372:255–57.

Calvert, S. E., T. F. Pedersen, and R. E. Karlin. 2001. Geochemical and isotopic evidence for post-glacial paleoceanographic changes in Saanich Inlet, British Columbia. *Marine Geology* 174:287–305.

Canfield, D. E. 2001. Biogeochemistry of sulfur isotopes. *Reviews in Mineralogy and Geochemistry* 43:607–36.

Canfield, D. E. 1998. A new model for Proterozoic ocean chemistry. *Nature* 396:450–53.

Chang, V. T-C., R. J. P. Williams, A. Makishima, N. S. Belshawl, and R. K. O'Nions. 2004. Mg and Ca isotope fractionation during $CaCO_3$ biomineralisation. *Biochemical and Biophysical Research Communications* 323:79–85.

Clementz, M. T., P. Holden, and P. L. Koch. 2003. Are calcium isotopes a reliable monitor of trophic level in marine settings? *International Journal of Osteoarchaeology* 13:29–36.

Coleman, M. L., M. Ader, S. Chaudhuri, and J. D. Coates. 2003. Microbial isotopic fractionation of perchlorate chlorine. *Applied Environmental Microbiology* 69:4997–5000.

Collier, R. W. 1985. Molybdenum in the northeast Pacific Ocean. *Limnology and Oceanography* 30:1351–54.

Cooper, W. C., and J. R. Glover. 1974. The toxicology of selenium and its compounds. In *Selenium*, eds. R. A. Zingaro and W. C. Cooper, 654–74. New York: Van Nostrand Reinhold.

Coplen, T. B. 1996. Atomic weights of the elements 1995. *Pure and Applied Chemistry* 68:2339–59.

Coplen, T. B., J. K. Bohlke, P. De Bievre, T. Ding, N. E. Holden, J. A. Hopple, H. R. Krouse, et al. 2002. Isotope-abundance variations of selected elements: (IUPAC Technical Report). *Pure and Applied Chemistry* 74:1987–2017.

Croal, L. R., C. M. Johnson, B. L. Beard, and D. K. Newman. 2004. Iron isotope fractionation by Fe(II)-oxidizing photoautotrophic bacteria. *Geochimica Cosmochimica Acta* 68:1227–42.

Crosby, H. A., C. M. Johnson, E. E. Roden, and B. L. Beard. 2005. Fe(II)-Fe(III) electron/atom exchange as a mechanism for Fe isotope fractionation during dissimilatory iron oxide reduction. *Environmental Science & Technology* 39:6698–704.

Crosby, H. A., E. E. Roden, C. M. Johnson, and B. L. Beard. 2007. The mechanisms of iron isotope fractionation produced during dissimilatory Fe(III) reduction by *Shewanella putrefaciens* and *Geobacter sulfurreducens*. *Geobiology* 5:169–89.

Crusius, J., S. Calvert, T. Pedersen, and D. Sage. 1996. Rhenium and molybdenum enrichments in sediments as indicators of oxic, suboxic and sulfidic conditions of deposition. *Earth and Planetary Science Letters* 145:65–78.

Dauphas, N., and O. Rouxel. 2006. Mass spectrometry and natural variations of iron isotopes. *Mass Spectrometry Reviews* 25:515–50.

Dauphas, N., M. V. Zuilen, M. Wadhwa, A. M. Davis, B. Marty, and P. E. Janney. 2004. Clues from Fe isotope variations on the origin of early Archean BIFs from Greenland. *Science* 306:2077–80.

Davidson, G. R., and R. L. Bassett. 1993. Application of boron isotopes for identifying contaminants such as fly-ash leachate in groundwater. *Environmental Science & Technology* 27:172–76.

Dean, W. E., J. V. Gardner, and D. Z. Piper. 1997. Inorganic geochemical indicators of glacial-interglacial changes in productivity and anoxia on the California continental margin. *Geochimica Cosmochimica Acta* 61:4507–18.

Dean, W. E., D. Z. Piper, and L. C. Peterson. 1999. Molybdenum accumulation in Cariaco basin sediment over the past 24 k.y.: A record of water-column anoxia and climate. *Geology* 27:507–10.

De La Rocha, C., and D. J. DePaolo. 2000. Isotopic evidence for variations in the marine calcium cycle over the Cenozoic. *Science* 289:1176–78.

DePaolo, D. J. 2004. Calcium isotopic variations produced by biological, kinetic, radiogenic, and nucleosynthetic processes. In *Geochemistry of non-traditional stable isotopes*, eds. C. M. Johnson, B. L. Beard, and F. Albarède, 255–88. Washington DC: The Mineralogical Society of America.

Dideriksen, K., J. A. Baker, and S. L. S. Stipp. 2006. Iron isotopes in natural carbonate minerals determined by MC-ICP-MS with a ^{58}Fe-^{54}Fe double spike. *Geochimica Cosmochimica Acta* 70:118–32.

Dideriksen, K., J. A. Baker, and S. L. S. Stipp. 2008. Equilibrium Fe isotope fractionation between inorganic aqueous Fe(III) and the siderophore complex, Fe(III)-desferrioxamine B. *Earth and Planetary Science Letters* 269:280–90.

Douglas, D. J., and S. C. Tanner. 1998. Fundamental considerations in ICP-MS. In *Inductively coupled plasma mass spectrometry*, ed. A. Montaser, 615–79. New York: Wiley-VCH.

Drenzek, N. J., T. I. Eglinton, C. O. Wirsen, N. C. Sturchio, L. J. Heraty, K. R. Sowers, Q. Z. Wu, H. D. May, and C. M. Reddy. 2004. Invariant chlorine isotopic signatures during microbial PCB reductive dechlorination. *Environmental Pollution* 128:445–48.

Eggenkamp, H. G. M. 1994. The geochemistry of chlorine isotopes. PhD diss., Utrecht University.

Ehrlich, H. L. 1996. *Geomicrobiology*. New York: Marcel Dekker.

Eisenhauer, A., T. F. Nägler, P. Stille, J. Kramers, N. Gussone, B. Bock, J. Fietzke, D. Hippler, and A-D. Schmitt. 2004. Proposal for international agreement on Ca notation resulting from discussions at workshops on stable isotope measurements held in Davos (Goldschmidt 2002) and Nice (EGS-AGU-EUG 2003). *Geostandards and Geoanalytical Research* 28:149–51.

Ellis, A. S., T. M. Johnson, and T. D. Bullen. 2002. Chromium isotopes and the fate of hexavalent chromium in the environment. *Science* 295:2060–62.

Ellis, A. S., T. M. Johnson, and T. D. Bullen. 2004. Using chromium stable isotope ratios to quantify Cr (VI) reduction: Lack of sorption effects. *Environmental Science & Technology* 38:3604–7.

Ellis, A. S., T. M. Johnson, T. D. Bullen, and M. J. Herbel. 2003. Stable isotope fractionation of selenium by natural microbial consortia. *Chemical Geology* 195:119–29.

Emerson, D. 2000. Microbial oxidation of Fe(II) and Mn(II) at circumneutral pH. In *Environmental metal-microbe interactions*, ed. D. R. Lovley, 31–52. Washington, DC: ASM Press.

Emerson, S. R., and S. S. Huested. 1991. Ocean anoxia and the concentrations of molybdenum and vanadium in seawater. *Marine Chemistry* 34:177–96.

Emmanuel S., Y. Erel, A. Matthews, and N. A. Teutsch. 2005. A preliminary mixing model for Fe isotopes in soils. *Chemical Geology* 222:23–34.

Erickson, B. E., and G. R. Helz. 2000. Molybdenum (VI) speciation in sulfidic waters: Stability and lability of thiomolybdates. *Geochimica Cosmochimica Acta* 64:1149–58.

Ewing, S. A., W. Yang, D. J. DePaolo, G. Michalski, C. Kendall, B. W. Stewart, M. Thiemens, and R. Amundson. 2008. Non-biological fractionation of stable Ca isotopes in soils of the Atacama Desert, Chile. *Geochimica Cosmochimica Acta* 72:1096–1110.

Fantle, M. S., and D. J. DePaolo. 2004. Iron isotopic fractionation during continental weathering. *Earth and Planetary Science Letters* 228:547–62.

Fantle, M. W., and D. J. DePaolo. 2005. Variations in the marine Ca cycle over the past 20 million years. *Earth and Planetary Science Letters* 237:102–17.

Fantle, W. M., and D. J. DePaolo. 2007. Ca isotopes in carbonate sediment and pore fluid from ODP Site 870A: The Ca^{2+}(aq)-calcite equilibrium fractionation factor and calcite recrystallization rates in Pleistocene sediments. *Geochimica Cosmochimica Acta* 71:2524–46.

Farkaš, J., F. Böhm, K. Wallmann, J. Blenkinsop, A. Eisenhauer, R. van Geldern, A. Munnecke, S. Voigt, and J. Veizer. 2007. Calcium isotope record of Phanerozoic oceans: Implications for chemical evolution of seawater and its causative mechanisms. *Geochimica Cosmochimica Acta* 71:5117–34.

Fehr, M. A., P. S. Andersson, U. Hålenius, and C-M. Mörth. 2008. Iron isotope variations in Holocene sediments of the Gotland Deep, Baltic Sea. *Geochimica Cosmochimica Acta* 72:807–26.

Fietzke, J., A. Eisenhauer, N. Gussone, B. Bock, V. Liebetrau, Th. F. Nägler, H. J. Spero, J. Bijma, and C. Dullo. 2004. Direct measurement of $^{44}Ca/^{40}Ca$ ratios by MC-ICP-MS using the cool plasma technique. *Chemical Geology* 206:11–20.

Fischer, A. J., E. A. Rowan, and R. F. Spalding. 1987. VOCs in groundwater influenced by large-scale withdrawals. *Groundwater* 2:407–13.

Flesch, G. D., A. R. Anderson Jr., and H. J. Svec. 1973. A secondary isotopic standard for $^6Li/^7Li$ determinations. *International Journal of Mass Spectrometry and Ion Processes* 12:265–72.

Frausto da Silva, J. J. R., and R. J. P. Williams. 2001. *The biological chemistry of the elements: The inorganic chemistry of life*. Oxford: Clarendon Press.

Fredrickson, J. K., J. M. Zachara, D. W. Kennedy, H. L. Dong, T. C. Onstott, N. W. Hinman, and S. M. Li. 1998. Biogenic iron mineralization accompanying the dissimilatory reduction of hydrous ferric oxide by a groundwater bacterium. *Geochimica Cosmochimica Acta* 62:3239–57.

Ganther, H. E. 1974. The biochemistry of selenium. In *Selenium*, eds. R. A. Zingaro and W. C. Cooper, 546–614. New York: Van Nostrand Reinhold Co.

Gates, W. P., H. T. Wilkinson, and J. W. Stucki. 1993. Swelling properties of microbially reduced ferruginous smectite. *Clays and Clay Minerals* 41:360–64.

Gellenbeck, D. J., and T. D. Bullen. 1991. Isotopic characterization of strontium and lithium in potential sources of ground-water contamination in Central Arizona. *EOS, Transactions, American Geophysical Union* 72:178.

Ghiorse, W. C. 1989. Manganese and iron as physiological electron donors and acceptors in aerobic-anaerobic transition zones. In *Microbial Mats*, ed. C. Y. Rosenberg, 163–79, Washington, DC: ASM Press.

Gordon, G. W., T .W. Lyons, G. L. Arnold, J. Role, B. B. Sageman, and A. D. Anbar. 2009. When do black shales tell molydenum isotope tales. *Geology* 37:535–538.

Graham, S., N. Pearson, S. Jackson, W. Griffin, and S. Y. O'Reilly. 2004. Tracing Cu and Fe from source to porphyry: In situ determination of Cu and Fe isotope ratios in sulfides from the Grasberg Cu-Au deposit. *Chemical Geology* 207:147–69.

Griffith, E. M., A. Paytan, R. Kozdon, A. Eisenhauer, and A. C. Ravelo. 2008. Influences on the fractionation of calcium isotopes in planktonic foraminifera. *Earth and Planetary Science Letters* 268:124–36.

Guelke, M., and F. von Blanckenburg. 2007. Fractionation of stable iron isotopes in higher plants. *Environmental Science & Technology* 41:1896–901.

Gulick, R. W., M. W. Lechevallier, and T. S. Barhorst. 2001. Occurrence of perchlorate in drinking water sources. *Journal of the American Water Works Association* 93:66–76.

Gussone, N., A. Eisenhauer, A. Heuser, M. Dietzel, B. Bock, F. Böhm, H. J. Spero, D. W. Lea, J. Bijma, and T. F. Nägler. 2003. Model for kinetic effects on calcium isotope fraction-ation ($\delta^{44}Ca$) in inorganic aragonite and cultured planktonic foraminifera. *Geochimica Cosmochimica Acta* 67:1375–82.

Gussone, N., A. Eisenhauer, R. Tiedemann, G. H. Haug, A. Heuser, B. Bock, Th. F. Nägler, and A. Müeller. 2004. Reconstruction of Caribbean Sea surface temperature and salinity fluctuations in response to the Pliocene closure of the central American gateway and radiative forcing, using $\delta^{44/40}Ca$, $\delta^{18}O$ and Mg/Ca ratios. *Earth and Planetary Science Letters* 227:201–14.

Hagiwara, Y. 2000. Selenium isotope ratios in marine sediments and algae. A reconnaissance study. M.S. diss., University of Illinois.

Heidenreich, J. E., and M. H. Thiemens. 1983. The non mass-dependent oxygen isotope effect in the electrodissociation of carbon dioxide. A step toward understanding NoMaD chemistry. *Geochimica Cosmochimica Acta* 49:1303–6.

Heising, S., L. Richter, W. Ludwig, and B. Schink. 1999. *Chlorobium ferrooxidans* sp nov., a phototrophic green sulfur bacterium that oxidizes ferrous iron in coculture with a "Geospirillum" sp strain. *Archives of Microbiology* 172:116–24.

Heising, S., and B. Schink, 1998. Phototrophic oxidation of ferrous iron by a *Rhodomicrobium vannielii* strain. *Microbiology-UK* 144:2263–69.

Helz, G. R., C. V. Miller, J. M. Charnock, J. F. W. Mosselmans, R. A. D. Pattrick, C. D. Garner, and D. J. Vaughn. 1996. Mechanism of molybdenum removal from the sea and its concentration in black shales: EXAFS evidence. *Geochimica Cosmochimica Acta* 60:3631–42.

Heraty, L. J., M. E. Fuller, L. Huang, T. Abrajano, and N. C. Sturchio. 1999. Isotopic frac-tionation of carbon and chlorine by microbial degradation of dichloromethane. *Organic Geochemistry* 30:793–99.

Herbel, M. J., T. M. Johnson, R. S. Oremland, and T. D. Bullen. 2000. Selenium stable iso-tope fractionation during bacterial dissimilatory reduction of selenium oxyanions. *Geochimica Cosmochimica Acta* 64:3701–9.

Herbel, M. J., T. M. Johnson, K. K. Tanji, S. Gao, and T. D. Bullen. 2002. Selenium stable isotope ratios in agricultural drainage water systems of the western San Joaquin Valley, CA. *Journal of Environmental Quality* 31:1146–56.

Hill, P. S., and E. A. Schauble. 2008. Modeling the effects of bond environment on equilibrium iron isotope fractionation in ferric aquo-chloro complexes. *Geochimica Cosmochimica Acta* 72:1939–58.

Hippler, D., A. Eisenhauer, and T. F. Nägler. 2006. Tropical Atlantic SST history inferred from Ca isotope thermometry over the last 140ka. *Geochimica Cosmochimica Acta* 70:90–100.

Hippler, D., A-D. Schmitt, N. Gussone, A. Heuser, P. Stille, A. Eisenhauer, and T. F. Nägler. 2003. Calcium isotopic compositions of various reference materials and seawater. *Geostandards Newsletter* 27:13–19.

Hogan, J. F., and J. D. Blum. 2003. Boron and lithium isotopes as groundwater tracers: A study at the Fresh Kills Landfill, Staten Island, New York, USA. *Applied Geochemistry* 18:615–27.

Holt, B. D., N. C. Sturchio, T. A. Abrajano, and L. J. Heraty. 1997. Conversion of chlorinated volatile organic compounds to carbon dioxide and methyl chloride for isotopic analysis of carbon and chlorine. *Analytical Chemistry* 69:2727–33.

Huh, Y., L. H. Chan, L. Zhang, and J. M. Edmond. 1998. Lithium and its isotopes in major world rivers: Implications for weathering and the oceanic budget. *Geochimica Cosmochimica Acta* 62:2039–51.

Icopini, G. A., A. D. Anbar, S. S. Ruebush, M. Tien, and S. L. Brantley. 2004. Iron isotope fractionation during microbial reduction of iron: The importance of adsorption. *Geology* 32:205–8.

Immenhauser, A., T. F. Näegler, T. Steuber, and D. Hippler. 2005. A critical assessment of mollusk $^{18}O/^{16}O$, Mg/Ca and $^{44}Ca/^{40}Ca$ ratios as proxies for Cretaceous seawater temperature seasonality. *Palaeogeography Palaeoclimatology Palaeoecology* 215:221–37.

Jacobson, A. D., and C. Holmden. 2008. $\delta^{44}Ca$ evolution in a carbonate aquifer and its bearing on the equilibrium isotope fractionation for calcite. *Earth and Planetary Science Letters* 270:349–53.

Jendrzejewski, N., H. G. M. Eggenkamp, and M. L. Coleman. 1997. Sequential determination of chlorine and carbon isotopic composition in single microliter samples of chlorinated solvent. *Analytical Chemistry* 69:4259–66.

Jendrzejewski, N., H. G. M. Eggenkamp, and M. L. Coleman. 2001. Characterization of chlorinated hydrocarbons from chlorine and carbon isotopic compositions: Scope of application to environmental problems. *Applied Geochemistry* 16:1021–31.

Johnson, C. M., and B. L. Beard. 1999. Correction of instrumentally produced mass fractionation during isotopic analysis of Fe by thermal ionization mass spectrometry. *International Journal of Mass Spectrometry* 193:870–99.

Johnson, C. M., B. L. Beard, N. J. Beukes, C. Klein, and J. M. O'Leary. 2003. Ancient geochemical cycling in the Earth as inferred from Fe isotope studies of banded iron formations from the Transvaal Craton. *Contributions to Mineralogy and Petrology* 144:523–47.

Johnson, C. M., B. L. Beard, and E. E. Roden. 2008. The iron isotope fingerprints of redox and biogeochemical cycling in modern and ancient Earth. *Annual Review of Earth and Planetary Sciences* 36:457–93.

Johnson, C. M., B. L. Beard, E. E. Roden, D. K. Newman, and K. H. Nealson. 2004. Isotopic constraints on biogeochemical cycling of Fe. In *Geochemistry of non-traditional stable isotopes*, eds. C. M. Johnson, B. L. Beard, and F. Albarède, 359–408. Washington, DC: The Mineralogical Society of America.

Johnson, C. M., E. E. Roden, S. A. Welch, and B. L. Beard. 2005. Experimental constraints on Fe isotope fractionation during magnetite and Fe carbonate formation coupled to dissimilatory hydrous ferric oxide reduction. *Geochimica Cosmochimica Acta* 69:963–93.

Johnson, C. M., J. L. Skulan, B. L. Beard, H. Sun, K. H. Nealson, and P. S. Braterman. 2002. Isotopic fractionation between Fe(III) and Fe(II) in aqueous solutions. *Earth and Planetary Science Letters* 195:141–53.

Johnson, T. M., and T. D. Bullen. 2003. Selenium isotope fractionation during reduction by Fe(II)-Fe(III) hydroxide-sulfate (green rust). *Geochimica Cosmochimica Acta* 67:413–19.

Johnson, T. M., and T. D. Bullen. 2004. Mass-dependent fractionation of selenium and chromium isotopes in low-temperature environments. In *Geochemistry of non-traditional stable isotopes*, eds. C. M. Johnson, B. L. Beard, and F. Albarède, 289–317. Washington DC: The Mineralogical Society of America.

Johnson, T. M., T. D. Bullen, A. S. Ellis, E. Sikora, and J. Kitchen. 2005. Cr isotopes as indicators of Cr(VI) reduction and contaminant sources. *Geochimica Cosmochimica Acta* 69:A204.

Johnson, T. M., T. D. Bullen, and P. T. Zawislanski. 2000. Selenium stable isotope ratios as indicators of sources and cycling of selenium: Results from the northern reach of San Francisco Bay. *Environmental Science & Technology* 34:2075–79.

Johnson, T. M., M. J. Herbel, T. D. Bullen, and P. T. Zawislanski. 1999. Selenium isotope ratios as indicators of selenium sources and oxyanion reduction. *Geochimica Cosmochimica Acta* 63:2775–83.

Kasemann, S. A., C. J. Hawkesworth, A. R. Prave, A. E. Fallick, and P. N. Pearson. 2005. Boron and calcium isotope composition in Neoproterozoic carbonate rocks from Nambia: Evidence for extreme environmental change. *Earth and Planetary Science Letters* 231:73–86.

Kasemann, S. A., D. N. Schmidt, P. N. Pearson, and C. J. Hawkesworth. 2008. Biological and ecological insights into Ca isotopes in planktic foraminifers as a paleotemperature proxy. *Earth and Planetary Science Letters* 271:292–302.

Kaufmann, R. S. 1984. Chlorine in ground water: Stable isotope distribution. PhD diss., University of Arizona.

Kaufmann, R. S., A. Long, H. Bentley, and S. Davis. 1984. Natural chlorine isotope variations. *Nature* 309:338–40.

Kim, J., H. L. Dong, J. Seabaugh, S. W. Newell, and D. D. Eberl. 2004. Role of microbes in the smectite-to-illite reaction. *Science* 303:830–32.

Komor, S. C. 1997. Boron contents and isotopic compositions of hog manure, selected fertilizers, and water in Minnesota. *Journal of Environmental Quality* 26:1212–22.

Kostka, J. E., and K. H. Nealson. 1995. Dissolution and reduction of magnetite by bacteria. *Environmental Science & Technology* 29:2535–40.

Kostka, J. E., J. W. Stucki, K. H. Nealson, and J. Wu. 1996. Reduction of structural Fe(III) in smectite by a pure culture of *Shewanella putrefaciens* strain MR-1. *Clays and Clay Minerals* 44:522–29.

Krouse, H. R., and H. G. Thode. 1962. Thermodynamic properties and geochemistry of isotopic compounds of selenium. *Canadian Journal of Chemistry* 40:367–75.

Lee, T., and F. Tera. 1986. The meteoritic chromium isotopic composition and limits for radioactive Mn-53 in the early Solar-System. *Geochimica Cosmochimica Acta* 50:199–206.

Lemarchand, D., G. J. Wasserburg, and D. A. Papanastassiou. 2004. Rate-controlled calcium isotope fractionation in synthetic calcite. *Geochimica Cosmochimica Acta* 68:4665–78.

Lemly, A. D. 1985. Toxicology of selenium in a freshwater reservoir: Implications for environmental hazard evaluation and safety. *Ecotoxicology and Environmental Safety* 10:314–38.

Lemly, A. D. 1998. Pathology of selenium poisoning in fish. In *Environmental chemistry of selenium*, eds. W. T. Frankenberger Jr. and R. A. Engberg, 281–96. New York: Marcel Dekker.

Liermann, .L. J., R. L. Guynn, A. Anbar, and S. L. Brantley. 2005. Production of a molybdophore during metal-targeted dissolution of silicates by soil bacteria. *Chemical Geology* 220:285–302.

Long, A. E., C. J. Eastoe, R. S. Kaufmann, J. G. Martin, L. Wirt, and J. B. Finley. 1993. High-precision measurement of chlorine stable isotope ratios. *Geochimica Cosmochimica Acta* 57:2907–12.

Lovley, D. R. 1987. Organic-matter mineralization with the reduction of ferric iron: A review. *Geomicrobiology Journal* 5:375–99.

Lovley, D. R. 1991. Dissimilatory Fe(III) and Mn(IV) reduction. *Microbiology Reviews* 55:259–87.

Lovley, D. R., and E. J. P. Phillips. 1988. Novel mode of microbial energy-metabolism: Organic-carbon oxidation coupled to dissimilatory reduction of iron or manganese. *Applied Environmental Microbiology* 54:1472–80.

Lovley, D. R., J. F. Stolz, G. L. Nord, and E. J. P. Phillips. 1987. Anaerobic production of magnetite by a dissimilatory iron-reducing microorganism. *Nature* 330:252–54.

Lyons, T. W., J. P. Werne, D. J. Hollander, and R. W. Murray. 2003. Contrasting sulfur geochemistry and Fe/Al and Mo/Al ratios across the last oxic-to-anoxic transition in the Cariaco Basin, Venezuela. *Chemical Geology* 195:131–57.

Magenheim, A. J., A. J. Spivack, P. J. Michael, and J. M. Gieskes. 1995. Chlorine stable isotope composition of the oceanic crust: Implications for Earth's distribution of chlorine. *Earth and Planetary Science Letters* 131:427–32.

Magenheim, A. J., A. J. Spivack, C. Volpe, and B. Ransom. 1994. Precise determination of stable chlorine isotopic ratios in low-concentration natural samples. *Geochimica Cosmochimica Acta* 58:3117–21.

Malinovsky, D., I. Rodushkin, D. C. Baxter, J. Ingri, and B. Öhlander. 2005. Molybdenum isotope ratio measurements on geological samples by MC-ICPMS. *International Journal of Mass Spectrometry* 245:94–107.

Mandernack, K. W., D. A. Bazylinski, W. C. Shanks, and T. D. Bullen. 1999. Oxygen and iron isotope studies of magnetite produced by magnetotactic bacteria. *Science* 285:1892–96.

Maréchal, C., P. Télouk, and F. Albarède. 1999. Precise analysis of copper and zinc isotopic compositions by plasma-source mass spectrometry. *Chemical Geology* 156:251–73.

Marriott, C. S., G. M. Henderson, N. S. Belshaw, and A. W. Tudhope. 2004. Temperature dependence of δ^7Li and δ^{44}Ca and Li/Ca during growth of calcium carbonate. *Earth and Planetary Science Letters* 222:615–24.

Matthews, A., X.-K. Zhu, and R. K. O'Nions. 2001. Kinetic iron stable isotope fractionation between iron (-II) and (-III) complexes in solution. *Earth and Planetary Science Letters* 192:81–92.

McManus, J., T. F. Nägler, C. Siebert, C. G. Wheat, and D. E. Hammond. 2002. Oceanic molybdenum isotope fractionation: Diagenesis and hydrothermal ridge-flank alteration. *Geochemistry Geophysics Geosystems* 3:1078, doi:10.1029/2002GC000356.

McManus, J., C. Siebert, R. Poulson, T. Nägler, W. M. Berelson, and S. Severmann. 2005. Molybdenum and molybdenum isotope diagenesis in continental margin settings: Geochemical balance and paleoproxy implications. *Geochimica Cosmochimica Acta* 69:A576.

Monaster, A., M. G. Minnich, H. Liu, A. G. T. Gustavsson, and R. F. Browner. 1998a. Fundamental aspects of sample introduction in ICP spectrometry. In *Inductively coupled plasma mass spectrometry*, ed. A. Montaser, 335–420. New York: Wiley-VCH.

Monaster, A., M. G. Minnich, J. A. McLean, H. Liu, J. A. Caruso, and C. W. McLeod. 1998b. Sample introduction in ICPMS. In *Inductively coupled plasma mass spectrometry*, ed. A. Montaser, 83–264. New York: Wiley-VCH.

Morford, J. L., and S. Emerson. 1999. The geochemistry of redox sensitive trace metals in sediments. *Geochimica Cosmochimica Acta* 63:1735–50.

Morris, A. W. 1975. Dissolved molybdenum and vanadium in the northeast Atlantic Ocean. *Deep-Sea Research Part A-Oceanographic Research Papers* 22:49–54.

Myers, C. R., and K. H. Nealson. 1988. Bacterial manganese reduction and growth with manganese oxide as the sole electron-acceptor. *Science* 240:1319–21.

Nägler, T. F., A. Einsenhauer, A. Müller, C. Hemleben, and J. Kramers. 2000. The δ^{44}Ca-temperature calibration on fossil and cultured *Globigerinoides sacculifer*: New tool for reconstruction of past sea surface temperatures. *Geochemistry Geophysics Geosystems* 1:1052, doi:10.1029/2000GC000091.

Nägler, T. F., M. M. Mills, and C. Siebert. 2004. Biological fractionation of Mo isotopes during N-2 fixation by *Trichodesmium* sp. IMS 101. *Geochimica Cosmochimica Acta* 68:A364.

Nägler, T. F., C. Siebert, H. Luschen, and M. E. Bottcher. 2005. Sedimentary Mo isotope record across the Holocene fresh-brackish water transition of the Black Sea. *Chemical Geology* 219:283–95.

Neal, A. L., K. M. Rosso, G. G. Geesey, Y. A. Gorby, and B. J. Little. 2003. Surface structure effects on direct reduction of iron oxides by *Shewanella oneidensis*. *Geochimica Cosmochimica Acta* 67:4489–4503.

Nealson, K. H. 1983. The microbial iron cycle. In *Microbial geochemistry,* ed. W. Krumbein, 159–90. Boston, MA: Blackwell Scientific.

Nealson, K. H., and C. R. Myers. 1990. Iron reduction by bacteria: A potential role in the genesis of banded iron formations. *American Journal of Science* 290A:35–45.

Nealson, K. H., and D. Saffarini. 1994. Iron and manganese in anaerobic respiration: Environmental significance, physiology, and regulation. *Annual Review of Microbiology* 48:311–43.

Nriagu, J. O., and E. Niebor. 1988. *Chromium in the natural and human environments.* New York: John Wiley and Sons.

Numata, M., N. Nakamura, and T. Gamo. 2001. Precise measurement of chlorine stable isotopic ratios by thermal ionization mass spectrometry. *Geochemistry Journal* 35:89–100.

Numata, M., N. Nakamura, H. Koshikawa, and Y. Terashima. 2002. Chlorine isotope fractionation during reductive dechlorination of chlorinated ethenes by anaerobic bacteria. *Environmental Science & Technology* 36:4389–94.

Ottonello, G., and M. V. Zuccolini. 2005. *Ab-initio* structure, energy and stable Cr isotopes equilibrium fractionation of some geochemically relevant H-O-Cr-Cl complexes. *Geochimica Cosmochimica Acta* 69:851–74.

Paneth, P. 2003. Chlorine kinetic isotope effects on enzymatic dehalogenations. *Accounts of Chemical Research* 36:120–26.

Pearce, C. R., A. S. Cohen, A. L. Coe, and K. W. Burton. 2008. Molybdenum isotope evidence for global ocean anoxia coupled with perturbations to the carbon cycle during the early Jurassic. *Geolog* 36:231–234.

Pietruszka, A. J., R. J. Walker, and P. A. Candela. 2006. Determination of mass-dependent molybdenum isotopic variations by MC-ICP-MS: An evaluation of matrix effects. *Chemical Geology* 225:121–36.

Poitrasson, F., A. N. Halliday, D.-C. Lee, S. Levasseur, and N. Teutsch. 2004. Iron isotope differences between Earth, Moon, Mars, and Vesta as possible records of contrasted accretion mechanisms. *Earth and Planetary Science Letters* 223:253–66.

Polyakov, V. B., R. N. Clayton, J. Hoita, and S. D. Mineev. 2007. Equilibrium iron isotope fractionation factors of minerals: Reevaluating from the data of nuclear inelastic resonant X-ray scattering and Mössbauer spectroscopy. *Geochimica Cosmochimica Acta* 71:3833–46.

Polyakov, V. B., and S. D. Mineev. 2000. The use of Mössbauer spectroscopy in stable isotope geochemistry. *Geochimica Cosmochimica Acta* 64:849–65.

Poulson, R. L., C. M. Johnson, and B. L. Beard. 2005. Evidence of Fe isotope exchange at the nanoparticulate ferrihydrite surface. *American Mineralogist* 90:758–63.

Poulson, R. L., C. Siebert, J. McManus, and W. M. Berelson. 2006. Authigenic molybdenum isotope signatures in marine sediments. *Geology* 34:617–20.

Qi, H. P., T. B. Coplen, Q. Z. Wang, and Y. H. Wang. 1997. Unnatural isotopic composition of lithium reagents. *Analytical Chemistry* 69:4076–78.

Raiswell, R., and T. F. Anderson. 2005. Reactive iron enrichment in sediments deposited beneath euxinic bottom waters: Constraints on supply by shelf recycling. In *Mineral deposits and earth evolution*, eds. I. MacDonald, A. J. Boyce, I. B. Butler, R. J. Herrington, and D. A. Poyla, 248:179–94. London: The Geological Society.

Rashid, K., and H. R. Krouse. 1985. Selenium isotopic fractionation during SeO_3 reduction to Se^0 and H_2Se. *Canadian Journal of Chemistry* 63:3195–99.

Reddy, C. M., L. J. Heraty, B. D. Holt, N. C. Sturchio, T. I. Eglinton, N. J. Drenzek, L. Xu, J. L. Lake, and K. A. Maruya. 2000. Stable chlorine isotopic compositions of aroclors and aroclor-contaminated sediments. *Environmental Science & Technology* 34:2866–70.

Rees, C. B., and H. G. Thode. 1966. Selenium isotope effects in the reduction of sodium selenite and of sodium selenate. *Canadian Journal of Chemistry* 44:419–27.

Richet, P., Y. Bottinga, and M. Javoy. 1977. A review of hydrogen, carbon, nitrogen, oxygen, sulphur, and chlorine stable isotope fractionation among gaseous molecules. *Annual Review of Earth and Planetary Sciences* 5:65–110.

Roden, E. E., M. R. Leonardo, and F. G. Ferris. 2002. Immobilization of strontium during iron biomineralization coupled to dissimilatory hydrous ferric oxide reduction. *Geochimica Cosmochimica Acta* 66:2823–39.

Roden, E. E., and D. R. Lovley. 1993. Dissimilatory Fe(III) reduction by the marine microorganism *Desulfuromonas acetoxidans*. *Applied Environmental Microbiology* 59:734–42.

Rodriguez-Navarro, A. B., K. F. Gaines, C. S. Romanek, and G. R. Masson. 2002. Mineralization of clapper rail eggshell from a contaminated salt marsh system. *Archives of Environmental Contamination and Toxicology* 43:449–60.

Rodriguez-Navarro, A. B., C. S. Romanek, P. Alvarez-Lloret, and K. F. Gaines. 2006. Effect of *in ovo* exposure to PCBs and Hg on Clapper Rail bone mineral chemistry from a contaminated salt marsh in coastal Georgia. *Environmental Science & Technology* 40:4936–42.

Rodushkin, I., A. Stenberg, H. Andren, D. Malinovsky, and D. C. Baxter. 2004. Isotopic fractionation during diffusion of transition metals in solution. *Analytical Chemistry* 76:2148–51.

Roe, J. E., A. D. Anbar, and J. Barling. 2003. Nonbiological fractionation of Fe isotopes: Evidence of an equilibrium isotope effect. *Chemical Geology* 195:69–85.

Rouxel, O. J., A. Bekker, and K. J. Edwards. 2005. Iron isotope constraints on the Archean and Paleoproterozoic ocean redox state. *Science* 307:1088–91.

Rouxel, O., N. Dobbek, J. Ludden, and Y. Fouquet. 2003. Iron isotope fractionation during oceanic crust alteration. *Chemical Geology* 202:155–82.

Rouxel, O., Y. Fouquet, and J. N. Ludden. 2004. Subsurface processes at the Lucky Strike hydrothermal field, mid-Atlantic Ridge: Evidence from sulfur, selenium, and iron isotopes. *Geochimica Cosmochimica Acta* 68:2295–311.

Rouxel, O., J. Ludden, J. Carginan, L. Marin, and Y. Fouquet. 2002. Natural variations of Se isotopic composition determined by hydride generation multiple collector inductively coupled plasma mass spectrometry. *Geochimica Cosmochimica Acta* 66:3191–99.

Rouxel, O., W. C. Shanks III, W. Bach, and K. J. Edwards. 2008. Integrated Fe- and S-isotope study of seafloor hydrothermal vents at East Pacific Rise 9-10°N. *Chemical Geology* 252:214–27.

Russell, W. A., D. A. Papanastassiou, and T. A. Tombrello. 1978. Ca isotope fractionation on the Earth and other solar system materials. *Geochimica Cosmochimica Acta* 42:1075–90.

Sageman, B. B., A. E. Murphy, J. P. Werne, C. A. V. Straeten, D. J. Hollander, and T. W. Lyons. 2003. A tale of shales: The relative roles of production, decomposition, and dilution in the accumulation of organic-rich strata, Middle-Upper Devonian, Appalachian basin. *Chemical Geology* 195:229–73.

Schauble, E. A. 2004. Applying stable isotope fractionation theory to new systems. In *Geochemistry of non-traditional stable isotopes*, eds. C. M. Johnson, B. L. Beard, and F. Albarède, 65–111. Washington DC: The Mineralogical Society of America.

Schauble, E. A., G. R. Rossman, and H. P. Taylor. 2001. Theoretical estimates of equilibrium Fe-isotope fractionations from vibrational spectroscopy. *Geochimica Cosmochimica Acta* 65:2487–97.

Schauble, E. A., G. R. Rossman, and H. P. Taylor Jr. 2003. Theoretical estimates of equilibrium chlorine-isotope fractionations. *Geochimica Cosmochimica Acta* 67:3267–81.

Schauble, E. A., G. R. Rossman, and H. P. Taylor Jr. 2004. Theoretical estimates of equilibrium chromium-isotope fractionations. *Chemical Geology* 205:99–114.

Schmitt, A. D., F. Chabaux, and P. Stille. 2003a. Variations of the $^{44}Ca/^{40}Ca$ ratio in seawater during the past 24 million years: Evidence from $\delta^{44}Ca$ and $\delta^{18}O$ in Miocene phosphates. *Geochimica Cosmochimica Acta* 67:2607–14.

Schmitt, A. D., F. Chabaux, and P. Stille. 2003b. The calcium riverine and hydrothermal isotopic fluxes and the oceanic calcium mass balance. *Earth and Planetary Science Letters* 213:503–18.

Schmitt, A. D. S., and P. Stille. 2005. The source of calcium in wet atmospheric deposits: Ca-Sr isotope evidence. *Geochimica Cosmochimica Acta* 69:3463–68.

Schuessler, J. A., R. Schoenberg, H. Behrens, and F. von Blanckenburg. 2007. The experimental calibration of the iron isotope fractionation factor between pyrrohtite and peralkaline rhyolitic melt. *Geochimica Cosmochimica Acta* 71:417–33.

Severmann, S., C. M. Johnson, B. L. Beard, and J. McManus. 2006. The effect of early diagenesis on the Fe isotope composition of porewaters and authigenic minerals in continental margin sediments. *Geochimica Cosmochimica Acta* 70:2006–22.

Severmann, S., T. W. Lyons, A. Anbar, J. McManus, and G. Gordon. 2008. Modern iron isotope perspective on benthic iron shuttle and the redox evolution of ancient oceans. *Geology* 36:487–90.

Shahar, A., E. D. Young, and C. E. Manning. 2008. Equilibrium high-temperature Fe isotope fractionation between fayalite and magnetite: An experimental calibration. *Earth and Planetary Science Letters* 268:330–38.

Sharma, M., M. Polizzotto, and A. D. Anbar. 2001. Iron isotopes in hot springs along the Juan de Fuca Ridge. *Earth and Planetary Science Letters* 194:39–51.

Shouakar-Stash, O., R. J. Drimmie, and S. K. Frape. 2005. Determination of inorganic chlorine stable isotopes by continuous flow isotope ratio mass spectrometry. *Rapid Communications in Mass Spectrometry* 19:121–27.

Shouakar-Stash, O., S. K. Frape, and R. J. Drimmie. 2003. Stable hydrogen, carbon and chlorine isotope measurements of selected chlorinated organic solvents. *Journal of Contaminant Hydrology* 60:211–28.

Siebert, C., J. D. Kramers, T. Meisel, P. Morel, and T. F. Nägler. 2005. PGE, Re-Os, and Mo isotope systematics in Archean and early Proterozoic sedimentary systems as proxies for redox conditions of the early Earth. *Geochimica Cosmochimica Acta* 69:1787–1801.

Siebert, C., J. McManus, A. Bice, R. Poulson, and W. M. Berelson. 2006. Molybdenum isotope signatures in continental margin marine sediments. *Earth and Planetary Science Letters* 241:723–33.

Siebert, C., T. F. Nägler, and J. D. Kramers. 2001. Determination of molybdenum isotope fractionation by double-spike multicollector inductively coupled plasma mass spectrometry. *Geochemistry Geophysics Geosystems* 2:1032, doi:10.1029/2000GC000124.

Siebert, C., T. F. Nägler, F. von Blanckenburg, and J. D. Kramers. 2003. Molybdenum isotope records as a potential new proxy for paleoceanography. *Earth and Planetary Science Letters* 211:159–71.

Sime, N. G., C. L. De La Rocha, and A. Galy. 2005. Negligible temperature dependence of calcium isotope fractionation in 12 species of planktonic foraminifera. *Earth and Planetary Science Letters* 232:645–53.

Sime, N. G., C. L. De La Rocha, E. T. Tipper, A. Tripati, A. Galy, and M. J. Bickle. 2007. Interpreting the Ca isotope record of marine biogenic carbonates. *Geochimica Cosmochimica Acta* 71:3979–89.

Skorupa, J. P. 1998. Selenium poisoning of fish and wildlife in nature: Lessons from twelve real-world-examples. In *Environmental chemistry of selenium*, eds. W. T. Frankenberger Jr. and R. A. Engberg, 315–54. New York: Marcel Dekker.

Skulan, J. L., B. L. Beard, and C. M. Johnson. 2002. Kinetic and equilibrium Fe isotope fractionation between aqueous Fe(III) and hematite. *Geochimica Cosmochimica Acta* 66:2995–3015.

Skulan, J., and D. J. DePaolo. 1999. Calcium isotope fractionation between soft and mineralized tissues as a monitor of calcium use in vertebrates. *Proceedings of the National Academy of Sciences* 96:13709–713.

Skulan, J., D. J. DePaolo, and T. L. Owens. 1997. Biological control of calcium isotopic abundances in the global calcium cycle. *Geochimica Cosmochimica Acta* 61:2505–10.

Soudry, D., I. Segal, Y. Nathan, C. R. Glenn, L. Halicz, Z. Lewy, and D. L. VonderHaar. 2004. $^{44}Ca/^{42}Ca$ and $^{143}Nd/^{144}Nd$ isotope variations in Cretaceous-Eocene Tethyan francolites and their bearing on phosphogenesis in the southern tethys. *Geology* 32:389–92.

Staubwasser, M., F. von Blanckenburg, and R. Schoenberg. 2006. Iron isotopes in the early marine diagenetic iron cycle. *Geology* 34:629–32.

Stewart, M. A. 2000. Geochemistry of dikes and lavas from Hess Deep: Implications for crustal construction processes beneath mid-ocean ridges and the stable-chlorine isotope geochemistry of mid-ocean ridge basalt glasses. PhD diss., Duke University.

Stewart, M. A., and A. J. Spivak. 2004. The stable-chlorine isotope compositions of natural and anthropogenic materials. In *Geochemistry of non-traditional stable isotopes*, eds. C. M. Johnson, B. L. Beard, and F. Albarède, 231–54. Washington DC: The Mineralogical Society of America.

Stiefel, E. I. 1997. Chemical keys to molybdenum enzymes. *Journal of the Chemical Society-Dalton Transactions* 3915–23.

Straub, K. L., M. Benz, and B. Schink. 2001. Iron metabolism in anoxic environments at near neutral pH. *FEMS Microbiology Ecology* 34:181–86.

Sturchio, N. C., J. L. Clausen, L. J. Heraty, L. Huang, B. D. Holt, and T. A. Abrajano. 1998. Chlorine isotope investigation of natural attenuation of trichloroethene in an aerobic aquifer. *Environmental Science & Technology* 32:3037–42.

Sturchio, N. C., P. B. Hatzinger, M. D. Arkins, C. Suh, and L. J. Heraty. 2003. Chlorine isotope fractionation during microbial reduction of perchlorate. *Environmental Science & Technology* 37:3859–63.

Taylor, P. D. P., R. Maeck, and P. De Bièvre. 1992. Determination of the absolute isotopic composition and atomic weight of a reference sample of natural iron. *International Journal of Mass Spectrometry and Ion Processes* 121:111–25.

Taylor, P. D. P., R. Maeck, F. Hendricks, and P. De Bièvre. 1993. The gravimetric preparation of synthetic mixtures of iron isotopes. *International Journal of Mass Spectrometry and Ion Processes* 128:91–97.

Taylor, S. R., and S. M. McLennan. 1985. *The continental crust: Its composition and evolution.* Boston, MA: Blackwell.

Teutsch, N., U. von Grunten, D. Porcelli, O. A. Cirpka, and A. N. Halliday. 2005. Adsorption as a cause for iron isotope fractionation in reduced groundwater. *Geochimica Cosmochimica Acta* 69:4175–85.

Thompson, A., J. Ruiz, O. A. Chadwick, M. Titus, and J. Chorover. 2007. Rayleigh fractionation of iron isotopes during pedogenesis along a climate sequence of Hawaiian basalt. *Chemical Geology* 238:72–83.

Tipper, E. T., A. Galy, and M. J. Bickle. 2006. Riverine evidence for a fractionated reservoir of Ca and Mg on the continents: Implications for the oceanic Ca cycle. *Earth and Planetary Science Letters* 247:267–79.

Tomascak, P. B. 2004. Developments in the understanding and application of lithium isotopes in the earth and planetary sciences. In *Geochemistry of non-traditional stable isotopes*, eds. C. M. Johnson, B. L. Beard, and F. Albarède, 153–95. Washington, DC: The Mineralogical Society of America.

Tomascak, P. B., E. J. Krogstad, K. L. Prestegaard, and R. J. Walker. 1995a. Li isotope variability in groundwaters from Maryland. *EOS, Transactions, American Geophysical Union* 76:S122.

Tomascak, P. B., E. J. Krogstad, K. L. Prestegaard, and R. J. Walker. 1995b. Li isotope geochemistry of stormwater from Maryland. *EOS, Transactions, American Geophysical Union* 76:F207.

Tossell, J. A. 2005. Calculating the partitioning of the isotopes of Mo between oxic and sulfidic species in aqueous solution. *Geochimica Cosmochimica Acta* 69:2981–93.

Urbansky, E. T. 2002. Perchlorate as an environmental contaminant. *Environmental Science and Pollution Research* 9:187–92.

Urey, H. C. 1947. The thermodynamic properties of isotopic substances. *Journal of Chemical Society* 562–81.

Urey, H. C., and L. J. Greiff. 1935. Isotopic exchange equilibria. *Journal of the American Chemical Society* 57:321–27.

van Warmerdam, E. M., S. K. Frape, R. Aravena, R. J. Drimmie, H. Flatt, and J. A. Cherry. 1995. Stable chlorine and carbon isotope measurements of selected chlorinated organic solvents. *Applied Geochemistry* 10:547–52.

Vengosh, A., K. G. Heumann, S. Juraske, and R. Kasher. 1994. Boron isotope application for tracing sources of contamination in groundwater. *Environmental Science & Technology* 28:1968–74.

Wachsmann, M., and K. G. Heumann. 1992. Negative thermal ionization mass spectrometry of main group elements, part 2. 6th group: Sulfur, selenium, and tellurium. *International Journal of Mass Spectrometry and Ion Processes* 114:209–20.

Walczyk, T., and F. von Blanckenburg. 2002. Natural iron isotope variations in human blood. *Science* 295:2065–66.

Wasylenki, L. E., A. D. Anbar, and G. W. Gordon. 2006a. Temperature dependence of Mo isotope fractionation during adsorption to δ-MnO_2: Implications for the paleoredox proxy. *Geochimica Cosmochimica Acta* 70:A691.

Wasylenki, L. E., R. D. Mathur, L. J. Liermann, A. D. Anbar, and S. L. Brantley. 2006b. Toward multi-element isotopic biosignatures: Experimental investigation of microbial metal assimilation. *Geochimica Cosmochimica Acta* 70:A691.

Webster, C. L. 1972. Selenium isotope analysis and geochemical applications. PhD diss., Colorado State University.

Wedepohl, K. H. 1970. *The handbook of geochemistry.* Berlin: Springer-Verlag.

Welch, S. A., B. L. Beard, C. M. Johnson, and P. S. Braterman. 2003. Kinetic and equilibrium Fe isotope fractionation between aqueous Fe(II) and Fe(III). *Geochimica Cosmochimica Acta* 67:4231–50.

Werne, J. P., B. B. Sageman, T. W. Lyons, and D. J. Hollander. 2002. An integrated assessment of a "type euxinic" deposit: Evidence for multiple controls on black shale deposition in the Middle Devonian Oatka Creek Formation. *American Journal of Science* 302:110–43.

Weyer, S., A. D. Anbar, G. P. Brey, C. Münker, K. Mezger, and A. B. Woodland. 2005. Iron isotope fractionation during planetary differentiation. *Earth and Planetary Science Letters* 240:251–64.

Weyer, S., and J. Schwieters. 2003. High precision Fe isotope measurements with high mass resolution MC-ICPMS. *International Journal of Mass Spectrometry* 226:355–68.

Wheat, C. G., and M. J. Mottl. 2000. Composition of pore and spring waters from Baby Bare: Global implications of geochemical fluxes from a ridge flank hydrothermal system. *Geochimica Cosmochimica Acta* 64:629–42.

Widdel, F., S. Schnell, S. Heising, A. Ehrenreich, B. Assmus, and B. Schink. 1993. Ferrous iron oxidation by anoxygenic phototrophic bacteria. *Nature* 362:834–36.

Wiederhold, J. G., S. M. Kraemer, N. Teutsch, P. M. Borer, A. N. Halliday, and R. Kretzschmar. 2006. Iron isotope fractionation during proton-promoted, ligand-controlled, and reductive dissolution of goethite. *Environmental Science & Technology* 40:3787–93.

Wiederhold, J. G., N. Teutsch, S. M. Kraemer, A. N. Halliday, and R. Kretzschmar. 2007a. Iron isotope fractionation during pedogenesis in redoximorphic soils. *Soil Science Society of America Journal* 71:1840–50.

Wiederhold, J. G., N. Teutsch, S. M. Kraemer, A. N. Halliday, and R. Kretzschmar. 2007b. Iron isotope fractionation in oxic soils by mineral weathering and podzolization. *Geochimica Cosmochimica Acta* 71:5821–33.

Wiegand, B. A., O. A. Chadwick, P. M. Vitousek, and J. L. Wooden. 2005. Ca cycling and isotopic fluxes in forested ecosystems in Hawaii. *Geophysical Research Letters* 32:L11404.

Wieser, M. E., and J. R. DeLaeter. 2003. A preliminary study of isotope fractionation in molybdenites. *International Journal of Mass Spectrometry and Ion Processes* 225:177–83.

Wiesli, R. A., B. L. Beard, and C. M. Johnson. 2004. Experimental determination of Fe isotope fractionation between aqueous Fe(II), siderite, and "green rust" in abiotic systems. *Chemical Geology* 211:343–62.

Xiao, Y.-K., and C.-G. Zhang. 1992. High precision isotopic measurement of chlorine by thermal ionization mass spectrometry of the Cs_2Cl^+ ion. *International Journal of Mass Spectrometry and Ion Processes* 116:183–92.

Yamaguchi, K. E., C. M. Johnson, B. L. Beard, and H. Ohmoto. 2005. Biogeochemical cycling of iron in the Archean-Paleoproterozoic Earth: Constraints from iron isotope variations in sedimentary rocks from the Kaapvaal and Pilbara Cratons. *Chemical Geology* 218:135–69.

Yarincik, K. M., R. W. Murray, T. W. Lyons, L. C. Peterson, and G. H. Haug. 2000. Oxygenation history of bottom waters in the Cariaco Basin, Venezuela, over the past 578,000 years: Results from redox-sensitive metals (Mo, V, Mn, and Fe). *Paleoceanography* 15:593–604.

Zachara, J. M., R. K. Kukkadapu, J. K. Fredrickson, Y. A. Gorby, and S. C. Smith. 2002. Biomineralization of poorly crystalline Fe(III) oxides by dissimilatory metal reducing bacteria (DMRB). *Geomicrobiology Journal* 19:179–207.

Zhang, C. L., J. Horita, D. R. Cole, J. Z. Zhou, D. R. Lovley, and T. J. Phelps. 2001. Temperature-dependent oxygen and carbon isotope fractionations of biogenic siderite. *Geochimica Cosmochimica Acta* 65:2257–71.

Zhang, C. L., H. Vali, C. S. Romanek, T. J. Phelps, and S. V. Liu. 1998. Formation of single-domain magnetite by a thermophilic bacterium. *American Mineralogist* 83:1409–18.

Zheng, Y., A. van Geen, R. F. Anderson, J. V. Gardner, and W. E. Dean. 2000. Intensification of the northeast Pacific oxygen minimum zone during the Bolling-Allerod warm period. *Paleoceanography* 15:528–36.

Zhu, P., and J. D. MacDougall. 1998. Calcium isotopes in the marine environment and the oceanic calcium cycle. *Geochimica Cosmochimica Acta* 62:1691–98.

Index